Instructors: Student Success Starts with You

Tools to enhance your unique voice

Want to build your own course? No problem. Prefer to use an OLC-aligned, prebuilt course? Easy. Want to make changes throughout the semester? Sure. And you'll save time with Connect's auto-grading too.

65%
Less Time Grading

Study made personal

Incorporate adaptive study resources like SmartBook® 2.0 into your course and help your students be better prepared in less time. Learn more about the powerful personalized learning experience available in SmartBook 2.0 at **www.mheducation.com/highered/connect/smartbook**

Laptop: McGraw Hill; Woman/dog: George Doyle/Getty Images

Affordable solutions, added value

Make technology work for you with LMS integration for single sign-on access, mobile access to the digital textbook, and reports to quickly show you how each of your students is doing. And with our Inclusive Access program you can provide all these tools at a discount to your students. Ask your McGraw Hill representative for more information.

Padlock: Jobalou/Getty Images

Solutions for your challenges

A product isn't a solution. Real solutions are affordable, reliable, and come with training and ongoing support when you need it and how you want it. Visit **www.supportateverystep.com** for videos and resources both you and your students can use throughout the semester.

Checkmark: Jobalou/Getty Images

Students: Get Learning that Fits You

Effective tools for efficient studying

Connect is designed to help you be more productive with simple, flexible, intuitive tools that maximize your study time and meet your individual learning needs. Get learning that works for you with Connect.

Study anytime, anywhere

Download the free ReadAnywhere app and access your online eBook, SmartBook 2.0, or Adaptive Learning Assignments when it's convenient, even if you're offline. And since the app automatically syncs with your Connect account, all of your work is available every time you open it. Find out more at www.mheducation.com/readanywhere

> *"I really liked this app—it made it easy to study when you don't have your textbook in front of you."*
>
> - Jordan Cunningham,
>   Eastern Washington University

Calendar: owattaphotos/Getty Images

Everything you need in one place

Your Connect course has everything you need—whether reading on your digital eBook or completing assignments for class, Connect makes it easy to get your work done.

Learning for everyone

McGraw Hill works directly with Accessibility Services Departments and faculty to meet the learning needs of all students. Please contact your Accessibility Services Office and ask them to email accessibility@mheducation.com, or visit www.mheducation.com/about/accessibility for more information.

Top: Jenner Images/Getty Images, Left: Hero Images/Getty Images, Right: Hero Images/Getty Images

THE
UNFINISHED
NATION

THE UNFINISHED NATION

A Concise History of the American People
Volume 1: To 1877

Tenth Edition

ALAN BRINKLEY
Columbia University

JOHN M. GIGGIE
University of Alabama

ANDREW J. HUEBNER
University of Alabama

THE UNFINISHED NATION: A CONCISE HISTORY OF THE AMERICAN PEOPLE
VOLUME 1: TO 1877, TENTH EDITION

Published by McGraw Hill LLC, 1325 Avenue of the Americas, New York, NY 10019. Copyright © 2022 by McGraw Hill LLC. All rights reserved. Printed in the United States of America. Previous editions © 2019, 2016, and 2014. No part of this publication may be reproduced or distributed in any form or by any means, or stored in a database or retrieval system, without the prior written consent of McGraw Hill LLC, including, but not limited to, in any network or other electronic storage or transmission, or broadcast for distance learning.

Some ancillaries, including electronic and print components, may not be available to customers outside the United States.

This book is printed on acid-free paper.

1 2 3 4 5 6 7 8 9 LCR 26 25 24 23 22 21

ISBN   978-1-264-30925-2 (bound edition)
MHID   1-264-30925-2 (bound edition)
ISBN   978-1-264-30928-3 (loose-leaf edition)
MHID   1-264-30928-7 (loose-leaf edition)

Senior Portfolio Manager: *Jason Seitz*
Product Development Manager: *Dawn Groundwater*
Senior Product Developer: *Lauren A. Finn*
Senior Marketing Manager: *Michael Gedatus*
Content Project Managers: *Sherry Kane/Vanessa McClune*
Senior Buyer: *Susan K. Culbertson*
Designer: *Beth Blech*
Content Licensing Specialist: *Sarah Flynn*
Cover Image: *1817 White House: Library of Congress, Prints & Photographs Division [LC-DIG-ppmsca-09502]; White House Front Lawn: Jill Braaten/McGraw Hill; and Blueprint Grid: McGraw Hill.*
Compositor: *Aptara®, Inc.*

All credits appearing on page or at the end of the book are considered to be an extension of the copyright page.

Library of Congress Cataloging-in-Publication Data

Names: Brinkley, Alan, author. | Giggie, John M. (John Michael), 1965- author. | Huebner, Andrew J., author.
Title: The unfinished nation : a concise history of the American people / Alan Brinkley, Columbia University ; John Giggie, University of Alabama ; Andrew Huebner, University of Alabama.
Other titles: Concise history of the American people
Description: Tenth edition. | [Dubuque, Iowa] : McGraw Hill Education, [2022] | Includes index.
Identifiers: LCCN 2021026391 (print) | LCCN 2021026392 (ebook) | ISBN 9781264309252 (v. 1 ; hardcover) | ISBN 9781264309283 (v. 1 ; spiral bound) | ISBN 9781264309306 (v. 2 ; hardcover) | ISBN 9781264309313 (v. 2 ; spiral bound) | ISBN 9781260726831 (hardcover) | ISBN 9781264309214 (spiral bound) | ISBN 9781265360917 (hardcover) | ISBN 9781264309207 (ebook) | ISBN 9781264309245 (ebook other)
Subjects: LCSH: United States—History—Textbooks.
Classification: LCC E178.1 .B827 2022 (print) | LCC E178.1 (ebook) | DDC 973—dc23
LC record available at https://lccn.loc.gov/2021026391
LC ebook record available at https://lccn.loc.gov/2021026392

The Internet addresses listed in the text were accurate at the time of publication. The inclusion of a website does not indicate an endorsement by the authors or McGraw Hill LLC, and McGraw Hill LLC does not guarantee the accuracy of the information presented at these sites.

mheducation.com/highered

BRIEF CONTENTS

PREFACE xvii

1 THE COLLISION OF CULTURES 1
2 TRANSPLANTATIONS AND BORDERLANDS 25
3 SOCIETY AND CULTURE IN PROVINCIAL AMERICA 55
4 THE EMPIRE IN TRANSITION 82
5 THE AMERICAN REVOLUTION 106
6 THE CONSTITUTION AND THE NEW REPUBLIC 134
7 THE JEFFERSONIAN ERA 155
8 EXPANSION AND DIVISION IN THE EARLY REPUBLIC 185
9 JACKSONIAN AMERICA 202
10 AMERICA'S ECONOMIC REVOLUTION 227
11 COTTON, SLAVERY, AND THE OLD SOUTH 253
12 ANTEBELLUM CULTURE AND REFORM 271
13 THE IMPENDING CRISIS 295
14 THE CIVIL WAR 319
15 RECONSTRUCTION AND THE NEW SOUTH 350

APPENDIX A-1
GLOSSARY G
INDEX I-1

CONTENTS

PREFACE xvii

1 THE COLLISION OF CULTURES 1

AMERICA BEFORE COLUMBUS 2
  The Peoples of the Precontact Americas 2
  The Growth of Civilizations: The South 3
  The Civilizations of the North 4

EUROPE LOOKS WESTWARD 6
  Commerce and Sea Travel 6
  Christopher Columbus 7
  The Spanish Empire 9
  Northern Outposts 11
  Biological and Cultural Exchanges 11
  Africa and America 17

THE ARRIVAL OF THE ENGLISH 18
  Incentives for Colonization 19
  The First English Settlements 20
  The French and the Dutch in America 22

Consider the Source: Bartolomé de Las Casas, "Of the Island of Hispaniola" (1542) 10

Debating the Past: Why Do Historians So Often Differ? 14

America in the World: The International Context of the Early History of the Americas 16

CONCLUSION 23
KEY TERMS/PEOPLE/PLACES/EVENTS 24
RECALL AND REFLECT 24

Don Mammoser/Shutterstock

2 TRANSPLANTATIONS AND BORDERLANDS 25

THE EARLY CHESAPEAKE 26
  Colonists and Native Peoples 26
  Reorganization and Expansion 27
  Slavery and Indenture in the Virginia Colony 29
  Bacon's Rebellion 30
  Maryland and the Calverts 32

THE GROWTH OF NEW ENGLAND 33
  Plymouth Plantation 33
  The Massachusetts Bay Experiment 35
  The Expansion of New England 35
  King Philip's War 37

THE RESTORATION COLONIES 40
  The English Civil War 40
  The Carolinas 41
  New Netherland, New York, and New Jersey 42
  The Quaker Colonies 43

BORDERLANDS AND MIDDLE GROUNDS 44
  The Caribbean Islands 44
  Slaveholder and Enslaved in the Caribbean 45
  The Southwest Borderlands 46
  The Southeast Borderlands 47
  The Founding of Georgia 47
  Middle Grounds 49

THE DEVELOPMENT OF EMPIRE 51
  The Dominion of New England 52
  The "Glorious Revolution" 52

Consider the Source: Cotton Mather on the Recent History of New England (1692) 38

Debating the Past: Native Americans and the Middle Ground 50

CONCLUSION 53
KEY TERMS/PEOPLE/PLACES/EVENTS 54
RECALL AND REFLECT 54

Universal History Archive/Universal Images Group/Getty Images

• ix

3 SOCIETY AND CULTURE IN PROVINCIAL AMERICA 55

THE COLONIAL POPULATION 56
- Indentured Servitude 56
- Birth and Death 57
- Medicine in the Colonies 57
- Women and Families in the Colonies 60
- The Beginnings of Slavery in English America 62
- Changing Sources of European Immigration 63

THE COLONIAL ECONOMIES 65
- Slavery and Economic Life 65
- Industry and Its Limits 65
- The Rise of Colonial Commerce 66
- The Rise of Consumerism 68

PATTERNS OF SOCIETY 69
- Southern Communities 69
- Northern Communities 70
- Cities 73

AWAKENINGS AND ENLIGHTENMENTS 74
- The Pattern of Religions 74
- The Great Awakening 75
- The Enlightenment 76
- Literacy and Technology 76
- Education 78
- The Spread of Science 79
- Concepts of Law and Politics 79

Consider the Source: Gottlieb Mittelberger, the Passage of Indentured Servants (1750) 58

Debating the Past: The Witchcraft Trials 72

CONCLUSION 80
KEY TERMS/PEOPLE/PLACES/EVENTS 81
RECALL AND REFLECT 81

Bettmann/Getty Images

4 THE EMPIRE IN TRANSITION 82

LOOSENING TIES 83
- A Decentralized Empire 83
- The Colonies Divided 83

THE STRUGGLE FOR THE CONTINENT 84
- New France and the Iroquois Nation 84
- Anglo-French Conflicts 85
- The Great War for Empire 85

THE NEW IMPERIALISM 89
- Burdens of Empire 90
- The British and Native Americans 90
- Battles over Trade and Taxes 92

STIRRINGS OF REVOLT 93
- The Stamp Act Crisis 93
- Internal Rebellions 96
- The Townshend Program 96
- The Boston Massacre 97
- The Philosophy of Revolt 97
- Sites of Resistance 99
- The Tea Excitement 99

COOPERATION AND WAR 102
- New Sources of Authority 102
- Lexington and Concord 102

America in the World: The First Global War 86

Consider the Source: Benjamin Franklin, Testimony Against the Stamp Act (1766) 94

Patterns of Popular Culture: Taverns in Revolutionary Massachusetts 100

CONCLUSION 104
KEY TERMS/PEOPLE/PLACES/EVENTS 104
RECALL AND REFLECT 105

Library of Congress Prints and Photographs Division [LC-USZC4-5315]

5 THE AMERICAN REVOLUTION 106

THE STATES UNITED 107
Defining American War Aims 107
The Declaration of Independence 110
Mobilizing for War 111

THE WAR FOR INDEPENDENCE 112
New England 112
The Mid-Atlantic 113
Securing Aid from Abroad 115
The South 116
Winning the Peace 117

WAR AND SOCIETY 120
Loyalists and Religious Groups 120
The War and Slavery 121
Native Americans and the Revolution 122
Women's Rights and Roles 123
The War Economy 125

MPI/Hulton Archive/Getty Images

THE CREATION OF STATE GOVERNMENTS 125
The Principles of Republicanism 125
The First State Constitutions 126
Revising State Governments 126

THE SEARCH FOR A NATIONAL GOVERNMENT 127
The Confederation 127
Diplomatic Failures 128
The Confederation and the Northwest 128
Native Americans and the Western Lands 130
Debts, Taxes, and Daniel Shays 130

Debating the Past: The American Revolution 108

America in the World: The Age of Revolutions 118

Consider the Source: The Correspondence of Abigail Adams on Women's Rights (1776) 124

CONCLUSION 132
KEY TERMS/PEOPLE/PLACES/EVENTS 132
RECALL AND REFLECT 133

6 THE CONSTITUTION AND THE NEW REPUBLIC 134

FRAMING A NEW GOVERNMENT 135
Advocates of Reform 135
A Divided Convention 136
Compromise 137
The Constitution of 1787 137

ADOPTION AND ADAPTATION 141
Federalists and Antifederalists 141
Completing the Structure 142

FEDERALISTS AND REPUBLICANS 143
Hamilton and the Federalists 144
Enacting the Federalist Program 144
The Republican Opposition 145

ESTABLISHING NATIONAL SOVEREIGNTY 146
Securing the West 146
Maintaining Neutrality 147

THE DOWNFALL OF THE FEDERALISTS 150
The Election of 1796 150
The Quasi War with France 150
Repression and Protest 151
The "Revolution" of 1800 152

Debating the Past: The Meaning of the Constitution 138

Consider the Source: Washington's Farewell Address, *American Daily Advertiser*, September 19, 1796 148

CONCLUSION 153
KEY TERMS/PEOPLE/PLACES/EVENTS 154
RECALL AND REFLECT 154

National Archives and Records Administration

7 THE JEFFERSONIAN ERA 155

THE RISE OF CULTURAL NATIONALISM 156
Educational and Literary Nationalism 156
Medicine and Science 157
Cultural Aspirations of the New Nation 158
Religion and Revivalism 158

STIRRINGS OF INDUSTRIALISM 160
Technology in the United States 160
Transportation Innovations 163
Country and City 166

JEFFERSON THE PRESIDENT 166
The Federal City and the "People's President" 166
Dollars and Ships 168
Conflict with the Courts 168

DOUBLING THE NATIONAL DOMAIN 169
Jefferson and Napoleon 169
The Louisiana Purchase 171
Exploring the West 171
The Burr Conspiracy 175

EXPANSION AND WAR 175
Conflict on the Seas 176
Impressment 176
"Peaceable Coercion" 177

Bettmann/Getty Images

Native Americans and the British 178
Tecumseh and the Prophet 179
Florida and War Fever 179

THE WAR OF 1812 180
Battles with the Nations 180
Battles with the British 181
The Revolt of New England 182
The Peace Settlement 183

America in the World: The Global Industrial Revolution 162

Patterns of Popular Culture: Horse Racing 164

Consider the Source: Thomas Jefferson to Meriwether Lewis (1803) 172

CONCLUSION 183
KEY TERMS/PEOPLE/PLACES/EVENTS 184
RECALL AND REFLECT 184

8 EXPANSION AND DIVISION IN THE EARLY REPUBLIC 185

STABILIZING ECONOMIC GROWTH 186
The Government and Economic Growth 186
Transportation 187

EXPANDING WESTWARD 188
Westward Migration 188
White Settlers in the Old Northwest 188
The Plantation System in the Old Southwest 189
Trade and Trapping in the Far West 189
Eastern Images of the West 190

THE "ERA OF GOOD FEELINGS" 191
The End of the First Party System 191
John Quincy Adams and Florida 191
The Panic of 1819 192

SECTIONALISM AND NATIONALISM 193
The Missouri Compromise 193
Marshall and the Court 195

The Court and Native Peoples 197
The Latin American Revolution and the Monroe Doctrine 198

THE REVIVAL OF OPPOSITION 199
The "Corrupt Bargain" 199
The Second President Adams 200
Jackson Triumphant 200

Consider the Source: Thomas Jefferson Reacts to the Missouri Compromise (1820) 194

CONCLUSION 201
KEY TERMS/PEOPLE/PLACES/EVENTS 201
RECALL AND REFLECT 201

Yale University Art Gallery

CONTENTS · xiii

9 JACKSONIAN AMERICA 202

THE RISE OF MASS POLITICS 203
Expanding Democracy 203
Tocqueville and *Democracy in America* 205
The Legitimization of Party 205
President of the Common People 207

"OUR FEDERAL UNION" 208
Calhoun and Nullification 208
The Rise of Van Buren 209
The Webster-Hayne Debate 209
The Nullification Crisis 210

THE REMOVAL OF NATIVE AMERICANS 210
White Attitudes toward Native Peoples 210
The "Five Civilized Tribes" 211
Trail of Tears 213
The Meaning of Removal 214

JACKSON AND THE BANK WAR 215
Biddle's Institution 215
The "Monster" Destroyed 216

THE CHANGING FACE OF AMERICAN POLITICS 217
Democrats and Whigs 218

POLITICS AFTER JACKSON 219
Van Buren and the Panic of 1837 219
The Log Cabin Campaign 220
The Frustration of the Whigs 224
Whig Diplomacy 224

Debating the Past: Jacksonian Democracy 206

Consider the Source: Letter from Chief John Ross to the Senate and House of Representatives (1836) 212

Patterns of Popular Culture: The Penny Press 222

CONCLUSION 225
KEY TERMS/PEOPLE/PLACES/EVENTS 225
RECALL AND REFLECT 226

Yale University Art Gallery

10 AMERICA'S ECONOMIC REVOLUTION 227

THE CHANGING AMERICAN POPULATION 228
Population Trends 228
Urban Growth and Immigration, 1840-1860 229
The Rise of Nativism 230

TRANSPORTATION AND COMMUNICATIONS REVOLUTIONS 231
The Canal Age 231
The Early Railroads 232
The Triumph of the Rails 233
The Telegraph 234
New Technology and Journalism 236

COMMERCE AND INDUSTRY 236
The Expansion of Business, 1820-1840 236
The Emergence of the Factory 236

Universal History Archive/Universal Images Group/Getty Images

Advances in Technology 237
Rise of the Industrial Ruling Class 238

MEN AND WOMEN AT WORK 238
Recruiting a Native Workforce 238
The Immigrant Workforce 239
The Factory System and the Artisan Tradition 241
Fighting for Control 242

PATTERNS OF SOCIETY 242
The Rich and the Poor 242
Social and Geographical Mobility 243
Middle-Class Life 244
The Changing Family 245
The "Cult of Domesticity" 246
Leisure Activities 246

THE AGRICULTURAL NORTH 248
Northeastern Agriculture 248
The Old Northwest 249
Rural Life 250

Consider the Source: Handbook to Lowell (1848) 240

CONCLUSION 251
KEY TERMS/PEOPLE/PLACES/EVENTS 251
RECALL AND REFLECT 252

11 COTTON, SLAVERY, AND THE OLD SOUTH 253

THE COTTON ECONOMY 254
The Rise of King Cotton 254
Southern Trade and Industry 257

SOUTHERN WHITE SOCIETY 257
The Planter Class 259
The "Southern Lady" 260
Beneath the Planter Class 260

SLAVERY: THE "PECULIAR INSTITUTION" 261
Slavery and Punishment 261
Life under Slavery 261
Slavery in the Cities 265
Free Blacks 265
The Slave Trade 266

BLACK CULTURE UNDER SLAVERY 267
Religion 267
Language and Song 268
Family 268
Resistance 269

Consider the Source: Senator James Henry Hammond Declares, "Cotton Is King" (1858) 258

Debating the Past: Analyzing Slavery 262

CONCLUSION 270
KEY TERMS/PEOPLE/PLACES/EVENTS 270
RECALL AND REFLECT 270

MPI/Archive Photos/Getty Images

12 ANTEBELLUM CULTURE AND REFORM 271

THE ROMANTIC IMPULSE 272
Nationalism and Romanticism in American Painting 272
An American Literature 273
Literature in the Antebellum South 273
The Transcendentalists 274
The Defense of Nature 276
Visions of Utopia 276
Redefining Gender Roles 277
The Mormons 277

REMAKING SOCIETY 279
Revivalism, Morality, and Order 279
Health, Science, and Phrenology 279
Medical Science 280
Education 281
Rehabilitation 282
The Rise of Feminism 282
Struggles and Successes of Black Women 283

THE CRUSADE AGAINST SLAVERY 286
Early Opposition to Slavery 286
Black Abolitionists 286
Garrison and Abolitionism 290
Anti-Abolitionism 290
Abolitionism Divided 291

Consider the Source: Declaration of Sentiments and Resolutions, Seneca Falls, New York (1848) 284

America in the World: The Abolition of Slavery 288

Patterns of Popular Culture: Sentimental Novels 292

CONCLUSION 294
KEY TERMS/PEOPLE/PLACES/EVENTS 294
RECALL AND REFLECT 294

Bettmann/Getty Images

13 THE IMPENDING CRISIS 295

LOOKING WESTWARD 296
Manifest Destiny 296
Americans in Texas 297
Oregon 298
The Westward Migration 298

EXPANSION AND WAR 300
The Democrats and Expansion 300
The Southwest and California 301
The Mexican War 302

THE SECTIONAL DEBATE 304
Slavery and the Territories 304
The California Gold Rush 305
Rising Sectional Tensions 307
The Compromise of 1850 308

THE CRISES OF THE 1850s 309
The Uneasy Truce 309
"Young America" 309
Slavery, Railroads, and the West 310
The Kansas–Nebraska Controversy 310
"Bleeding Kansas" 311
The Free-Soil Ideology 312
The Pro-Slavery Argument 313

Buchanan and Depression 313
The *Dred Scott* Decision 314
Deadlock over Kansas 315
The Emergence of Lincoln 315
John Brown's Raid 316
The Election of Lincoln 316

Consider the Source: Wilmot Proviso (1846) 306

CONCLUSION 317
KEY TERMS/PEOPLE/PLACES/EVENTS 318
RECALL AND REFLECT 318

Library of Congress, Prints and Photographs Division Washington, D.C. 20540 USA [LC-USZ62-11138]

14 THE CIVIL WAR 319

THE SECESSION CRISIS 320
The Withdrawal of the South 320
The Failure of Compromise 321
The Opposing Sides 321
Going to War 323

THE MOBILIZATION OF THE NORTH 324
Economic Nationalism 324
Raising the Union Armies 325
Wartime Politics 326
The Politics of Emancipation 327
Black Americans and the Union Cause 329
Women, Nursing, and the War 330

THE MOBILIZATION OF THE SOUTH 331
The Confederate Government 331
Money and Manpower 332
Economic and Social Effects of the War 333

STRATEGY AND DIPLOMACY 333
The Commanders 333
The Role of Sea Power 334

Library of Congress Prints and Photographs Division [LC-USZ61-903]

Europe and the Disunited States 336

CAMPAIGNS AND BATTLES 337
The Technology of War 337
The Opening Clashes, 1861 338
The Western Theater 338
The Virginia Front, 1862 339
The Progress of the War 341
1863: Year of Decision 341
The Last Stage, 1864–1865 345

Consider the Source: Ordinances of Secession (1860/1861) 322

Consider the Source: Letter from a Refugee (1862) 326

Consider the Source: The Emancipation Proclamation (1863) 328

CONCLUSION 348
KEY TERMS/PEOPLE/PLACES/EVENTS 349
RECALL AND REFLECT 349

15 RECONSTRUCTION AND THE NEW SOUTH 350

THE PROBLEMS OF PEACEMAKING 351
The Aftermath of War and Emancipation 351
Competing Notions of Freedom 351
Plans for Reconstruction 353
The Death of Lincoln 355
Johnson and "Restoration" 357

RADICAL RECONSTRUCTION 358
The Black Codes 359
The Fourteenth Amendment 359
The Congressional Plan 361
The Impeachment of Andrew Johnson 362

THE SOUTH IN RECONSTRUCTION 363
Politics 363
Education 364
Landownership and Tenancy 364

THE GRANT ADMINISTRATION 366
The Soldier President 366
The Grant Scandals 367
The Greenback Question 367
Republican Diplomacy 367

THE ABANDONMENT OF RECONSTRUCTION 368
The Southern States for Southern Whites 368
Waning Northern Commitment 369
The Compromise of 1877 369
The Legacy of Reconstruction 370

THE NEW SOUTH 371
The "Redeemers" 371
Industrialization and the New South 371
Black Americans and the New South 372
The Lost Cause 373
The Birth of Jim Crow 374

Consider the Source: Southern Blacks Demand Federal Aid (1865) 354

Debating the Past: Reconstruction 356

Consider the Source: Mississippi Black Codes (1865) 360

Patterns of Popular Culture: The Minstrel Show 376

CONCLUSION 378
KEY TERMS/PEOPLE/PLACES/EVENTS 378
RECALL AND REFLECT 378

Corbis/Getty Images

APPENDIX A-1
The Declaration of Independence A-2
The Constitution of the United States A-6
GLOSSARY G
INDEX I-1

PREFACE

The title *The Unfinished Nation* is meant to suggest several things. It is a reminder of the exceptional diversity of the United States—of the degree to which, despite all the many efforts to build a single, uniform definition of the meaning of American nationhood, that meaning remains contested. It is a reference to the centrality of change in American history—to the ways in which the nation has continually transformed itself and continues to do so in our own time. And it is also a description of the writing of American history itself—of the ways in which historians are engaged in a continuing, ever-unfinished process of asking new questions.

Like any history, *The Unfinished Nation* is a product of its time and reflects the views of the past that historians of recent generations have developed. The writing of our nation's history—like our nation itself—changes constantly. It is not, of course, the past that changes. Rather, historians adjust their perspectives and priorities, ask different kinds of questions, and uncover and incorporate new historical evidence. There are now, as there have always been, critics of changes in historical understanding who argue that history is a collection of facts and should not be subject to "interpretation" or "revision." But historians insist that history is not simply a collection of facts. Names and dates and a record of events are only the beginning of historical understanding. Writers and readers of history interpret the evidence before them, and inevitably bring to the task their own questions, concerns, and experiences.

This edition continues the evolution of the *The Unfinished Nation* as authors John M. Giggie and Andrew J. Huebner build upon this canonical text, with a focus on making history relatable and accessible to today's students. John M. Giggie is a historian of race and religion, Andrew J. Huebner is a historian of war and society, and both more generally study and teach American social and cultural history. Their interests join and complement Alan Brinkley's expansive base of knowledge in the history of American politics, society, and culture. Alan's scholarship inspired John and Andrew as graduate students and they are honored to continue the work of *The Unfinished Nation*. They endeavor to bring their own scholarly interests and sensitivities to an already vibrant, clear, concise, and balanced survey of American history. The result, we hope, is a text that explores the great range of ideas, institutions, individuals, and events that make up the fabric of society in the United States.

It is a daunting task to attempt to convey the history of the United States in a single book, and the tenth edition of *The Unfinished Nation* has, as have all previous editions, been carefully written and edited to keep the book as concise and readable as possible. It features most notably an enlarged focus on the history of Native Americans, the experiences of enslaved peoples in the United States, the ever-shifting political landscape with its associated opportunities and challenges, the Civil War and Reconstruction periods, the struggles and successes of Black Americans since the Civil War, and dramatic political and economic change in the twenty-first century, including discussion of the COVID-19 pandemic. Across these subjects, we recognize that to understand the full complexity of the American past it is necessary to understand both the forces that divide Americans and the forces that draw them together. Thus we've sought to explore the development of foundational ideals like democracy and equality as well as the ways that our nation's fulfillment of those ideals remains, like so much else, unfinished.

Paired with Connect History, a digital assignment and assessment platform, instructors and students utilizing *The Unfinished Nation* are able to accomplish more in less time. Among other resources, Connect History offers interactive map assignments and tools to strengthen critical reading and writing skills.

AMERICA'S HISTORY IS STILL UNFOLDING

Is American History finished? Not yet! *The Unfinished Nation* shows that as more details are uncovered, dates may not change—but perceptions and reality definitely can. The United States and its history are in a constant state of change.

Just like the United States, this edition evolves, benefiting from the voices of John M. Giggie and Andrew J. Huebner, whose expertise sheds light on perspectives that shape an examination of the past. Their aim is to help students ask new questions. By doing so, students find their own answer to the question: is American History finished?

PRIMARY SOURCES HELP STUDENTS THINK CRITICALLY ABOUT HISTORY

Primary sources help students think critically about history and expose them to contrasting perspectives of key events. The Tenth Edition of *The Unfinished Nation* provides three different ways to use primary source documents in your course.

Power of Process is a critical thinking tool for reading and writing about primary sources. As part of Connect History, Power of Process contains a database of over 400 searchable primary sources in addition to the capability for instructors to upload their own sources. Instructors can then select a series of strategies for students to use to analyze and comment on a source. The Power of Process framework helps students develop essential academic skills such as understanding, analyzing, and synthesizing readings and visuals such as maps, leading students toward higher order thinking and writing.

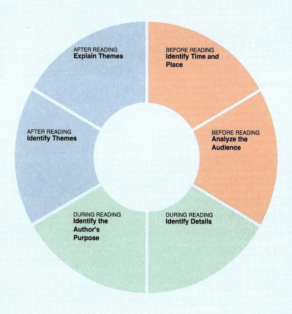

The Power of Process landing page makes it easy for instructors to find pre-populated documents or to add their own.

Features that offer contrasting perspectives or showcase historical artifacts. Within the print or eBook, the Tenth Edition of *The Unfinished Nation* offers the following features:

CONSIDER THE SOURCE

In every chapter, Consider the Source features guide students through careful analysis of historical documents and prompt them to closely examine the ideas expressed, as well as the historical circumstances. Among the classic sources included are Cotton Mather on New England society, Benjamin Franklin's testimony against the Stamp Act, Abigail Adams' letter to her husband regarding women's rights, and the Emancipation Proclamation. Concise introductions provide context, and concluding questions prompt students to understand, analyze, and evaluate each source.

DEBATING THE PAST

Debating the Past essays introduce students to the contested quality of much of the American past and provide a sense of the evolving nature of historical scholarship. From examining specific differences in historical understandings of the Constitution, to exploring the causes of the Civil War, these essays familiarize students with the interpretive character of historical understanding.

AMERICA IN THE WORLD

America in the World essays focus on specific parallels between American history and those of other nations and demonstrate the importance of the many global influences on the American story. Topics such as the global Industrial Revolution and the abolition of slavery provide concrete examples of the connections between the history of the United States and the history of other nations.

PATTERNS OF POPULAR CULTURE

Patterns of Popular Culture essays bring fads, crazes, hangouts, hobbies, and entertainment into the story of American history, encouraging students to expand their definition of what constitutes history and gain a new understanding of what popular culture reveals about a society.

 Select primary source documents that meet the unique needs of your course. No two history courses are the same. Using McGraw Hill Education's Create allows you to quickly and easily create custom course materials with cross-disciplinary content and other third-party sources.

- **CHOOSE YOUR OWN CONTENT:** Create a book that contains only the chapters you want, in the order you want. Create will even renumber the pages for you!
- **ADD READINGS:** Use our American History Collections to include primary sources, or Taking Sides: Annual Editions. Add your own original content, such as syllabus or History major requirements!
- **CHOOSE YOUR FORMAT:** Print or eBook? Softcover, spiral-bound, or loose-leaf? Black-and-white or color? Perforated, three-hole punched, or regular paper?
- **CUSTOMIZE YOUR COVER:** Pick your own cover image and include your name and course information right on the cover. Students will know they're purchasing the right book—and using everything they purchase!
- **REVIEW YOUR CREATION:** When you are all done, you'll receive a free PDF review copy in just minutes! To get started, go to create.mheducation.com and register today.

WRITING ASSIGNMENT

McGraw Hill's new Writing Assignment Plus tool delivers a learning experience that improves students' written communication skills and conceptual understanding with every assignment.

Assign, **monitor**, and **provide feedback** on writing more efficiently and grade assignments within McGraw Hill Connect®. Writing Assignment Plus gives you time-saving tools with a just-in-time basic writing and originality checker.

Features include:

- Grammar/writing checking with McGraw Hill learning resources.
- Originality checker with McGraw Hill learning resources.
- Streamlined tools for faculty to make grading writing easier.
- Writing stats that identify common student issues.
- Rubric building and scoring.
- Ability to assign draft and final deadline milestones.
- Tablet-readiness, with tools for all learners.

MAP TOOLS TO PROMOTE STUDENT LEARNING

Using Connect History and more than 100 maps, students can learn the course material more deeply and study more effectively than ever before.

Interactive maps give students a hands-on understanding of geography. *The Unfinished Nation* offers over 30 interactive maps that support geographical as well as historical thinking. These maps appear in both the eBook and Connect History exercises. For some interactive maps, students click on the boxes in the map legend to see changing boundaries, visualize migration routes, or analyze war battles and election results. With others, students manipulate a slider to help them better understand change over time. New interactive maps feature advanced navigation features, including zoom, as well as audio and textual animation.

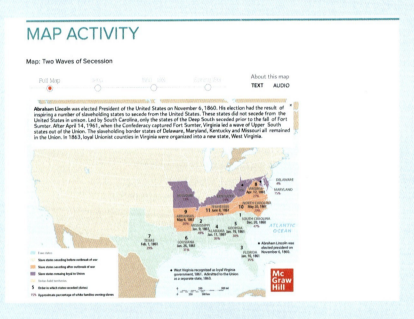

SMARTBOOK

Available within Connect History, SmartBook has been updated with improved learning objectives to ensure that students gain foundational knowledge while also learning to make connections to help them formulate a broader understanding of historical events. SmartBook 2.0 personalizes learning to individual student needs, continually adapting to pinpoint knowledge gaps and to focus learning on topics that need the most attention. Study time is more productive and, as a result, students are better prepared for class and coursework. For instructors, SmartBook 2.0 tracks student progress and provides insights that can help guide teaching strategies.

CONTEXTUALIZE HISTORY

Help students experience history in a whole new way with our Podcast Assignments. We've gathered some of the most interesting and popular history podcasts currently available and built assignable questions around them. These assignments allow instructors to bring greater context and nuance to their courses while engaging students through the storytelling power of podcasts.

INSTRUCTOR RESOURCES

The Unfinished Nation offers an array of instructor resources for the U.S. history course:

Instructor's Manual The Instructor's Manual provides a wide variety of tools and resources for presenting the course, including learning objectives and ideas for lectures and discussions.

Test Bank By increasing the rigor of the test bank development process, McGraw Hill has raised the bar for student assessment. Each question has been tagged for level of difficulty, Bloom's taxonomy, and topic coverage. Organized by chapter, the questions are designed to test factual, conceptual, and higher-order thinking.

Test Builder Available within Connect, Test Builder is a cloud-based tool that enables instructors to format tests that can be printed and administered within a Learning Management System. Test Builder offers a modern, streamlined interface for easy content configuration that matches course needs without requiring a download.

Test Builder enables instructors to

- Access all test bank content from a particular title.
- Easily pinpoint the most relevant content through robust filtering options.
- Manipulate the order of questions or scramble questions and/or answers.
- Pin questions to a specific location within a test.
- Determine your preferred treatment of algorithmic questions.
- Choose the layout and spacing.
- Add instructions and configure default settings.

PowerPoint The PowerPoint presentations highlight the key points of the chapter and include supporting visuals. All slides are WCAG compliant.

Remote Proctoring and Browser-Locking Capabilities. Remote proctoring and browser-locking capabilities, hosted by Proctorio within Connect, provide control of the assessment environment by enabling security options and verifying the identity of the student. Seamlessly integrated within Connect, these services allow instructors to control students' assessment experience by restricting browser activity, recording students' activity, and verifying students are doing their own work. Instant and detailed reporting gives instructors an at-a-glance view of potential academic integrity concerns, thereby avoiding personal bias and supporting evidence-based claims.

CHAPTER-BY-CHAPTER CHANGES

We have extensively revised the narrative and features in this Tenth Edition to bring in new scholarship, particularly as it relates to the experiences and perspectives of Native Americans, Black Americans, and women throughout American history. Revisions and updates in every chapter reflect the most recent scholarship as well as the advice of our panel of reviewers. Following are the major changes organized by chapter:

Chapter 1, The Collision of Cultures
- Revised content addressing political and cultural achievements of indigenous societies before the arrival of Europeans.
- Updated material on Henry Hudson's travels.

Chapter 2, Transplantations and Borderlands
- Expanded discussion of Anne Hutchinson's social role and theology.
- Revised content on William Penn and the Pennsylvania Quakers, with a focus on relationships with native peoples.
- New material on the Barbados Slave Code of 1661.

Chapter 3, Society and Culture in Provincial America
- New and revised material pertaining to early medicine in the American colonies.
- Revised material pertaining to early industry in the colonies.
- Updated discussion of the experiences of enslaved people in the Southern communities.
- Revised material on the economies and social patterns in Northern settlements, including discussion of enslaved people living in Northern colonies.
- Revised content on the religious heritage of enslaved Africans.

Chapter 4, The Empire in Transition
- Updated treatment of native resistance to European powers.
- Revised discussion of the different approaches taken by British and French colonies in North American, particularly in regards to relationships with natives peoples.

Chapter 5, The American Revolution
- Updates regarding the 1619 project in the Debating the Past feature on the American Revolution.
- Revised material on the role of enslaved people in the Revolutionary War as well as the way that war affected the lives of enslaved people.

Chapter 6, The Constitution and the New Republic
This chapter features substantial reworking of the coverage of the Constitution, slavery, and the rancor of early American politics. Specific updates include

- Revised material on slavery and the Constitution, including significant revisions in the Debating the Past feature.
- Expanded discussion of the system of checks and balances.
- Revised material on the Bill of Rights.
- Revised material on Hamilton's approach to the national economy.
- New material on the emergence of a two-party system.
- Updated treatment of the Alien and Sedition Acts.

Chapter 7, The Jeffersonian Era
This chapter has been significantly revised to reflect the newest scholarship on the Jeffersonian period. Particular attention has been paid to westward expansion, violence against and dispossession of Native Americans, and the War of 1812. Other updates include

- Clarifications in discussion of the impact of the cotton gin on the slavery system.
- Expanded material on the Louisiana Purchase.

Chapter 8, Expansion and Division in the Early Republic
- Expanded discussion of the impacts of westward migration on indigenous societies during the early republican period.
- Revised material pertaining to the plantation system in the Old Southwest.
- New content on Jackson's activities during the Seminole War.
- Revised material on the Missouri Compromise.
- Updated discussion of the effect of the Marshall Court's rulings on native peoples.
- New material on the formation of the second two-party system.
- New material pertaining to the "corrupt bargain."

Chapter 9, Jacksonian America
- Expanded material on the Dorr Rebellion.
- Updated discussion in the Debating the Past feature reflecting the most recent scholarship pertaining to Jacksonian democracy.
- Significant updates and revisions pertaining to the forced removal of Native Americans during the Jackson presidency.
- Revised content on the philosophies and approaches of the Democrats and the Whigs.

Chapter 10, America's Economic Revolution
- Expanded discussion of immigration and urban growth.
- New material on the growth of the railroads.
- New material on women's early efforts to unionize.
- New material on class conflict.

Chapter 11, Cotton, Slavery, and the Old South
- Expanded and revised material on the cotton economy.
- Significantly revised material on Southern white society, including the roles of women, the class divide among white Southerners, and the commitment to slavery as the foundation of the economy.
- Significantly revised material on slavery in the American South, including updates to the Debating the Past feature to reflect recent scholarship, expanded material on the reliance on punishment as a way of managing enslaved individuals, new content on the diets and daily lives of enslaved people, and a revised discussion of the experiences of those sold through the slave trade.
- Significantly rewritten content on Black culture under slavery, with new or revised material on religion, language, song, family, and means of resistance.

Chapter 12, Antebellum Culture and Reform
- Revised discussion of the importance of the Hudson River school.
- New content on Southern writers.
- Revised material on the development, culture, and theology of the Shakers.
- Revised content on health care in the antebellum period, including new material on the racist applications of phrenology.
- Rewritten discussion of prison reform efforts.
- Updated material on early opposition to slavery.
- New and revised material on the Southern response to the abolition movement.

Chapter 13, The Impending Crisis
- Rewritten material pertaining to the concept of manifest destiny.
- Updated material on John Brown and "Bleeding Kansas."

Chapter 14, The Civil War
This chapter has been significantly rewritten and revised. Changes include
- New material reflecting latest scholarship on the question of why the South seceded
- New Consider the Source feature "Ordinances of Secession (1860/1861)."
- New material on the differences in the ways that Southerners and Northerners viewed the Civil War while it was being fought, as well as material on the way that enslaved people and free Blacks viewed the Civil War.
- Revised discussion of political changes in the North associated with the War.
- New Consider the Source feature "Letter from a Refugee (1862)."
- New Consider the Source feature "The Emancipation Proclamation (1863)."
- New and revised content on the experiences of Black soldiers who fought for the Union.
- New and revised content on women's roles during the Civil War in both the North and the South.
- New material on the politic effects of the Conscription Act.
- New material on the skills, strategies, and personalities of key Southern and Northern military leaders.
- New material on the cultural and class background of Civil War soldiers in both the Union and Confederate armies.
- Updated material on the impact of military technology on Civil War battles.

Chapter 15, Reconstruction and the New South

This chapter has been significantly rewritten and revised. Changes include

- Updated content on the number of causalities caused by the Civil War.
- Clarified discussion of competing notions of freedom held by white Southerners and Black Americans.
- New material on the development of Black higher education.
- Updated and revised material on the scholarship pertaining to Reconstruction in the Debating the Past box, with a focus on the contributions of gender historians.
- New material on the significance and limitations of the Fourteenth Amendment.
- New material on Black politics during Reconstruction.
- New material on the growth and activities of the Ku Klux Klan.
- New section on the growth and cultural significance of Lost Cause mythology in the South.
- Revised treatment of the birth of Jim Crow.

ACKNOWLEDGMENTS

We would like to express our deep appreciation to the following faculty members who contributed to the development of *The Unfinished Nation, Tenth Edition*:

Charles Adams, *North Central Texas College*
Alana Aleman, *Lone Star College*
Shelly Bailess, *Liberty University*
Sheryl Ballard, *Houston Community College*
Kelly Cantrell, *East Mississippi Community College*
John Carr Shanahan, *University of Texas at San Antonio*
Caitlin Curtis, *Liberty University*
Roger Hardaway, *Northwestern Oklahoma State University*
Sandra Harvey, *Lone Star College*
Raymond Hylton, *Virginia Union University*
Bradley Keefer, *Kent State University*
Frederic Krome, *University of Cincinnati*
Bob Miller, *University of Cincinnati*
Carey Roberts, *Liberty University*
James Thomas, *Houston Community College*
Shawna Williams, *Houston Community College*

ABOUT THE AUTHORS

ALAN BRINKLEY (1949-2019) was the Allan Nevins Professor of History at Columbia University. He served as university provost at Columbia from 2003 to 2009. He authored works such as *Voices of Protest: Huey Long, Father Coughlin, and the Great Depression*, which won the 1983 National Book Award; *American History: Connecting with the Past*; *The End of Reform: New Deal Liberalism in Recession and War*; *Liberalism and Its Discontents*; *Franklin D. Roosevelt*; and *The Publisher: Henry Luce and His American Century*. He served as board chair of the National Humanities Center, board chair of the Century Foundation, and a trustee of Oxford University Press. He was also a member of the Academy of Arts and Sciences. In 1998-1999 he was the Harmsworth Professor of History at Oxford University, and in 2011-2012 the Pitt Professor at the University of Cambridge. He won the Joseph R. Levenson Memorial Teaching Award at Harvard and the Great Teacher Award at Columbia. He was educated at Princeton and Harvard.

JOHN M. GIGGIE is associate professor of history and African American studies at the University of Alabama where he also serves as Director of the Summersell Center for the Study of the South. He is the author of *After Redemption: Jim Crow and the Transformation of African American Religion in the Delta, 1875-1917*, editor of *America Firsthand*, editor of *Faith in the Market: Religion and the Rise of Commercial Culture* and co-editor of *Dixie Great War: World War I and the American South*. He is a series editor for Religion and Culture at the University of Alabama Press. In 2020, Prof. Giggie taught the first Black history course offered daily for an entire year at an Alabama public school. He is co-founder of the West Side Scholars Academy, a middle school summer enrichment program that focuses on local civil rights history. He is managing a research study of lynching in Alabama and preparing a book on civil rights protests in West Alabama. He was educated at Amherst College and Princeton University.

ANDREW J. HUEBNER is associate professor of history at the University of Alabama. He is the author of *Love and Death in the Great War* (2018) and *The Warrior Image: Soldiers in American Culture from the Second World War to the Vietnam Era* (2008). He is co-editor of *Dixie's Great War* (2020) as well as two other forthcoming edited volumes on the subject of war and society in the United States. In 2017, he was named an Organization of American Historians (OAH) Distinguished Lecturer. He received his PhD from Brown University.

THE UNFINISHED NATION

1 | THE COLLISION OF CULTURES

AMERICA BEFORE COLUMBUS
EUROPE LOOKS WESTWARD
THE ARRIVAL OF THE ENGLISH

LOOKING AHEAD

1. How did the societies of native people in the South differ from those in the North in the precontact period (before the arrival of the Europeans)?
2. What effects did the arrival of Europeans have on the native peoples of the Americas?
3. How did patterns of settlement differ within the Americas?

THE DISCOVERY OF THE AMERICAS did not begin with Christopher Columbus. It began many thousands of years earlier, when human beings first crossed into the new continents and began to people them. By the end of the fifteenth century, when the first important contact with Europeans occurred, the Americas were already home to millions of men and women.

These ancient civilizations experienced many changes and many catastrophes during their long history. But likely none was as tragically transforming as the arrival of Europeans. In the first violent years of Spanish and Portuguese exploration, the impact of the new arrivals was profound. European invaders brought with them diseases (most notably smallpox) previously unknown to native peoples to which they had no immunity. The result was a demographic disaster that killed millions of people, weakened existing societies, and greatly aided the Spanish and Portuguese in their rapid and devastating takeover of empires that had existed long before they arrived.

But the European immigrants were never able to eliminate the influence of the indigenous peoples (whom they came to call "Indians"). In their many interactions, whether beneficial or ruinous, these very different civilizations shaped one another, learned from one another, and changed one another forever.

TIME LINE

AMERICA BEFORE COLUMBUS

While relatively little is known about the first peoples in the Americas, archaeologists continue to discover ancient artifacts that provide new information about them.

THE PEOPLES OF THE PRECONTACT AMERICAS

For many decades, scholars believed that all early migrations into the Americas came from humans crossing an ancient land bridge over the Bering Strait into what is now Alaska, approximately 11,000 years ago. The migrations were probably a result of the development of new stone tools—spears and other hunting implements—used to pursue the large animals that crossed between Asia and North America. All of these land-based migrants are thought to have come from a Mongolian stock related to that of modern-day Siberia. Scholars refer to these migrants as the **"Clovis" people**, so named for a town in New Mexico where archaeologists first discovered evidence of their tools and weapons in the 1930s.

More recent archaeological evidence, however, suggests that not all the early migrants to the Americas came across the Bering Strait. Some migrants from Asia appear to have settled as far south as modern-day Chile and Peru even before people began moving into North America by land. These first South Americans may have come by sea, using boats.

This new information suggests that the early population of the Americas was more diverse and more scattered than scholars previously assumed. Recent DNA evidence has identified a possible early population group that does not seem to have come from Asia. This suggests that thousands of years before Columbus, there may have been some migration from Europe.

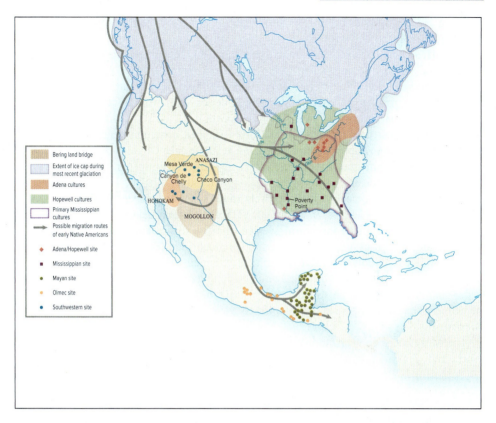

NORTH AMERICAN MIGRATIONS This map tracks some of the very early migrations into, and within, North America in the centuries preceding contact with Europe. It shows the now-vanished land bridge between Siberia and Alaska over which thousands, perhaps millions, of migrating people passed into the Americas. It also shows the locations of some of the earliest settlements in North America. • *What role did the extended glacial field in what is now Canada play in residential patterns in the ancient American world?*

The *Archaic period* is a scholarly term for the early history of humans in America, beginning around 8000 B.C. In the first part of this period, most humans supported themselves through hunting and gathering, using the same stone tools that earlier Americans had brought with them. Later in the Archaic period, population groups began to expand their activities and develop new tools, such as nets and hooks for fishing, traps for smaller animals, and baskets for gathering berries, nuts, seeds, and small plants. Still later, some groups began to farm. Farming, of course, requires people to stay in one place. In agricultural areas, the first sedentary settlements slowly began to form, creating the basis for larger civilizations.

THE GROWTH OF CIVILIZATIONS: THE SOUTH

The most elaborate early civilizations emerged in South and Central America and in Mexico. In Peru, the Incas created the largest empire in the Americas, stretching almost 2,000 miles along western South America. The Incas developed a complex administrative state, an irrigation system, and a large network of paved roads that welded together the populations of many peoples under a single government.

Organized societies emerged around 10,000 B.C. in **Mesoamerica**, a region comprising Mexico and much of Central America. The Olmec people, whose roots trace back to between 1600 and 1500 B.C., were the first complex society in the region. A more sophisticated culture grew up in parts of Central America and in the Yucatán peninsula of Mexico, in an area known as Maya. Mayan civilization, which stretched back to 1800 B.C. and was at its most powerful about A.D. 300, developed a written language, a numerical system similar to the Arabic numeral system, an accurate calendar, an advanced agricultural system, and important trade routes into other areas of the continents.

Gradually, the societies of the Maya region were superseded by other Mesoamerican groups, who have become known collectively (and somewhat inaccurately) as the Aztecs. They called themselves Mexica. In about A.D. 1325, the Mexicas built the city of Tenochtitlán on a large island in a lake in central Mexico, the site of present-day Mexico City. With a population as high as 100,000 by A.D. 1500, Tenochtitlán featured large and impressive public buildings, schools that all male children attended, an organized military, a medical system, and an enslaved workforce drawn from conquered peoples. It was a city built over water and featuring a sophisticated water navigation system, much like Venice, Italy, but larger. The Mexicas gradually established their dominance over almost all of central Mexico.

The Mesoamerican civilizations were for many centuries the center of civilized life in North and Central America—the hub of culture and trade.

THE CIVILIZATIONS OF THE NORTH

The peoples north of Mexico developed less elaborate but still substantial civilizations. Inhabitants of the northern regions of the continent subsisted on combinations of hunting, gathering, and fishing. They included the Inuit of the Arctic Circle, who fished and hunted seals; big-game hunters of the northern forests, who led nomadic lives based on the pursuit of moose and caribou; nations of the Pacific Northwest, who relied heavily on salmon fishing and who created substantial permanent settlements along the coast; and groups spread through relatively arid regions of the Far West, who built successful communities based on fishing, hunting small game, and gathering edible plants.

Other societies in North America were primarily agricultural. Among the most developed were those in the Southwest. Between A.D. 900 and 1150, the ancient Pueblo people developed a thriving center of culture and commerce in Chaco Canyon, in modern-day northwestern New Mexico. At its apex, Chaco Canyon boasted a population of 15,000, 12 towns, and 200 villages—one of the largest of which was Pueblo Bonita. Composed of sandstone, timber, and adobe, Pueblo Bonita soared five stories high and had 600 rooms. There would not be another structure of this size in North America until the 1880s. At roughly the same period, the Hopis lived in small masonry villages, farmed corn, and developed an elaborate irrigation system, a ceremonial culture, and a trade network stretching across what is now Arizona. And the Zunis, based in the desert areas of present-day Arizona, Utah, and New Mexico, built large stone and adobe villages centered on a plaza, created elaborate pottery, and farmed corn and other grains.

The eastern third of what is now the United States—much of it covered with forests and inhabited by the Woodland Indians—had the greatest food resources of any area of the continent. Most of the many peoples of this region engaged in farming, hunting, gathering,

HOW THE EARLY NORTH AMERICANS LIVED This map shows the various ways in which the native peoples of North America supported themselves before the arrival of European civilization. They survived largely on the resources available in their immediate surroundings. Note, for example, the reliance on the products of the sea by those living along the northern coastlines of the continent, and the way in which groups in relatively inhospitable climates in the North—where agriculture was difficult—relied on hunting large game. • *What different kinds of farming would have emerged in the very different climates of the agricultural regions shown on this map?*

and fishing simultaneously. In the South there were permanent settlements and large trading networks based on the corn, legumes, and squash grown in the rich lands of the Mississippi River valley. **Cahokia**, a trading center located near present-day St. Louis, had a population of 40,000 at its peak in A.D. 1200. Residents traded not only their crops but also their locally made hand tools and pottery. Occupying six square miles, Cahokia was the largest and most populous urban center north of Tenochtitlán and would remain so until Philadelphia in 1780.

The agricultural societies of the Northeast were more mobile. Farming techniques there were designed to exploit the land quickly rather than to develop permanent settlements. Many of the nations living east of the Mississippi River were loosely linked together by

(Don Mammoser/Shutterstock)

PUEBLO VILLAGE OF THE SOUTHWEST

common linguistic roots. The largest of these language groups consisted of the Algonquian, who lived along the Atlantic seaboard from Canada to Virginia; the Iroquois Confederacy, which was centered in what is now upstate New York; and the Muskogean, which consisted of the peoples in the southernmost regions of the eastern seaboard.

Most indigenous societies were matrilineal, meaning that family association and clan membership flowed through the mother's heritage. In contrast, in Europe ancestral descent typically followed paternal lines. All Native American groups assigned women the majority of work to care for children, prepare meals, and gather certain foods. But the allocation of other tasks varied from one society to another. In the case of the Hopi, women and men shared cultural authority. Women assumed leadership roles in the household, economy, and social system; men tended to predominate in religion and politics. Yet women reserved the power to negate or renegotiate trade or land deals forged by men if they deemed them unjust or imbalanced.

EUROPE LOOKS WESTWARD

Europeans were almost entirely unaware of the existence of the Americas before the fifteenth century. A few early wanderers—Leif Eriksson, an eleventh-century Norse seaman, and others—had glimpsed parts of the eastern Atlantic on their voyages. But even if their discoveries had become common knowledge (and they did not), there would have been little incentive for others to follow. Europe in the Middle Ages (roughly A.D. 500-1500) was too weak, divided, and decentralized to inspire many great ventures. By the end of the fifteenth century, however, conditions in Europe had changed and the incentive for overseas exploration had grown.

COMMERCE AND SEA TRAVEL

Two important social changes encouraged Europeans to look toward new lands. The first was the significant growth in Europe's population in the fifteenth century. The Black Death, a catastrophic epidemic of the bubonic plague that began in Constantinople in 1347, had killed more than a third of the people on the Continent (according to some estimates). But a century and a half later, the population had rebounded. With that growth came a reawakening of commerce. A new merchant class was emerging to meet the rising demand for goods from abroad. As trade increased, and as advances in navigation made long-distance sea travel more feasible, interest in expanding trade grew even more quickly. The second change was the emergence of new governments that were more united and powerful than the feeble political entities of the feudal past. In the western areas of Europe in particular, strong new monarchs were eager to enhance the commercial development of their nations.

Above all, Europeans who craved commercial glory had dreamed of trade with the East. It was not a new dream. In the early fourteenth century, Marco Polo and other adventurers had returned from Asia bearing exotic spices, cloths, and dyes and even more exotic tales. Yet for two centuries, that trade had been limited by the difficulties of the long overland journey to the Asian courts. By the mid-fourteenth century, talk of finding a faster, safer sea route to East Asia had began.

The Portuguese were the preeminent maritime power in the fifteenth century, largely because of Prince Henry the Navigator, who devoted much of his life to the promotion of exploration. In 1486, after Henry's death, the Portuguese explorer Bartholomeu Dias rounded the southern tip of Africa (the Cape of Good Hope). In 1497–1498, Vasco da Gama proceeded all the way around the cape to India. But the Spanish, not the Portuguese, were the first to encounter the *New World,* the term Europeans applied to the ancient lands previously unknown to them.

CHRISTOPHER COLUMBUS

Christopher Columbus was born and reared in Genoa, Italy. He spent his early seafaring years in the service of the Portuguese, stoking his ambitions of undertaking a great voyage of discovery. By the time he was a young man, he believed he could reach East Asia by sailing west, across the Atlantic, rather than east, around Africa. Columbus thought the world was far smaller than it actually is. He also was convinced that the Asian continent extended farther eastward than it actually does. Most important, he did not realize that anything lay to the west between Europe and the lands of Asia.

Columbus failed to enlist the leaders of Portugal to back his plan, so he turned instead to Spain. The marriage of Spain's two most powerful regional rulers, Ferdinand of Aragon and Isabella of Castile, had produced the strongest and most ambitious monarchy in Europe. Columbus appealed to Queen Isabella for support for his proposed westward voyage, and in 1492, she agreed. Commanding ninety men and three ships—the *Niña,* the *Pinta,* and the *Santa María*—Columbus left Spain in August 1492 and sailed west into the Atlantic. Ten weeks later, he sighted land and assumed he had reached an island off Asia. In fact, he had landed in the modern-day Bahamas. When he pushed on and encountered what we now call Cuba, he assumed he had reached Japan. He returned to Spain, bringing with him several captured native people as evidence of his achievement. (He called the indigenous people "Indians" because he believed they were from the East Indies in the Pacific.)

But Columbus did not bring back news of the great khan's court in China or any samples of the fabled wealth of the Indies. And so a year later he tried again, only this time with a much larger expedition. As before, he headed into the Caribbean, discovering several other islands and leaving a small and short-lived **colony** on Hispaniola. On a third voyage, in 1498, he finally reached the mainland and cruised along the northern coast of South America. He then realized, for the first time, that he had encountered not a part of Asia but a separate continent.

Columbus ended his life in obscurity. Ultimately, he was even unable to give his name to the land he had revealed to the Europeans. That distinction went instead to a Florentine merchant, Amerigo Vespucci, who wrote a series of vivid descriptions of the lands he visited on a later expedition to the New World and helped popularize the idea that the Americas were new continents.

Partly as a result of Columbus's initiative, Spain began to devote greater resources and energy to maritime exploration. In 1513, the Spaniard Vasco de Balboa crossed the Isthmus

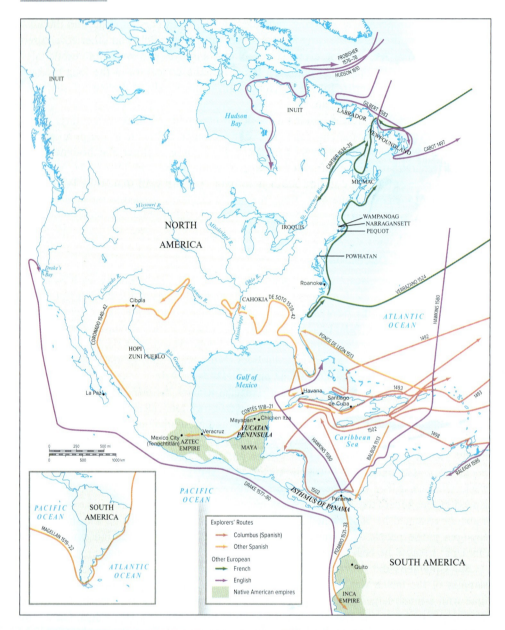

EUROPEAN EXPLORATION AND CONQUEST, 1492–1583 This map shows the many voyages of exploration to, and conquest of, North America launched by Europeans in the late fifteenth and sixteenth centuries. Note how Columbus and the Spanish explorers who followed him tended to move quickly into the lands of Mexico, the Caribbean, and Central and South America, while the English and French explored the northern territories of North America. In all cases they encountered native peoples, whose roots trace back centuries before the arrival of the Europeans. • *What factors might have led these various nations to explore and colonize different areas of the New World?*

of Panama and became the first known European to gaze westward upon the great ocean that separated America from China. Seeking access to that ocean, Ferdinand Magellan, a Portuguese in Spanish employ, found the strait that now bears his name at the southern end of South America, struggled through the stormy narrows and into the ocean (so calm by contrast that he christened it the *Pacific*), and then proceeded to the Philippines. There Magellan died in a conflict with local indigenous people, but his expedition went on to complete the first known circumnavigation of the globe (1519-1522). By 1550, Spaniards had explored the coasts of North America as far north as Oregon in the west and Labrador in the east.

THE SPANISH EMPIRE

In time, Spanish explorers in the New World stopped thinking of America simply as an obstacle to their search for a route to Asia and began instead to view it as a possible source of wealth itself. The Spanish claimed for themselves the whole of the New World, except for a large part of the east coast of South America (today's Brazil) that was reserved by a papal decree for the Portuguese.

In 1518, Hernando Cortés, who had been an unsuccessful Spanish government official in Cuba for fourteen years, led a small military expedition of about 600 men against the Aztecs in Mexico and their powerful emperor, Montezuma, after hearing stories of their great treasures. Moving his warriors through Mexico, he befriended a native group that he labeled the Tlaxcalans, who were rivals of the Aztecs and would become crucial military allies. Approaching Tenochtitlán, Cortés benefited from perfect timing. His arrival seemed to fulfill a popular Aztec prophecy that claimed the god Quetsalcoatl was to return to Earth. The Aztecs mistook Cortés and his fighters—mysterious light skinned men—as divine company and greeted them as honored figures. Cortés, with the support of the Tlaxcalans, quickly took control of the city. Key to his success was the use of steel swords, lances with iron or steel points, body armor that repelled or blunted arrows, and a type of early musket called harquebus—all weapons unknown to the Aztecs. An Aztec counter-rebellion soon restored them to power. But not for long.

A smallpox epidemic, begun when a Spanish soldier died from the disease while in Tenochtitlán, spread among the Aztecs and gutted the population. When Cortés re-attacked, again with the backing of the Tlaxcalans, he fought a depleted people. Even more significantly, he employed a series of new and aggressive military tactics—blocking delivery of food and water to the city, choking off canals, destroying aqueducts—that brought the city to its knees after 75 days. Cortés laid claim to Tenochtitlán, ruthlessly destroying temples and homes and establishing himself as one of the most brutal of the Spanish **conquistadores** (conquerors). Twenty years later, Francisco Pizarro overpowered the Incas in Peru and opened the way for other Spanish advances into South America.

The first Spanish settlers in America were interested largely in exploiting the American stores of gold and silver, and they were fabulously successful. For 300 years, beginning in the sixteenth century, the mines of Spanish America yielded more than ten times as much gold and silver as all the rest of the world's mines combined. Before long, however, most Spanish settlers in America traveled to the New World for other reasons. Many went in hopes of profiting from agriculture. They helped establish elements of European civilization permanently in America. Other Spaniards—priests, friars, and missionaries—went to America to spread Catholicism; through their efforts, the influence of the Catholic Church ultimately extended throughout South and Central America and Mexico. They sometimes evangelized with an iron fist, forcing whole families to forsake their sacred beliefs and practices, be baptized, and adopt the teachings of the Catholic Church or face physical punishment and even death. Yet one of the first friars to work in the colonies, **Bartolomé de Las Casas**, fought for the fair treatment of native peoples by the Spanish as part of his ministry. (See "Consider the Source: Bartolomé de Las Casas, 'Of the Island of Hispaniola.'")

CONSIDER THE SOURCE

BARTOLOMÉ DE LAS CASAS, "OF THE ISLAND OF HISPANIOLA" (1542)

Bartolomé de Las Casas, a Dominican friar from Spain, was an early European settler of the West Indies. He devoted much of his life to describing the culture of native peoples and chronicling the many abuses they suffered at the hands of their colonizers. This excerpt is from a letter he addressed to Spain's Prince Philip.

God has created all these numberless people to be quite the simplest, without malice or duplicity, most obedient, most faithful to their natural Lords, and to the Christians, whom they serve; the most humble, most patient, most peaceful and calm, without strife nor tumults; not wrangling, nor querulous, as free from uproar, hate and desire of revenge as any in the world. . . . Among these gentle sheep, gifted by their Maker with the above qualities, the Spaniards entered as soon as they knew them, like wolves, tigers and lions which had been starving for many days, and since forty years they have done nothing else; nor do they afflict, torment, and destroy them with strange and new, and divers kinds of cruelty, never before seen, nor heard of, nor read of. . . .

The Christians, with their horses and swords and lances, began to slaughter and practice strange cruelty among them. They penetrated into the country and spared neither children nor the aged, nor pregnant women, nor those in child labour, all of whom they ran through the body and lacerated, as though they were assaulting so many lambs herded in their sheepfold. They made bets as to who would slit a man in two, or cut off his head at one blow: or they opened up his bowels. They tore the babes from their mothers' breast by the feet, and dashed their heads against the rocks. Others they seized by the shoulders and threw into the rivers, laughing and joking, and when they fell into the water they exclaimed: "boil body of so and so!" They spitted the bodies of other babes, together with their mothers and all who were before them, on their swords.

They made a gallows just high enough for the feet to nearly touch the ground, and by thirteens, in honor and reverence of our Redeemer and the twelve Apostles, they put wood underneath and, with fire, they burned the Indians alive.

They wrapped the bodies of others entirely in dry straw, binding them in it and setting fire to it; and so they burned them. They cut off the hands of all they wished to take alive, made them carry them fastened on to them, and said: "Go and carry letters": that is; take the news to those who have fled to the mountains.

They generally killed the lords and nobles in the following way. They made wooden gridirons of stakes, bound them upon them, and made a slow fire beneath; thus the victims gave up the spirit by degrees, emitting cries of despair in their torture.

UNDERSTAND, ANALYZE, & EVALUATE

1. How did Bartolomé de Las Casas characterize the indigenous people of Hispaniola? How do you think they would have responded to this description?
2. What metaphor did Las Casas use to describe the native peoples and where does this metaphor come from?
3. What role did Las Casas expect the Spaniards to play on Hispaniola? What did they do instead?

Source: MacNutt, Francis Augustus, *Bartholomew de Las Casas: His Life, His Apostolate, and His Writings*. New York: G.P. Putnam's Sons, 1909, 14.

By the end of the sixteenth century, the Spanish Empire included the Caribbean islands, Mexico, and southern North America. It also spread into South America and included what is now Chile, Argentina, and Peru. In 1580, when the Spanish and Portuguese monarchies temporarily united, Brazil came under Spanish jurisdiction as well.

Northern Outposts

In 1565, the Spanish established the fort of St. Augustine in Florida, their first permanent settlement in what is now the United States. It was little more than a small military outpost. A more substantial colonizing venture began in the Southwest in 1598, when Don Juan de Oñate traveled north from Mexico with a party between 600 and 700, claimed for Spain some of the lands of the Pueblo in what is now New Mexico, and began to establish a colony. It was a bloody affair. In October 1898, the Acoma Pueblos refused to turn over food to Oñate's soldiers and, in a small battle, killed as many as 13 of them, including Oñate's nephew. In January of the next year, Oñate ordered retribution. His men lay siege to the Acoma village, killing at least 800. They enslaved all survivors older than 12 years for a period of 20 years and cut off the right foot of all men of fighting age.

Oñate granted **encomiendas** (the right to exact tribute and labor from native peoples on large tracts of land) to favored Spaniards. In 1609, Spanish colonists founded Santa Fe. By 1680, there were over 2,000 Spanish colonists living among about 30,000 Pueblos. The economic heart of the colony was cattle and sheep, raised on the *ranchos* that stretched out around the small towns Spanish settlers established.

Part of the Spanish expansion in the North included converting native peoples to Catholicism. As in the South, it met with uneven results. Many native peoples simply rejected the attempt, mixed the precepts and practices of their own faith with Catholicism, or only selectively adopted Catholic rituals and teachings. At other times native peoples and Spanish officials differed over what constituted conversion. Matters came to a head in 1680, when Spanish priests and the colonial government tried to suppress native rituals. In response, **Popé**, a Pueblo religious leader, led an uprising that killed hundreds of European settlers, captured Santa Fe, and drove the Spanish from the region. Ironically, the rebellion was so widespread and included so many different indigenous groups that the native revolutionaries used Spanish as their common language in order to communicate with one other. Twelve years later, however, the Spanish returned and crushed a last revolt in 1696.

Many Spanish colonists now realized that they could not hope to prosper in New Mexico while in constant conflict with a native population that greatly outnumbered them. Although the Spanish intensified their assimilation efforts, they also now permitted the Pueblos to own land. They stopped commandeering Pueblo labor, and they tolerated the survival of tribal religious rituals. There was significant intermarriage between Europeans and native women. By 1750, the Spanish population had grown to about 4,000. The Pueblo population had declined (through disease, war, and migration) to about 13,000—less than half what it had been in 1680. New Mexico had by then become a reasonably stable, but still weak and isolated, outpost of the Spanish Empire.

Biological and Cultural Exchanges

European and native cultures never entirely merged in the Spanish Empire. Nevertheless, the arrival of whites launched a process of interaction between diverse peoples that left

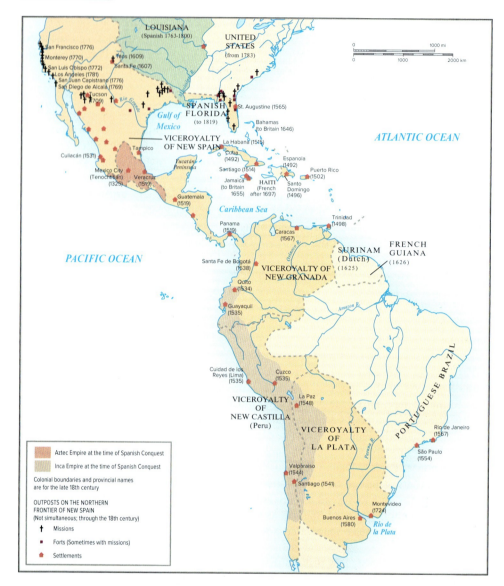

SPANISH AMERICA From the time of Columbus's initial voyage in 1492 until the mid-nineteenth century, Spain was the dominant colonial power in the New World. From the southern regions of South America to the northern regions of the Pacific Northwest, Spain controlled one of the world's vastest empires. Note how much of the Spanish Empire was simply grafted upon the earlier empires of native peoples—the Inca in what is today Chile and Peru and the Aztec across much of the rest of South America, Mexico, and the Southwest of what is now the United States. • *What characteristics of Spanish colonization would account for their preference for already settled regions?*

no one unchanged. That Europeans were exploring the Americas at all was a result of early contacts with the native peoples, from whom they had learned of the rich deposits of gold and silver. From then on, the history of the Americas became one of increasing levels of exchanges—some beneficial, others catastrophic—among different peoples and cultures.

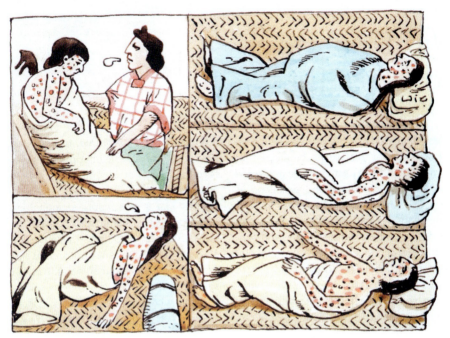

(Dorling Kindersley/Getty Images)

SMALLPOX AMONG THE AZTECS This illustration by a Spanish missionary in the fifteenth century depicts victims of smallpox in various stages of the disease, which was introduced to the Americas by Europeans.

The first and perhaps most profound result of this exchange was the importation of European diseases to the New World. It would be difficult to exaggerate the consequences of the exposure of Native Americans to such illnesses as influenza, measles, typhus, and above all smallpox. Although historians have debated the question of how many people lived in the Americas before the arrival of Europeans, it is estimated that millions died. (See "Debating the Past: Why Do Historians So Often Differ?") The high fatality rates were due partly to the way that native cultures traditionally cared for the very ill. They tended to surround the sick with constant companions and visitors as a way to encourage healing—a practice that inadvertently helped spread the highly contagious diseases they were encountering for the first time. Unlike in Europe, where experience with the bubonic plague had taught the benefits of isolating the infected, there was no notion of quarantine among native societies of the Americas. In some areas, native populations were virtually wiped out within a few decades of their first contact with whites. On Hispaniola, where Columbus had landed in the 1490s, the native population quickly declined from approximately one million to about five hundred. In the Maya area of Mexico, as much as 95 percent of the population perished within a few years of the native peoples' first contact with the Spanish. Still, not everyone died, not every community was ravaged, and some rebuilt over time. And many of the nations north of Mexico were spared the worst of the epidemics. But for other areas of the New World, this was a disaster at least as grave as, and in some places far worse than, the Black Death that had killed over one-third of the population of Europe two centuries before. Some Europeans, watching this biological catastrophe, saw it as clear evidence of God's will that they should dominate the New World—and its native population.

DEBATING THE PAST

Why Do Historians So Often Differ?

Early in the twentieth century, when the professional study of history was still relatively new, many historians believed that questions about the past could be answered with the same certainty and precision as questions in more-scientific fields. By sifting through available records, using precise methods of research and analysis, and producing careful, closely argued accounts of the past, they believed they could create definitive histories that would survive without controversy. Scholars who adhered to this view believed that real knowledge can be derived only from direct, scientific observation of clear "fact." They were known as "positivists."

A vigorous debate continues to this day over whether historical research can ever be truly objective. Almost no historian any longer accepts the positivist claim that history could ever be an exact science. Disagreement about the past is, in fact, at the heart of the effort to understand history. Critics of contemporary historical scholarship often denounce the way historians are constantly revising earlier interpretations. Some denounce the act of interpretation itself. History, they claim, is "what happened," and historians should "stick to the facts."

Historians, however, continue to differ with one another both because the facts are seldom as straightforward as their critics claim and because facts by themselves mean almost nothing without an effort to assign meaning to them. Some historical facts, of course, are not in dispute. Everyone agrees, for example, that the Japanese bombed Pearl Harbor on December 7, 1941, and that Abraham Lincoln was elected president in 1860. But many other facts are much harder to determine—among them, for example, the question of how large the American population was before the arrival of Columbus, or how many enslaved workers resisted slavery. This sounds like a reasonably straightforward question, but it is almost impossible to answer with any certainty—because the records of slave resistance are spotty and the definition of "resistance" is a matter of considerable dispute.

Even when a set of facts is clear and straightforward, historians disagree—sometimes quite radically—over what they mean. Those disagreements can be the result of political and ideological disagreements. Some of the most vigorous debates in recent decades have been between scholars who believe that economic interests and class divisions are the key to understanding the past, and those who believe that ideas and culture are at least as important as material interests. Debates can also occur over differences in methodology—between those who believe that quantitative studies can answer important historical questions and those who believe that other methods come closer to the truth.

Most of all, historical interpretation changes in response to the time in which it is written. Historians may strive to be objective in their work, but no one can be entirely free from the assumptions and political concerns of the present. In the 1950s, the omnipresent shadow of the Cold War shaped histories of Communist countries. The civil rights movements prompted scholars to reconsider what they knew about the lives and achievements of Black Americans, women, Hispanics, and gays and lesbians. The rise of postcolonial societies pushed historians to reexamine assumptions built into the telling of the rise and fall of empires—that they were the products of an elite cadre of men—and rethink the role of workers and the less powerful in influencing the course of events. The "cultural turn" at the end of the twentieth century placed a newfound stress on examining how various forces of culture—gender, sexuality, race, ethnicity, language—deeply affected the ways in which people experienced and understood the world. Its effects are still rippling through the academy, asking historians to ever widen their lens of analysis when seeking to explain people's motivations and actions.

Historians regularly debate over which types of interpretation come closest to capturing the truth of the past with no clear-cut consensus likely to come into focus any time soon. Such debate, though, is a sign of the health of the profession. Scholars need to constantly revisit how they talk about the past and be challenged to defend their decisions in order to make sure they are capturing the full range of human experience when writing their histories. Indeed, understanding the past is a forever continuing—and forever contested—process. •

UNDERSTAND, ANALYZE, & EVALUATE

1. What are some of the reasons historians so often disagree?
2. Is there ever a right or wrong in historical interpretation? What value might historical inquiry have other than reaching a right or wrong conclusion?
3. If historians so often disagree, how should a student of history approach historical content? How might disagreement expand our understanding of history?

Not all aspects of the exchange were disastrous to native peoples. The Europeans introduced important new crops (among them sugar and bananas), domestic livestock (cattle, pigs, and sheep), and, perhaps most significantly, the horse, which gradually became central to the lives of many native peoples and transformed their societies. Less beneficially, the transfer of European grass seed and the grazing and feeding habits of European animals devastated local flora.

The exchange was at least as important (and often more advantageous) to the Europeans. In both North and South America, the arriving peoples learned from native cultures new agricultural techniques appropriate to the demands of the new land. They discovered new crops—above all maize (corn), which Columbus took back to Europe from his first trip to America. Such foods as squash, pumpkins, beans, sweet potatoes, tomatoes, peppers, and potatoes also found their way into European diets.

In South America, Central America, and Mexico, Europeans and native groups lived in intimate, if unequal, contact with one another. Many native people gradually came to speak Spanish or Portuguese, but they created a range of dialects fusing the European languages with elements of their own. European men outnumbered European women by at least ten to one. Intermarriage—often forced—became frequent between Spanish immigrants and native women. Before long, the population of the colonies came to be dominated (numerically, at least) by people of mixed race, or **mestizos**.

Virtually all the enterprises of the Spanish and Portuguese colonists depended on native workforces. In some places, indigenous people were sold into slavery. More often, colonists used a coercive (or "indentured") wage system, under which native people worked in the mines and on the plantations under duress for fixed periods. That was not, in the end, enough to meet the labor needs of the colonists. As early as 1502, European settlers began importing enslaved laborers from Africa.

AMERICA IN THE WORLD

The International Context of the Early History of the Americas

Most Americans understand that our nation of late has become intimately bound up with the rest of the world—that we live in what many call the "age of **globalization**." But few extend that idea backward in time and consider how the story of America before Columbus and the effort by European powers to settle it was also part of a global current of ideas and events. Indeed, until recently historians typically studied these early chapters from the nation's past mostly in isolation from larger world events and non-European societies. By contrast today, scholars of early American history now examine what happened in the New World from a broadly international perspective.

That perspective is often called the "**Atlantic World**" and it explores the intermingling of peoples from Africa, Europe, and the Americas and the profound effects of those interactions. The phrase has a long intellectual genealogy, stretching back to the foundational work of C. L. R. James, W. E. B. Du Bois, Frantz Fanon, and Eric Williams. They demonstrated that the origins of the New World were deeply enmeshed in the practice and institution of slavery, on the one hand, and that African (and later African American) culture lay at the root of the evolution of culture in the Americas, on the other.

The idea of an Atlantic World rests in part on the obvious connections between western Europe and the Spanish, British, French, and Dutch colonies in North and South America. All the early European civilizations of the Americas were part of a great imperial project launched by the major powers of Europe. The European immigrations to the Americas beginning in the sixteenth century, the advance of slavery and the introduction of it in the New World, the defeat and devastation of native populations, the creation of European agricultural and urban settlements, and the imposition of imperial regulations on trade, commerce, landowning, and political life—all of these forces reveal the influence of Old World **imperialism** on the history of the New World.

But the expansion of empires is only one part of the creation of the Atlantic World. At least equally important—and closely related—is the expansion of commerce from Europe and Africa to the Americas. Although some northern and southern Europeans traveled to the New World in search of religious freedom, or to escape oppression, or to search for adventure, the great majority were in search of economic opportunity. Not surprisingly, therefore, their settlements in the Americas were almost from the start intimately connected to Europe through the growth of commerce between them and to Africa through the capture and import of enslaved workers. This international commercial dynamic between America and Europe was responsible not just for the growth of trade, but also for the increases in migration over time—as the demand for labor in the New World drew more and more settlers from the Old World. Commerce was also a principal reason for the rise of slavery in the Americas, and for the growth of the slave trade between European America and Africa.

Religion was also a powerful force influencing migration to the New World and shaping human interactions there. Depending on the decade, some Europeans—Puritans, Anabaptists—relocated in part to escape persecution for their principles. At other times, Catholics and members of the Church of England built settlements to win converts and extend their religious empires. Significantly, European transplants had to come to terms with the religion of the native people they encountered, which led to a variety of responses: indifference, evangelism, repression, or the growth of hybrid sacred practices and convictions. Adding to the mix were enslaved African, who brought their own indigenous religions. They found themselves the subjects of intense and sometimes brutal proselytizing attempts by Europeans, which met with only uneven success. Some enslaved workers adopted the faith of their owners. But African American religion as a whole generally emerged as a series of spiritual beliefs and rituals that mixed African, European, and sometimes indigenous beliefs. It also influenced the religion of Europeans and (to a lesser extent) native peoples, particularly in the evolution of their public revivals and preaching traditions in the New World.

The early history of the Americas was also closely bound up with the intellectual life of northern and southern Europe and Latin America. The Enlightenment—the cluster of ideas that emerged in the seventeenth and eighteenth centuries emphasizing the power of human reason—moved quickly to the Americas, producing intellectual ferment throughout the New World. Thinkers from Britain and Spain, for example, stressed the sanctity of individual rights, the proper nature and role of representative government, and the fairness of law that eventually undergirded the American Revolution, the Haitian Revolution, and Latin American revolutions of the eighteenth century. Scientific and technological knowledge—another product of the Enlightenment—traveled constantly across the Atlantic and back. Americans borrowed industrial technology from Britain. Europe acquired much of its early knowledge of electricity from experiments done in America. But the Enlightenment was only one part of the continuing intellectual connections within the Atlantic World, connections that spread artistic, scholarly, and political ideas widely through the lands bordering the ocean.

Instead of thinking of the early history of what became the United States simply as the story of the growth of thirteen small colonies along the Atlantic seaboard of North America, the idea of the Atlantic World encourages us to think of early American history as a vast pattern of exchanges and interactions—trade, migration, religious and intellectual exchange, and many other relationships—among all the societies bordering the Atlantic: northern and southern Europe, western Africa, the Caribbean, and North and South America. •

UNDERSTAND, ANALYZE, & EVALUATE

1. What is the Atlantic World?
2. What has led historians to begin studying the idea of an Atlantic World?

Africa and America

Over one-half of all those enterering the New World between 1500 and 1800 were Africans, sent against their will. Most came from West and Central Africa.

Europeans and white Americans came to portray African society as primitive. But most Africans were, in fact, highly civilized peoples with well-developed economies and political systems. The residents of the Gold Coast had substantial commercial contact with the Mediterranean world—trading ivory, gold, and enslaved labor for finished goods—and, largely as a result, became early converts to Islam. After the collapse of the ancient kingdom of Ghana around A.D. 1100, they created the even larger empire of Mali, whose trading center at Timbuktu became fabled as a learned meeting place of the peoples of many lands. In West Central Africa, the Kingdom of Kongo flourished. It was a regional center for trade, where residents sold goods they manufactured, such as pottery and copper and iron goods. By early 1500, the majority of the ruling class had converted to Catholicism and the Kingdom was sending a formal emissary to the Vatican. By the end of the sixteenth century its population was nearly 500,000.

As in many indigenous societies in America, African families tended to be matrilineal. Women played a major role, often the dominant role, in trade. In many areas, they were also the principal farmers while the men hunted, fished, raised livestock, fought battles; in these areas women chose their own leaders to make decisions and policies for the community as a whole. Everywhere women managed child care and food preparation.

Small elites of priests and nobles stood at the top of many African societies. Most people belonged to a large middle group of farmers, traders, crafts workers, and others. At the bottom of society were enslaved men and women, not all of them African, who were put into bondage after being captured in wars, because of criminal behavior, or as a result of unpaid debts. Those enslaved in Africa were generally in bondage for a fixed term, and in the meantime they retained certain legal protections (including the right to marry). Children did not inherit their parents' condition of bondage.

The African slave trade long preceded European settlement in the New World. As early as the eighth century, West Africans began selling small numbers of enslaved workers to traders from the Mediterranean and later to the Portuguese. In the sixteenth century, however, the market for enslaved labor increased dramatically as a result of the growing European demand for sugarcane. The small areas of sugar cultivation in the Mediterranean could not meet popular demand, and production soon spread to new areas: to the island of Madeira off the African coast, which became a Portuguese colony, and not long thereafter (still in the sixteenth century) to the Caribbean islands and Brazil. Sugar was a labor-intensive crop, and the demand for African workers in these new areas of cultivation was high. At first the slave traders were overwhelmingly Portuguese. By the seventeenth century, though, the Dutch had won control of most of the market. And in the eighteenth century, the English dominated it. By 1700, slavery had spread well beyond its original locations in the Caribbean and South America and into the English colonies to the north. The relationship among European, African, and native peoples—however unequal—reminds us of the global context to the history of America. (See "America in the World: The International Context of the Early History of the Americas.")

THE ARRIVAL OF THE ENGLISH

England's first documented contact with the New World came only five years after Spain's. In 1497, John Cabot (like Columbus, a native of Genoa) sailed to the northeastern coast of North America on an expedition sponsored by King Henry VII, in an unsuccessful search for a northwest passage through the New World to the Orient. But nearly a century passed before the English made any serious efforts to establish colonies in America.

Significantly, England's first experience with colonization came not in the New World but in neighboring Ireland. The English had long laid claim to the island, but only in the late sixteenth century did serious efforts at colonization begin. The long, brutal process by which the English attempted to subdue the Irish created an important assumption about colonization: the belief that settlements in foreign lands must retain a rigid separation from the native populations. Unlike the Spanish in America, the English in Ireland tried to build a separate society of their own, peopled with emigrants from England itself. They would take that concept with them to the New World.

INCENTIVES FOR COLONIZATION

Interest in **colonization** grew in part as a response to social and economic problems in sixteenth-century England. The English people faced frequent and costly European wars as well as almost constant religious strife within their own land. Many suffered, too, from harsh economic changes in their countryside. Because the worldwide demand for wool was growing rapidly, landowners were converting their land from fields for crops to pastures for sheep. The result was a reduction in the amount of land available for growing food. England's food supply declined at the same time that the English population was growing—from 3 million in 1485 to 4 million in 1603. To some of the English, the New World began to seem attractive because it offered something that was growing scarce in England: land.

At the same time, new merchant capitalists were prospering by selling the products of England's growing wool-cloth industry abroad. At first, most exporters did business almost entirely as individuals. In time, however, merchants formed companies, whose **charters** from the king gave them monopolies for trading in particular regions. Investors in these companies often made fantastic profits, and they were eager to expand their trade.

Central to this trading drive was the emergence of a new concept of economic life known as **mercantilism**. Mercantilism rested on the belief that one person or nation could grow rich only at the expense of another, and that a nation's economic health depended, therefore, on selling as much as possible to foreign lands and buying as little as possible from them. The principles of mercantilism spread throughout Europe in the sixteenth and seventeenth centuries. One result was the increased attractiveness of acquiring colonies, which became the source of raw materials and a market for the colonizing power's goods.

In England, the mercantilistic program thrived at first on the basis of the flourishing wool trade with the European continent, and particularly with the great cloth market in Antwerp. In the 1550s, however, that glutted market began to collapse, and English merchants had to look elsewhere for overseas trade. Some English believed colonies would solve their problems.

There were also religious motives for colonization—a result of the **Protestant Reformation**. Protestantism began in Germany in 1517, when Martin Luther challenged some of the basic practices and beliefs of the Roman Catholic Church. Luther quickly won a wide following among ordinary men and women in northern Europe. When the pope excommunicated him in 1520, Luther began leading his followers out of the Catholic Church entirely.

The Swiss theologian John Calvin went even further in rejecting the Catholic belief that human behavior could affect an individual's prospects for salvation. Calvin introduced the doctrine of predestination. God "elected" some people to be saved and condemned others to damnation; each person's destiny was determined before birth, and no one could change that predetermined fate. But those who accepted Calvin's teachings came to believe that the way they led their lives might reveal to them their chances of salvation. A wicked or useless existence would be a sign of damnation, saintliness and diligence possibly signs of grace. The new creed spread rapidly throughout northern Europe.

In 1529, King Henry VIII of England, angered by the refusal of the pope to grant him a divorce from his Spanish wife, broke England's ties with the Catholic Church and established himself as the head of the Christian faith in his country. This was known as the English Reformation. After Henry's death, his Catholic daughter, Queen Mary, restored England's allegiance to Rome and persecuted Protestants. When Mary died in 1558, her half sister, **Elizabeth I**, became England's sovereign and once again severed the nation's connection with the Catholic Church, this time for good.

To many English people, however, the new Church of England was not reformed enough. They clamored for changes that would "purify" the church and quickly became known as **Puritans**. Most only wanted to simplify worship and reform the leadership of the church. Their frustration mounted steadily as political and ecclesiastical authorities refused to respond to their demands.

Puritan discontent grew rapidly after the death of Elizabeth, the last of the Tudors, and the accession of James I, the first of the Stuarts, in 1603. Convinced that kings ruled by divine right, James quickly antagonized the Puritans by resorting to illegal and arbitrary taxation, favoring English Catholics in the granting of charters and other favors, and supporting "high-church" forms of ceremony, meaning a strong stress on traditional and very formal liturgical practices. By the early seventeenth century, some Puritans were beginning to look for places of refuge outside the kingdom.

THE FIRST ENGLISH SETTLEMENTS

The first permanent English settlement in the New World was established at Jamestown, in Virginia, in 1607. But for nearly thirty years before that, English merchants and adventurers had been engaged in a series of failed efforts to create colonies in America.

Through much of the sixteenth century, the English had harbored mixed feelings about the New World. They were intrigued by its possibilities, but they were also fearful of Spain, which remained the dominant force in America. In 1588, King Philip II of Spain sent one of the largest military fleets in the history of warfare—the Spanish Armada—across the English Channel to attack England itself. The smaller English fleet, taking advantage of its greater maneuverability, defeated the armada and, in a single stroke, ended Spain's domination of the Atlantic. This great shift in naval power caused English interest in colonizing the New World to grow quickly.

The pioneers of English colonization were Sir Humphrey Gilbert and his half brother Sir Walter Raleigh—both veterans of earlier colonial efforts in Ireland. In 1578, Gilbert obtained from Queen Elizabeth a six-year patent granting him the exclusive right "to inhabit and possess any remote and heathen lands not already in the possession of any Christian prince." Five years later, after several setbacks, he led an expedition to Newfoundland, looking for a good place to build a profitable colony. A storm sank his ship, and he was lost at sea. The next year, Sir Walter Raleigh secured his own six-year grant from the queen and sent a small group of men on an expedition to explore the North American coast. When they returned, Raleigh named the region they had explored Virginia, in honor of Elizabeth, who was known as the "Virgin Queen."

In 1585, Raleigh recruited his cousin, Sir Richard Grenville, to lead a group of men to the island of **Roanoke**, off the coast of what is now North Carolina, to establish a colony. Grenville deposited the settlers on the island, destroyed a native village as retaliation for a minor theft, and returned to England. The following spring, with long-overdue supplies and reinforcements from England, Sir Francis Drake unexpectedly arrived in Roanoke. The dispirited colonists boarded his ships and left.

THE COLLISION OF CULTURES • 21

(Photo12/Universal Images Group/Getty Images)

ROANOKE A drawing by one of the colonists in the ill-fated Roanoke expedition of 1585 became the basis for this engraving by Theodor de Bry, published in England in 1590. A small European ship approaches the island of Roanoke, in the center. The wreckage of several larger vessels farther out to sea suggests the danger of the journey while the presence of native settlements on the mainland and on Roanoke itself reflects the contact between two different cultures to come.

Raleigh tried again in 1587, sending an expedition to Roanoke carrying ninety-one men, seventeen women, and nine children. The settlers attempted to take up where the first group of colonists had left off. John White, the commander of the expedition, returned to England after several weeks, in search of supplies and additional settlers. Because of a war with Spain, he was unable to return to Roanoke for three years. When he did, in 1590, he found the island deserted, with no clue to the fate of the settlers other than the cryptic inscription "Croatoan" carved on a post.

The Roanoke disaster marked the end of Sir Walter Raleigh's involvement in English colonization of the New World. No later colonizers would receive grants of land in America as vast or undefined as those Raleigh and Gilbert had acquired. Yet the colonizing impulse remained very much alive. In the early years of the seventeenth century, a group of London merchants decided to renew the attempt at colonization in Virginia. A rival group of merchants, from the area around Plymouth, was also interested in American ventures and was sponsoring voyages of exploration farther north. In 1606, James I issued a new charter, which divided North America between the two groups. The London group got the exclusive right to colonize the south, and the Plymouth merchants received the same right in the north. Through the efforts of these and other companies, the first enduring English colonies would soon be established in North America.

THE FRENCH AND THE DUTCH IN AMERICA

English settlers in North America encountered not only native groups but also other Europeans who were, like them, driven by mercantilist ideas. There were scattered North American outposts of the Spanish Empire and, more important, there were French and Dutch settlers who were also vying for a stake in the New World.

In the early sixteenth century, eager to discover new trade routes across the Atlantic and locate a new corridor to the Pacific, the French King, Francis I, turned to Giovanni da Verrazzano, an Italian explorer. After rough seas forced him to abort his maiden voyage in 1523, Verrazano set sail the next year and successfully landed at Cape Fear. He charted his way north along the Atlantic Coast, including stops in New York Bay, Long Island, Narraganset Bay, Cape Cod, and finally Newfoundland. Crafting detailed maps and providing accounts of his interactions with native people, Verrazano laid the pathway for future generations of European explorers.

Nearly 40 years later, in 1562, Frenchman Jean Ribault established a small settlement he called Charlesfort in present-day Parris Island, South Carolina. Poor leadership, inadequate supplies, and a lack of cooperation with native residents ushered in its demise after only a year. In 1564, Rene Goulaine de Laudonniere, an officer in Ribault's original force, built Fort Caroline, near what is now Jacksonville, Florida. It too nearly collapsed within a year for similar reasons, but a fortuitous stop-over by an English ship allowed residents to trade for much needed supplies. Fort Caroline quickly became entangled in larger territorial conflicts between the French and Spanish, who sacked the fort in 1565, killed most of its residents, and built their own fortification, Fort San Mateo. It lasted until 1569, when a vengeful French force burned it to the ground.

Samuel de Champlain founded the first permanent French settlement in North America at Quebec in 1608, less than a year after the English settled Jamestown. Central to its success was Champlain's winning effort to form strong political partnerships with the Montagnais, Algonquin, and Huron, even going to war with them against the Iroquois. These bonds facilitated the expansion of the French fur trade in the region. Champlain also promoted close interaction between *coureurs de bois*—male French fur traders and trappers—and indigenous peoples as a way of knitting together the different cultures. The *coureurs de bois* settled deep in the region, learned native languages and customs, and sometimes intermarried. They enlarged the network of fur trading, which helped open the way for French agricultural estates (or *seigneuries*) along the St. Lawrence River and for the development of trade and military centers at Quebec and Montreal.

Jesuit missionaries spread French influence as well. These members of the male-only Catholic Society of Jesus arrived in 1634 and five years later founded the Sainte-Marie-aux-Huron, which as the name suggests was meant to encourage cooperation and conversion among the Huron. The Jesuits soon expanded the mission, adding a farm, hospital, mill, and church to introduce the native people to their faith, way of life, and skills like blacksmithing. The Jesuits learned the local tongue and customs in an attempt to build trust and collaboration and eventually launched new missions in the region. While their work certainly enhanced relationships with the native population, the Jesuits faced limits in what they could accomplish. The Huron, like indigenous people from other parts of the New World, often challenged attempts to make them into Catholics and farm and live like Frenchmen. Some flat-out resisted. Others mixed traditional religion practices with Catholicism, creating a new faith hybrid.

The Dutch, too, established a presence in North America. Holland in the early seventeenth century was one of the leading trading nations of the world, and its commerce moved to America in the seventeenth century. In 1609, Henry Hudson, an English explorer in the employ of the Dutch East Indies Company, steered his ship across the Atlantic in pursuit of a new route to Asia. Instead he made landfall at Newfoundland and then traveled south. He eventually sailed into the waters surrounding Manhattan Island, which led to the formation of the Dutch colony, New Netherland, in what is now part of New York City. From it, Dutch trappers moved into the interior toward the Appalachian Mountains and built a profitable trade in furs.

The Dutch, like other European powers, controlled a broad global empire of commerce and colonization. In addition to North America, the Dutch settled portions of the Caribbean and South America. They built settlements in Sint Maarten in 1618. St. Croix in 1625, Bonaire and Curacao in 1634, and Sint Eustatius in 1636. For 24 years, from 1630 to 1654, Holland also controlled a giant swath of northeastern Brazil, representing half of all European settlements there at the time. They introduced sugar and the business of sugar trading to Barbados, an English colony.

CONCLUSION

The lands that Europeans eventually named the Americas were the home of many millions of people before the arrival of Columbus. Many had migrated from Asia thousands of years earlier. These pre-Columbian Americans spread throughout the Western Hemisphere and eventually created great civilizations. Among the most notable of them were the Inca in Peru and the Maya and Aztec in Mexico. In the regions north of what was later named the Rio Grande, the human population was smaller and the civilizations were less advanced than they were farther south. Even so, North American native peoples created a cluster of civilizations that thrived and expanded. They included the Mississippian peoples, notably the Chickasaw, Cherokee, Choctaw, Muscogee, Creek, Houma, Seminole, and Tunica-Biloxis; as well as the Pueblo of the modern American Southwest and the Algonquian who dwelled mostly in contemporary New England and eastern Canada.

In the century after European contact, these native populations suffered catastrophes that all but destroyed many of the civilizations they had built: brutal invasions by Spanish and Portuguese conquistadores and a series of plagues inadvertently imported by Europeans. By the middle of the sixteenth century, the Spanish and Portuguese—no longer faced with large-scale and effective resistance from the native populations—had largely established colonial control over all of South America and much of North America.

In the parts of North America that would eventually become the United States, the European presence was for a time much less powerful. The Spanish established an important northern outpost in what is now New Mexico, a society in which Europeans and indigenous people lived closely together, though on unequal terms. On the whole, however, native North Americans remained largely undisturbed by Europeans until English, French, and Dutch migrations began in the early seventeenth century.

KEY TERMS/PEOPLE/PLACES/EVENTS

Atlantic World 16
Bartolomé de Las Casas 9
Cahokia 5
charter 19
Christopher Columbus 7
Clovis people 2
colonization 19

colony 7
conquistador 9
Elizabeth I 20
encomienda 11
globalization 16
imperialism 16
mercantilism 19

Mesoamerica 4
mestizo 15
Popé 11
Protestant Reformation 19
Puritans 20
Roanoke 20

RECALL AND REFLECT

1. Why did European countries seek to establish settlements in the New World?
2. How did indigenous women occupy different social roles and exercise different social responsibilities than European women?
3. What was the response of native peoples to the efforts by Europeans to settle near them?
4. What role did disease play in the settlement of the New World?

Design element: Stars and Stripes: McGraw Hill Education.

2 | TRANSPLANTATIONS AND BORDERLANDS

THE EARLY CHESAPEAKE
THE GROWTH OF NEW ENGLAND
THE RESTORATION COLONIES
BORDERLANDS AND MIDDLE GROUNDS
THE DEVELOPMENT OF EMPIRE

LOOKING AHEAD

1. Why did the English seek to settle in North America? What were their various motivations?
2. How did Native Americans affect the early history of English settlements?
3. How did English colonies differ from another? That is, how did settlements in, say, Virginia differ from the Massachusetts Bay colony?

THE FIRST PERMANENT ENGLISH SETTLEMENTS were small, fragile communities, generally unprepared for the hardships they were to face. Seeking to secure a greater degree of control over their lives and improve their futures, immigrants from the British Isles found a world populated by Native American nations; by colonists, explorers, and traders from Spain, France, and the Netherlands; and by immigrants from other parts of Europe and, soon, enslaved people from Africa. American society was from the beginning a fusion of many cultures in which disparate people coexisted, often quite violently.

All of English North America was, in effect, a borderland during the early years of colonization. Through much of the seventeenth century, English colonies both relied on and did battle with native peoples and struggled with challenges from other Europeans in their midst. Eventually, however, some areas of English settlement—most notably the growing communities along the eastern seaboard—managed to dominate their own regions, marginalizing or expelling indigenous inhabitants and other challengers. In these eastern colonies, the English created significant towns and cities; constructed political, religious, and educational institutions; and built productive agricultural systems. They also instituted slavery here as they did in every colony.

TIME LINE

- **1607** Jamestown founded
- **1619** First recorded enslaved African in Virginia
- **1620** Pilgrims found Plymouth Colony
- **1622** Powhatan attack Virginia
- **1624** Dutch settle Manhattan
- **1630** Puritans establish Massachusetts Bay colony
- **1634** Maryland founded
- **1636** Roger Williams founds Rhode Island
- **1637** Anne Hutchinson expelled from Massachusetts Bay colony; Pequot War
- **1663** Carolina chartered
- **1664** English capture New Netherland
- **1675** King Philip's War
- **1676** Bacon's Rebellion
- **1681** Pennsylvania chartered
- **1686** Dominion of New England
- **1688** Glorious Revolution
- **1732** Georgia chartered

THE EARLY CHESAPEAKE

Once James I had issued his 1606 charters, the London Company moved quickly and decisively to launch a colonizing expedition headed for Virginia. A party of 144 men aboard three ships—the *Godspeed*, the *Discovery*, and the *Susan Constant*—set sail for America on December 19, 1606. Only 104 men survived the journey. They reached the North American coast in late April 1607, sailed into the Chesapeake and up a river they named the James, in honor of their king. They established their colony, **Jamestown**, on a peninsula in the river on May 24. They chose this inland setting because they believed it would provide a measure of comfort and security.

COLONISTS AND NATIVE PEOPLES

For nearly a decade, Jamestown was a tiny colony teetering on the brink of collapse. During this time local indigenous people were more powerful than the English. Coastal Virginia was the home of numerous native nations, many of whom joined forces to form the Powhatan Confederacy, named after its chief, Wahunsonacock, otherwise known as Powhatan. Composed of at least 30 Algonquian-speaking peoples, the Confederacy occupied a broad territory in what is now southern Maryland, the Chesapeake Bay, and the Virginia coast. Major groups included the Arrohateck, the Appamattuck, the Pamunkey, the Chickahominy, and the Mattapony. Chief **Powhatan** initially viewed the English as simply another group among many in his Confederacy, one that posed no immediate threat and could be removed at any time.

The early history of Jamestown certainly gave him little reason to change his mind. The English newcomers were vulnerable to local diseases, particularly malaria, which was especially virulent along the marshy rivers where they had chosen to build their homes. They also faced ordeals that were to a large degree of their own making. They spent more time searching for gold and other exports than growing enough food to be self-sufficient. And they could create no real community without women, who had not been recruited for the expedition. Within a few months after the first colonists arrived in Virginia, only 38 or the 144 men who had sailed to America were alive, the rest killed by diseases and famine.

Jamestown's early survival required the British immigrants to admit their mistakes and learn from the Powhatan. This was not easy for the settlers, because they believed that English civilization, with its oceangoing vessels, muskets, and other advanced weaponry, was greatly superior. Yet native agricultural techniques were far better adapted to the soil and climate of Virginia than those of English origin. The Powhatan were settled farmers whose villages were surrounded by neatly ordered fields. They grew a variety of crops—beans, pumpkins, vegetables, and above all maize (corn). Some of their farmlands stretched over hundreds of acres and supported substantial populations. The colony's leader, twenty-seven-year-old Captain **John Smith**, convinced the colonists to swallow their pride and learn from the locals what and when to plant and harvest, how to make dugout canoes and navigate local waterways, and where to hunt and fish. They also traded extensively with native inhabitants for food. Only steady help from the Powhatan and Smith's efforts to impose work and order on the community kept Jamestown from completely dying out.

Reorganization and Expansion

As Jamestown struggled to survive, the London Company (now renamed the Virginia Company) was already dreaming of bigger things. In 1609, it obtained a new charter from the king, which increased its power and enlarged its territory. It offered stock in the company to planters who were willing to migrate at their own expense. And it provided free passage to Virginia for poorer people who would agree to serve the company for seven years. In the spring of 1609, two years after the first arrival of the English, a fleet of nine vessels was dispatched to Jamestown with approximately 600 people, including some women and children.

Disaster quickly followed. One of the Virginia-bound ships was lost at sea in a hurricane. Another ran aground in the Bermuda islands and was unable to sail for months. Many of the new settlers succumbed to fevers before the first winter came. And the winter of 1609–1610 was especially severe and came to be known as "starving time." By then as well the native people began to realize that the colonists were a possible threat and blocked the English from moving inland. Barricaded in a small palisade, unable to hunt or cultivate food, the settlers lived on what they could find: "dogs, cats, rats, snakes, toadstools, horsehides," and even "the corpses of dead men," as one survivor recalled. When the migrants who had run aground in Bermuda finally arrived in Jamestown the following May, they found only about 60 emaciated people still alive. The new arrivals scooped up the survivors, put them on their ship, and set sail for England. As they proceeded down the James, they met an English ship coming up the river—part of a fleet bringing supplies and the colony's first governor, Lord De La Warr. The departing settlers agreed to return to Jamestown. Relief expeditions soon began to arrive, and the effort to turn a profit in Jamestown resumed.

New settlements soon began lining the river above and below Jamestown. The immigrants learned about a new crop from the native people—tobacco, which was already popular among the Spanish colonies to the south. It was also being imported to Europe and becoming much sought-after by the citizenry, so much so that King James I began to worry about its potential health hazards. John Rolfe, who had arrived in Jamestown in 1610, began experimenting with tobacco in 1612 but needed help to grow it on a large scale. Drawing on indigenous expertise and planting seeds grown in the West Indies, he developed tobacco as Virginia's first profitable crop. His success encouraged other planters to raise tobacco up and down the James River and eventually to move deeper inland, intruding more and more into the native farmlands.

As Jamestown expanded and slowly developed a profitable economy, residents attempted to drive away local Powhatan inhabitants. During two years of bloody raids, settlers captured Chief Powhatan's young daughter, Pocahontas, in spring 1613. She was probably about 18 years old at the time. Ironically, only several years earlier she had played a key role in mediating differences between her people and the Europeans. Powhatan refused to ransom her. During her years of captivity living among the English, Pocahontas learned their language, customs, and religion, converted to Christianity and took (or was given) the name "Rebecca" upon her baptism. In April 1614 she married John Rolfe and a year later had a son, Thomas, with him. She likely was a key source for Rolfe of information about growing tobacco. While historians generally agree about the course of her life after becoming a prisoner, they are unclear about why Pocahontas pursued this path. Was her conversion genuine or undertaken simply to fulfill the legal requirement for an indigenous woman to marry an Englishman? And was Rolfe's marital proposal primarily political, despite his professed affection for her, as suggested in a letter he penned to Sir Dale describing his motivations as being "for the good of this plantation, for the honour of our country, for the glory of God, for my own salvation, and for the converting to the true knowledge

(Carol M. Highsmith/Library of Congress, Prints and Photographs Division [LC-DIG-highsm-13340])

POCAHONTAS In this reproduction of an eighteenth-century painting based itself on a 1616 engraving by Simon van de Passe, Pocahontas is presented to English society as one of its own. She is represented in expensive and high-status European dress—neck collar, tailored clothing—and holding Ostrich feathers, symbols of nobility. Her name is inscribed twice, first as Matoaka, a family name given to her as a baby, and then Rebecca, the name she adopted upon converting to Christianity.

of God and Jesus Christ, an unbelieving creature, namely Pokahuntas [sic]. To whom my hearty and best thoughts are, and have a long time been so entangled, and enthralled in so intricate a labyrinth." Regardless, the marriage ushered in a period of uneasy détente between the English and the Powhatan that lasted several years.

This relatively peaceful period accelerated the development of a tobacco economy, which in turn created a heavy demand for labor and land. To entice new workers to the colony, the Virginia Company established what it called the **headright system**. Headrights were fifty-acre grants of land. Those who already lived in the colony received two headrights apiece. Each new settler received a single headright for himself or herself. This system encouraged family groups to migrate together, since the more family members who traveled to America, the more land the family would receive. In addition, anyone who paid for the passage of immigrants to Virginia would receive an extra headright for each arrival. As a result, some colonists were quickly able to assemble large plantations, establishing domain over land used by native inhabitants for generations to hunt, fish, and farm.

The Virginia Company also transported ironworkers and other skilled crafts workers to Virginia to diversify the economy. In 1619, it sent 100 English women to become the wives of male colonists. It also promised white male colonists the full rights of Englishmen, an end to strict and arbitrary rule, and even a share in self-government. On July 30, 1619, delegates from the various communities met as the **House of Burgesses**, the first elected legislature within what was to become the United States.

During these years, Pocahontas and Rolfe had moved to England. Arriving in June 1616, she soon became the subject of popular fascination—the native woman who had converted to Christianity and married an English subject. Her unusual status sparked popular debate about the possibility of assimilating indigenous inhabitants and other non-English into English culture. The couple embarked on a return voyage to Virginia in March 1617. Pocahontas fell gravely ill shortly after embarking, forcing the ship to land near Gravesend on the Thames River, where she died. After burying her, Rolfe continued his travels to Virginia.

A year after Pocahontas's death her father, Chief Powhatan, died. His passing created a vacuum of leadership among the Powhatan, which was eventually filled by his brother, Opechancanough, who sought to stop English encroachment and force them to depart the region. On a March morning in 1622, Powhatan men called on white settlements as if to offer goods for sale then suddenly attacked. They killed between 350 and 400 whites of both sexes and all ages before retreating. Although they killed about one-quarter of the total population of Jamestown, the Powhatan were not seeking to eliminate all settlers; they did not practice what would become known as "total warfare." Instead, they intended to deliver a powerful and graphic message to scare all newcomers into leaving immediately. It did not work, however, and instead set off a series of conflicts that, after more than twenty years, resulted in the defeat of the Powhatan.

SLAVERY AND INDENTURE IN THE VIRGINIA COLONY

In late August of 1619, John Rolfe recorded that "20 and odd Negroes" arrived at Jamestown aboard a Dutch ship. It was actually the British war vessel, the *White Lion,* which had recently raided a Portuguese slave ship for its human cargo. Nevertheless, Rolfe provided the first documented instance of Africans arriving in North America, though the Spanish in the South had brought some earlier. Historians are uncertain if English colonists in Jamestown initially viewed the Africans as servants, to be held for a term of years and then freed, or as enslaved. Likely it was the former, as the majority of laborers at this time were "bonded" to a master or employer for a fixed period of time. Within about ten years, however, English colonists noted that it was "customary practice to hold some Negroes in a form of life service." But they also indicated that other Africans worked for a period of time after which they were freed. This variability would not last long. Virginians began to depend on African laborers to farm tobacco and the judicial system began to codify what Blacks could and could not do. In 1639, a law forbade them from owning arms; in 1640, Virginia courts condemned a Black runaway servant, John Punch, to "serve his said master . . . for the time of his natural Life"; and most importantly, in 1662, the Virginia General Assembly declared a "Negro women's children to serve according to the condition of the mother." The 1662 code made it clear that children born to enslaved mothers were to be enslaved themselves, for life. Even if the father was a free person, Black or white, a child's status mirrored that of the mother.

The number of Blacks living in Virginia was fairly small during the colony's earliest decades, estimated to be about 23 in 1625 and 300 in 1648. Providing most of the labor to grow the colony at that time were **indentured servants**—who were primarily white English immigrants who inked a contract or an "indenture" that bound them to work for a set period of time for a person or institution in exchange for travel costs to Virginia and all living expenses once there. As tobacco farms increased in size and number and the need for laborers to work them rose accordingly, though, owners slowly turned toward owning and using enslaved Blacks as opposed to white indentured servants. They bought increasing numbers of enslaved workers, and the enslaved population in Virginia rose steeply, to 3,000 by 1680 and 10,000 by 1704.

During these decades, the Virginia Company in London became defunct. In 1624, James I revoked the company's charter, and the colony came under the control of the crown, where it would remain until 1776. The colony, but not the company, had survived—but at a terrible cost. In Virginia's first seventeen years, more than 8,500 white settlers had arrived in the colony, and nearly 80 percent of them had died. Countless native inhabitants had died as well, and slavery became part of the colony.

Bacon's Rebellion

For more than thirty years, one man—Sir **William Berkeley**, the royal governor of Virginia—dominated the politics of the colony. He took office in 1642 at the age of 36 and with but one brief interruption remained in control of the government until 1677. In his first years as governor, he helped open up the interior of Virginia by sending explorers across the Blue Ridge Mountains and crushing a 1644 uprising of Native Americans, who agreed to a treaty ceding to England most of the territory east of the mountains and establishing a new boundary, west of which white settlement would be prohibited. The rapid growth of the Virginia population made this agreement difficult to keep. Between 1640 and 1660, Virginia's population rose from 8,000 to over 40,000. By 1652, English settlers had established three counties in the territory set aside by the treaty for native inhabitants.

In the meantime, Berkeley was expanding his own powers. By 1670, the vote for delegates to the House of Burgesses, once open to all white men, was restricted to landowners. Elections were rare, and the same burgesses, representing the established planters of the eastern (or tidewater) region of the colony, remained in office year after year. The more recent settlers on the frontier were underrepresented.

Popular resentment of the power of the governor and the Tidewater aristocrats grew steadily in the newly settled lands of the West (often known as the "backcountry"). In 1676, it bubbled over into a major conflict, dubbed **Bacon's Rebellion** after its leader Nathaniel Bacon, a distant relative of Governor Berkeley himself. Bacon had a good farm in the West and a seat on the governor's council. But he believed he and other members of the new backcountry gentry were being treated unfairly. Specifically, Bacon opposed the governor's refusal to allow white settlers to move farther west and take more land, in defiance of the boundary established by treaty. Adding to Bacon's bitterness was Berkeley's control of the lucrative fur trade, which Bacon desperately wanted to profit from.

The turbulence in Virginia reflected not just the tension between Berkeley and Bacon, both of them frontier aristocrats. It was also a result of the consequences of the indentured servant system. By the 1670s, many poor young men—predominantly white but some Black—had finished their term as indentures and found themselves without a home or any money. Many of them began moving around the colony, sometimes working, sometimes begging, sometimes stealing. This large, landless itinerant group were quickly drawn into what became Bacon's Rebellion.

In 1675, a major conflict erupted in the West between English settlers and native inhabitants. As the fighting escalated, Bacon and other concerned landholders demanded that the governor send the militia. When Berkeley refused, Bacon responded by offering to organize a volunteer army of backcountry men who would do their own fighting. Berkeley rejected that offer too. Bacon simply ignored him and launched a series of vicious but unsuccessful pursuits of native opponents.

When Berkeley heard of the unauthorized military effort, he proclaimed Bacon and his men to be rebels. Confused and angry, Bacon now took aim at the governor and all elites, whom he openly criticized for lacking empathy and support for the lower classes. He attracted a range of volunteers and built an alliance of Black and white workers as well as enslaved Africans, who asked for an end to their bondage in exchange for their allegiance. Bacon led his troops, numbering about 500, east to Jamestown twice. The first time he won a temporary pardon from the governor; the second time, after the governor repudiated the first agreement, Bacon burned much of the city and drove the governor into exile. But then Bacon died suddenly of dysentery, and Berkeley regained control. Among the last to surrender were some 400 indentured servants and 80 enslaved people. In 1677, the Native Americans reluctantly signed a new treaty that opened new lands to white settlement.

Bacon's Rebellion was part of a continuing struggle to define the native and white spheres of influence in Virginia. It also revealed the bitterness of the competition among rival white leaders as well as the potential for instability in the colony's large population of free landless men, both Black and white, and enslaved people. Most notably, the uprising forced landed elites in both eastern and western Virginia to recognize a common interest in quelling social unrest from below and shoring up their cultural dominance. As a result, they began to reduce their dependence on indentured laborers and rely more heavily on enslaved workers. Enslaved Blacks, unlike white indentured servants, did not need to be released after a fixed term and therefore did not threaten to become an unstable, landless class. Elite whites, perhaps predictably, now sought more enslaved workers, preferring those shipped straight from Africa, hoping that their presumed lack of knowledge about English culture would make them less likely to partner in any rebellion with white workers and servants. To minimize any chance of cooperation between Blacks and whites, they also enlisted poor whites to organize slave patrols. And, most effectively, they created new laws. In 1705 the Virginia General Assembly passed the Virginia Slave Codes, which formally defined the enslaved as property or "real estate" and endowed slaveholders with near limitless power over enslaved workers. The new **slave codes** declared that "All Negro, mulatto and Indian slaves within this dominion . . . shall be held to be real estate. If any slave resist his master . . . correcting such slave, and shall happen to be killed in such correction . . . the master shall be free of all punishment . . . as if such accident never happened." They further specified that an enslaved worker needed a written pass to leave the slaveholder's home or plantation, that robbery would be met with sixty lashes and time in the stocks, where his or her ears would be loped off, and that associating with whites without official sanction would result in a whipping, branding, or maiming of some sort. These codes laid the foundation for the practice of slavery for generations to come.

Maryland and the Calverts

The Maryland colony ultimately came to look much like Virginia, but its origins were quite different. **George Calvert**, the first Lord Baltimore, envisioned establishing a colony in America both as a great speculative venture in real estate and as a refuge for English Catholics like himself. Calvert died while he was still negotiating with the king in London for a charter to establish such a colony in the Chesapeake region. But in 1632, his son **Cecilius Calvert**, the second Lord Baltimore, received the charter.

THE GROWTH OF THE CHESAPEAKE, 1607–1750 This map shows the political forms of European settlement in the region of Chesapeake Bay in the seventeenth and early eighteenth centuries. Note the several different kinds of colonial enterprises: the royal colony of Virginia, controlled directly by the English crown after the failure of the early commercial enterprises there; and the proprietary regions of Maryland, northern Virginia, and North Carolina, which were under the control of powerful English aristocrats. • *Why were most settlements, regardless of who founded them, found near bodies of water? What does it suggest about the type of economic activities each engaged in?*

Lord Baltimore remained in England, but he named his brother, Leonard Calvert, governor of the colony. In March 1634, two ships—the *Ark* and the *Dove*—bearing Calvert along with 200 or 300 other colonists entered the Potomac River, turned into one of its eastern tributaries, and established the village of St. Mary's on a high, dry bluff. Neighboring Native Americans, in particular the Susquehannocks, befriended the settlers and provided them with temporary shelter and corn.

The Calverts needed to attract thousands of settlers to Maryland if their expensive colonial venture was to pay. They quickly went about encouraging the immigration of Protestants as well as their fellow English Catholics. As non-Catholics began to arrive in ever greater number, however, the Calverts realized that Catholics would always be a minority in the colony and so they adopted a policy in 1649 that they hoped would ensure a degree of acceptance among different sacred traditions: the "Act Concerning Religion." Dubbed the Maryland Toleration Act, it decreed religious toleration among all resident Christians. Still, politics in Maryland remained plagued for years with chronic tensions, and at times violence, between the Catholic minority and the growing Protestant majority.

At the insistence of the first settlers, the Calverts agreed in 1635 to the calling of a representative assembly—the House of Delegates. But they retained absolute authority to distribute land as he wished; and since Lord Baltimore granted large estates to his relatives and to other English aristocrats, a distinct upper class soon established itself. By 1640, a severe labor shortage forced a modification of the land-grant procedure; and Maryland, like Virginia, adopted a headright system—a grant of 100 acres to each male settler, another 100 for his wife and each servant, and 50 for each of his children. But the great landlords of the colony's earliest years remained powerful. Like Virginia, Maryland became a center of tobacco cultivation; planters worked their land with the aid, first, of indentured servants imported from England and then, beginning late in the seventeenth century, of enslaved labor from Africa.

THE GROWTH OF NEW ENGLAND

The northern regions of English North America were slower to attract settlers than those in the South. That was in part because the Plymouth Company was not able to mount a successful colonizing expedition after receiving its charter in 1606. It did, however, sponsor other explorations. Captain John Smith, after departing from Jamestown, made an exploratory journey for the Plymouth merchants, wrote an enthusiastic pamphlet about the lands he had seen, and called them New England.

Plymouth Plantation

A discontented congregation of Puritan Separatists in England—those disenchanted with the Church of England and seeking to "separate" from it—established the first enduring European settlement in New England. In 1608, a congregation of Separatists from the English hamlet of Scrooby began emigrating quietly (and illegally), a few at a time, to Leyden, Holland, where they believed they could enjoy freedom of worship. As foreigners in Holland, they had to work at unskilled and poorly paid jobs. They also watched with alarm as their children began to adapt to Dutch society and drift away from their church. Finally, some of the Separatists decided to move again, this time across the Atlantic; there, they hoped to create a stable, protected community where they could spread "the gospel of the Kingdom of Christ in those remote parts of the world."

In 1620, leaders of the Scrooby group obtained permission from the Virginia Company to settle in Virginia. The "Pilgrims," as they saw themselves, sailed from Plymouth, England, in September 1620 on the *Mayflower*; thirty-five "saints" (the Puritan Separatists) and sixty-seven "strangers" (people who were not part of the congregation) were aboard. In November, after a long and difficult voyage, they sighted land—the shore of what is now Cape Cod. That had not been their destination, but it was too late in the year to sail farther south. So the Pilgrims chose a site for their settlement in the area just north of the cape, a place John Smith had labeled "Plymouth" on a map he had drawn during his earlier exploration of New England. Because Plymouth lay outside the London Company's territory, the settlers were not bound by the company's rules. While still aboard ship, the saints in the group drew up an agreement, the **Mayflower Compact**, to establish a government for themselves. Then, on December 21, 1620, they stepped ashore.

The Pilgrims' first winter was a difficult one. Half the colonists perished from malnutrition, disease, and exposure. But the colony survived, in large part because of crucial assistance from the Wampanoags. Significantly, that help was not immediately forthcoming from the Wampanoags nor did it betoken their complete trust in the newcomers. Earlier English ships had anchored in the area for supplies, captured native people, and sold them into slavery. An epidemic had broken out before 1615 that killed thousands of Wampanoags and left them vulnerable to attack by their rivals, the Pequot and the Narragansett. Eventually the Wampanoags reached out to the immigrants in large part to form an alliance that might help them in any future conflict with their enemies. They now traded furs with the colonists and showed them how to cultivate corn and how to hunt wild animals for meat. The first autumn harvest, the settlers invited some Wampanoags to join them in a feast, the original Thanksgiving. But the relationship between the settlers and the local Native Americans was always strained. Thirteen years after the Pilgrims arrived, a devastating smallpox epidemic wiped out much of the indigenous population around Plymouth.

New colonists arrived slowly but steadily from England and in a decade the population reached 300. The people of **Plymouth Plantation** chose as their governor William Bradford, who ruled successfully for many years. The Pilgrims could not create rich farms on the sandy and marshy soil around Plymouth, but they developed a profitable trade in fish and furs.

(Alexander Sviridov/Shutterstock)

PLYMOUTH PLANTATION (Re-creation)

The Massachusetts Bay Experiment

In 1628, another group of Puritan merchants began organizing a new colonial venture in America. They obtained a grant of land in New England for most of the area now comprising Massachusetts and New Hampshire. They acquired a charter from the king—now Charles I, who had inherited the throne at the death of his father, James I, in 1625—that allowed them to create the **Massachusetts Bay Company** and to establish a colony in the New World. Some members of the Massachusetts Bay Company wanted to create a refuge in New England for Puritans. They bought out the interests of company members who preferred to stay in England, and the new owners elected a governor, **John Winthrop**. They then sailed for New England in 1630. With 17 ships and 1,000 people, it was the largest single migration of its kind in the seventeenth century. Winthrop carried with him the charter of the Massachusetts Bay Company, which decreed that the colonists would be responsible to no company officials in England.

This migration quickly produced several settlements. The port of Boston became the capital, but in the course of the next decade colonists established many new towns in eastern Massachusetts: Charlestown, Newtown (later renamed Cambridge), Roxbury, Dorchester, Watertown, Ipswich, Concord, Sudbury, and others.

Governor Winthrop and the other founders believed they were building a holy commonwealth, a model—a "city upon a hill"—for the corrupt world to see and emulate. They built a **theocracy**, a society in which the church was almost indistinguishable from the state. Ironically, these Puritans who had sought religious freedom in England now created a society that brooked no tolerance of religious or political dissent. Their strict rules and laws soon produced a wave of critics who would eventually leave and found new colonies.

Like other new settlements, the Massachusetts Bay colony had early difficulties. During the first winter (1629-1630), nearly 200 people died and many others decided to leave. But the colony soon grew and prospered. The nearby Pilgrims and Wampanoags helped with food and advice. Incoming settlers brought needed tools and other goods. The prevalence of families in the colony helped establish a feeling of commitment to the community and a sense of order among the settlers, and it also ensured that the population would reproduce itself.

The Expansion of New England

It did not take long for British settlement to begin moving outward from Massachusetts Bay. Some migrated in search of soil more productive than the stony land around Boston. Others left because of the oppressiveness of the church-dominated government of Massachusetts.

The Connecticut River valley, about 100 miles west of Boston, began attracting English families as early as the 1630s because of its fertile lands and its distance from Massachusetts Bay. In 1635, Thomas Hooker, a minister of Newtown (now Cambridge), defied the Massachusetts government, led his congregation southwest, and established the town of Hartford. Four years later, the people of Hartford and two other newly founded towns nearby adopted a constitution known as the Fundamental Orders of Connecticut, which created an independent colony with a government similar to that of Massachusetts Bay, but which gave a larger proportion of the men the right to vote and hold office. Women were barred from voting, as they were virtually everywhere in the colonies.

Another Connecticut colony grew up around New Haven on the Connecticut coast. Unlike Hartford, the Fundamental Articles of New Haven (1639) established a Bible-based government even stricter than that of Massachusetts Bay. New Haven remained independent until 1662, when a royal charter officially gave the Hartford colony jurisdiction over the New Haven settlements.

European settlement in what is now Rhode Island was a result of the vigorous religious and political dissent of **Roger Williams**, a controversial young minister who lived for a time in Salem, Massachusetts. Williams was a confirmed Separatist who argued that the Massachusetts church should abandon all allegiance to the Church of England and permit its citizens to worship as they saw fit. More radically, he proclaimed that the land the colonists were occupying belonged to the local Native Americans. Not surprisingly the colonial government voted to deport him, but he escaped before they could force him to leave. During the winter of 1635–1636, Williams took refuge with the Narragansett; the following spring he bought a tract of land from them and, with a few followers, created the town of Providence. He also learned their language and called for the Bible to be translated into the Narragansett's native tongue. In 1644, after obtaining a charter from Parliament, he established a government similar to that of Massachusetts but without any ties to a specific church. That year he famously made the case for a strict separation between church and state, writing "An enforced uniformity of religion throughout a nation or civil state, confounds the civil and religious, denies the principles of Christianity and civility, and that Jesus Christ has come in the flesh." For a time, Rhode Island was the only colony in which all believers, including Jews, Quakers, and Baptists, could worship without interference.

Another challenge to the established religious order in Massachusetts Bay came from **Anne Hutchinson**, a learned and charismatic woman from a substantial Boston family who pushed the limits of religious orthodoxy and the domestic roles of women. She sparked the **Antinomian**, or free grace, controversy. Specifically, she challenged the clerical doctrine of the covenant of grace, which held that when individuals sincerely confessed belief in Christ, then God promised them salvation. The "elect" then demonstrated their special status by performing good works. Hutchinson differed sharply with this orthodoxy, claiming that one's works and words had no bearing at all on the state of one's salvation. The only matter of significance was whether God selected a person for eternal life or not; what that person did or did not do in the course of daily life mattered none in proving God's favor. She was a popular voice, often hosting groups of women in her home. Hutchinson's teachings threatened the spiritual authority of established clergy, who spoke on the presumption that they were already saved. As her influence grew and as she began to deliver open attacks on members of the clergy, the Massachusetts hierarchy mobilized to stop her. They would not tolerate any threats to their spiritual authority, especially coming from a woman occupying a social role—public spokesperson, theological critic—typically reserved for men at this time.

In 1637, Hutchinson was convicted of heresy and sedition and was banished. With her family and some of her followers, she fled to a point on Narragansett Bay not far from Providence. Later she moved south to New York, where in 1643 she and her family died during a Native American uprising.

New Hampshire and Maine were established in 1629 by two English proprietors. But few settlers moved into these northern regions until the religious disruptions in Massachusetts Bay. In 1639, John Wheelwright, a disciple of Anne Hutchinson, led some of his fellow dissenters to Exeter, New Hampshire. Others soon followed. New Hampshire became a separate colony in 1679. Maine remained a part of Massachusetts until 1820.

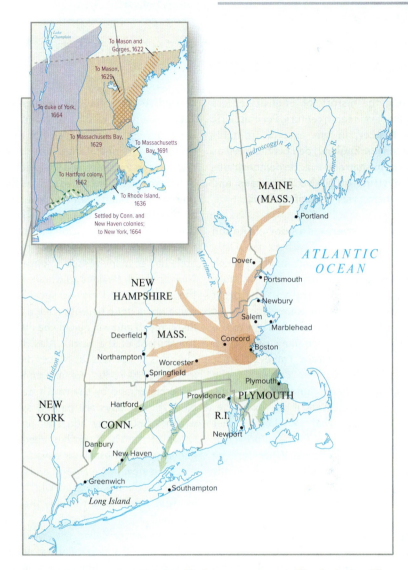

THE GROWTH OF NEW ENGLAND, 1620–1750 The European settlement of New England, as this map reveals, traces its origins primarily to two small settlements on the Atlantic Coast. The first was the Pilgrim settlement at Plymouth, which began in 1620 and spread out through Cape Cod, southern Massachusetts, and the islands of Martha's Vineyard and Nantucket. The second, much larger settlement began in Boston in 1630 and spread rapidly through western Massachusetts, north into New Hampshire and Maine, and south into Connecticut. • *Why would the settlers of Massachusetts Bay have expanded so much more rapidly and expansively than those of Plymouth?*

KING PHILIP'S WAR

In 1637, hostilities broke out between English settlers in the Connecticut Valley and the Pequot people of the region. The **Pequot War** resulted in the near elimination of local Native Americans. But the bloodiest and most prolonged encounter between whites and native people in the seventeenth century was the fourteen-month conflict that whites called **King Philip's War** and that ultimately killed more than 3,000 native people and 1,000 colonists.

CONSIDER THE SOURCE

COTTON MATHER ON THE RECENT HISTORY OF NEW ENGLAND (1692)

Reviewing the history of English settlers during the seventeenth century, Puritan cleric Cotton Mather, in this excerpt from his history of New England, depicts the Devil as the root of mishap and evil. He demonstrates a real mistrust of Native Americans and portrays them as servants of the Devil.

I believe there never was a poor plantation more pursued by the wrath of the Devil than our poor New England; and that which makes our condition very much the more deplorable is that the wrath of the great God himself at the same time also presses hard upon us. It was a rousing alarm to the Devil when a great company of English Protestants and Puritans came to erect evangelical churches in a corner of the world where he had reigned without any control for many ages; and it is a vexing eyesore to the Devil that our Lord Christ should be known and owned and preached in this howling wilderness. Wherefore he has left no stone unturned, that so he might undermine this plantation and force us out of our country.

First, the Indian Powwows used all their sorceries to molest the first planters here; but God said unto them, "Touch them not!" Then, seducing spirits came to root in this vineyard, but God so rated them off that they have not prevailed much farther than the edges of our land. After this, we have had a continual blast upon some of our principal grain, annually diminishing a vast part of our ordinary food. Herewithal, wasting sicknesses, especially burning and mortal agues, have shot the arrows of death in at our windows. Next, we have had many adversaries of our own language, who have been perpetually assaying to deprive us of those English liberties in the encouragement whereof these territories have been settled. As if this had not been enough, the Tawnies among whom we came have watered our soil with the blood of many hundreds of our inhabitants. Desolating fires also have many times laid the chief treasure of the whole province in ashes. As for losses by sea, they have been multiplied upon us; and particularly in the present French War, the whole English nation have observed that no part of the nation has proportionately had so many vessels taken as our poor New England. Besides all which, now at last the devils are (if I may so speak) in person come down upon us, with such a wrath as is justly much and will quickly be more the astonishment of the world. Alas, I may sigh over this wilderness, as Moses did over his, in Psalm 90.7, 9: "We are consumed by thine anger, and by thy wrath we are troubled: All our days are passed away in thy wrath." And I may add this unto it: the wrath of the Devil too has been troubling and spending of us all our days....

Let us now make a good and a right use of the prodigious descent which the Devil in great wrath is at this day making upon our land. Upon the death of a great man once, an orator called the town together, crying out, "Concurrite cives, dilapsa cunt vestra moenia!" That is, "Come together neighbors, your town walls are fallen down!" But such is the descent of the Devil at this day upon our selves that I may truly tell you, the walls of the whole world are broken down! The usual walls of defense about mankind have such a gap made in them that the very devils are broke in upon us to seduce the souls, torment the bodies, sully the credits, and consume the estates of our neighbors, with impressions both as real and as furious as if the invisible world were becoming incarnate on purpose for the vexing of us....

In as much as the devil is come down in great wrath, we had need labor, with all the care and speed we can, to divert the great wrath of Heaven from coming at the same time upon us. The God of Heaven has with long and loud admonitions been calling us to a reformation of our provoking evils as the only way to avoid that wrath of his which does not only threaten but consume us. It is because we have been deaf to those calls that we are now by a provoked God laid open to the wrath of the Devil himself.

Source: Mather, Cotton, *The Wonders of the Invisible Word*. Boston, 1692, 41–43, 48; cited in Richard Godbeer, *The Salem Witch Hunt: A Brief History with Documents*. Boston: Bedford St. Martin's, 48–49.

UNDERSTAND, ANALYZE, & EVALUATE

1. According to Cotton Mather, what particular hardships did the colonists suffer?
2. What did Mather mean when he wrote that "now at last the devils [have descended] in person"?
3. What deeper explanation did Cotton Mather offer for New England's crisis? What response did he suggest?

In June 1675, the Wampanoag, led by their chief **Metacom** whom the English called "King Philip," joined with the Nipmuck, Pocumtuck, and Narragansett in an attempt to drive out the English settlers and resist any further encroachments upon their land. Stoking native resentment were efforts by the colonists to force them to recognize English sovereignty. Not all native groups followed Metacom's lead, however, and some actually backed the English, including the Nauset, Mohegan, and Pequot. Still, Metacom and his warriors terrorized a string of Massachusetts towns for over a year. In early 1676 the tide began to turn when white settlers forged a new alliance with a group of Mohawk allies, who soon ambushed Metacom and beheaded him. Without Metacom, the fragile coalition of native groups collapsed and the white settlers were quickly able to crush the uprising. In the aftermath, some survivors fled to Canada. Colonists seized many of those who surrendered and sold them into slavery in the West Indies. The native nations who had originally led the uprising were left greatly weakened as military powers and would never again be able to mount such a major assault against the English settlers.

The conflicts between native peoples and settlers were crucially affected by earlier exchanges of technology between the English and Native Americans. In particular, Native Americans made effective use of a relatively new European weapon that they had acquired from the English: the flintlock rifle. It replaced the earlier staple of colonial musketry, the matchlock rifle, which proved too heavy, cumbersome, and inaccurate to be effective. The matchlock had to be steadied on a fixed object and ignited with a match before firing. The flintlock could be held up without external support and fired without a match.

Despite rules forbidding colonists to instruct native people on how to use and repair the weapons, they learned to handle the rifles and repair them very effectively on their own. In King Philip's War, the very high casualties on both sides were partly a result of the use of these more advanced rifles.

The violence of the war, the settlers' insatiable appetite for land, and their uneven respect for indigenous culture affected their understanding of their own recent history. Leaders increasingly portrayed native people as "heathens" and barbarians. Religious officials in particular came to consider local Native Americans as a threat to their hopes of creating a godly community in the New World. (See "Consider the Source: Cotton Mather on the Recent History of New England.")

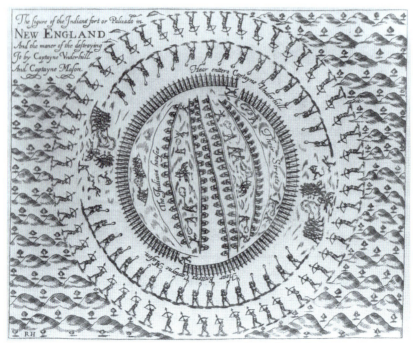

(Library of Congress, Prints and Photographs Division [LC-USZ62-32055])

A PEQUOT VILLAGE DESTROYED An English artist drew this view of a fortified Pequot village in Connecticut surrounded by English soldiers and their allies from other indigenous nations during the Pequot War in 1637. The invaders massacred more than 600 residents of the settlement.

THE RESTORATION COLONIES

For nearly thirty years after Lord Baltimore received the charter for Maryland in 1632, no new English colonies were established in America. England was dealing with troubles of its own at home.

The English Civil War

After Charles I dissolved Parliament in 1629 and began ruling as an absolute monarch, he alienated a growing number of his subjects. Desperately in need of money, Charles called Parliament back into session in 1640 and asked it to levy new taxes. But he antagonized the members by dismissing them twice in two years; and in 1642, members of Parliament organized a military force, sparking the English Civil War.

The conflict between the Cavaliers (the supporters of the king) and the Roundheads (the forces of Parliament, who were largely Puritans) lasted seven years. In 1649, the Roundheads defeated the king's forces and shocked all of Europe by beheading the monarch. The stern Roundhead leader Oliver Cromwell assumed the position of "protector." During his reign, Crowell looked westward to expand his influence. He authorized the ambitious Western Design—a vast expedition designed to wrest Jamaica, Puerto Rico, Cuba, and Hispaniola from the Spanish. He earnestly believed that God had called him to engage the Spanish and seize the land for England, but he met with patchy success. Indeed, his plans failed everywhere except Jamaica, which would eventually become an important engine of the sugar and slave economy in the English colonies in the eighteenth century.

When Cromwell died in 1658, his son and heir proved unable to maintain his authority. Two years later, Charles II, the son of the executed king, returned from exile and seized the throne, in what became known as the Restoration. Among the results of the Restoration was the swift resumption of colonization in America. Charles II rewarded faithful courtiers with grants of land in the New World, and in the twenty-five years of his reign he issued charters for four additional colonies: Carolina, New York, New Jersey, and Pennsylvania.

Charles II supported religious toleration—and would have allowed Catholicism again in England, to the dismay of many Protestants. Parliament refused to back him, however. The prudent Charles was wise enough not to fight for the right of Catholics to worship openly lest it cost him his throne. But he made a private agreement with Louis XIV of France that he would become a Catholic—which he did only on his deathbed. His son, James II, faced a hostile Parliament that suspected him of Catholic allegiances.

THE CAROLINAS

In charters issued in 1663 and 1665, Charles II awarded joint title to eight proprietors. They received a vast territory stretching south from Virginia to the Florida peninsula and west to the Pacific Ocean. Like Lord Baltimore, they received almost kingly powers over their grant, which they prudently called Carolina (a name derived from the Latin word for "Charles"). They reserved tremendous estates for themselves and distributed the rest through a headright system similar to those in Virginia and Maryland. Although committed Anglicans themselves, the proprietors promised religious freedom to all members of all Christian faiths. They also created a representative assembly. They hoped to attract settlers from the existing American colonies and to avoid the expense of financing expeditions from England.

Their initial efforts to profit from settlement in Carolina, however, failed. One proprietor, Anthony Ashley Cooper, tried a new course of action. He convinced the other proprietors to finance expeditions to Carolina from England, the first of which set sail with 300 people in the spring of 1670. The 100 people who survived the voyage established a settlement at Port Royal on the Carolina coast. Ten years later, they founded a city at the junction of the Ashley and Cooper Rivers, which in 1690 became the colonial capital. They called it Charles Town (later renamed Charleston).

With the aid of the English philosopher John Locke, Cooper (now the earl of Shaftesbury) drew up the Fundamental Constitution for Carolina in 1669. It divided the colony into counties of equal size and divided each county into equal parcels. It also established a social hierarchy with the proprietors themselves (who were to be known as "seigneurs") at the top, a local aristocracy (consisting of lesser nobles known as "landgraves" or "caciques") below them, and then ordinary settlers ("leet-men"). At the bottom of this stratified society would be poor whites, who would have few political rights, and enslaved Africans. Proprietors, nobles, and other landholders would have a voice in the colonial parliament in proportion to the size of their landholdings.

In reality, Carolina developed along lines quite different from the carefully ordered vision of Cooper and Locke. For one thing, the northern and southern regions of settlement were widely separated and socially and economically distinct from each other. The northern settlers were mainly backwoods farmers. In the south, fertile lands and the good harbor at Charles Town promoted a more prosperous economy and a more stratified, aristocratic society. Settlements grew up rapidly along the Ashley and Cooper Rivers and colonists established a flourishing trade, particularly (beginning in the 1670s) in rice. The development of the rice economy depended heavily on the role of enslaved Africans. It was they who brought the experience and knowledge of how to grow the crop. They grew rice for themselves and then for the slaveholders, who eventually forced their enslaved workers to raise it in large-scale production.

Southern Carolina very early developed commercial ties to the large (and overpopulated) European colony on the Caribbean island of Barbados. During the first ten years of settlement, most of the new residents in Carolina were Barbadians, some of whom established themselves as substantial landlords. African slavery had taken root in Barbados earlier than in any of the mainland colonies, and the white Caribbean migrants—tough, uncompromising profit seekers—established a similar slave-based plantation society in Carolina.

Carolina was one of the most divided English colonies in America. There were tensions between the small farmers of the Albemarle region in the North and the wealthy planters in the South. And there were conflicts between the rich Barbadians in southern Carolina and the smaller landowners around them. After Lord Shaftesbury's death, the proprietors proved unable to establish order. In 1719, the colonists seized control of the colony from them. Ten years later, the king divided the region into two royal colonies, North Carolina and South Carolina.

New Netherland, New York, and New Jersey

In 1664, Charles II granted his brother James, the Duke of York, all the territory lying between the Connecticut and Delaware Rivers. This land, however, was also claimed by the Dutch. The growing conflict between the English and the Dutch was part of a larger commercial rivalry between the two nations throughout the world. The English particularly resented the Dutch presence in America, because it served as a wedge between the northern and southern English colonies and provided bases for Dutch smugglers evading English custom laws. Months after James received the grant, an English fleet under the command of Richard Nicolls put in at New Amsterdam, the capital of the Dutch colony of New Netherland, and extracted a surrender from the governor, Peter Stuyvesant. Several years later, in 1673, the Dutch reconquered and briefly held their old provincial capital. But they lost it again, this time for good, in 1674.

The Duke of York renamed his territory New York. It contained not only Dutch and English but also Scandinavians, Germans, French, and a large number of Africans (brought in as enslaved workers by the Dutch West India Company), as well as members of several different indigenous groups. James wisely made no effort to impose his own Roman Catholicism on the colony. He delegated powers to a governor and a council but made no provision for representative assemblies.

Slavery was different in New York, in part because of its Dutch roots. In the 1640s, the Dutch West India Company responded to a petition for freedom from enslaved laborers by granting them "half freedom," in which they were technically manumitted but their children were not and they were required to pay a fee to the Company every year. The Company allowed enslaved laborers to work for wages and also awarded them land, between 2 and 18 acres, in a watery, hilly region about a mile north from the city, in part to create buffer between native inhabitants and white settlers. Formerly enslaved men and women eventually inhabited over 130 acres in what was New York's first free Black community.

Power in New York was concentrated in the hands of English and Dutch landlords, wealthy fur traders, and the duke's political appointees. By 1685, when the Duke of York ascended the English throne as James II, New York contained about four times as many people (around 30,000) as it had twenty years before.

Shortly after James received his charter, he gave a large part of the land south of New York to a pair of political allies, both Carolina proprietors, Sir John Berkeley and Sir George Carteret. Carteret named the territory New Jersey. The venture in New Jersey generated few profits and in 1674 Berkeley sold his half interest. The colony was divided into two jurisdictions, East Jersey and West Jersey, which squabbled with each other until 1702, when the two halves of the colony were again joined. New Jersey, like New York, was a colony of enormous ethnic and religious diversity and the weak colonial government made few efforts to impose strict control over the fragmented society. Unlike New York, New Jersey developed no important class of large landowners.

THE QUAKER COLONIES

Pennsylvania was born out of the efforts of a dissenting English Protestant sect, the Society of Friends., who wished to find a home for their own distinctive social order. The Society began in the mid-seventeenth century under the leadership of George Fox, a Nottingham shoemaker, and Margaret Fell. Their followers came to be known as **Quakers** (from Fox's instruction to them to "tremble at the name of the Lord"). Unlike the Puritans, Quakers rejected the concept of predestination (that God foreordained who would go to heaven and to hell) and original sin (that all people were born sinful). Every person, they believed, had divinity within themselves from birth and needed only learn to cultivate it; all could attain salvation.

The Quakers had no formal church government and no paid clergy; in their worship they spoke up one by one as the spirit moved them. Disregarding distinctions of gender and class, they addressed one another with the terms *thee* and *thou,* words commonly used in other parts of British society only in speaking to servants and social inferiors. Confirmed pacifists, they would not take part in wars. Until the mid-eighteenth century, though, Quakers did not oppose slavery and indeed some held enslaved workers themselves. (In Barbados, many did.) Unpopular in England, the Quakers began looking to America for asylum. A few migrated to New England or Carolina, but most Quakers wanted a colony of their own. As members of a despised sect, however, they could not get the necessary royal grant without the aid of someone influential at the court.

Fortunately for the Quaker cause, a number of wealthy and prominent men had converted to the faith. One of them was **William Penn**, an outspoken evangelist who had been in prison several times. Penn worked with George Fox on plans for a Quaker colony in America. When Penn's father died in 1681, Charles II settled a large debt he had owed to the older Penn by making an enormous land grant of territory between New York and Maryland to the son. At the king's insistence, the territory was to be named Pennsylvania, after Penn's late father.

Penn soon made Pennsylvania the best-known and most cosmopolitan of all the British colonies in America. Pennsylvania prospered from the outset because of Penn's successful recruiting and advertising, his careful planning, and the region's mild climate and fertile soil. Penn sailed to Pennsylvania in 1682 to oversee the laying out of the city he named Philadelphia ("Brotherly Love" in Greek) between the Delaware and Schuylkill Rivers.

Penn's relatively good relations with Native Americans were a result in large part of his religious beliefs. Besides extending Quakerism's rejection of war or physical violence to formal relationships with Native Americans, Penn generally worked to respect indigenous culture. He recognized native claims to the land in the province and was usually scrupulous in reimbursing native nations for their land when it was taken. In later years, however, the relationships between the British residents of Pennsylvania and the indigenous inhabitants were not always so peaceful.

By the late 1690s, some residents of Pennsylvania were beginning to resist the nearly absolute power of the proprietor. Pressure from these groups grew to the point that in 1701, shortly before he departed for England for the last time, Penn agreed to a Charter of Liberties for the colony. The charter established a representative assembly (consisting, alone among the British colonies, of only one house) that greatly limited the authority of the proprietor. The charter also permitted "the lower counties" of the colony to establish their own representative assembly. The three lower counties did so in 1703 and as a result became, in effect, a separate colony—Delaware—although until the American Revolution it continued to have the same governor as Pennsylvania.

BORDERLANDS AND MIDDLE GROUNDS

The English colonies clustered along the Atlantic seaboard of North America eventually united, expanded, and became the beginnings of a powerful nation. But in the seventeenth and early eighteenth centuries, their future was not at all clear. In those years, they were small, frail settlements surrounded by other competing colonies and by native nations. The British Empire in North America was, in fact, much smaller and weaker than the great Spanish Empire to the south, and in many ways weaker than the enormous French Empire to the north.

The continuing contests for control of North America were most clearly visible in areas around the borders of British settlement—the Caribbean and the northern, southern, and western borders of the coastal colonies. In the borderland regions societies emerged that were very different from those in the British seaboard colonies—these areas have been described as **middle grounds**, in which diverse civilizations encountered one another and, for a time at least, shaped one another.

THE CARIBBEAN ISLANDS

The Chesapeake was the site of the first permanent English settlements in the North American continent. Throughout the first half of the seventeenth century, however, the most important destinations for British immigrants were the islands of the Caribbean and the northern way station of Bermuda. Far smaller in size than the colonies would eventually become, they still were home to more than half of English migrants to the New World in the early seventeenth century.

Before the arrival of Europeans, most of the Caribbean islands had substantial native populations. But beginning with Christopher Columbus's first visit in 1492, and accelerating after 1496, these populations were severely weakened and reduced by European epidemics.

The Spanish Empire claimed title to all the islands in the Caribbean, but Spain created substantial settlements only in the largest of them: Cuba, Hispaniola, and Puerto Rico. English, French, and Dutch traders began settling on some of the smaller islands early in the sixteenth century, despite the Spanish claim to them. After Spain and the Netherlands went to war in 1621 (distracting the Spanish navy and leaving the British in the Caribbean relatively unmolested), the pace of English colonization increased. Its most important settlements were on St. Kitts (1623), Barbados (1627), Antigua (1632), and Jamaica (1655).

In their first years in the Caribbean, English settlers experimented unsuccessfully with tobacco and cotton. They soon discovered that the most lucrative crop was sugar, for which there was a substantial and growing market in Europe. Sugarcane could also be distilled into rum, for which there was also a booming market abroad. Unsurprisingly, planters quickly devoted almost all of their land to sugarcane.

Because sugar was a labor-intensive crop, British planters found it necessary to import laborers. As in the Chesapeake, they began by bringing indentured servants from England but they soon came to prefer enslaved Africans, over whom they could exert greater control for longer periods. By midcentury, the English planters in the Caribbean were relying more and more on an enslaved African workforce, which soon substantially outnumbered them. By the early eighteenth century, there were four times as many enslaved Africans as there were white settlers on Barbados and there were African majorities on all English Leeward Islands.

SLAVEHOLDER AND ENSLAVED IN THE CARIBBEAN

Fearful of revolts by the enslaved, whites in the Caribbean monitored their labor forces closely and often harshly. Indeed, planters paid little attention to the welfare of their workers. Many concluded that it was cheaper to buy new enslaved laborers periodically than to protect the well-being of those they already owned, and it was not uncommon for slaveholders to literally work enslaved laborers to death. Few enslaved Africans lasted more than a decade in the brutal Caribbean working environment—they were either sold to planters in North America or died. Even whites, who worked far less hard than did the enslaved, often succumbed to the harsh climate; most died before the age of forty.

Establishing a stable society and culture was extremely difficult in such harsh and even deadly conditions. Still, English Leeward Islands landowners in the Caribbean islands built a range of intuitions that replicated those found in their homeland—government, courts, churches, and juridical codes. Even after many returned home to escape the harsh conditions, the institutions they birthed continued to shape society. The Barbados' slave code of 1661, for example, influenced the practice of slavery in the Caribbean for decades. It empowered slaveholders to treat enslaved workers as they saw fit, including inflicting severe punishments, with little fear of legal penalty. It stated that for "any Negro or other slave under punishment by his master . . . no person whatsoever shall be liable to any fine therefore." The code went further, however, specifying in exacting detail the type of pain to be inflicted on those who were "disobedient." "If any Negro or slave whatsoever shall offer any violence to any Christian by striking or the like, such Negro or slave shall for his or her first offence be severely whipped by the Constable. For his second offence of that nature he shall be severely whipped, his nose slit, and be burned in some part of his face with a hot iron." The code also required slaveholders to provide at least one set of clothes to every enslaved worker every year, but specified not much else; it said nothing about providing health care, housing, food, or a period of rest.

Despite living in subhuman conditions, enslaved Africans developed a range of social practices that gave them a measure of comfort and relief and nurtured an African-Caribbean culture at least partly independent of the slaveholders' control. They sustained African religious and social traditions and blended them with local customs to create new signature expressions of faith. They strove to build families, even though many were destroyed by death or the slave trade. And they challenged their lowly status by resisting—often quietly and in hidden ways—the circumstances of their lives. They sometimes sabotaged crops, poisoned seeds, set small fires to the harvest, broke hoes and shovels, slowed down their pace of work, or feigned illness to stay out of the fields. Occasionally they revolted, as in 1675 and 1692 in Barbados. While none of these efforts overturned the slave system, they signaled a determination to buck the system and to build a life of their own making.

The Caribbean settlements were the most important colonies for Britain in the seventeenth and early eighteenth centuries, a vital center in the Atlantic trading world. They were the key source for sugar and rum and a market for goods made in the mainland colonies and in England. More importantly, they were the first principal source of enslaved Africans for the mainland colonies.

THE SOUTHWEST BORDERLANDS

By the end of the seventeenth century, the Spanish had established a sophisticated and impressive empire. Compared to all but a few English settlers in North America, Spanish residents in the Southwest enjoyed much greater prosperity. Their capital, Mexico City, was the most dazzling metropolis in the Americas.

The principal Spanish colonies north of Mexico—Florida, Texas, New Mexico, Arizona, and California—were less important economically to the empire. They attracted religious minorities, Catholic missionaries, and independent ranchers fleeing the heavy hand of imperial authority. Spanish troops defended them, but they remained weak and peripheral. An exception was New Mexico, the most prosperous and populous of these Spanish outposts. By the end of the eighteenth century, New Mexico had a non-native population of over 10,000—the largest European settlement west of the Mississippi and north of Mexico—and was steadily expanding through the region.

The Spanish began to colonize California once they realized that other Europeans—among them English merchants and French and Russian trappers—were beginning to establish a presence in the region. Formal Spanish settlement of California began in the 1760s, when the governor of Baja California was ordered to create outposts of the empire farther north. Soon a string of missions, forts (or *presidios*), and trading communities were springing up along the Pacific Coast, beginning with San Diego and Monterey in 1769 and eventually San Francisco (1776), Los Angeles (1781), and Santa Barbara (1786). They sought to control a local native population that had as many as 300 separate groups and nations, most notably Tipai, Ipai, Luisena, Gabrielino, Chumash, and Ohlone.

As was historically the case when Europeans moved into occupied land, the arrival of the Spanish in California had a devastating effect on the local population, who died in great numbers from the diseases the colonists brought with them. As the new settlements spread, the Spanish, here as elsewhere in the Americas, insisted that the remaining Native Americans convert to Catholicism. Once again they met with mixed results, with some native people adopting Catholic beliefs and rituals, others rejecting them, and still others creating a hybrid with their traditional convictions and practices. The Spanish colonists were also intent on creating a prosperous agricultural economy. Abetting them was the encomienda program, implemented by the Spanish crown, which authorized colonists to force indigenous people to work for them and pay a yearly tribute or tax. Not all native people complied, of course, and some ran away or fought back.

The Spanish considered the greatest threat to the northern borders of their empire to be the growing ambitions of the French. In the 1680s, French explorers traveled down the Mississippi Valley to the mouth of the river and claimed those lands for France in 1682. They called the territory Louisiana in honor of King Louis XIV. Fearful of French incursions farther west, the Spanish began to fortify their claim to Texas by establishing new forts, missions, and settlements there, including San Fernando (later San Antonio) in 1731.

The Southeast Borderlands

The southeastern areas of what is now the United States posed a direct challenge to English ambitions in North America. After Spain claimed Florida in the 1560s, missionaries and traders began moving northward into Georgia and westward into what is now known as the Florida panhandle. Some ambitious Spaniards began to dream of expanding their empire still farther north, into what became the Carolinas and beyond. The founding of Jamestown in 1607 dampened those hopes and replaced them with fears. The Spaniards worried that the English colonies would threaten their existing settlements in Florida and Georgia. As a result, the Spanish built forts in both regions to defend themselves against the increasing English presence. Throughout the eighteenth century, the area between the Carolinas and Florida was the site of continuing tension and frequent conflict—between the Spanish and the English and, to a lesser degree, between the Spanish and the French, who were threatening Spain's northwestern borders with settlements in Louisiana and in what is now Alabama.

There was no formal war between England and Spain in these years, but that did not mean there was an absence of hostilities between them in the Southeast. English pirates continually harassed the Spanish settlements and, in 1668, actually sacked St. Augustine. The English encouraged native groups in Florida to rise up against the Spanish missions. The Spanish offered freedom to enslaved Africans owned by English settlers in the Carolinas if they agreed to convert to Catholicism. About 100 Africans accepted the offer, and the Spanish later organized some of them into a military regiment to defend the northern border of New Spain. By the early eighteenth century, the constant fighting in the region had driven almost all the Spanish out of Florida except for settlers in St. Augustine on the Atlantic Coast and Pensacola on the Gulf Coast.

Eventually, after more than a century of conflict in the southeastern borderlands, the English prevailed—acquiring Florida in the aftermath of the Seven Years' War (known in America as the French and Indian War) and rapidly populating it with settlers from their colonies to the North. Before that point, however, protecting the southern boundary of the British Empire in North America was a continual concern to the British and contributed in crucial ways to the founding of the colony of Georgia.

The Founding of Georgia

Georgia—the last English colony to be established in what would become the United States—was founded to create a military barrier against Spanish lands on the southern border of English America. It was also designed to provide a refuge for the impoverished, a place where British men and women without prospects at home could begin anew. Its founders, led by General **James Oglethorpe**, served as unpaid trustees of a society created to serve the needs of the British Empire.

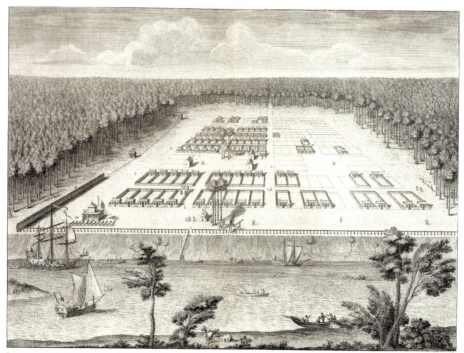

(Universal History Archive/Universal Images Group/Getty Images)

SAVANNAH IN 1734 This view of Savannah by an English artist shows the intensely orderly character of early settlement in the Georgia colony. As the colony grew, its residents gradually abandoned the plan created by Oglethorpe and his fellow trustees.

Oglethorpe, himself a veteran of the most recent Spanish wars with Britain, was keenly aware of the military advantages of an English colony south of the Carolinas. Yet his interest in settlement rested even more on his philanthropic commitments. As head of a parliamentary committee investigating English prisons, he had been appalled by the plight of honest debtors rotting in confinement. Such prisoners, and other poor people in danger of succumbing to a similar fate, could, he believed, become the farmer–soldiers of the new colony in America.

In 1732, King George II granted Oglethorpe and his fellow trustees control of the land between the Savannah and Altamaha Rivers. Their colonization policies reflected the vital military purposes of the colony. They limited the size of landholdings to make the settlement compact and easier to defend against Spanish and native attacks. They excluded Africans, free or enslaved; Oglethorpe feared that Black labor would produce internal revolts and that disaffected enslaved workers might turn to the Spanish as allies. The trustees strictly regulated trade with Native Americans, again to limit the possibility of sparking conflict and pushing indigenous nations into a wartime alliance with the Spanish. They also excluded Catholics for fear they might collude with their coreligionists in the Spanish colonies to the south.

Oglethorpe himself led the first colonial expedition to Georgia, which built a fortified town at the mouth of the Savannah River in 1733 and later constructed additional forts south of the Altamaha. In the end, only a few debtors were released from jail and sent to Georgia. Instead, the trustees brought hundreds of impoverished tradesmen and artisans from Britain and Scotland and many religious refugees from Switzerland and Germany. Among the immigrants was a small group of Jews. British settlers made up a lower proportion of the European population of Georgia than of any other English colony.

Oglethorpe (whom some residents of Georgia began calling "our perpetual dictator") created almost constant dissension and conflict through his heavy-handed regulation of the colony. He also suffered military disappointments, such as leading a failed 1740 assault on the Spanish outpost at St. Augustine, Florida. Gradually, as the threats from Spain receded, he lost his grip on the colony, which over time became more like the rest of British North America, with an elected legislature that loosened the restrictions on settlers. Georgia grew more slowly than other southern colonies and developed along lines roughly similar to those of South Carolina.

Middle Grounds

The struggle for the North American continent was not just a conflict among competing European empires. It was also a series of contests among the many different peoples who shared the continent—the Spanish, English, French, Dutch, and other colonists, on one hand, and the many Native American nations with whom they shared the continent, on the other.

In no part of the Americas did colonial settlers quickly establish their dominance, subjugating and displacing native groups with dispatch. Instead, the balance of power was in constant flux; never was colonial rule inevitable. Along the western borders of English settlement, in particular, Europeans and Native Americans lived together in regions in which neither side was able to establish clear and persistent advantage. In these middle grounds, the two populations—despite frequent conflicts—carved out ways of living together, with each side making concessions to the other. (See "Debating the Past: Native Americans and the Middle Ground.") To be sure, settlers came to power faster in some colonies, like Virginia and parts of New England, but even here Native Americans often continued to live among them and contest their control.

It is important to recognize that these western areas tended to fall along the peripheries of European empires, in which the influence of formal colonial governments was at times minimal. European settlers, and the soldiers scattered in forts throughout these regions to protect them, were unable to displace the native inhabitants. So they had to negotiate their own relationships with native nations. In those relationships, Europeans found themselves obligated to adapt to indigenous expectations often as much as Native Americans had to adapt to European ones.

In the seventeenth century, before many English settlers had entered the interior, the French were particularly adept at creating successful associations with Native Americans. French missionaries strove to learn local languages and customs and work within relationships between and among native groups. Fur traders practiced similar habits, welcoming the chance to attach themselves to—even to marry within—local nations. They also recognized the importance of treating indigenous leaders with respect and channeling gifts and tributes through them. But by the mid-eighteenth century, French influence in the interior was in decline, and British settlers gradually became the dominant European group. Eventually, the British learned the lessons that the French had long ago absorbed—that simple commands and raw force were ineffective in creating a workable long-term relationship with Native Americans; that they too had to learn to deal with indigenous leaders through gifts and ceremonies and mediation. In large western regions—especially those around the

DEBATING THE PAST

Native Americans and the Middle Ground

For many generations, historians chronicling the westward movement of European settlement in North America portrayed Native Americans largely as weak and inconvenient obstacles swept aside by the inevitable progress of "civilization." Indigenous people were presented either as murderous savages or as relatively docile allies of white people, but rarely as important actors of their own. Francis Parkman, the renowned nineteenth-century American historian, described Native Americans as a civilization "crushed" and "scorned" by the march of European powers in the New World. Many subsequent historians departed little from his assessment.

In the past half century, historians have challenged this traditional view first by examining how white civilization victimized native peoples. Gary Nash's *Red, White, and Black* (1974) was an early important presentation of this approach, as was Ramon Guttierez's *When Jesus Came, the Corn Mothers Went* (1991). They, and other scholars, rejected the optimistic, progressive view of white triumph over adversity and presented, instead, a picture of white brutality and futile native resistance, ending in defeat.

Subsequently, scholars saw Native Americans and Euro-Americans as uneasy partners in the shaping of a new society in which, for a time at least, both were a vital part. Richard White's influential 1991 book, *The Middle Ground*, was among the first significant statements of this view. White examined the culture of the Great Lakes region in the eighteenth century, in which the Algonquian created a series of complex trading and political relationships with French, English, and American settlers and travelers. In this "borderland" between the growing European settlements in the east and the still largely intact indigenous civilizations farther west, a new kind of hybrid society emerged in which many cultures intermingled. James Merrell's *Into the American Woods* (1999) contributed further to this new view of collaboration by examining the world of negotiators and go-betweens along the western Pennsylvania frontier in the seventeenth and eighteenth centuries. Like White, he emphasized the complicated blend of European and Native American diplomatic rituals that allowed both groups to conduct business, make treaties, and keep the peace.

Daniel Richter extended the idea of a middle ground further in two important books: *The Ordeal of the Long-house* (1992) and *Facing East from Indian Country* (2001). Richter demonstrates that the Iroquois Confederacy was an active participant in the power relationships in the Hudson River basin; in his later book, he tells the story of European colonization from the Native American perspective, revealing how Western myths of "first contact" such as the story of John Smith and Pocahontas look entirely different when seen through the eyes of Native Americans, who remained in many ways the more powerful of the two societies in the seventeenth century.

Building on Richter but firmly centering the power of Native American societies in the story of European colonization, Kathleen Duval (*The Native Ground: Indians and Colonists in the Heart of the Continent,* 2008) demonstrates how indigenous societies often determined the character of relationships with colonial powers. Indeed, she argues that it was Native Americans who drew Europeans into their practices of diplomacy, warfare, family, agriculture, and gender. Similarly, Pekka Hamalainen, in *The Comanche Empire* (2008), details how the Comanche controlled vast swaths of the American Southwest, besting European colonial powers through their military and economic power throughout the eighteenth and early nineteenth centuries. He portrays the Comanche as building a successful empire that unraveled more because of internal mistakes and divisions than European conquest. And Michael Witgen, in *An Infinity of Nations: How the Native New World Shaped Early North America* (2013), studies how the Anishinaabe and Dakota of the Great Lakes and Northern Plains controlled trade and diplomacy in these regions despite the incursions of Europeans. Even as they interacted with European agents, they were far from conquered or absorbed. Instead, they tended to control the patterns of interaction and cooperation and nurtured an independent culture for generations.

UNDERSTAND, ANALYZE, & EVALUATE

1. How have historians' views of Native Americans and their role in the European colonization of North America changed over time?
2. Why do you think scholars changed their portrayal of native peoples in early America so radically?

Great Lakes—they established a precarious peace with native inhabitants that lasted for several decades.

But as the English (and after 1776 American) presence in the region grew, the balance of power between Europeans and Native Americans shifted. Newer settlers had difficulty adapting to the complex rituals that the earlier migrants had developed. The stability of the relationships between native peoples and whites deteriorated. By the early nineteenth century, the middle ground had collapsed, replaced by a European world in which Native Americans were ruthlessly subjugated and eventually removed. Nevertheless, for a considerable period of early American history the story of the relationship between whites and indigenous inhabitants was not simply a story of conquest and subjugation, but relied on accommodation and tolerance.

THE DEVELOPMENT OF EMPIRE

The British colonies in America had begun as separate projects, and for the most part they grew up independent of one another and subject to only nominal control from London. But by the mid-seventeenth century, the growing commercial success of the colonial ventures was producing pressure in England for a more uniform structure to the empire.

The English government began trying to regulate colonial trade in the 1650s, when Parliament passed laws to keep Dutch ships out of the British colonies. Later, Parliament passed three important **Navigation Acts**. The first of them, in 1660, closed the colonies to all trade except that carried by English ships. The English also required that tobacco and other items be exported from the colonies only to England or to English possessions. The second act, in 1663, required that all goods sent from Europe to the colonies pass through England on the way, where they would be subject to English taxation. The third act, in 1673, imposed duties on the coastal trade among the English colonies, and it provided for the appointment of customs officials to enforce the Navigation Acts. These acts formed the legal basis of England's regulation of the colonies for a century.

The Dominion of New England

Before the creation of the Navigation Acts, all the colonial governments except that of Virginia had operated largely independently of the crown, with governors chosen by the proprietors or by the colonists themselves and with powerful representative assemblies. Officials in London recognized that to increase their control over the colonies and enforce the new laws, they would have to increase English authority.

In 1675, the king created a new body, the Lords of Trade, to make recommendations for imperial reform. In 1679, the king moved to increase his control over Massachusetts. He stripped it of its authority over New Hampshire and chartered a separate, royal colony there whose governor he would himself appoint. And in 1684, citing the colonial assembly's defiance of the Navigation Acts, he revoked the Massachusetts charter.

Charles II's brother, James II, who succeeded him to the throne in 1685, went further. He created a single **Dominion of New England**, which combined the government of Massachusetts with the governments of the rest of the New England colonies and later with those of New York and New Jersey as well. He appointed a single governor, Sir Edmund Andros, to supervise the entire region from Boston. Andros's rigid enforcement of the Navigation Acts and his brusque dismissal of the colonists' claims to the "rights of Englishmen" made him highly unpopular.

The "Glorious Revolution"

James II, unlike his father, was openly Catholic. He made powerful enemies when he appointed fellow Catholics to high offices. The restoration of Catholicism in England led to popular fears that the Vatican and the pope would soon overtake the country, with the king's support. At the same time, James II tried to control Parliament and the courts with an iron fist and rule as an absolute monarch. By 1688, the opposition to the king was so great that Parliament voted to force out James II. His daughter, Mary II, and her husband, William of Orange, of the Netherlands—both Protestants—replaced James II to reign jointly. James II went to Ireland, raised an army, and fought William but lost. He eventually left the country and spent the rest of his life in France. No Catholic monarch has reigned since. This coup came to be known as the "Glorious Revolution."

When Bostonians heard of the overthrow of James II, they arrested and imprisoned the unpopular Andros. The new sovereigns in England abolished the Dominion of New England and restored separate colonial governments. In 1691, however, they combined Massachusetts with Plymouth and made it a single, royal colony. The new charter restored the colonial assembly, but it gave the crown the right to appoint the governor. It also replaced church membership with property ownership as the basis for voting and officeholding.

Andros had been governing New York through a lieutenant governor, Captain Francis Nicholson, who enjoyed the support of the wealthy merchants and fur traders of the province. Other, less-favored colonists had a long list of grievances against Nicholson and his allies. The leader of the New York dissidents was **Jacob Leisler**, a German merchant. In May 1689, when news of the Glorious Revolution and the fall of Andros reached New York, Leisler raised a militia, captured the city fort, drove Nicholson into exile, and proclaimed himself the new head of government in New York. For two years, he tried in vain to stabilize his power in the colony amid fierce factional rivalry. In 1691, when William and Mary appointed a new governor, Leisler briefly resisted. He was convicted of treason and executed. Fierce rivalry between what became known as the "Leislerians" and the "anti-Leislerians" dominated the politics of the colony for years thereafter.

In Maryland, many people wrongly assumed that their proprietor, the Catholic Lord Baltimore, who was living in England, had sided with the Catholic James II and opposed William and Mary. So in 1689, an old opponent of the proprietor's government, the Protestant John Coode, led a revolt that drove out Lord Baltimore's officials and resulted in Maryland's establishment as a royal colony in 1691. The colonial assembly then established the Church of England as the colony's official religion and excluded Catholics from public office. Maryland became a proprietary colony again in 1715, after the fifth Lord Baltimore joined the Anglican Church.

The Glorious Revolution of 1688 in Britain touched off revolutions, mostly bloodless, in several colonies. Under the new king and queen, the representative assemblies that had been abolished were revived and the scheme for colonial unification from above was abandoned. But the Glorious Revolution in America did not stop the reorganization of the empire. The new governments that emerged in America actually increased the crown's potential authority. As the first century of English settlement in America came to its end, the colonists were becoming more a part of the imperial system than ever before.

CONCLUSION

The English colonization of North America was part of a larger effort by several European nations to expand the reach of their increasingly commercial societies. Indeed, for many years, the British Empire in America was among the smallest and weakest of the imperial ventures there, overshadowed by the French to the north and the Spanish to the south.

In the English colonies along the Atlantic seaboard, new agricultural and commercial societies gradually emerged—those in the South centered on the cultivation of tobacco and rice and were heavily reliant on enslaved labor; those in the northern colonies centered on more traditional food crops and relied less on enslaved labor. Substantial trading centers emerged in such cities as Boston, New York, Philadelphia, and Charles Town, and a growing proportion of the population became prosperous and settled in these increasingly complex communities. By the early eighteenth century, English settlement had spread from northern New England (in what is now Maine) south into Georgia.

But this growing English presence coexisted with, and often was in conflict with, other Europeans—most notably the Spanish and the French—in certain areas of North America. In these borderlands, societies did not assume the settled, prosperous form that was slowly taking root in the Tidewater and New England. Borderland colonies were raw, sparsely populated settlements in which Europeans, including over time increasing numbers of English, had to learn to accommodate not only one another but also the Native Americans with whom they shared these interior lands. By the middle of the eighteenth century, there was a significant European presence across a broad swath of North America—from Florida to Maine, and from Texas to Mexico to California. No European power, however, controlled any major part of these large geographic regions. Yet changes were under way within the British Empire that would soon lead to its dominance through a much larger area of North America.

KEY TERMS/PEOPLE/PLACES/EVENTS

Anne Hutchinson 36
Antinomianism 36
Bacon's Rebellion 30
Cecilius Calvert 32
Dominion of New
  England 52
George Calvert 32
headright system 29
indentured servants 30
Jacob Leisler 52
James Oglethorpe 47

Jamestown 26
John Smith 27
John Winthrop 35
King Philip's War 37
Massachusetts Bay
  Company 35
Mayflower Compact 34
Metacom 39
middle grounds 44
Navigation Acts 51
Pequot War 37

Plymouth Plantation 34
Powhatan 26
Quakers 43
Roger Williams 36
slave codes 31
theocracy 35
Virginia House of
  Burgesses 29
William Berkeley 30
William Penn 43

RECALL AND REFLECT

1. Compare patterns of colonization between the Spanish and the English. What similarities do you see? What differences?
2. How did the institution of slavery differ between Virginia and the Caribbean? What accounts for these differences?
3. How did the relationships between European settlers and Native Americans along the Atlantic seaboard differ from those found in the interior regions near and around what we now call the Great Lakes?
4. How did the Glorious Revolution in England affect England's North American colonies?

Design element: Stars and Stripes: McGraw Hill Education.

3 | SOCIETY AND CULTURE IN PROVINCIAL AMERICA

THE COLONIAL POPULATION
THE COLONIAL ECONOMIES
PATTERNS OF SOCIETY
AWAKENINGS AND ENLIGHTENMENTS

LOOKING AHEAD

1. What accounted for the rapid increase in the colonial population in the seventeenth century?
2. Why did African slavery expand so quickly in the late seventeenth century?
3. How did religion shape and influence colonial society?

MOST PEOPLE IN ENGLAND believed that the English colonies were outposts of the English world. And so did most English immigrants who came to North America. It is certainly true that as the colonies grew and became more prosperous, they came to closely resemble English society. To be sure, some of the early settlers had come to America to escape what they considered English tyranny. But by the early eighteenth century, many colonists considered themselves British just as much as the men and women in Britain itself did.

However, the colonies were actually quite different from England and from one other. What distinguished them from England was not simply landscape and climate but also the constant engagement with Native Americans, experimentation with new systems of local government, attempts to establish religious orthodoxy and the rebellions occasioned with them, and efforts to learn about and raise new crops. African laborers and enslaved workers were stitched into the fabric of colonial life almost from the start. Indeed, the English colonies would eventually become the destination for millions of forcibly transplanted Africans. The area that would become the United States was a magnet for immigrants from many lands other than England: Scotland, Ireland, the European continent, eastern Russia, and the Spanish and French Empires already established in America. Indeed, part of the story of the development of the English colonies is just how distinctive they were becoming from England itself.

TIME LINE

1636
America's first college, Harvard, founded

1639
First printing press in colonies begins operation

1685
Huguenots migrate to America

1692
Salem witchcraft trials conclude

1697
Importations of enslaved Africans increase

1720
Cotton Mather starts smallpox inoculation

1734
Great Awakening begins

1739
Stono rebellion
Great Awakening intensifies
Stono slave rebellion

Zenger trial

1740s
Indigo production begins

THE COLONIAL POPULATION

After uncertain beginnings, the non-indigenous population of English North America grew rapidly and substantially through continued immigration, importation of enslaved Africans, and natural increase. By the late seventeenth century, Europeans and Africans outnumbered Native Americans along the Atlantic Coast.

A few of the early settlers were members of the English upper classes, but most were English laborers. Some came independently, such as the religious dissenters in early New England. But in the Chesapeake, at least three-fourths of the immigrants in the seventeenth century arrived as indentured servants.

INDENTURED SERVITUDE

The system of temporary (or "indentured") servitude developed out of practices in England. Most indentured servants were young men who bound themselves to masters for fixed terms of servitude (usually four to five years) in exchange for passage to America, food, and shelter. Their passage was typically a terrible trial of want and hunger. (See "Consider the Source: Gottlieb Mittelberger, the Passage of Indentured Servants.") Male indentured servants were supposed to receive clothing, tools, and occasionally land upon completion of their service. In reality, however, many left service with nothing. Most women indentures—who constituted roughly one-fourth of the total in the Chesapeake—worked as domestic servants and were expected to marry when their term of servitude expired.

By the late seventeenth century, the indentured servant population had become one of the largest elements of the colonial population and was creating serious social problems. Some former indentures managed to establish themselves successfully as farmers, tradespeople, or artisans, and some of the women married propertied men. But many found themselves without land, without employment, without families, and without prospects. As a result, there emerged in some areas, particularly in the Chesapeake, a large floating population of predominantly young single men who were a source of social unrest, prompting elites to consider enslaved Africans as a better, more dependable and controllable form of laborer.

Shortly after the arrival of enslaved Africans, planters began to view them as critical to their economic successes. Most importantly, they saw great benefit in having a permanent labor population without hope of freedom, consigned to a life of work and servitude and—planters assumed—accepting of their plight. Accelerating efforts to import more enslaved labor were a series of economic and demographic changes. Beginning in the 1670s, a decrease in the birthrate in England and an improvement in economic conditions reduced the pressures on laboring men and women to emigrate; correspondingly, the flow of indentured servants to North America slowly declined. Those who did make the voyage as indentures generally avoided the southern colonies, where prospects for advancement were slim.

Birth and Death

Immigration remained for a time the greatest source of population growth in the colonies. But the most important long-range factor in the increase of the colonial population was its ability to reproduce itself. Improvement in the reproduction rate began in New England and the mid-Atlantic colonies, where natural increase was the most important source of population growth after the 1650s. The New England population more than quadrupled through reproduction alone in the second half of the seventeenth century. This rise was a result not only of families having large numbers of children, it was also because life expectancy in New England was unusually high.

The South was very different. The high death rates in the Chesapeake region did not begin to decline to levels found elsewhere until the mid-eighteenth century. Throughout the seventeenth century, the average life expectancy for European men in the region was just over forty years, and for women slightly less. One in four white children died in infancy, and half died before the age of twenty. Children who survived infancy often lost one or both of their parents before reaching maturity. Widows, widowers, and orphans thus formed a substantial proportion of the white Chesapeake population. Only after settlers developed immunity to local diseases (particularly malaria) did life expectancy increase significantly. Population growth was substantial in the region, but it was largely a result of immigration.

The natural increases in the population in the seventeenth and eighteenth centuries reflected an improving balance between men and women in the colonies. In the early years of settlement, more than three-quarters of the white population of the Chesapeake consisted of men. In New England, which from the beginning had attracted more families than the southern colonies, 60 percent of the inhabitants were still male in 1650. But as more women began to arrive to the colonies and birthrates increased, the ratio of men to women became less unequal. Throughout the colonial period, the population almost doubled every twenty-five years. By 1775, the non-indigenous population of the colonies was over 2 million.

Medicine in the Colonies

Death rates were very high for any person living in the colonial era. They reflected the state of medicine at the time, which was often unhelpful in providing succor to the sick

CONSIDER THE SOURCE

GOTTLIEB MITTELBERGER, THE PASSAGE OF INDENTURED SERVANTS (1750)

Gottlieb Mittelberger, a German laborer, traveled to Philadelphia in 1750 and chronicled his voyage.

Both in Rotterdam and in Amsterdam the people are packed densely, like herrings so to say, in the large sea-vessels. One person receives a place of scarcely 2 feet width and 6 feet length in the bedstead, while many a ship carries four to six hundred souls; not to mention the innumerable implements, tools, provisions, water-barrels and other things which likewise occupy such space.

On account of contrary winds it takes the ships sometimes 2, 3, and 4 weeks to make the trip from Holland to . . . England. But when the wind is good, they get there in 8 days or even sooner. Everything is examined there and the custom-duties paid, whence it comes that the ships ride there 8, 10 or 14 days and even longer at anchor, till they have taken in their full cargoes. During that time every one is compelled to spend his last remaining money and to consume his little stock of provisions which had been reserved for the sea; so that most passengers, finding themselves on the ocean where they would be in greater need of them, must greatly suffer from hunger and want. Many suffer want already on the water between Holland and Old England.

When the ships have for the last time weighed their anchors near the city of Kaupp [Cowes] in Old England, the real misery begins with the long voyage. For from there the ships, unless they have good wind, must often sail 8, 9, 10 to 12 weeks before they reach Philadelphia. But even with the best wind the voyage lasts 7 weeks.

But during the voyage there is on board these ships terrible misery, stench, fumes, horror, vomiting, many kinds of sea-sickness, fever, dysentery, headache, heat, constipation, boils, scurvy, cancer, mouth rot, and the like, all of which come from old and sharply salted food and meat, also from very bad and foul water, so that many die miserably.

Add to this want of provisions, hunger, thirst, frost, heat, dampness, anxiety, want, afflictions and lamentations, together with other trouble, as . . . the lice abound so frightfully, especially on sick people, that they can be scraped off the body. The misery reaches the climax when a gale rages for 2 or 3 nights and days, so that every one believes that the ship will go to the bottom with all human beings on board. In such a visitation the people cry and pray most piteously.

Children from 1 to 7 years rarely survive the voyage. I witnessed . . . misery in no less than 32 children in our ship, all of whom were thrown into the sea. The parents grieve all the more since their children find no resting-place in the earth, but are devoured by the monsters of the sea.

That most of the people get sick is not surprising, because, in addition to all other trials and hardships, warm food is served only three times a week, the rations being very poor and very little. Such meals can hardly be eaten, on account of being so unclean. The water which is served out of the ships is often very black, thick and full of worms, so that one cannot drink it without loathing, even with the greatest thirst. Toward the end we were compelled to eat the ship's biscuit which had been spoiled long ago; though in a whole biscuit there was scarcely a piece the size of a dollar that had not been full of red worms and spiders' nests. . . .

At length, when, after a long and tedious voyage, the ships come in sight of land, so that the promontories can be seen, which the people were so eager and anxious to see, all creep from below on deck to see the land from afar and they weep for joy, and pray and sing, thanking and praising God. The sight of the land makes the people on board the ship, especially the sick and the half dead, alive again, so that their hearts leap within them; they shout and rejoice, and are content to bear their misery in patience, in the hope that they may soon reach the land in safety. But alas!

When the ships have landed at Philadelphia after their long voyage, no one is permitted to leave them except those who pay for their passage or can give good security; the others, who cannot pay, must remain on board the ships till they are purchased, and are released from the ships by their purchasers. The sick always fare the worst, for the healthy are naturally preferred and purchased first; and so the sick and wretched must often remain on board in front of the city for 2 or 3 weeks, and frequently die, whereas many a one, if he could pay his debt and were permitted to leave the ship immediately, might recover and remain alive.

The sale of human beings in the market on board the ship is carried out thus: Every day Englishmen, Dutchmen and High-German people come from the city of Philadelphia and other places, in part from a great distance, say 20, 30, or 40 hours away, and go on board the newly arrived ship that has brought and offers for sale passengers from Europe, and select among the healthy persons such as they deem suitable for their business, and bargain with them how long they will serve for their passage money, which most of them are still in debt for. When they have come to an agreement, it happens that adult persons bind themselves in writing to serve 3, 4, 5 or 6 years for the amount due by them, according to their age and strength. But very young people, from 10 to 15 years, must serve till they are 21 years old.

Many parents must sell and trade away their children like so many head of cattle; for if their children take the debt upon themselves, the parents can leave the ship free and unrestrained; but as the parents often do not know where and to what people their children are going, it often happens that such parents and children, after leaving the ship, do not see each other again for many years, perhaps no more in all their lives. . . . It often happens that whole families, husband, wife and children, are separated by being sold to different purchasers, especially when they have not paid any part of their passage money.

When a husband or wife has died at sea, when the ship has made more than half of her trip, the survivor must pay or serve not only for himself or herself but also for the deceased.

When both parents have died over halfway at sea, their children, especially when they are young and have nothing to pawn or pay, must stand for their own and their parents' passage, and serve till they are 21 years old. When one has served his or her term, he or she is entitled to a new suit of clothes at parting; and if it has been so stipulated, a man gets in addition a horse, a woman, a cow. When a serf has an opportunity to marry in this country, he or she must pay for each year which he or she would have yet to serve, 5 or 6 pounds. •

UNDERSTAND, ANALYZE, & EVALUATE

1. What hardships did passengers suffer at sea? What relief could they hope for upon reaching Philadelphia?
2. Explain the different purchase agreements between passengers and masters. How did the death of a family member affect a passenger's indenture contracts?
3. What do the ordeals of indentured servants tell us about prospects in Europe? What do they tell us about the concept of liberty in the colonies?

and healing the very ill. Unaware of bacteria, many communities were plagued with infectious diseases transmitted by garbage or unclean water. Physicians had little or no understanding of infection and sterilization and their surgeries regularly left patients with severe complications or even killed them. Not surprisingly, many women and infants died from infections and complications contracted during childbirth.

Because of the limited extent of formal medical knowledge, and the lack of regulations for any practitioners at the time, it was relatively easy for people to practice medicine, sometimes without much professional training. One group who benefited from this lack of educational and legal standards were women, who established themselves in considerable numbers as midwives. Midwives assisted women in childbirth, but they also dispensed other vital medical advice. Their success reflected their skill and compassion as health-care providers and the fact that they generally treated friends and neighbors—a local and familiar focus that was not usual for male doctors. As midwives' popularity grew, male doctors felt threatened and slowly attempted to drive them out, a goal largely accomplished by the early nineteenth century.

Midwives and doctors alike practiced medicine on the basis of the prevailing assumptions of their time, most of them derived from the theory of "humoralism" popularized by the famous second-century Roman physician Galen. Galen argued that the human body was governed by four "humors" that were lodged in four bodily fluids: yellow bile (or "choler"), black bile ("melancholy"), blood, and phlegm. In a healthy body, the four humors existed in balance. Illness represented an imbalance and suggested the need for removing whatever fluid was causing the imbalance. This was the reasoning behind the principal medical techniques of the seventeenth century: purging, expulsion, and bleeding. Bleeding was practiced mostly by male physicians. Midwives favored "pukes" and laxatives.

The great majority of early Americans, however, had little contact with physicians, or even midwives, and dealt with illness on their own. The assumption that treating illness was the exclusive province of trained professionals, so much a part of modern times, was still far off in the distance.

That seventeenth-century medicine rested so much on ideas produced 1,400 years before is evidence of how little support there was for the scientific method in England and America at the time. Bleeding, for example, had been in use for hundreds of years, during which time there had been no evidence at all that it helped people recover from illness; indeed, there was considerable evidence that bleeding could do great harm. But what would seem in later eras to be the simple process of testing scientific assumptions was not yet a common part of Western thought. That was one reason that the birth of the Enlightenment in the late seventeenth century—with its faith in human reason and its belief in the capacity of individuals and societies to create better lives—was important not just to politics but also to science.

WOMEN AND FAMILIES IN THE COLONIES

White women lived under the principle of coverture, in which they had their legal rights assumed by their husbands upon marriage. Whereas an adult unmarried woman could own property and enter into contracts on her own, though often under the care of her father, once she married she lost such legal rights. Widows had a considerable amount of power because in most colonies they could inherit and hold property following their husbands' deaths. High death rates meant that a number of women gained control of property, making them highly desirable mates for men seeking wives.

Because there were many more men than women in seventeenth-century America, few women remained unmarried for long. The average white European woman in America married for the first time at twenty or twenty-one years of age. Because of the large numbers of indentured servants who were forbidden to marry until their terms of service expired, and because female indentured servants frequently had their terms of service extended, premarital sexual relationships were not uncommon. Children born out of wedlock to indentured white women were often taken from their mothers at a young age and were themselves bound as indentured servants.

White women in the Chesapeake could anticipate a life consumed with childbearing. The average wife experienced pregnancies every two years. Those who lived long enough bore an average of eight children apiece (up to five of whom typically died in infancy or early childhood). Since childbirth was one of the most frequent causes of female death, many women did not survive to see their children grow to maturity. Those who did, however, were often widowed, since they were usually much younger than their husbands.

In New England, where many more immigrants arrived with family members and where death rates declined more quickly, family structure was much more stable than in the Chesapeake. As in the Chesapeake, women married young, began producing children early, and continued to do so well into their thirties. In contrast to their southern counterparts, however, northern children were more likely to survive, and their families were more likely to remain intact. Fewer New England women became widows, and those who did generally lost their husbands later in life.

The longer life span in New England meant that parents continued to control their children longer than did parents in the South. Few sons and daughters could choose a spouse entirely independently of their parents' wishes. Men tended to rely on their fathers for land to cultivate. Women needed dowries from their parents if they were to attract desirable husbands.

(Bettmann/Getty Images)

LIFE IN THE AMERICAN COLONIES This colored engraving shows an idealized domestic life for white Americans during the eighteenth century. Depicted are family members at work in their cozy surroundings. The industriousness they are exhibiting was a virtue of the era.

The Beginnings of Slavery in English America

The demand for enslaved Africans to supplement the indentured or free labor force existed almost from the first moments of settlement. For a time, however, enslaved workers were hard to find. Not until the mid-seventeenth century, when a substantial commerce in enslaved labor grew up between the Caribbean islands and the southern colonies, did enslaved Africans become generally available in North America. Just how slavery actually took root and spread has been a source of endless debate among historians.

The rising demand for enslaved labor in North America beginning in the late seventeenth century helped expand the transatlantic slave trade. Before it ended in the nineteenth century, it was responsible for the forced immigration of as many as 11 million Africans to the Americas and the Caribbean. In the flourishing slave markets on the African coast, native chieftains brought members of rival tribes captured in western and central Africa to the ports. A small number were also captured in raids by European slave traders.

After they were captured and marched to ports along the west African coast, terrified Africans were then tightly packed into the dark, filthy holds of ships for the horrors of the **middle passage**—the long transatlantic journey to the Americas or the Caribbean. The journey took three to four months, during which time up to 600 Africans were chained together in columns deep in the bowels of the ship or stuffed onto shelves lining the hull. So cramped were the quarters that most could not stand up. Men were kept apart from women and children. Food and fresh air were scarce. Many died, their corpses dumped overboard.

Olaadah Equiano, an African from Eboe (present-day Nigeria), was seized by slave traders at the age of 11 in 1745 and sent to the West Indies. He later escaped and penned an autobiography detailing his life, including the middle passage. "I was soon put down under the decks. . . . The closeness of the place, the heat of the climate, added to the number in the ship, which was so crowded that each had scarcely room to turn himself, almost suffocated us. This produced copious perspirations, so that the air soon became unfit for respiration, from a variety of loathsome smells, and brought on a sickness among the slaves, of which many died." Upon arrival in the New World, enslaved workers like Equiano were auctioned off to white landowners and transported, frightened and bewildered, to their new homes.

Most enslaved Africans shipped to the New World landed not in an English colony but the Caribbean islands, Brazil, or territories of the Spanish Empire. From there enslaved workers were sometimes purchased and transported by traders to North America. But not until the 1670s did slave traders start importing enslaved labor directly from Africa to North America. Even then, the flow remained small for a time, mainly because a single group, the Royal African Company of England, monopolized the trade and kept prices high and supplies low. Indeed, only 5 to 7 percent of all enslaved Africans were ever sent directly to English North America.

A turning point in the history of the Black population in North America was 1697, the year rival traders broke the Royal African Company's monopoly. With the trade now open to competition, prices fell and the number of Africans greatly increased. In 1700, about 25,000 enslaved Africans lived in English North America. Heavily concentrated in a few southern colonies at this time, they were already beginning to outnumber whites in some areas. Sixty years later, the number of enslaved Africans in the English mainland colonies had ballooned to approximately a quarter of a million, the majority of whom lived in the South.

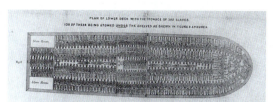

(Library of Congress, Prints & Photographs Division [LC-USZ62-44000])

AFRICAN SLAVE TRADE This image is taken from a plate from British author Amelia Opie's poem, *Black Slaves in the Hold of the Slave Ship: or How to Make Sugar*, published in London in 1826. Opie's poem depicts the life of an African who was captured by slave traders, transported to the West Indies on a slave ship, and forced to live a brutal life working on sugar plantations.

Initially, the legal and social status of the enslaved laborers was fluid. In some areas, white and Black laborers worked together on terms of relative equality. Some enslaved Blacks were treated much like white hired servants, and some were freed after a fixed term of servitude. But white society eventually determined that slavery promised the most reliable and pliable labor force and beginning in the late seventeenth century began to pass a series of restrictive slave codes. In 1662, Virginia declared that slavery followed the condition of the mother, meaning that children of enslaved women were themselves enslaved. Two years later Maryland passed a law stipulating that any free-born woman who married an enslaved man became enslaved herself. In 1667, Virginia reversed an earlier law and declared that any enslaved person who underwent the rite of Christian baptism was still enslaved. And in 1712 South Carolina announced that "all negro[e]s, mulattoes, mestizo[s] or Indians, which at any time heretofore have been sold, or now are held or taken to be, or hereafter shall be bought and sold for slaves, are hereby declared slaves; and they, and their children, are hereby made and declared slaves."

CHANGING SOURCES OF EUROPEAN IMMIGRATION

The most distinctive and enduring feature of the American population was that it brought together peoples of many different geographic origins, ethnic groups, and nationalities. North America was home to a highly diverse population. The British colonies were the home to Native Americans, English immigrants, forcibly imported Africans, and a wide range of other European groups. Among the earliest European immigrants were the French Calvinists (known as Huguenots). The Edict of Nantes of 1598 had assured them freedom of religion in France. But in 1685 the Edict was revoked, driving about 10,000 Huguenots to North America. Germany had similar laws banning Protestantism, driving many Germans to America where they settled in Pennsylvania. They came to be known as the "Pennsylvania Dutch," a corruption of the German term for their nationality, *Deutsch*. Frequent wars in Europe drove many other immigrants to the American colonies.

The most numerous of the newcomers were the so-called **Scotch-Irish**—Scotch Presbyterians who had settled in northern Ireland (in the province of Ulster) in the early seventeenth century. Most of the Scotch-Irish in America pushed out to the western edges of European settlement and occupied land without much regard for who actually claimed to own it.

There were also immigrants from Scotland itself and from southern Ireland. The Irish migrated steadily over a long period. Some abandoned their Roman Catholic religion and much of their ethnic identity after they arrived in America.

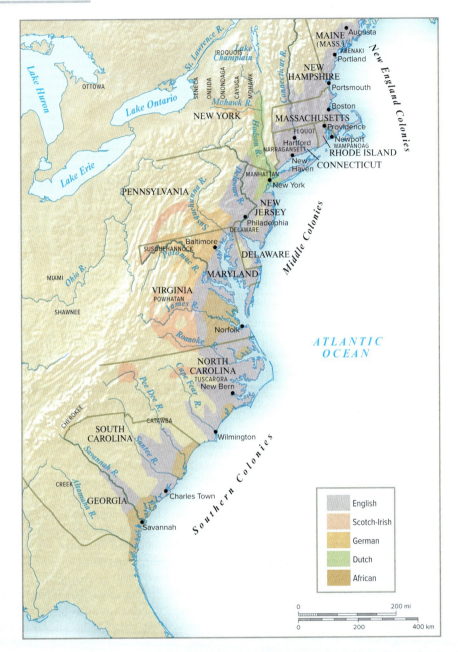

POPULATIONS LIVING IN COLONIAL AMERICA, 1760 Even though the entire Atlantic seaboard of what is now the United States had become a series of British colonies by 1760, the nonnative population consisted of people from many places. As this map reveals, English settlers dominated most of the regions of North America. But note the large areas of German settlement in the western Chesapeake and Pennsylvania; the swath of Dutch settlement in New York and New Jersey; the Scotch-Irish regions in the western regions of the South; and the large areas in which enslaved Africans were becoming the majority of the population. Note too the presence of multiple native nations along the seaboard and interior lands, which predated the influx of Europeans. They played a vital role in the evolution of the European colonies, sometimes as allies and other times as enemies, but always as a key force shaping colonial culture. • *What aspects of the history of these colonies help explain their ethnic composition?*

THE COLONIAL ECONOMIES

Farming, hunting, and fishing dominated most areas of European settlement, as well as long-established Native American communities in North America, throughout the seventeenth and eighteenth centuries. Even so, no colony was alike. Each developed its own economic focus and character—though all eventually incorporated slavery into the routines of daily life.

SLAVERY AND ECONOMIC LIFE

In every colony, enslaved labor was essential to economic productivity. Enslaved workers performed different jobs under different conditions depending on their colony of residence, but they were an integral and very visible part of every local culture. In Virginia, where tobacco was the dominant crop, planters responded to the rising demand from markets in the colonies and Europe by bringing in more enslaved labor to work large plantations. By the mid-1700s, nearly 150,000 enslaved workers lived in Virginia. South Carolina and Georgia relied on rice production, since the low-lying coastline with its many tidal rivers made it possible to create rice paddies that could be flooded and drained. Rice cultivation was so difficult and unhealthy that many white workers simply refused to perform it, forcing planters in South Carolina and Georgia to grow dependent on enslaved labor. African workers were highly valued because many had lived and worked in rice-producing regions of west Africa and were experts in cultivation techniques and harvesting strategies. In South Carolina in 1765, Blacks, nearly all of whom were enslaved, outnumbered whites 90,000 to 40,000, and the port of Charleston imported more enslaved workers than any other city in the colonies.

There were fewer enslaved workers in the North, in large part because of the lack of plantation-based economies dominated by a single crop. Enslaved men and women in Massachusetts, for example, worked on farms that raised a broad variety of crops, and many served as domestics and tradesmen. Massachusetts was the center of the slave trade for New England and was home to about 4,500 enslaved people in 1754. The largest slaveholding state in New England, though, was Connecticut. In 1774, nearly 6,500 enslaved workers lived there and about one-half of all ministers, lawyers, judges, and public officials held one or more enslaved workers.

INDUSTRY AND ITS LIMITS

In northern New England, colder weather and hard, rocky soil made it difficult for colonists to develop the kind of large-scale commercial farming system that Southerners were creating. Conditions for agriculture were better in southern New England and the middle colonies, where the soil was more fertile and the weather more temperate. New York, Pennsylvania, and the Connecticut River valley were the chief suppliers of wheat to much of New England, parts of the South, and the Caribbean. Even in this region, however, a commercial economy emerged alongside the agricultural one.

Almost every colonist engaged in a certain amount of industry at home, and these efforts occasionally provided families with surplus goods they could trade or sell. Beyond this domestic production, craftsmen and artisans established themselves in colonial towns as cobblers, blacksmiths, rifle-makers, cabinetmakers, silversmiths, and printers. In some areas, entrepreneurs harnessed water power to run small mills for grinding grain, processing cloth, or milling lumber. And in several coastal areas, large-scale shipbuilding operations began to flourish.

The first attempt to establish a significant metals industry in the colonies was an ironworks established in Saugus, Massachusetts, in 1646. The Saugus Ironworks produced pots, anvils, and nails for twenty-four years before financial problems forced it to close its doors. Later metalwork operations were more successful. The largest industrial enterprise anywhere in English North America was the ironworks of the German ironmaster Peter Hasenclever in northern New Jersey. Founded in 1764 with British capital, it employed several hundred laborers. There were other, smaller ironmaking enterprises in every northern colony and in several southern colonies as well. Even so, these and other growing industries did not immediately become the basis for the kind of explosive industrial growth that Great Britain experienced in the late eighteenth century—in part because English parliamentary regulations such as the Iron Act of 1750 restricted metal processing in the colonies. Similar prohibitions limited the manufacture of woolens, hats, and other goods. But the biggest obstacles to industrialization in America were an inadequate labor supply, a small domestic market, and inadequate transportation facilities and energy supplies.

More important than manufacturing were industries that took advantage of the natural resources of the continent. By the mid-seventeenth century, the flourishing fur trade of earlier years was in decline. Taking its place were lumbering, mining, and fishing. These industries provided commodities that could be exported to England in exchange for manufactured goods. And they helped produce the most distinctive feature of the northern economy: a thriving commercial class.

Technological progress, however, did not reach all colonists, even in the North. Up to half of all farmers did not own a plow, even less a wagon. Substantial numbers of households lacked pots and kettles for cooking and only about half owned guns. Few owned candles because they were either unable to afford candle molds or wax or lacked access to commercially produced candles. The low levels of ownership of these and other elementary tools were not because such things were difficult to make, but because most Americans remained too poor or too isolated to be able to obtain them.

THE RISE OF COLONIAL COMMERCE

Perhaps the most remarkable feature of colonial commerce was that it was able to survive at all. American merchants faced bewildering obstacles and lacked so many of the basic institutions of trade that they managed to stay afloat only with great difficulty and through considerable ingenuity. The colonies had almost no gold or silver, and their paper currency was not acceptable as payment for goods from abroad. For many years, colonial merchants had to rely on barter or on money substitutes such as beaver skins, rice, sugar, or tobacco.

A second obstacle was lack of information about the supply and demand of goods and services. Traders had no way of knowing what they would find in foreign ports. American colonial vessels sometimes stayed at sea for years, journeying from one port to another and trading one commodity for another in a desperate attempt to turn a profit. There was also an enormous number of small, fiercely competitive companies, which made the problem of organizing the system of commerce even more acute.

Nevertheless, commerce in the colonies grew. There was elaborate trade among the colonies themselves and with the West Indies. The mainland colonies offered their Caribbean trading partners rum, agricultural products, meat, and fish. The islands offered sugar, molasses, and at times enslaved labor in return. There was also trade with England, continental Europe, and the west coast of Africa. This commerce has often been described, somewhat inaccurately, as the **triangular trade**, suggesting a neat and orderly process by which merchants carried rum and other goods from New England to Africa, exchanged

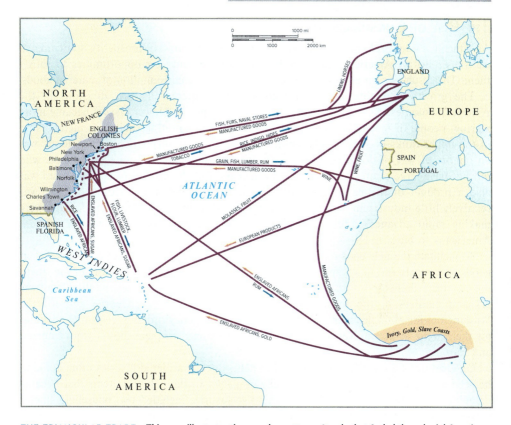

THE TRIANGULAR TRADE This map illustrates the complex pattern of trade that fueled the colonial American economy in the seventeenth and eighteenth centuries. A simple explanation of this trade is that the American colonies exported raw materials (agricultural products, furs, and others) to Britain and Europe and imported manufactured goods in return. While that explanation is not untrue it is far from complete, largely because the Atlantic trade was not a simple exchange between America and Europe, but a complex network of exchanges involving the Caribbean, Africa, and the Mediterranean. Note the important exchanges between the North American mainland and the Caribbean islands, the important trade between the American colonies and Africa, and the wide range of European and Mediterranean markets in which Americans were active. Not shown on this map, but also very important to colonial commerce, was a large coastal trade among the various regions of British North America. • *Why did the major ports of trade emerge almost entirely in the northern colonies?*

their merchandise for enslaved workers, whom they then transported to the West Indies (hence the term *middle passage* for the dreaded journey—it was the second of the three legs of the voyage), and then exchanged enslaved men and women for sugar and molasses, which they then shipped back to New England to be distilled into rum. In reality, the so-called triangular trade in rum, enslaved labor, and sugar was a complicated maze of highly diverse trade routes. Out of this risky set of operations emerged a group of adventurous entrepreneurs who by the mid-eighteenth century were beginning to constitute a distinct merchant class. The British Navigation Acts protected them from foreign competition in the colonies. They had ready access to the market in England for such colonial products as furs, timber, and American-built ships. But they also developed markets illegally outside the British Empire—in the French, Spanish, and Dutch West Indies—where they could often get higher prices for their goods than in the British colonies.

The Rise of Consumerism

As affluent residents of the colonies grew in number, the growing prosperity and commercialism of British America created both new appetites and new opportunities to satisfy them. The result was an emerging preoccupation with the consumption of material goods.

The growth of eighteenth-century consumerism increased class divisions in the American colonies. As the difference between the upper and lower classes became more glaring, people of means became more intent on demonstrating their own membership in the upper ranks of society. The ability to purchase and display consumer goods was an important way of doing so, particularly for wealthy people in cities and towns, who did not have large estates to indicate their success. But the growth of consumerism was also a product of the early stages of the Industrial Revolution. Although there was relatively little industry in America in the eighteenth century, England and Europe were making rapid advances and producing more and more affordable goods for affluent Americans to buy.

To facilitate the new consumer appetites, merchants and traders began advertising their goods in journals and newspapers. Agents of urban merchants—the ancestors of the traveling salesman—fanned out through the countryside, attempting to interest wealthy landowners and planters in the luxury goods now available to them. George and Martha Washington, for example, spent considerable time and money ordering elegant furnishings for their home at Mount Vernon, goods that were shipped to them mostly from England and Europe.

One feature of a consumer society is that items once considered luxuries often come to be seen as necessities once they are readily available. In the colonies, items that became commonplace after having once been expensive luxuries included tea, household linens, glassware, manufactured cutlery, crockery, furniture, among many other things. Another result of consumerism is the association of material goods—of the quality of a person's home and possessions and clothing, for example—with virtue and "refinement." The ideal of the cultivated "gentleman" and the gracious "lady" became increasingly powerful throughout the colonies in the eighteenth century. In part that meant striving to become educated

(Fototeca Gilardi/Hulton Archive/Getty Images)

THE BEGINNINGS OF A CONSUMER SOCIETY: In this idealized image of the colonial home, wealth and status are communicated through the display of manufactured furniture, braided rugs, and brass candlestick holders as well as the white woman tucked in the corner, presumably a mother or daughter, who is wearing clean, expensive clothing and embroidering by herself.

and "refined" in speech and behavior. Wealthy Americans read books on manners and fashion. They smoked tobacco from pipes. They bought magazines about London society. And they strove to develop themselves as witty and educated conversationalists. They also commissioned portraits of themselves and their families, devoted large portions of their homes to entertainment, built shelves and cases in which they could display fashionable possessions, constructed formal gardens, and lavished attention on their wardrobes and hairstyles.

PATTERNS OF SOCIETY

Although there were sharp social distinctions in the colonies, the well-defined and deeply entrenched class system of England failed to reproduce itself in America. Aristocracies emerged, to be sure, but they tended to rely less on landownership than control of a substantial workforce, and they were generally less secure and less powerful than their English counterparts. More than in England, white people in America lived with opportunities for social mobility—both up and down. There were also new forms of community in America, and they varied greatly from one region to another.

Southern Communities

The developing plantation system of the American South produced one form of community. The first plantations emerged in the tobacco-growing areas of Virginia and Maryland. Some of the early planters became established aristocrats with vast estates, but most did not. On the whole, seventeenth-century colonial plantations were rough and relatively small. In the early days in Virginia, they were little more than crude clearings where landowners and indentured servants worked side by side in conditions so harsh that death was an everyday occurrence. Most landowners lived in rough cabins or houses, with their servants or enslaved workers nearby. The economy of the plantation was precarious. Planters could not control their markets, so even the largest plantations were constantly at risk. When prices fell, planters faced the prospect of ruin.

Enslaved Black people, of course, lived very differently. On the smaller farms with only a handful of enslaved workers, it was not always possible for a rigid separation to develop between whites and Blacks. But by the early eighteenth century, over three-fourths of all enslaved men and women lived on plantations with at least ten enslaved laborers, and nearly one-half lived in communities of fifty enslaved workers or more. In those settings, enslaved people were more easily able to develop aspects of culture that reflected their own beliefs and values and testified to their humanity. Although whites seldom encouraged formal marriages among enslaved laborers, many Blacks themselves developed strong and elaborate family structures. There were also distinctive forms of slave religion, which variously blended Christianity with African folklore and sacred practices and became a central element in the emergence of Black culture.

Nevertheless, Black society was subject to constant intrusions from and interaction with white society. Domestic enslaved workers, for example, were often isolated from their own community. Enslaved women regularly faced sexual assault from slaveholders and overseers; the mixed-race children of these unions, though, were rarely recognized by their white fathers. On some plantations, enslaved workers were treated with a modicum of compassion and fairness, but this was rare and never lasted very long. More typically enslaved people encountered physical brutality and occasionally even sadism, against which they were virtually powerless.

Enslaved men and women challenged their status in ways large and small. The most serious example in the colonial period was the **Stono Rebellion** in South Carolina in 1739, during which about 100 enslaved people banded together, seized weapons, killed several whites, and attempted to escape south to Florida. The uprising was quickly crushed and most participants were executed. A more frequent form of resistance was running away, sometimes to nearby Native American nations in the hope of finding freedom there. Some native groups accepted the runaways, but others practiced slavery themselves and held as slaves Africans or native from other groups themselves. More often, runaways were caught and returned to the slaveholders before they could reach a protective community. Subtler, often undetected forms of resistance were practiced within the confines of slavery as enslaved people evaded or defied slaveholders' wishes through lying, cheating, stealing, and foot-dragging.

Most enslaved people, male and female, worked as field hands. But on the larger plantations that aspired to genuine self-sufficiency, some enslaved workers learned trades and crafts: blacksmithing, carpentry, shoemaking, spinning, weaving, sewing, midwifery, and others. These skilled craftspeople were at times hired out to other planters. Some set up their own establishments in towns or cities and shared their profits with the slaveholders. A few were able to buy their freedom.

Northern Communities

It is important to note that enslaved men and women in the North experienced much of the same degradation and humiliation as in the South. While fewer in number, they still experienced similar levels of violence, barriers to freedom, and white presumptions about their unfitness for citizenship and divine appointment as permanent laborers. Few towns or cities in New England were without enslaved Blacks, who worked in the fields, in homes, and in shops and barns.

The characteristic social unit in New England was not the isolated farm or the large plantation but the town. In the early years of colonization, each new settlement drew up a **covenant** binding all residents tightly together both religiously and socially. Colonists laid out a village, with houses and a meetinghouse arranged around a shared pasture, or "common." They divided up the outlying fields and woodlands among the residents; the size and location of a family's field depended on the family's numbers, wealth, and social station. Most lived near each other.

Once a town was established, residents held a yearly "town meeting" to decide important questions and choose a group of "selectmen," who ran the town's affairs. Participation in the meeting was generally restricted to adult white males who were members of the local church. Residents of the town were normally required to attend church services.

New Englanders did not adopt the English system of primogeniture—the passing of all property to the firstborn son. Instead, a father divided up his land among all his sons. His control of this inheritance gave him great power over the family. Often a son would reach his late twenties before his father would allow him to move into his own household and work his own land. Even then, sons would usually continue to live in close proximity to their fathers.

The early Puritan community was a tightly knit organism. But as the years passed and the communities grew, social strains began to affect their communal structure. This was

AFRICAN POPULATION AS A PROPORTION OF TOTAL POPULATION, CA. 1775 This map illustrates the parts of the colonies in which enslaved workers made up a large proportion of the population—in some areas, a majority. The enslaved population was smallest in the western regions of the southern colonies and in the area north of the Chesapeake, although there remained a significant African population in parts of New Jersey and New York (some enslaved, some free). • *What explains the dense concentration of enslaved workers in certain areas?*

partly due to the increasing commercialization of New England society. It was also partly due to population growth. In the first generations, fathers often controlled enough land to satisfy the needs of themselves and their sons. After several generations, however, there was often too little land to go around, particularly in communities surrounded by other

DEBATING THE PAST

The Witchcraft Trials

The witchcraft trials of the 1690s—which began in Salem, Massachusetts, and spread to other areas of New England—have been the stuff of popular legend for centuries. They have also engaged the interest of generations of historians, who have tried to explain why these seventeenth-century Americans became so committed to the belief that some of their own neighbors were agents of Satan. Although there have been many explanations of the witchcraft phenomenon, some of the most important in recent decades have focused on the central place of women in the story.

Through the first half of the twentieth century, most historians dismissed the witchcraft trials as "hysteria," prompted by the intolerance and rigidity of Puritan society. This interpretation informed the most prominent popular portrayal of witchcraft in the twentieth century: Arthur Miller's play *The Crucible*, first produced in 1953, which was clearly an effort to use the Salem trials as a comment on the great anticommunist frenzy of his own era. But at almost the same time, Perry Miller, the renowned scholar of Puritanism, argued in a series of important studies that belief in witchcraft was not a product of simple public excitement or intolerance but a widely shared part of the religious worldview of the seventeenth century. To the Puritans, witchcraft seemed not only plausible but scientifically rational.

A new wave of interpretation of witchcraft began in the 1970s, with the publication of *Salem Possessed* (1976), by Paul Boyer and Stephen Nissenbaum. Their examination of the town records of Salem in the 1690s led them to conclude that the witchcraft controversy was a product of class tensions between the poorer, more marginal residents of one part of Salem and the wealthier, more privileged residents of another. These social tensions, which could not find easy expression on their own terms, led some poorer Salemites to lash out at their richer neighbors by charging them, or their servants, with witchcraft. A few years later, John Demos, in *Entertaining Satan* (1983), examined witchcraft accusations in a larger area of New England and similarly portrayed them as products of displaced anger about social and economic grievances that could not be expressed otherwise. Demos provided a far more complex picture of the nature of these grievances than had Boyer and Nissenbaum, but like them, saw witchcraft as a symptom of a persistent set of social and psychological tensions.

At about the same time, a number of scholars were beginning to look at witchcraft through the scholarly lens of gender. Famously, Carol Karlsen's *The Devil in the Shape of a Woman* (1987) demonstrated through intensive scrutiny of records across New England that a disproportionate number of those accused of witchcraft were property-owning widows or unmarried women—in other words, women who did not fit comfortably into the normal pattern of male-dominated families. Karlsen concluded that such women were vulnerable to these accusations because they seemed threatening to people (including many women) who were accustomed to women as subordinate members of the community. Mary Beth Norton's *In the Devil's Snare* (2002) placed the witchcraft trials in the context of other events of their time—and in particular the terrifying upheavals and dislocations that the Indian Wars of the late seventeenth century created in Puritan communities. In the face of this crisis, in which refugees from King William's War were fleeing towns destroyed by native combatants and flooding Salem and other eastern towns, fear and social instability grew. Accusations of witchcraft and public trials and executions helped publicize and shore up social norms.

The witchcraft trials reflected a greater-than-normal readiness to connect aberrant behavior—such as the actions of independent or powerful women—to supernatural causes. The result was a wave of deadly witchcraft accusations. •

UNDERSTAND, ANALYZE, & EVALUATE

1. How did the Salem witchcraft trials reflect attitudes toward women and the status of women in colonial New England?
2. Why were colonial New Englanders willing to believe accusations of witchcraft about their fellow colonists?

towns. As a result, in many communities groups of younger residents broke off and moved elsewhere to form towns of their own.

The tensions building in Puritan communities could produce dramatic events. One example was the widespread excitement in the 1680s and 1690s over accusations of witchcraft—the human exercise of satanic powers—in New England. The most famous outbreak was in Salem, Massachusetts. Fear of the devil's influences spread quickly throughout the town, and hundreds of people, were accused of witchcraft. (See "Debating the Past: The Witchcraft Trials.") Between 140 and 150 people were accused of witchcraft and at least twenty were ultimately put to death before the **Salem witchcraft trials** finally ended in 1692. The vast majority of the accused and executed were women.

The Salem experience was not unique. Accusations of witchcraft popped up in many New England towns in the early 1690s, centering mostly on women who were pushing the boundaries of society. Research into the background of accused "witches" reveals that most were middle-aged women, often widowed, with few or no children. Some were of low social position, were often involved in domestic conflicts, had frequently been accused of crimes, and were considered abrasive by their neighbors. Still others were women who, through inheritance or hard work, had come into possession of substantial property of their own and thus challenged the power of men in the community.

In addition to reflecting bubbling social anxieties in New England towns, the witchcraft controversies signaled the highly religious character of New England societies. New Englanders believed in the power of Satan and the ongoing struggle between good and evil. Belief in witchcraft was not a superstition embraced only by the marginal but a common feature of mainstream Puritan religious conviction.

CITIES

In the 1770s, the two largest colonial ports—Philadelphia and New York—had populations of 28,000 and 25,000, respectively, which made them larger than most English urban centers of their time. Boston (16,000), Charles Town (later Charleston), South Carolina (12,000), and Newport, Rhode Island (11,000), were also substantial communities by the standards of the day.

Colonial cities served as trading centers for the farmers of their regions, as markets for international commerce, and locales where thousands of enslaved laborers were bought and sold. They were the centers of what industry existed in the colonies. They were also the locations of the most advanced schools and sophisticated cultural activities and of shops where imported goods could be bought. In addition, they were communities with urban social problems: crime, vice, pollution, traffic. Unlike most smaller towns, cities set up constables' offices and fire departments and developed systems for supporting the urban poor, whose numbers became especially large in times of economic crisis.

Finally, cities were places where new ideas could circulate and be discussed. There were newspapers, books, and other publications from abroad, and hence new intellectual influences. The taverns and coffeehouses of cities provided forums in which people could gather and debate the issues of the day. That is one reason why the Revolutionary crisis, when it began to build in the 1760s and 1770s, originated in the cities.

AWAKENINGS AND ENLIGHTENMENTS

Intellectual life in colonial America revolved around the conflict between the traditional emphasis on a personal God deeply involved in individual lives and the new spirit of the Enlightenment, which stressed the importance of science and human reason. The old views placed a high value on a stern moral code in which intellect was less important than faith. The Enlightenment suggested that people had substantial control over their own lives and societies.

THE PATTERN OF RELIGIONS

Religious toleration flourished in parts of America to a degree unmatched in any European nation. Settlers in America brought with them so many different religious practices that it proved almost impossible to impose a single religious code on any large area for a substantial period of time.

Although the Church of England was legally established as the official faith in Virginia, Maryland, New York, the Carolinas, and Georgia, many colonialists chose not to strictly abide by the laws and customs. Enforcement was uneven and lessened over time. Even in New England, where the Puritans had originally believed that they were all part of a single faith, there was a growing tendency in the eighteenth century for different congregations to affiliate with different denominations. In parts of New York and New Jersey, Dutch settlers had established their own Calvinist denomination, the Dutch Reformed. American Baptists developed a great variety of offshoots and shared the belief that rebaptism, usually by total immersion, was necessary when believers reached maturity. But while some Baptists remained Calvinists (believers in predestination), others came to believe in salvation by free will.

Protestants extended toleration to one another more readily than they did to Roman Catholics. New Englanders, in particular, viewed their Catholic neighbors in New France (Canada) not only as commercial and military rivals but also as dangerous agents of Rome. In most of the English colonies, however, Roman Catholics were too few to cause serious conflict. They were most numerous in Maryland, where they numbered 3,000. Perhaps for that reason they suffered the most persecution in that colony. After the overthrow of the original proprietors in 1691, Catholics in Maryland not only lost their political rights but also were forbidden to hold religious services except in private houses.

Jews in provincial America totaled no more than about 2,000 at any time. The largest community lived in New York City. Smaller groups settled in Newport and Charles Town, and there were scattered Jewish families in all the colonies. Nowhere could they vote or hold office. Only in Rhode Island could they practice their religion openly.

Enslaved Africans brought their own religious heritage. Though from diverse religious environments in West and western Central Africa, they generally shared a central belief in a Supreme Being or Creator and a pantheon of lesser divinities, whom they appeased and sought favor from through prayer, song, dance, and sacrifice. They aimed to create and sustain a harmonious bond with nature and supernatural beings, including not only gods but also spirits and deceased family ancestors. Many strove to continue traditional practices in their new locations but faced stern scrutiny and even hostility. Slaveholders regularly compelled the enslaved to adopt their own sacred beliefs, resulting in the creation of hybrid faiths that combined African religions with Christianity and Judaism or a practice of worshipping in secret, out of sight and earshot of whites.

Enslaved people from the Kingdom of Kongo, because of early contact with the Portuguese, tended to be Catholic while those from the Senegambia region often included Muslims. As many as 10 percent of enslaved Africans brought to the colonies were Muslim, but they left only traces of their faith. Like other Africans, they often took to worshiping clandestinely or integrating their beliefs with the slaveholder's principles. Ayuba Suleimon Diallo, born in 1700 into a noble family in Bondu (now Senegal), was captured in 1730, packed on a slave ship, and sold in Annapolis, Maryland, where he worked for two years as a tobacco hand. He ran away, was captured, and placed in jail. There he became known as a devout Muslim of royal lineage whose story of bondage won the sympathy of the Royal African Company, which freed him with the hope he might be of service to them in his native country. He later published an autobiography. African Muslim names appear on muster rolls in the Revolutionary War, such as Yusef ben Ali, Bampett Muhamad, and Joseph Sabo. And in 1777 Thomas Jefferson, arguing for an expansive view of religious tolerance in Virginia and quoting John Locke, wrote that "neither Pagan nor Mahamedan [Muslim] nor Jew ought to be excluded from the civil rights for the Commonwealth because of his religion."

THE GREAT AWAKENING

By the beginning of the eighteenth century, some Americans were growing troubled by the apparent decline in religious piety in their society. The movement of the population westward and the wide scattering of settlements had caused many communities to lose touch with organized religion. The rise of commercial prosperity created a more secular outlook in urban areas. The progress of science and free thought caused at least some colonists to doubt traditional religious beliefs.

Concerns about weakening piety surfaced as early as the 1660s in New England, where the Puritan oligarchy warned of a decline in the power of the church. Ministers preached sermons of despair (known as **jeremiads**), deploring the signs of waning piety. By the standards of other societies or other eras, the Puritan faith remained remarkably strong. But to New Englanders, the "declension" of religious piety seemed a serious problem. By the early eighteenth century, similar concerns about declining piety were emerging in other regions and among members of other faiths. The result was the first great American revival: the **Great Awakening**.

The Great Awakening began in earnest in the 1730s and reached its climax in the 1740s. It was potentially a subversive force in society, challenging traditions of power and deference.

The rhetoric of the revival emphasized the potential for every person to break away from the constraints of the past and start anew in his or her relationship with God. Such beliefs reflected in part the desires of many people to break away from their families or communities and start a new life. Not surprisingly, then, the revival had particular appeal to women (the majority of converts) and to younger sons of the third or fourth generation of settlers—those who stood to inherit the least land and who faced the most uncertain futures. Enslaved men and women flocked to hear this message of a new community as well, and even participated, when allowed, in public services.

Powerful **evangelists** from England helped spread the revival. John and Charles Wesley, the founders of Methodism, visited Georgia and other colonies in the 1730s. George Whitefield, a powerful open-air preacher from England, made several evangelizing tours through the colonies and drew massive crowds. He spoke in every colony and multiple times in Massachusetts and Connecticut—so many times, in fact, that it was estimated that every resident heard him preach at least once. But the outstanding preacher of the Great Awakening was the New England Congregationalist **Jonathan Edwards**. From his pulpit in Northampton, Massachusetts, Edwards attacked the new doctrines of easy salvation for all. He preached anew the traditional Puritan ideas of the absolute sovereignty of God, predestination, and salvation by God's grace alone. His vivid descriptions of hell terrified his listeners.

The Great Awakening led to the division of existing congregations (between "New Light" revivalists and "Old Light" traditionalists) and to the founding of new ones. It also affected areas of society outside of the churches. Some of the revivalists denounced book learning as a hindrance to salvation. But other evangelists saw education as a means of furthering religion, and they founded or led schools for the training of New Light ministers.

THE ENLIGHTENMENT

The Great Awakening caused one great cultural upheaval in the colonies. The **Enlightenment** caused a very different one. The Enlightenment was the product of scientific and intellectual discoveries in Europe in the seventeenth century—discoveries that revealed the "natural laws" that regulated the workings of nature. The new scientific knowledge encouraged many thinkers to begin celebrating the power of human reason and to argue that rational thought, not just religious faith, could create progress and advance knowledge in the world.

In celebrating reason, the Enlightenment encouraged men and women to look to themselves and their own intellect—not just to God—for guidance as to how to live their lives and shape their societies. It helped produce a growing interest in education and a heightened concern with politics and government.

In the early seventeenth century, Enlightenment ideas in America were largely borrowed from Europe—from such thinkers as Francis Bacon and John Locke of England, Baruch Spinoza of Amsterdam, and René Descartes and Jean-Jacques Rousseau of France. Later, however, Americans such as Benjamin Franklin, Thomas Paine, Thomas Jefferson, and James Madison made their own important contributions to Enlightenment thought.

LITERACY AND TECHNOLOGY

White male Americans achieved a high degree of literacy in the eighteenth century. By the time of the Revolution, well over one-half of all white men could read and write. The literacy rate for women lagged behind until the nineteenth century. While opportunities

> Poor Richard, 1743.
>
> AN
>
> # Almanack
>
> For the Year of Chrift
>
> # 1743,
>
> Being the Third after LEAP YEAR.
>
> | And makes fince the Creation | Years |
> |---|---|
> | By the Account of the Eaftern *Greeks* | 7251 |
> | By the Latin Church, when ☉ ent. ♈ | 6942 |
> | By the Computation of *W. W.* | 5752 |
> | By the *Roman* Chronology | 5692 |
> | By the *Jewifh* Rabbies | 5504 |
>
> *Wherein is contained,*
>
> The Lunations, Eclipfes, Judgment of the Weather, Spring Tides, Planets Motions & mutual Afpects, Sun and Moon's Rifing and Setting, Length of Days, Time of High Water, Fairs, Courts, and obfervable Days.
>
> Fitted to the Latitude of Forty Degrees, and a Meridian of Five Hours Weft from *London,* but may without fenfible Error, ferve all the adjacent Places, even from *Newfoundland* to *South-Carolina.*
>
> By *RICHARD SAUNDERS*, Philom.
>
> PHILADELPHIA:
> Printed and fold by *B. FRANKLIN,* at the New Printing-Office near the Market.

(Library of Congress, Prints and Photographs Division [LC-USZ62-58189])

GUIDE TO THE SEASONS Among their many purposes, almanacs sought to help farmers predict weather and plan for the demands of changing seasons.

for education beyond the primary level were scarce for men, they were almost nonexistent for women.

The large number of colonists who could read created a market for the first widely circulated publications in America other than the Bible: almanacs. By 1700, there were dozens, perhaps hundreds, of almanacs circulating throughout the colonies and even in the sparsely settled lands to the west. Most families had at least one. Almanacs provided medical advice, navigational and agricultural information, practical wisdom, humor, and predictions about the future—most famously, predictions about weather patterns for the coming year, which many farmers used as the basis for decisions about crops, even though the predictions were notoriously unreliable. The most famous almanac in eighteenth-century America was *Poor Richard's Almanac,* published by Benjamin Franklin in Philadelphia.

The wide availability of reading material in colonial America by the eighteenth century was a result of the spread of printing technology. The first printing press began operating in the colonies in 1639, and by 1695 there were more towns in America with printers than there were in England. At first, many of these presses did not get very much use. Over time, however, the rising literacy of the society created a demand for books, pamphlets, and almanacs that the presses rushed to fill.

The first newspaper in the colonies, *Publick Occurrences,* was published in Boston in 1690 using a relatively advanced printing facility. It was the first step toward what would eventually become a large newspaper industry. One reason the Stamp Act of 1765, which imposed a tax on printed materials, created such a furor was that printing technology had by then become central to colonial life.

EDUCATION

Even before Enlightenment ideas penetrated America, colonists placed a high value on formal education. Some families tried to teach their children to read and write at home, although the heavy burden of work in most agricultural households limited the time available for schooling. In Massachusetts, a 1647 law required that every town support a school; and a modest network of public schools emerged as a result. The Quakers and religious groups operated church schools, and in some communities widows or unmarried women conducted "dame schools" (all girl schools) in their homes. In cities, some master craftsmen set up evening schools for their apprentices.

African Americans had virtually no access to education. Occasionally a slaveholder would teach enslaved children to read and write; but as the slave system became more firmly entrenched, strong social (and ultimately legal) sanctions developed to discourage such efforts. Native Americans, too, remained largely outside the white educational system—to a large degree by choice. Some white missionaries and philanthropists established schools for Native Americans and helped create a small population of native people literate in spoken and written English.

Harvard, the first American college, was established in 1636 by Puritan theologians who wanted to create a training center for ministers. (The college was named for a Charlestown, Massachusetts, minister, John Harvard, who had left it his library and one-half of his estate.) In 1693, William and Mary College (named for the English king and queen) was established in Williamsburg, Virginia, by Anglicans. And in 1701, conservative Congregationalists, dissatisfied with the growing religious liberalism of Harvard, founded Yale (named for one of its first benefactors, Elihu Yale) in New Haven, Connecticut. Out of the Great Awakening emerged the College of New Jersey, founded in 1746 and known later as Princeton (after the town in which it was located); one of its first presidents was Jonathan Edwards. Despite the religious basis of these colleges, most of them offered curricula that included not only theology but also logic, ethics, physics, geometry, astronomy, rhetoric, Latin, Hebrew, and Greek. King's College, founded in New York City in 1754 and later renamed Columbia, was specifically devoted to the spread of secular knowledge. The Academy and College of Philadelphia, founded in 1755 and later renamed the University of Pennsylvania, was also a secular institution, established by a group of laymen under the inspiration of Benjamin Franklin.

After 1700, most colonial leaders received their entire education in America (rather than attending university in England, as had once been the case). But higher education remained available only to a few relatively affluent white men.

The Spread of Science

The clearest indication of the spreading influence of the Enlightenment in America was an increasing interest in scientific knowledge. Most of the early colleges established chairs in the natural sciences and introduced some of the advanced scientific theories of Europe, including Copernican astronomy and Newtonian physics, to their students. But the most vigorous promotion of science in these years occurred through the private efforts of amateurs and the activities of scientific societies. Leading merchants, planters, and even theologians became corresponding members of the Royal Society of London, the leading English scientific organization. Benjamin Franklin won international fame through his experiments with electricity. Particularly notable was his 1747 theory—and his 1752 demonstration, using a kite—that lightning and electricity were the same. (Previously, most scientists had believed that there were several distinct types of electricity.) His research on the way in which electricity could be "grounded" led to the development of the lightning rod, which greatly reduced fires and other damage to buildings during thunderstorms.

The high value that influential Americans were beginning to place on scientific knowledge was clearly demonstrated by the most controversial scientific experiment of the eighteenth century: inoculation against smallpox. The Puritan theologian **Cotton Mather** credited a man he had held in slavery, whom he had given the name Onesimus after the biblical slave who escaped from Philemon, for teaching him. In a 1716 letter to the Royal Society of London, Mather wrote that Onesimus, after contracting the disease, confided "he had undergone an Operation, which had given him something of ye Small-Pox, & would forever preserve him from it, adding, That it was often used among [Africans] and whoever had ye Courage to use it was forever free from ye Fear of the Contagion." Despite strong opposition, Mather urged inoculation on his fellow Bostonians during an epidemic in the 1720s. The results confirmed the effectiveness of the technique. Other theologians took up the cause, along with many physicians. By the mid-eighteenth century, inoculation had become a common medical procedure in America.

Concepts of Law and Politics

In law and politics, as in other parts of their lives, Americans in the seventeenth and eighteenth centuries believed that they were re-creating in the New World the practices and institutions of the Old World. But as in other areas, they created something very different. Although the American legal system adopted most of the essential elements of the English system, including such ancient rights as trial by jury, significant differences developed in court procedures, punishments, and the definition of crimes. In England, for example, a printed attack on a public official, whether true or false, was considered libelous. At the 1734-1735 trial of the New York publisher **John Peter Zenger**, the courts ruled that criticisms of the government were not libelous if factually true—a verdict that removed some colonial restrictions on the freedom of the press.

More significant for the future relationship between the colonies and England were differences emerging between the American and British political systems. Because the royal government was so far away, Americans created a group of institutions of their own that gave them a large measure of self-government. In most colonies, local communities grew accustomed to running their own affairs with minimal interference from higher authorities. The colonial assemblies came to exercise many of the powers that Parliament exercised in

(Fotosearch/Archive Photos/Getty Images)

COLONIAL PUNISHMENT American communities prescribed a wide range of punishments for misconduct and crime. Among the more common punishments were public humiliations—placing offenders in stocks, forcing them to wear badges of shame, or, as in this woodcut, binding them into a "ducking stool" and immersing them in water.

England. Provincial governors (appointed by the king after the 1690s) had broad powers on paper, but their actual influence was limited.

The result of all this was that the provincial governments became accustomed to acting more or less independently of Parliament, and a set of assumptions and expectations about the rights of the colonists took hold in America that was not shared by policymakers in England. These differences caused few problems before the 1760s, because the British did little to exert the authority they believed they possessed. But when, beginning in 1763, the English government began attempting to tighten its control over the American colonies, a great imperial crisis resulted.

CONCLUSION

Between the 1650s and the 1750s, the English colonies in America grew steadily in population, in the size of their economies, and in the sophistication and diversity of their cultures. Although most settlers in the 1750s still believed that they were fully a part of the British Empire, they were in fact living in a very different world.

Diversity and difference characterized individual colonies. They developed their own economies, systems of government, ideas about religious toleration, and rules governing interactions with Native Americans. What they shared was constant engagement with indigenous populations near the areas of their settlements. Those interactions varied from uneasy peace to outright hostility but always were part of each colony's experience. Also shared was a growing commitment to the enslavement of Africans or Black Americans. As increasing numbers of planters, farmers, landowners, merchants, ministers, and public officials determined that the presence of a slave class benefited them, colonial governments created slave codes and customs that birthed the colonial culture of human bondage. Many participated in the Great Awakening and embraced evangelical religion, leading to the trans-colonial spread of Baptist and Methodist churches. And most colonists shared a belief in certain basic principles of law and politics, which they considered embedded in the English constitution. Their interpretation of that constitution, however, was becoming increasingly different from that of the Parliament in England, laying the groundwork for future conflict.

KEY TERMS/PEOPLE/PLACES/EVENTS

Cotton Mather 79
covenant 70
Enlightenment 76
evangelist 76
Great Awakening 75
jeremiad 75
John Peter Zenger 79
Jonathan Edwards 76
middle passage 62
Salem witchcraft trials 73
Scotch-Irish 63
Stono Rebellion 70
triangular trade 66

RECALL AND REFLECT

1. How did patterns of family life and popular attitudes toward women differ in the northern and southern colonies?
2. How did the lives of enslaved Africans change over the course of the first century of slavery?
3. Who emigrated to North America in the seventeenth century, and why did they come?
4. How and why did life in the English colonies diverge from life in England?

Design element: Stars and Stripes: McGraw Hill Education.

4 THE EMPIRE IN TRANSITION

LOOSENING TIES
THE STRUGGLE FOR THE CONTINENT
THE NEW IMPERIALISM
STIRRINGS OF REVOLT
COOPERATION AND WAR

LOOKING AHEAD

1. How did the Seven Years' War change the balance of power in North America and throughout the world?
2. What policies did Parliament implement with regard to the colonies in the 1760s and 1770s, and why did Britain adopt these policies?
3. How did the colonists respond to Parliament's actions?

AS LATE AS THE 1750S, few Americans objected to their membership in the British Empire. The imperial system provided many benefits, and for the most part the British government left the colonies alone. Assemblies in those colonies passed laws, levied taxes, and otherwise strove to represent their white constituencies.

By the mid-1770s, the relationship between the American colonies and their British rulers was on the verge of unraveling. A global war between France and Britain had started in North America, and the colonists were thrust into the fight on Britain's side. Most indigenous groups, other than the Iroquois Confederacy, sided with the French. Britain's successes in that conflict left Native Americans divided and weakened, though it did not mark the end of their resistance to colonial encroachment in North America. Yet rather than uniting Britain and the colonists, the peace led to tensions, as London pressured the colonists to help pay for and otherwise contribute to the consolidation of empire. In the spring of 1775, the first shots were fired in a war that would ultimately win America its independence. How had it happened? And why so quickly?

LOOSENING TIES

In one sense, it had not happened quickly at all. Ever since the first days of English settlement, the ideas and institutions of the colonies had been diverging from those in Britain. In another sense, however, the revolutionary crisis emerged in response to relatively sudden changes in the administration of the empire. In 1763, the British government began to enforce a series of colonial policies that brought the differences between the two societies into sharp focus.

A Decentralized Empire

In the fifty years after the Glorious Revolution, the British Parliament established a growing supremacy over the king. Under Kings George I (1714–1727) and George II (1727–1760), the prime minister and his cabinet became the nation's real executives. Because these kings depended politically on the great merchants and landholders of Britain, they were less inclined than seventeenth-century monarchs to try to tighten control over the empire, which many merchants feared would disrupt profitable commerce with the colonies. As a result, administration of the colonies remained loose, decentralized, and inefficient. What was more, some men appointed to govern the colonies remained in Britain and hired substitutes to take their places in America.

The colonial assemblies, taking advantage of the weak imperial administration, had asserted their own authority to levy taxes, make appropriations, approve appointments, and pass laws. The assemblies came to look upon themselves as little parliaments, each practically as sovereign within its colony as Parliament itself was in Great Britain.

The Colonies Divided

Even so, the colonists continued to think of themselves as loyal British subjects. Many felt stronger ties to Great Britain (as it was called after a 1707 act of unification with

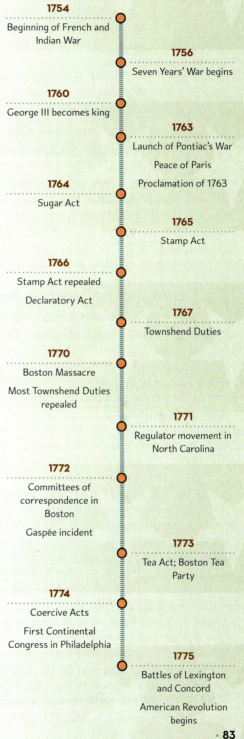

TIME LINE

1754 Beginning of French and Indian War

1756 Seven Years' War begins

1760 George III becomes king

1763 Launch of Pontiac's War
Peace of Paris
Proclamation of 1763

1764 Sugar Act

1765 Stamp Act

1766 Stamp Act repealed
Declaratory Act

1767 Townshend Duties

1770 Boston Massacre
Most Townshend Duties repealed

1771 Regulator movement in North Carolina

1772 Committees of correspondence in Boston
Gaspée incident

1773 Tea Act; Boston Tea Party

1774 Coercive Acts
First Continental Congress in Philadelphia

1775 Battles of Lexington and Concord
American Revolution begins

Scotland) than they did to the other American colonies. Although the colonies had slowly learned to cooperate with one another on such practical matters as intercolonial trade, road construction, and a colonial postal service, they remained reluctant to cooperate in larger ways, even when, in 1754, they faced a common threat from their old rivals, the French, and France's Native American allies. Delegates from Pennsylvania, Maryland, New York, and New England met in Albany in that year to negotiate a treaty with the Iroquois Confederacy. They tentatively approved a proposal by Benjamin Franklin to set up a "general government" to manage relations with Native Americans. War with the French and their native allies was already beginning when the **Albany Plan** was presented to the colonial assemblies. None approved it.

THE STRUGGLE FOR THE CONTINENT

The war that raged in North America through the late 1750s and early 1760s, which colonists called the **French and Indian War**, was the final stage in a long struggle among the three principal powers in northeastern North America: the British, the French, and the Iroquois. Two years into the conflict it expanded to Europe and beyond, where it became known as the **Seven Years' War**. The British victory in that struggle confirmed Britain's commercial supremacy and cemented its control over portions of North America.

NEW FRANCE AND THE IROQUOIS NATION

By the end of the seventeenth century, the French Empire in America was vast: it constituted the whole length of the Mississippi River and its delta (named Louisiana, after King Louis XIV) and the continental interior as far west as the Rocky Mountains and as far south as the Rio Grande. France claimed, in effect, the entire interior of the continent.

To secure their hold on these enormous claims, the French founded a string of communities, fortresses, missions, and trading posts. Would-be feudal lords established large estates *(seigneuries)* along the banks of the St. Lawrence River. On a high bluff above the river stood the fortified city of Quebec. Montreal to the south and Sault Sainte Marie and Detroit to the west marked the northern boundaries of French settlement. On the lower Mississippi there were plantations much like those in the southern colonies of British America, worked by enslaved Africans and owned by **Creoles** (people of European ancestry born in the Americas). New Orleans, founded in 1718 to service the French plantation economy, was soon as big as some of the larger cities of the Atlantic seaboard. Biloxi and Mobile to the east completed the string of French settlements.

Both the French and the British were aware that the battle for control of North America would be determined in part by who could best win the allegiance of native groups. The British—with their more advanced commercial economy—could usually offer Native Americans better and more plentiful goods. But the French offered tolerance. Unlike the British, who were much larger in number, the French settlers in the interior generally adjusted their own behavior to indigenous patterns. French fur traders frequently married native women and adopted native ways. Jesuit missionaries interacted comfortably with Native Americans and converted them to Catholicism by the thousands without challenging most of their social customs. By the mid-eighteenth century, therefore, the French had better and closer relations with most Native Americans of the interior than did the British.

The most powerful native group, however, had remained aloof from both sides. The **Iroquois Confederacy**—five nations (Mohawk, Seneca, Cayuga, Onondaga, and Oneida) that had formed a defensive alliance in the fifteenth century—had been the most powerful native presence in the Ohio Valley since the 1640s. Although the Iroquois claimed rights to the Valley, they maintained relations with the French and British and cemented their autonomy by trading successfully with both and astutely playing them against each other.

Anglo–French Conflicts

As long as peace and stability in the North American interior lasted, English and French colonists coexisted without serious difficulty. But after the Glorious Revolution in England, a complicated series of Anglo-French wars erupted in Europe and continued intermittently for nearly eighty years, creating important repercussions in America.

King William's War (1689-1697) produced only a few, indecisive clashes between the English and the French in northern New England. Queen Anne's War, which began in 1701 and continued for nearly twelve years, generated more substantial conflicts. The Treaty of Utrecht, which brought the conflict to a close in 1713, transferred substantial territory from the French to the British in North America, including Acadia (Nova Scotia) and Newfoundland. Two decades later, disputes over British trading rights in the Spanish colonies produced a conflict between Great Britain and Spain that soon grew into a much larger European war. The British colonists in America were drawn into the struggle, which they called King George's War (1744-1748). New Englanders captured the French bastion at Louisbourg on Cape Breton Island in Nova Scotia, but the peace treaty that finally ended the conflict forced them to abandon it.

In the aftermath of King George's War, relations among the British, French, and Iroquois in North America quickly deteriorated. The Iroquois granted trading concessions in the interior to British merchants for the first time. The French feared, probably correctly, that the British were using the concessions as a first step toward expansion into French lands. They began in 1749 to construct new fortresses in the Ohio Valley. The British responded by increasing their military forces and building fortresses of their own. The balance of power that the Iroquois had carefully maintained for so long rapidly disintegrated.

For the next five years, tensions between the British and French increased. In the summer of 1754, the governor of Virginia sent a militia force (under the command of an inexperienced young colonel, George Washington) into the Ohio Valley to challenge France's Fort Duquesne, on the site of what is now Pittsburgh. The colonel's men and an allied native force under the leader Tanaghrisson were met on the way by a French patrol, and the British-allied force killed a French officer and ten of his men. Washington built a crude stockade (**Fort Necessity**) not far from Fort Duquesne. After the Virginians staged an unsuccessful attack on a French detachment, the French countered with an assault on Fort Necessity, trapping Washington and his soldiers inside. After one-third of them died in the fighting, Washington surrendered. The clash marked the beginning of the French and Indian War.

The Great War for Empire

The French and Indian War lasted nearly nine years, and it moved through three distinct phases. The first phase—from Fort Necessity in 1754 until the expansion of the war to Europe in 1756—was primarily a local, North American conflict, but one that swept up native groups throughout the west. The Iroquois remained largely passive in the conflict. But virtually all the other nations sided with the French, though a few fought with the British and some with one side then the other. Native Americans tended to view these alliances as means for expelling one power or the other from their lands, not as endorsements of imperial presence

AMERICA IN THE WORLD

THE FIRST GLOBAL WAR

The French and Indian War in North America was part of a much larger conflict. Known in Europe as the Seven Years' War, it was one of the longest, most widespread, and most important wars in modern history. The war thrust Great Britain into conflicts across Europe and North America. Winston Churchill once wrote of it as the first "world war."

In North America, the war was a result of tensions along the frontiers of the British Empire. But it arose more broadly from larger conflicts among the great powers in Europe. It began in the 1750s with what historians have called a "diplomatic revolution." Well-established alliances between Britain and the Austro-Hungarian Empire and between France and Prussia collapsed, replaced by a new set of alliances setting Britain and Prussia against France and Austria. The instability that these changing alliances produced helped speed the European nations toward war.

The Austrian–British alliance collapsed because Austria suffered a series of significant defeats at the hands of the Prussians. To the British government, these failures suggested that the Austro-Hungarian Empire was now too weak to help Britain balance French power. As a result, Great Britain launched a search for new partnerships with the rising powers of northern Germany, Austria's enemies. In response, the Austrians sought an alliance with France to help protect them from the power of their former British allies. (One later result of this new alliance was the 1770 marriage of the future French king Louis XVI to the Austrian princess Marie Antoinette.) In the aftermath of these realignments, Austria sought again to defeat the Prussian-Hanover forces in Germany. In the process, Russia became concerned about the Austro-Hungarian Empire's possible dominance in central Europe and allied itself with the British and the Prussians. These complicated realignments eventually led to the Seven Years' War, which soon spread across much of the world. The war engaged not only most of the great powers in Europe, from Britain to Russia, but also the emerging colonial worlds—India, West Africa, the Caribbean, and the Philippines—as the powerful British navy worked to strip France, and eventually Spain, of its valuable colonial holdings.

The Seven Years' War was at heart a struggle for economic power. Colonial possessions, many European nations believed, were critical to their future wealth. The war's outcome affected not only the future of America but also the distribution of power throughout much of the world. It destroyed the French navy and much of the French Empire, and it elevated Great Britain to undisputed preeminence among the colonizing powers—especially when, at the conclusion of the war, India and all of eastern North America fell firmly under British control. The war also reorganized the balance of power in Europe, with Britain now preeminent among the great powers and Prussia (later to become the core of modern Germany) rapidly rising in wealth and military power.

The Seven Years' War was not only one of the first great colonial wars but also one of the last big wars of religion, and it extended the dominance of Protestantism in Europe. In what is now Canada, the war replaced French with British rule and thus replaced Catholic with Protestant dominance. The Vatican, no longer a military power itself, had relied on the great Catholic empires—Spain, France, and Austria-Hungary—as bulwarks of its power and influence. The shift of power toward Protestant governments in Europe and North America weakened the Catholic Church and reduced its geopolitical influence.

The conclusion of the Seven Years' War strengthened Britain and Germany and weakened France. But it did not provide any lasting solution to the rivalries among the great colonial powers. In North America, a dozen years after the end of the conflict, the American Revolution—which in many ways arose from the Seven Years' War—stripped the British Empire of one of its most important and valuable colonial appendages. By the time the American Revolution came to an end, the French Revolution had sparked another lengthy period of conflict, culminating in the Napoleonic Wars of the early nineteenth century, which once again redrew the map of Europe and, for a while, the world.

UNDERSTAND, ANALYZE, & EVALUATE

1. How did the Seven Years' War change the balance of power among the nations of Europe? Who gained and who lost in the war?
2. Why is the Seven Years' War described as one of the "most important wars in modern history"?

in North America. Combat engulfed western white settlements, native villages, and frontier forts, featuring clashes between European and colonial armies and native fighters arrayed across the alliances. By late 1755, many British settlers along the frontier had withdrawn east of the Allegheny Mountains to escape the hostilities.

The second phase of the struggle began in 1756, when the Seven Years' War expanded to Europe and beyond. (See "America in the World: The First Global War.") The fighting now spread to the West Indies, India, and Europe itself. But the principal struggle remained the war in North America, where so far Britain had suffered nothing but frustration and defeat. Beginning in 1757, **William Pitt**, the British secretary of state (and future prime minister), brought the war fully under British control. He planned military strategy, appointed commanders, and issued orders to the colonists. British commanders began forcibly enlisting colonists into the army, a practice known as **impressment**. Officers also seized supplies from local farmers and tradesmen and compelled colonists to offer shelter to British troops, all generally without compensation. The Americans resented these new impositions and firmly resisted them. By early 1758, the friction between the British authorities and the colonists was threatening to bring the war effort to a halt.

Beginning in 1758, Pitt initiated the third and final phase of the war by relaxing many of the policies that Americans had found obnoxious. He agreed to reimburse the colonists for all supplies requisitioned by the army. He returned control over recruitment to the colonial assemblies. And he dispatched large numbers of additional British troops to America, further strengthening the numerical advantage British colonists already enjoyed over the French. These moves turned the tide of battle in Great Britain's favor. After 1756, moreover, the French suffered from a series of poor harvests and were unable to sustain their early military successes.

By mid-1758, British regulars and colonial militias were seizing one French stronghold after another. Two British generals, Jeffrey Amherst and James Wolfe, captured the fortress at Louisbourg in July 1758. A few months later Fort Duquesne fell without a fight. The next year, at the end of a siege of Quebec, the army of General Wolfe struggled up a hidden ravine under cover of darkness, surprised the larger forces of the Marquis de Montcalm, and defeated them in a battle in which both commanders were killed. The dramatic fall of Quebec on September 13, 1759, marked the beginning of the end of the American phase

THE DEATH OF GENERAL WOLFE This engraving is based on a 1770 painting by Benjamin West of General James Wolfe, lying mortally wounded during the siege of Quebec in 1759. West took much dramatic license in the painting, positioning important and recognizable military figures around Wolfe who in fact were not present. He also depicted a Native American in the manner of a "noble savage," a stock figure in contemporary art and literature, simultaneously admired for his uncorrupted virtue but also set apart as irredeemably primitive. This portrayal also romanticized the connections between native nations and the British in the war. (Library of Congress Prints and Photographs Division [LC-DIG-pga-03470])

of the war. A year later, in September 1760, the French army formally surrendered to Amherst in Montreal. The hostilities finally ended in 1763, with the Peace of Paris, by which the French ceded to Great Britain some of their West Indian islands, most of their colonies in India and Canada, and all other French territory in North America east of the Mississippi. The French then turned over New Orleans and their claims west of the Mississippi to Spain, surrendering all title to the mainland of North America. Yet they kept hold of possessions in the Caribbean central to the economy of empire: Martinique, Guadeloupe, and Saint-Domingue (modern-day Haiti), the most profitable sugar colony in the world and home to hundreds of thousands of enslaved people.

The French and Indian War greatly expanded Britain's territorial claims in the New World. At the same time, the cost of the war greatly enlarged Britain's debt and substantially increased British resentment of the Americans. British leaders were contemptuous of the colonists for what they considered American military ineptitude during the war. They were angry that the colonists had made so few financial contributions to a struggle waged, they believed, largely for American benefit. And they were particularly bitter that colonial merchants had been selling food and other goods to the French in the West Indies throughout the conflict. All these factors combined to persuade many British leaders that a major reorganization of the empire would be necessary. London wanted increased authority over the colonies.

(Library of Congress Prints and Photographs Division [LC-USZC4-5315])

AN APPEAL FOR COLONIAL UNITY This sketch, one of the first American editorial cartoons, appeared in Benjamin Franklin's Philadelphia newspaper, the *Pennsylvania Gazette*, on May 9, 1754. It was meant to illustrate the need for intercolonial unity against the French and Native Americans, but later served as a revolutionary rallying cry.

The war had an equally profound effect on the American colonists. It was an experience that forced them, for the first time, to act in concert against a common foe. Yet resentments against British impressment and other wartime demands also mobilized common grievances against the government in London. The 1758 return of authority to the colonial assemblies seemed to many Americans to confirm the illegitimacy of British interference in local affairs. Thus Benjamin Franklin's famous woodcut of a divided snake—"Join, or Die"—appeared in 1754 to encourage cooperation with the British against the French and native opponents but later served to call for unity against Great Britain itself.

For native groups of the Ohio Valley, the British victory was disastrous. Many of the territorial spoils came out of indigenous land. Disease and starvation plagued indigenous peoples, and they held the British responsible. Those nations that had allied themselves with the French earned the enmity of the British. The Iroquois Confederacy, which had not allied with the French, fared only slightly better. British officials saw the passivity of the Iroquois during the war as evidence of duplicity. In the aftermath of the peace settlement, the fragile Iroquois alliance with the British quickly unraveled. Native Americans were increasingly divided and outnumbered, and would seldom again be in a position to deal with their European rivals on terms of military or political equality. But even before the war ended, a coalition under the leadership of a chief named **Pontiac** was planning a united rebellion against British rule. Native groups were to continue contesting the British for control of the Ohio Valley for another fifty years.

THE NEW IMPERIALISM

With the treaty of 1763, Great Britain found itself truly at peace for the first time in more than fifty years. As a result, the British government could now turn its attention to the organization of empire. Saddled with enormous debts from many years of war, Britain desperately needed new revenues. Responsible for vast holdings in the New World, the

Burdens of Empire

The experience of the French and Indian War should have suggested that increasing imperial control over the colonies would not be easy. Not only had the resentment of colonists forced Pitt to relax his policies in 1758, but the colonial assemblies continued to defy imperial trade regulations and other British demands. The most immediate problem for London, however, was its staggering war debt. Landlords and merchants in Britain were objecting strenuously to any further tax increases, and the colonial assemblies had repeatedly demonstrated their unwillingness to pay for the war effort. Many officials in Britain believed that only by taxing the Americans directly could the empire effectively meet its financial needs.

At this crucial moment in Anglo-American relations, the government of Great Britain saw the 1760 accession to the throne of **George III**. He was determined to reassert the authority of the monarchy, removing from power the relatively stable coalition of Whigs (opponents of absolute monarchy) that had governed for much of the century and replaced it with a new and very unstable coalition of his own. The weak new ministers that emerged as a result each lasted in office an average of only about two years.

The king had serious intellectual and psychological limitations. He suffered, apparently, from a rare mental disease that produced intermittent bouts of insanity. (Indeed, in the last years of his long reign he was, according to most accounts, unable to perform any official functions.) Yet even when George III was lucid, which was most of the time in the 1760s and 1770s, he was painfully immature and insecure. The king's personality, therefore, contributed both to the instability and to the rigidity of the British government during these critical years.

More directly responsible for the problems that soon emerged with the colonies, however, was **George Grenville**, whom the king made prime minister in 1763. Grenville shared the prevailing opinion within Britain that the colonists should be compelled to pay a part of the cost of defending and administering the empire.

The British and Native Americans

With the defeat of the French, settlers from the British colonies began immediately to move over the mountains into native lands in the upper Ohio Valley. An alliance of Ottawa, Potawatomi, and Ojibwe, under the Ottawa chieftain Pontiac, struck back. This "Three Fires Confederacy" maintained a long tradition of native resistance against European powers. Its motivations resembled, as well, the kind of war for independence the colonists would launch against the same British power twelve years later and which other groups (Apache, Comanche, Ute, Navajo) had waged against the Spanish.

Native Americans fighting loosely under Pontiac laid siege to Detroit and captured several British forts, at one point staging a ruse involving a stray lacrosse ball to infiltrate the garrison at Michilimackinac. Five hundred soldiers and 2,000 white settlers ended up dead in a region spanning from the Great Lakes to the Mississippi River to the Appalachians. The British determined to inflict horrific damage in return. Even as they negotiated, authorities at Fort Pitt gave blankets that had come from a smallpox hospital to a delegation from the Delaware Nation. The disease tore through that group the following summer.

The British government, fearing a disruption of western trade, issued the **Proclamation of 1763**, which forbade settlers to advance beyond the Appalachian Mountains. Many native groups supported the Proclamation as the best bargain available to them. The Cherokee, in particular, worked actively to hasten the drawing of the border, hoping finally to put an end to white encroachment onto their lands. But the much-hoped-for boundary failed to stop white settlers from moving back into lands farther into the Ohio Valley. Meanwhile the **Paxton Boys**, a band of Pennsylvania frontiersmen, massacred twenty Conestoga late in 1763. This was the sort of racial terror that animated native certainty that the British intended to "extirpate you from being a people," as one colonial official in 1764 told the Delaware, Shawnee, and Mingo.

Ultimately, white violence as well as illness, supply shortages, and internal divisions brought the native revolt to its end. Native Americans did win membership in trade alliances and promises that the British would enforce the boundary line. But in 1768, new agreements with the western nations pushed the border outward, and treaties failed to stop the white advance in any event. British settlers who had fought in the Seven Years' War were never going to give up lands they believed they had earned by blood. George Washington dismissed

COLONIAL CLAIMS IN 1763 This map shows the thirteen colonies at the end of the Seven Years' War. It shows the line of settlement established by the Proclamation of 1763 (the red line), as well as the extent of actual settlement in that year (the blue line). Note that in the middle colonies (North Carolina, Virginia, Maryland, and southern Pennsylvania), settlement had already reached the red line—and in one small area of western Pennsylvania moved beyond it—by the time of the Proclamation of 1763. Note also the string of forts established beyond the Proclamation line. • *How do the forts help explain the efforts of the British to restrict settlement? And what does the extent of actual settlement tell us?*

the Proclamation in 1767 as nothing more than "a temporary expedient to quiet the Minds of the Indians." What was more, just as colonists had fumed at being forced to support the war in the West in the 1750s, they now resented being taxed to bankroll the new British commitment to policing the imperial frontier.

Battles over Trade and Taxes

The Grenville ministry tried to increase its authority in the colonies in other ways as well. The **Sugar Act** of 1764 aimed to tighten British control over American trade with French and Spanish colonies through a series of tariffs and rules that would be more strictly enforced than in the past. The duty on French molasses from the Caribbean, for example, was reduced to discourage smugglers who had evaded paying the higher tax. But by lowering the tax and enforcing its payment, the British law aimed to raise revenue from its colony. In addition to regulating imports, the Sugar Act specified that colonists could export timber and iron only to Britain. It also established new vice-admiralty courts in America to try accused smugglers, thus cutting them off from sympathetic local juries. The Currency Act of 1764 required that the colonial assemblies stop issuing paper money.

Regular British troops were stationed permanently in America, and under the Mutiny Act of 1765 the colonists were required to help provision and maintain the army. Ships of the British navy patrolled American waters to search for smugglers. The customs service was reorganized and enlarged. Royal officials were required to take up their colonial posts in person instead of sending substitutes. Colonial manufacturing was restricted so that it would not compete with rapidly expanding industries in Great Britain.

At first, it was difficult for the colonists to resist these unpopular new laws. That was partly because Americans continued to harbor as many grievances against one another as they did against the authorities in London. In 1763, for example, the Paxton Boys descended on Philadelphia to demand tax relief and financial support for their violence against Native Americans. Colonial authorities conceded to their demands. In 1771, a small-scale civil war broke out in North Carolina when the "Regulators," farmers of the interior, organized and armed themselves to resist high taxes. The colonial governor appointed sheriffs to enforce the levies. An army of militiamen, most of them from the eastern counties, crushed the Regulator revolt.

The unpopularity of the Grenville program helped the colonists overcome their internal conflicts and led them to regard the policies from London as a threat to all Americans. Northern merchants would suffer from restraints on their commerce. The closing of the West to land speculation and fur trading enraged many colonists. Others were angered by the restriction of opportunities for manufacturing. Southern planters, in debt to British merchants, would be unable to ease their debts by speculating in western land. Small farmers would suffer from the abolition of paper money, which had been the source of most of their loans. Workers in towns faced the prospect of narrowing opportunities, particularly because of the restraints on manufacturing and currency. Everyone stood to suffer from increased taxes.

Most Americans soon found ways to live with the new British laws without terrible economic hardship. But their political grievances remained. Americans were accustomed to wide latitude in self-government. They believed that colonial assemblies had the sole right to control appropriations for the costs of government within the colonies. By attempting to raise extensive revenues directly from the public, the British government was challenging the basis of colonial political power.

STIRRINGS OF REVOLT

By the mid-1760s, a hardening of positions had begun in both Great Britain and America. The result was a progression of events that, more rapidly than imagined, diminished the British Empire in America.

The Stamp Act Crisis

Grenville could not have devised a better method for antagonizing and unifying the colonies than the **Stamp Act** of 1765. Unlike the Sugar Act of a year earlier, which affected only a few New England merchants, the new tax fell on everyone. It levied taxes on every printed document in the colonies: newspapers, almanacs, pamphlets, deeds, wills, licenses. British officials were soon collecting more than ten times as much revenue in America as they had been before 1763. More alarming than these taxes, however, was the precedent they seemed to create. In the past, taxes and duties on colonial trade had always been designed to regulate commerce. The Stamp Act, however, was clearly an attempt by Britain to raise revenue from the colonies without the consent of the colonial assemblies.

Few colonists believed that they could do anything more than grumble until the Virginia House of Burgesses roused Americans to action. The planter and lawyer **Patrick Henry** made a dramatic speech to the House in May 1765, concluding with a vague prediction that if present policies were not revised, George III, like earlier tyrants, might lose his head. Amid shocked cries of "Treason!" Henry introduced a set of resolutions (only some of which the assembly passed) declaring that Americans possessed the same rights as the British, especially the right to be taxed only by their own representatives; that Virginians should pay no taxes except those authorized by the Virginia assembly; and that anyone advocating the right of Parliament to tax Virginians should be deemed an enemy of the colony. Henry's resolutions were printed and circulated as the "**Virginia Resolves**."

In Massachusetts at about the same time, James Otis persuaded his fellow members of the colonial assembly to call an intercolonial congress to take action against the new tax. In October 1765, the Stamp Act Congress, as it was called, met in New York with delegates from nine colonies. In a petition to the British government, the congress denied that the colonies could rightfully be taxed except through their own provincial assemblies. Across the ocean, colonial agent **Benjamin Franklin** articulated such grievances before Parliament. (See "Consider the Source: Benjamin Franklin, Testimony against the Stamp Act.")

Meanwhile, in the summer of 1765, mobs were rising up in several colonial cities against the Stamp Act. The largest was in Boston, where men belonging to the newly organized **Sons of Liberty** terrorized stamp agents and burned stamps. The mob also attacked such supposedly pro-British aristocrats as the lieutenant governor, Thomas Hutchinson, who had privately opposed passage of the Stamp Act but who felt obliged to support it once it became law. Hutchinson's elegant house was pillaged and virtually destroyed.

The crisis finally subsided largely because Britain backed down. The authorities in London were less affected by the political protests than by economic pressure. Many New Englanders had stopped buying British goods to protest the Sugar Act of 1764, and the Stamp Act caused the boycott to spread. With pressure from British merchants, Parliament—under a new prime minister, the Marquis of Rockingham—repealed the unpopular law on March 18, 1766. To satisfy his strong and vociferous opponents, Rockingham also pushed through the Declaratory Act, which confirmed parliamentary authority over the colonies "in all

CONSIDER THE SOURCE

BENJAMIN FRANKLIN, TESTIMONY AGAINST THE STAMP ACT (1766)

In 1765 Parliament passed the first internal tax on the colonists, known as the Stamp Act. Benjamin Franklin was a colonial agent in London at the time, and as colonial opposition to the act grew, he found himself representing these views to the British government. In his testimony before Parliament he describes the role of taxes in Pennsylvania and the economic relationship between the colonies and the mother country.

Q. What is your name, and place of abode?
A. Franklin, of Philadelphia.
Q. Do the Americans pay any considerable taxes among themselves?
A. Certainly many, and very heavy taxes.
Q. What are the present taxes in Pennsylvania, laid by the laws of the colony?
A. There are taxes on all estates, real and personal; a poll tax; a tax on all offices, professions, trades, and businesses, according to their profits; an excise on all wine, rum, and other spirit; and a duty of ten pounds per head on all Negroes imported, with some other duties.
Q. For what purposes are those taxes laid?
A. For the support of the civil and military establishments of the country, and to discharge the heavy debt contracted in the last [Seven Years'] war. . . .
Q. Are not all the people very able to pay those taxes?
A. No. The frontier counties, all along the continent, have been frequently ravaged by the enemy and greatly impoverished, are able to pay very little tax. . . .
Q. Are not the colonies, from their circumstances, very able to pay the stamp duty?
A. In my opinion there is not gold and silver enough in the colonies to pay the stamp duty for one year.

Q. Don't you know that the money arising from the stamps was all to be laid out in America?
A. I know it is appropriated by the act to the American service; but it will be spent in the conquered colonies, where the soldiers are, not in the colonies that pay it.
Q. Do you think it right that America should be protected by this country and pay no part of the expense?
A. That is not the case. The colonies raised, clothed, and paid, during the last war, near 25,000 men, and spent many millions.
Q. Were you not reimbursed by Parliament?
A. We were only reimbursed what, in your opinion, we had advanced beyond our proportion, or beyond what might reasonably be expected from us; and it was a very small part of what we spent. Pennsylvania, in particular, disbursed about 500,000 pounds, and the reimbursements, in the whole, did not exceed 60,000 pounds. . . .
Q. Do you think the people of America would submit to pay the stamp duty, if it was moderated?
A. No, never, unless compelled by force of arms. . . .
Q. What was the temper of America towards Great Britain before the year 1763?
A. The best in the world. They submitted willingly to the government of the Crown, and paid, in all their courts, obedience to acts of Parliament. . . .
Q. What is your opinion of a future tax, imposed on the same principle with that of the Stamp Act? How would the Americans receive it?
A. Just as they do this. They would not pay it.
Q. Have not you heard of the resolutions of this House, and of the House of Lords, asserting the right of Parliament relating to America, including a power to tax the people there?

A. Yes, I have heard of such resolutions.

Q. What will be the opinion of the Americans on those resolutions?

A. They will think them unconstitutional and unjust.

Q. Was it an opinion in America before 1763 that the Parliament had no right to lay taxes and duties there?

A. I never heard any objection to the right of laying duties to regulate commerce; but a right to lay internal taxes was never supposed to be in Parliament, as we are not represented there.

Q. Did the Americans ever dispute the controlling power of Parliament to regulate the commerce?

A. No.

Q. Can anything less than a military force carry the Stamp Act into execution?

A. I do not see how a military force can be applied to that purpose.

Q. Why may it not?

A. Suppose a military force sent into America; they will find nobody in arms; what are they then to do? They cannot force a man to take stamps who chooses to do without them. They will not find a rebellion; they may indeed make one.

Q. If the act is not repealed, what do you think will be the consequences?

A. A total loss of the respect and affection the people of America bear to this country, and of all the commerce that depends on that respect and affection.

Q. How can the commerce be affected?

A. You will find that, if the act is not repealed, they will take very little of your manufactures in a short time.

Q. Is it in their power to do without them?

A. I think they may very well do without them.

Q. Is it their interest not to take them?

A. The goods they take from Britain are either necessaries, mere conveniences, or superfluities. The first, as cloth, etc., with a little industry they can make at home; the second they can do without till they are able to provide them among themselves; and the last, which are mere articles of fashion, purchased and consumed because of the fashion in a respected country; but will now be detested and rejected. The people have already struck off, by general agreement, the use of all goods fashionable in mourning.

Q. If the Stamp Act should be repealed, would it induce the assemblies of America to acknowledge the right of Parliament to tax them, and would they erase their resolutions [against the Stamp Act]?

A. No, never.

Q. Is there no means of obliging them to erase those resolutions?

A. None that I know of; they will never do it, unless compelled by force of arms.

Q. Is there a power on earth that can force them to erase them?

A. No power, how great so ever, can force men to change their opinions. . . .

Q. What used to be the pride of the Americans?

A. To indulge in the fashions and manufactures of Great Britain.

Q. What is now their pride?

A. To wear their old clothes over again, till they can make new ones.

UNDERSTAND, ANALYZE, & EVALUATE

1. What kind of taxes did colonists pay according to Franklin? What did the interviewer seem to think of the colonists' tax burden? What disagreements existed between Franklin and his interviewer on the purpose, legality, and feasibility of the stamp tax?
2. How did Franklin characterize the British–colonial relationship prior to 1763?
3. What colonial response to the Stamp Act and other "internal taxes" did Franklin predict? What, if anything, could Parliament do to enforce the colonists' compliance?

Source: *The Parliamentary History of England,* London, 1813, vol. XVI, 138–159; in Charles Morris, *The Great Republic by the Master Historians, vol. II.* R.S. Belcher Co., 1902.

cases whatsoever." But in their rejoicing over the Stamp Act repeal, most Americans paid little attention to this sweeping declaration of power.

Internal Rebellions

The conflicts with Britain were not the only uprisings emerging in the turbulent years of the 1760s. In addition to the Stamp Act crisis and other challenges to London, there were internal rebellions that had their roots in the class system in New York and New England. In the Hudson Valley in New York, great estates had grown up in which owners had rented out their land to small farmers. The revolutionary fervor of the time led many of these tenants to demand ownership of the land they worked. To emphasize their determination, they stopped paying rents.

The rebellion soon failed, but other challenges continued. In Vermont, which still was governed by New York, insurgent farmers challenged landowners (many of them the same owners whom tenants had challenged on the Hudson) by taking up arms and demanding ownership of the land they worked. Ethan Allen, later a hero of the Revolutionary War and himself a land speculator, took up the cause of the Green Mountain farmers and accused the landowners of trying to "enslave a free people." Allen eventually succeeded in making Vermont into a separate state, which broke up some of the large estates.

The Townshend Program

When the Rockingham government's policy of appeasement met substantial opposition in Britain, the king dismissed the ministry and replaced it with a new government led by the aging but still powerful William Pitt, who was now Lord Chatham. Chatham had in the past been sympathetic toward American interests. Once in office, however, he was at times so incapacitated by mental illness that leadership of his administration fell to the chancellor of the exchequer, Charles Townshend.

With the Stamp Act repealed, the greatest remaining American grievance involved the Mutiny (or Quartering) Act of 1765, which required colonists to shelter and supply British troops. Many colonists objected not so much to the actual burden as to its coercive character. The Massachusetts and New York assemblies went so far as to refuse to grant the mandated supplies to the troops.

Townshend responded in 1767 by disbanding the New York Assembly until the colonists agreed to obey the Mutiny Act. By singling out New York, he believed, he would avoid antagonizing all the colonies at once. He also imposed new taxes, known as the **Townshend Duties**, on various goods imported to the colonies from Great Britain—lead, paint, paper, and tea. Townshend assumed that since these were taxes purely on "external" transactions (imports from overseas), as opposed to the internal transactions the Stamp Act had taxed, the colonists would not object. But all the colonies resented the suspension of the New York Assembly, believing it to be a threat to every colonial government. And all the colonies rejected Townshend's careful distinction between external and internal taxation.

Townshend also established a board of customs commissioners in America. The new commissioners established their headquarters in Boston. They virtually ended smuggling in Boston, although smugglers continued to carry on a busy trade in other colonial seaports. The Boston merchants, angry that the new commission was diverting the lucrative smuggling trade elsewhere, helped organize a boycott of British goods that were subject to the Townshend Duties. Merchants in Philadelphia and New York joined them in a nonimportation agreement in 1768, and later some southern merchants and planters also agreed to cooperate. Throughout the colonies, American homespun and other domestic products became suddenly fashionable.

Late in 1767, Charles Townshend died. In March 1770, the new prime minister, Lord North, hoping to end the American boycott, repealed all the Townshend Duties except the tea tax.

The Boston Massacre

Before news of the repeal reached America, an event in Massachusetts inflamed colonial opinion. The harassment of the new customs commissioners in Boston had grown so intense that the British government had placed four regiments of regular troops in the city. Many of the poorly paid British soldiers looked for jobs in their off-duty hours and thus competed with local workers. Clashes between the two groups were frequent.

On the night of March 5, 1770, a mob of dockworkers, "liberty boys," and others began pelting the sentries at the customs house with rocks and snowballs. Hastily, Captain Thomas Preston of the British regiment lined up several of his men in front of the building to protect it. There was some scuffling, one of the soldiers was knocked down, and in the midst of it all, apparently, several British soldiers fired into the crowd, killing five people.

This murky incident, almost certainly the result of panic and confusion, was quickly transformed by local resistance leaders into the "**Boston Massacre**." It became the subject of such lurid (and inaccurate) accounts as the widely circulated pamphlet *Innocent Blood Crying to God from the Streets of Boston*. A famous engraving by Paul Revere portrayed the massacre as a calculated assault on a peaceful crowd. The British soldiers, tried before a jury of Bostonians and defended by future American president John Adams, were found guilty only of manslaughter and given token punishment. But colonial pamphlets and newspapers convinced many dissidents that the soldiers were guilty of murder.

The leading figure in fomenting public outrage over the Boston Massacre was the colonial official and political philosopher Samuel Adams, second cousin to John. Britain, he argued, had become a morass of sin and corruption; only in America did public virtue survive. In 1772, he proposed the creation of "**committees of correspondence**" in Boston to publicize grievances against Britain. Other colonies followed Massachusetts's lead, and a loose intercolonial network of political organizations was soon established that kept the spirit of dissent alive through the 1770s.

The Philosophy of Revolt

Although a superficial calm settled on the colonies after the Boston Massacre, the crises of the 1760s and early 1770s had helped arouse enduring challenges to British authority and produced powerful instruments for circulating colonial complaints. Yet revolutionary impulses rarely came down to simple arguments for democracy over monarchy, and they rarely pulled in a single direction. Some dissidents rejected or sought to modify British traditions of governance, others lobbied simply for Britain to leave the colonies alone, still others to break away from the monarchy altogether. Gradually these diverse voices would merge to form a political outlook in America that would serve to justify revolt, if not exactly a radical one.

British political philosophy, passed down and amended over time in the unwritten English constitution, called for distributing power among the three elements of society—the

(Barney Burstein/Corbis Historical/Getty Images)

THE BOSTON MASSACRE (1770), BY PAUL REVERE This sensationalized engraving of the conflict between British troops and Boston laborers is one of many important propaganda documents, by Revere and others, for the revolutionary cause in the 1770s. Among the victims of the massacre listed by Revere was Crispus Attucks, probably the first person of color to die in the struggle for American independence.

monarchy, the aristocracy, and the people—in order to prevent the exercise of unchecked authority. In the late seventeenth and early eighteenth centuries, colonial assemblies in North America resembled the elected and increasingly influential House of Commons in the British Parliament. Neither were particularly democratic, nor did leaders on either side of the Atlantic intend them to be. Though some colonial regions granted voting rights more broadly, most parts of British America and Britain proper endowed the vote rather sparsely to property holders. Through elections for the assemblies, those voters transferred authority over their lives to their representatives, who often went on to govern as they saw fit without much additional consultation from the public. By this political ideology, only independent, landowning men should vote (dependent men could be manipulated by those they depended upon), and only the best minds should hold public office. Two future presidents believed in this model: George Washington in his youth feared the ignorance of the "grazing multitude," John Adams the "common herd of mankind." Some American revolutionaries continued to harbor such suspicions of democracy and the masses up to 1776 and well beyond.

Some colonists opposed the privileges of hereditary aristocracy and the powers of the monarchy. But their galvanizing grievance by the 1770s, as the controversies over duties and quartering demonstrated, concerned matters of representation and **sovereignty**, or the authority to govern. Some colonists objected less to *how* they were governed than to *who* was governing them, or put another way, rejected British *practices* of governance more than their *principles*. This sort of frustration materialized in the slogan, "No taxation without representation." Whatever the nature of a tax, they said, it could not be levied without the consent of the colonists themselves. There were actually some supporters of colonial rights in Britain making such arguments on behalf of the colonists.

But to many other British observers and authorities, this clamor about representation made little sense. According to their constitutional theory, members of Parliament did not represent individuals or particular geographical areas. Instead, each member represented the interests of the whole nation and indeed the whole empire. The many boroughs of Britain that had no representative in Parliament, the whole of Ireland, and the colonies thousands of miles away—all were thus represented in Parliament at London, even though they elected no representatives of their own. This was the theory of "**virtual representation**." Americans, in fact, practiced the very same thing *within* their colonies, whereby assemblies did not reflect universal suffrage yet still claimed to represent their communities.

But for many colonists, the difference was the literal and figurative distance separating themselves from the men supposedly protecting their interests in Parliament. How could officials impose policies and taxes, said a religious leader in Georgia, on those "who never invested them with any such power"? Such thinkers may have believed in virtual representation across miles, but not continents and oceans. Soon enough, the king who shared power with Parliament similarly lost legitimacy. Indeed, this tradition, too, the colonists borrowed from the British, whose outburst of anti-monarchical dissent had separated King Charles I from his head over a century earlier.

SITES OF RESISTANCE

Colonists kept the growing spirit of resistance alive in many ways, but most of all through writing and talking. Dissenting leaflets, pamphlets, and books circulated widely through the colonies. In towns and cities, people gathered in churches, schools, town squares, and, above all, taverns to discuss politics.

Taverns were also places where resistance pamphlets and leaflets could be distributed and where meetings for the planning of protests and demonstrations could be held. Massachusetts had the most elaborately developed tavern culture, which was perhaps one reason why the spirit of resistance grew more quickly there than anywhere else. (See "Patterns of Popular Culture: Taverns in Revolutionary Massachusetts.")

America in the 1760s and early 1770s featured growing resentment about the continued enforcement of the Navigation Acts. Popular anger was visible in occasional acts of rebellion. At one point, colonists seized a British revenue ship on the lower Delaware River. In 1772, angry residents of Rhode Island boarded the British schooner *Gaspée,* set it afire, and sank it.

THE TEA EXCITEMENT

The revolutionary fervor of the 1760s intensified as a result of a new act of Parliament, one that involved the business of selling tea. In 1773, Britain's East India Company (on

PATTERNS OF POPULAR CULTURE

Taverns in Revolutionary Massachusetts

In colonial Massachusetts, as in many other American colonies in the 1760s and 1770s, taverns (or "public houses," as they were often known) were crucial to the development of popular resistance to British rule. The Puritan culture of New England created some resistance to taverns, and reformers tried to regulate or close them to reduce "public drunkenness," "lewd behavior," and "anarchy." But as the commercial life of the colonies expanded and more people began living in towns and cities, taverns became a central institution in American social life—and eventually in its political life as well.

Taverns were appealing, of course, because they provided alcoholic drinks in a culture where the craving for alcohol and the extent of drunkenness were very high. But taverns had other attractions as well. They were one of the few places where people could meet and talk openly in public; indeed, many colonists considered the life of the tavern as the only vaguely democratic experience available to them. The tavern was a mostly male institution, just as political life was considered a mostly male concern. Male camaraderie and political discourse fused together out of tavern culture.

As the revolutionary crisis deepened, taverns and pubs became central meeting places for cultivating resistance to British policies. Educated and uneducated men alike joined in animated discussions of events. The many who could not read learned about the contents of revolutionary pamphlets from listening to tavern conversations. They could join in the discussion of the new republican ideas emerging in America by participating in tavern celebrations of, for example, the anniversaries of resistance to the Stamp Act. Those anniversaries inspired elaborate toasts in public houses throughout the colonies.

In an age before wide distribution of newspapers, taverns and tavernkeepers were important sources of information about the political and social turmoil of the time. Taverns were also the settings for political events. In 1770, for example, a report circulated through the taverns of Danvers, Massachusetts, about a local man who was continuing to sell tea despite the colonial boycott. The Sons of Liberty brought the seller to the Bell Tavern and persuaded him to sign a confession and apology before a crowd of defiant men in the public room.

Almost all politicians who wanted any real contact with the public found it necessary to visit taverns in colonial Massachusetts. Samuel Adams spent considerable time in

(Library of Congress, Prints and Photographs Division [LC-USZC2-1367])

TAVERNS AND POLITICS The London Coffee House and other taverns were centers for pre-Revolutionary social and political life in colonial Philadelphia.

the public houses of Boston, where he sought to encourage resistance to British rule while taking care to drink moderately so as not to erode his stature as a leader. His cousin John Adams, although somewhat more skeptical of taverns and more sensitive to the vices they encouraged, also recognized their political value. In taverns, he once said, "bastards and legislatores are frequently begotten." •

UNDERSTAND, ANALYZE, & EVALUATE

1. Why were taverns so important in educating colonists about the relationship with Britain?
2. What gathering places today serve the same purposes as taverns did in colonial America?

---

the verge of bankruptcy) was sitting on large stocks of tea that it could not sell in Britain. In an effort to save the company, the government passed the **Tea Act** of 1773, which gave the company the right to export its merchandise directly to the colonies without paying any of the regular taxes that were imposed on colonial importers. The law provided no new tax on tea, but the original Townshend duty on the commodity survived, and the East India Company was now exempt from paying it. That meant cheaper tea for consumers, which Lord North had assumed would make the law welcome among the colonists.

But resistance leaders in America argued that the law, in effect, imposed an unfair tax on American merchants, who would be undersold by the East India Company and become disadvantaged in the colonial tea trade. The colonists responded by boycotting tea. Unlike earlier protests, most of which had involved relatively small numbers of people, the tea boycott mobilized large segments of the population. It also helped link the colonies together in a common experience of mass popular protest. Particularly important to the movement were the activities of colonial women, who led the boycott. The **Daughters of Liberty**—a recently formed women's organization—proclaimed, "rather than Freedom, we'll part with our Tea."

In the last weeks of 1773, with strong popular support, some colonial leaders made plans to prevent the East India Company from landing its cargoes. In Philadelphia and New York, determined colonists kept the tea from leaving the company's ships, and in Charles Town, South Carolina, they stored it away in a public warehouse. In Boston, local dissenters staged a spectacular drama. On the evening of December 16, 1773, three companies of fifty men each, masquerading as Mohawk, went aboard three ships, broke open the tea chests, and heaved them into the harbor. As the electrifying news of the **Boston Tea Party** spread, colonists in other seaports staged similar acts of resistance.

Parliament retaliated in four acts of 1774: closing the port of Boston, drastically reducing the powers of self-government in Massachusetts, permitting royal officers in America to be tried for crimes in other colonies or in Great Britain, and providing for the quartering of troops by the colonists. These **Coercive Acts** were more widely known in America as the "Intolerable Acts."

The Coercive Acts backfired. Far from isolating Massachusetts, they made the colony a martyr in the eyes of other colonies and sparked new resistance up and down the coast. Colonial legislatures passed a series of resolves supporting Massachusetts. Women's groups mobilized to extend the boycotts of British goods and to create substitutes for the tea, textiles, and other commodities they were shunning. In Edenton, North Carolina, fifty-one women signed an agreement in October 1774 declaring their "sincere adherence" to the anti-British resolutions of their provincial assembly and proclaiming their duty to do "every thing as far as lies in our power" to support the "publick good."

COOPERATION AND WAR

Beginning in 1765, colonial leaders developed a variety of organizations for converting popular discontent into action—organizations that in time formed the basis for an independent government.

New Sources of Authority

The passage of authority from the royal government to the colonists themselves began on the local level. In colony after colony, local institutions responded to the resistance movement by simply seizing authority. At times, entirely new institutions emerged.

The most effective of these new groups were the committees of correspondence. Massachusetts and Virginia and other colonies established these committees to foster continuous cooperation among them. After the royal governor dissolved the Virginia assembly in 1774, colonists met in the Raleigh Tavern at Williamsburg, declared that the Intolerable Acts menaced the liberties of every colony, and issued a call for a Continental Congress.

Delegates from all the colonies except Georgia were present when, in September 1774, the **First Continental Congress** convened in Philadelphia. They made five major decisions. First, they rejected a plan for a colonial union under British authority. Second, they endorsed a relatively moderate statement of grievances, which addressed the king as "Most Gracious Sovereign," but which also included a demand for the repeal of all oppressive legislation passed since 1763. Third, they approved a series of resolutions recommending that military preparations be made for defense against possible attack by British troops in Boston. Fourth, they agreed to a series of boycotts they hoped would stop all trade with Great Britain, and they formed a "Continental Association" to see that these agreements were enforced. Fifth, the delegates agreed to meet again the following spring.

During the winter, the Parliament in London debated proposals for placating the colonists, and early in 1775 Lord North finally won approval for a series of measures known as the Conciliatory Propositions. Parliament proposed that the colonies tax themselves at Parliament's demand. With this offer, Lord North hoped to separate the American moderates, whom he believed represented the views of the majority, from the extremist minority. But his offer was too little and too late. It did not reach America until after the first shots of war had been fired.

Lexington and Concord

For months, the farmers and townspeople of Massachusetts had been gathering arms and ammunition and preparing "minutemen" to fight on a moment's notice. The Continental Congress had approved preparations for a defensive war, and the citizen-soldiers waited only for an aggressive move by the British regulars in Boston.

There, General Thomas Gage, commanding the British garrison, considered his army too small to do anything without reinforcements. He resisted the advice of less cautious officers, who assured him that the Americans would back down quickly before any show of British force. When General Gage received orders to arrest the rebel leaders Sam Adams and the wealthy merchant John Hancock, known to be in the vicinity of Lexington, he still hesitated. But when he heard that the minutemen had stored a large supply of gunpowder in Concord (eighteen miles from Boston), he decided to act. On the night of April 18, 1775, he sent a detachment of about 1,000 men out toward Lexington, hoping to surprise the colonials and seize the illegal supplies without bloodshed.

But dissenters in Boston were watching the British movements closely, and during the night two horsemen, William Dawes and Paul Revere, rode out to warn the villages and farms. When the redcoats arrived in Lexington the next day, several dozen minutemen awaited them on the town common. Shots were fired and minutemen fell; eight were killed and ten wounded. Advancing to Concord, the British discovered that the Americans had hastily removed most of the powder supply. All along the road back to Boston, the British were harassed by the gunfire of farmers hiding behind trees, rocks, and stone walls. By the end of the day, the British had lost almost three times as many men as the Americans.

The first shot—the "shot heard 'round the world," as the poet Ralph Waldo Emerson later called it—had been fired. But who had fired first? According to one minuteman at Lexington, the British commander, Major Thomas Pitcairn, had shouted to the colonists on his arrival, "Disperse, ye rebels!" When they ignored him, he ordered his troops to fire. British officers and soldiers claimed that the minutemen had fired first. Whatever the

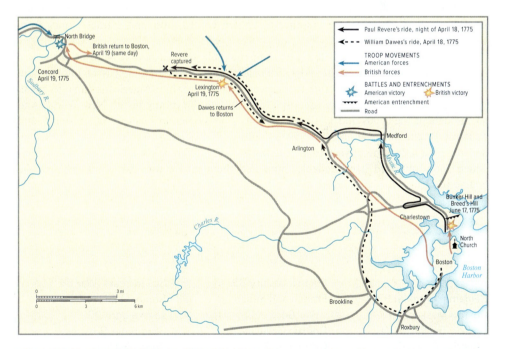

THE BATTLES OF LEXINGTON AND CONCORD, 1775 This map shows the fabled series of events that led to the first battle of the American Revolution. On the night of April 18, 1775, Paul Revere and William Dawes rode out from Boston to warn the outlying towns of the approach of British troops. Revere was captured just west of Lexington, but Dawes escaped and returned to Boston. The next morning, British forces moved out of Boston toward Lexington, where they met armed American minutemen on the Lexington common and exchanged fire. The British dispersed the Americans in Lexington. But they next moved on to Concord, where they encountered more armed minutemen, clashed again, and were driven back toward Boston. All along their line of march, they were harassed by riflemen. • *What impact did the Battles of Lexington and Concord have on colonial sentiment toward the British?*

truth, the rebels succeeded in circulating their account well ahead of the British version, adorning it with tales of British atrocities. The effect was to rally thousands of colonists to the rebel cause.

It was not immediately clear at the time that the skirmishes at Lexington and Concord were the first battles of a war. But whether people recognized it at the time or not, the American Revolution had begun.

CONCLUSION

When the Seven Years' War ended in 1763, it might have seemed reasonable to expect that relations between the British colonists in America and Great Britain itself would have been cemented more firmly than ever. But in fact, the resolution of that conflict altered the imperial relationship forever, in ways that ultimately drove Americans to rebel against British rule and begin a war for independence. To the British, the lesson of the French and Indian War was that the colonies in America needed firmer control from London. The empire was now much bigger, and it needed better administration. The war had produced great debts, and the Americans—among the principal beneficiaries of the war—should help pay them. And so for more than a decade after the end of the fighting, the British tried one strategy after another to tighten control over and extract money from the colonies.

To the colonists, this effort to tighten imperial rule seemed both a betrayal of the sacrifices they had made in the war and a challenge to their long-developing assumptions about the rights of British people to rule themselves. Gradually, white Americans came to see in the British policies evidence of a conspiracy to establish tyranny in the New World. And so throughout the 1760s and 1770s, the colonists developed an ideology of resistance and defiance. By the time the first shots were fired in the American Revolution in 1775, Britain and America had come to view each other as two very different societies. Their differences, which soon appeared irreconcilable, propelled them into a war that would change the course of history for both sides.

KEY TERMS/PEOPLE/PLACES/EVENTS

Albany Plan 84
Benjamin Franklin 93
Boston Massacre 97
Boston Tea Party 101
Coercive Acts 101
committees of correspondence 97
Creole 84
Daughters of Liberty 101
First Continental Congress 102

Fort Necessity 85
French and Indian War 84
George Grenville 90
George III 90
impressment 87
Iroquois Confederacy 84
Patrick Henry 93
Paxton Boys 91
Pontiac 89
Proclamation of 1763 90

Seven Years' War 84
Sons of Liberty 93
sovereignty 99
Stamp Act 93
Sugar Act 92
Tea Act 101
Townshend Duties 96
Virginia Resolves 93
virtual representation 99
William Pitt 87

RECALL AND REFLECT

1. Which Native Americans fought in the French and Indian War, and how did the war's outcome affect them? What about Native Americans who did not participate in the war?
2. How and why did the colonists' attitude toward Britain change from the time of the Seven Years' War to the beginning of the American Revolution?
3. What were the philosophical underpinnings of the colonists' revolt against Britain?
4. What did the slogan "No taxation without representation" mean, and why was it a rallying cry for the colonists?

5 | THE AMERICAN REVOLUTION

THE STATES UNITED
THE WAR FOR INDEPENDENCE
WAR AND SOCIETY
THE CREATION OF STATE GOVERNMENTS
THE SEARCH FOR A NATIONAL GOVERNMENT

LOOKING AHEAD

1. What were the military strategies (both British and American) of each of the three phases of the American Revolution? How successful were these strategies during each phase?
2. How did the American Revolution become an international conflict, not just a colonial war against the British?
3. How did the new national government of the United States reflect the principles of republicanism?

TWO STRUGGLES OCCURRED SIMULTANEOUSLY during the eight years of war that began in April 1775. The first was the military conflict with Great Britain. The second was a political conflict within America.

The military conflict, though modest compared to later wars, seemed to many participants an unusually brutal fight, pitting not only army against army but the civilian population against a powerful external force. The shift of the war from a traditional, conventional struggle to a new kind of conflict—a revolutionary war for liberation—is what made it possible for the United States to defeat the more powerful British.

At the same time, Americans were wrestling with the great political questions that the conflict necessarily produced: first, whether to demand independence from Britain; second, how to structure the new nation they had proclaimed; and third, how to deal with questions that the revolution had raised about slavery, the rights of Native Americans, the role of women, and the limits of religious tolerance in the new American society.

THE STATES UNITED

Although some Americans had long expected a military conflict with Britain, the actual beginning of hostilities in 1775 found the colonies generally unprepared for war against the world's greatest armed power.

Defining American War Aims

Three weeks after the Battles of Lexington and Concord, when the **Second Continental Congress** met in Philadelphia, delegates from every colony (except Georgia, which had not yet sent a representative) agreed to support the war. But they disagreed about its purpose. At one extreme was a group led by the Adams cousins (John and Samuel), Richard Henry Lee of Virginia, and others, who already favored independence. At the other extreme was a group led by such moderates as John Dickinson of Pennsylvania, who hoped for a quick reconciliation with Great Britain.

Most Americans believed at first that they were fighting not for independence but for a resolution of grievances against the British Empire. During the first year of fighting, however, many colonists began to change their minds. The costs of the war were so high that the original war aims began to seem too modest to justify them. Many colonists were enraged when the British began trying to recruit Native Americans, enslaved persons, and German mercenaries (the hated "**Hessians**"). Particularly galvanizing for slaveholders in the southern colonies was the royal governor of Virginia's announcement of late 1775 that enslaved people owned by rebels—*not* those enslaved by colonists loyal to the crown—could win freedom if they abandoned their owners and joined the British forces, a development known as **Lord Dunmore's Proclamation**. When the British government blockaded colonial ports and rejected all efforts at conciliation, many colonists concluded that independence was the only remaining option.

TIME LINE

1775
Second Continental Congress

Washington commands American forces

Lord Dunmore's Proclamation

1776
Paine's *Common Sense*

Declaration of Independence

Battle of Trenton

1777
Articles of Confederation adopted

British defeat at Saratoga

1778
French-American alliance

1781
Articles of Confederation ratified

Cornwallis surrenders at Yorktown

1783
Treaty of Paris

1784
Postwar depression begins

1786
Shays's Rebellion

1787
Northwest Ordinance

DEBATING THE PAST

THE AMERICAN REVOLUTION

Almost from the moment it ended, historians have debated the character, meaning, and origins of the American Revolution. For the first several generations after 1776, they developed what historians call a "whiggish" view of the rebellion. In this narrative, the colonists proceeded inexorably and with God's approval toward independence from Britain, scoring a preordained victory for Enlightenment ideals and progress, for liberty over tyranny. Later, in the decades before and during the Civil War, the perpetuation of this vision of the Revolution by George Bancroft and others served a contemporary need to emphasize American unity and greatness at a time of roiling division and sectional violence.

In the early twentieth century, historians downplayed the importance of ideology in the Revolution, attributing it, rather, to social and economic forces. Carl Becker, J. Franklin Jameson, Arthur M. Schlesinger, and other "progressive" historians, so named for the period in which they wrote, characterized the Revolution as a burst of radical, populist outrage against not just the monarchy or Parliament but against colonial elites with property, prestige, and power. Thus, said Becker (in 1909), there were really two revolutions, one against Britain, the other against colonial aristocrats, each animated by a different question: "The first was the question of home rule; the second was the question ... of who should rule at home."

Beginning in the 1950s, with the nation once again searching for unity in the Cold War era, a new generation of scholars began to reemphasize the role of consensus and ideology over class warfare and economic interests. Edmund S. Morgan (in 1956) argued that most eighteenth-century Americans shared common political principles and that the social and economic conflicts other historians had identified were not severe. Bernard Bailyn, in *The Ideological Origins of the American Revolution* (1967), found Revolutionary rhetoric rooted in deeply and widely held resentments against British imperial oppression. For Bailyn and for Pauline Maier in *From Resistance to Revolution* (1972), the independence movement still tipped radical but in its resentment of monarchical corruption and abuse of power rather than of elite property holders. Scholars of this period argued over what to name the ideologies that drove the rebellion, but they returned to the privileging of ideas over economic motivators.

By the late 1960s, for a new generation of historians, many influenced by the New Left, the pendulum was swinging back to class-based interpretations and to the internal struggles of the Revolutionary generation. Historians like Gary Nash in *The Urban Crucible* (1979) cited economic distress and the actions of mobs in colonial cities, the economic pressures on colonial merchants, and other changes in the character of American culture and society as critical prerequisites for the growth of the Revolutionary movement. Echoing Becker, Nash argued for two revolutions, one by common people against colonial elites and one by colonial elites bent on maintaining their status and power in a post-British environment.

According to many scholars up to the present day, colonial elites succeeded in reclaiming economic and political authority after the British left. These scholars point to the crafting of the Constitution, with its protections for the propertied white male elite and exclusion of everyone else from citizenship rights, as evidence for the victory of an essentially conservative

(MPI/Hulton Archive/Getty Images)

THE BRITISH SURRENDER This contemporary drawing depicts the formal surrender of British troops at Yorktown on October 19, 1781. Columns of American troops and a large French fleet flank the surrender ceremony, suggesting part of the reason for the British defeat. General Cornwallis, the commander of British forces in Virginia, did not himself attend the surrender. He sent a deputy in his place.

revolution. Along these same lines, work on women and people of color during the Revolution by Mary Beth Norton (*Liberty's Daughters*, 1980), Linda Kerber (*Women of the Republic*, 1980), and Sylvia R. Frey (*Water from the Rock*, 1991) turned attention to the role of those groups in the rebellion. They worked in auxiliary functions for the army or as wartime maintainers of home and industry, but also, in a deeper sense, as cultivators of civic virtue in the home, or "republican mothers." And they lamented that these contributions did not merit citizenship rights in the country's first constitutions.

The pendulum swung again with Gordon Wood, who argued in *The Radicalism of the American Revolution* (1992) for a *socially* radical revolution, whereby the Founders undermined time-worn social patterns of deference, patriarchy, and gender hierarchies. "Americans had become," Wood wrote, "almost overnight, the most liberal, the most democratic, the most commercially minded, and the most modern people in the world." The Revolution did not end slavery or the second-class status of women, he granted, but laid the foundations for those future transformations and should not have its radicalism undercut by those failures.

Others have found the picture Wood painted too rosy. They counter that a new democratic order was short-lived, or came much later, or was driven not by the framers but by societal actors left out of earlier histories of the Revolution. Edward Countryman (*A People in Revolution*, 1989), Woody Holton (*Forced Founders*, 1999; *Unruly Americans*, 2007), and T. H. Breen (*The Marketplace of Revolution*, 2004) argued for a radical Revolution that saw common people, for a time, shape the course of independence. These scholars see in the rebellion rhetorical groundwork for later change or the momentary ignition of possibilities for marginalized groups, but then a retrenchment of elite white rule.

Similarly, scholars of enslaved peoples, women, and Native Americans during the Revolution have tracked the contributions of these groups to the war as well as their appropriation of liberating rhetoric from Revolutionary political culture, but ultimately they have a bleak story to tell. Colin Calloway's *The American Revolution in Indian Country* (1995), Rosemarie Zagarri's *Revolutionary Backlash* (2007), Douglas Egerton's *Death or Liberty* (2009), Carroll Smith-Rosenberg's *This Violent Empire* (2012), and many others have located in the nation's birth foundational commitments to white supremacy, male dominance, and the destruction of indigenous peoples, despite various efforts by these groups to claim the Revolution's transformative potential for themselves. Such views found wide dissemination with the publication in 2019 of the 1619 Project in *The New York Times*, a collection of essays, teaching tools, and podcasts arguing for the centrality of racial oppression, and slavery in particular, to the nation's founding and history. For the contributing scholars and journalists, and much of the academic community, America's gradual (and still incomplete) inclusion of marginalized groups came in spite of, rather than because of, the intentions of the country's founders. •

UNDERSTAND, ANALYZE, & EVALUATE

1. In what way was the American Revolution an ideological struggle?
2. In what way was the American Revolution a social and economic conflict?
3. Was the Revolution a fundamentally liberating or constricting event for the nation's people?

Thomas Paine's impassioned pamphlet **Common Sense** crystallized these feelings in January 1776. Paine, who had emigrated from Britain less than two years before, sought to turn the anger of Americans toward parliamentary overreach as well as the British monarchy more broadly. It was simple common sense, Paine wrote, for Americans to break completely with a political system that could inflict such hardships on its own people. Written in plain terms and in English, rather than French or Latin like some treatises, *Common Sense* sold more than 100,000 copies in only a few months and helped build support for the idea of independence in the early months of 1776. (For more on the origins of the rebellion, see "Debating the Past: The American Revolution.")

THE DECLARATION OF INDEPENDENCE

In the meantime, the Continental Congress in Philadelphia was moving toward a complete break with Britain. At the beginning of the summer, it appointed a committee to draft a formal declaration of independence, and on July 2, 1776, it adopted a resolution: "That these United Colonies are, and, of right, ought to be, free and independent states; that they are absolved from all allegiance to the British crown, and that all political connexion between them and the state of Great Britain is, and ought to be, totally dissolved." Two days later, on July 4, Congress approved the **Declaration of Independence** itself, which provided formal justifications for this resolution.

The Declaration launched a period of energetic political innovation, as one colony after another reconstituted itself as a "state." By 1781, most states had produced written constitutions for themselves. At the national level, however, the process was more uncertain. In November 1777, finally, Congress adopted a plan for union, the **Articles of Confederation**. The document confirmed the existing system, weak and decentralized as it was.

Thomas Jefferson, a thirty-three-year-old Virginia lawyer and former member of the state House of Burgesses, wrote most of the Declaration. He had help from the Pennsylvania political theorist, inventor, and scientist Benjamin Franklin, and John Adams, a lawyer and Massachusetts delegate to the Second Continental Congress. The Declaration expressed concepts that had been circulating throughout the colonies over the previous few months in the form of at least ninety other, local "declarations of independence"—statements drafted up and down the coast by town meetings, artisan and militia organizations, county officials, grand juries, Sons of Liberty, and colonial assemblies. Jefferson borrowed heavily from these texts.

The final document had two parts. In the first, the Declaration restated the familiar contract theory of John Locke: that governments were formed to protect what Jefferson called "life, liberty and the pursuit of happiness." Although these lines have become the most famous part of the document, at the time, the second part loomed larger in the minds of the rebels. It listed the alleged crimes of the king, who, with the backing of Parliament, had violated his contract with the colonists and thus had forfeited all claim to their loyalty.

MOBILIZING FOR WAR

Financing the war was difficult. Congress had no authority to levy taxes on its own, and when it requisitioned money from the state governments, none contributed more than a small part of its expected share. Congress had little success borrowing from the public, since few Americans could afford to buy bonds. Instead, Congress issued paper money. Printing presses turned out enormous amounts of "Continental currency," and the states printed currencies of their own. The result, predictably, was soaring inflation, and Congress soon found the Continental currency was virtually worthless. Ultimately, Congress financed the war mostly by borrowing from other nations.

After a surge of revolutionary spirit in 1775, volunteer soldiers became scarce. States had to pay bounties or use a draft to recruit the needed men. At first, the militiamen remained under the control of their respective states. But Congress recognized the need for a centralized military command, and it created a Continental army with a single commander in chief: **George Washington**. An early advocate of independence with considerable military experience, Washington was admired, respected, and trusted by nearly all **American Patriots**, as supporters of independence came to be known. He took command of the new army in June 1775. With the aid of foreign military experts such as the Marquis de Lafayette from France and the Baron von Steuben from Prussia, he built a formidable force.

Though Britain successfully lured many enslaved people to join its side in exchange for freedom (including some owned by George Washington), several thousand Black men fought alongside the colonists, particularly in New England and other parts of the North. Almost all southern states refused to allow the enslaved to serve, even when Congress offered them money in exchange.

(©Anne S.K. Brown Military Collection, Brown University Library)

REVOLUTIONARY SOLDIERS Jean Baptist de Verger, a French officer serving in America during the Revolution, kept an illustrated journal of his experiences. Here he portrays four American soldiers carrying different kinds of arms: a Black infantryman with a light rifle, a musketman, a rifleman, and an artilleryman.

THE WAR FOR INDEPENDENCE

As the War for Independence began, the British seemed to have overwhelming advantages: the greatest navy and the best-equipped army in the world, the resources of an empire, a coherent structure of command. Yet the United States had advantages, too. Beginning in 1777, Americans received substantial aid from abroad. They were fighting on their own ground. They were more committed to the conflict than the British, who made a series of early miscalculations. The transformation of the war—through three regions—made it a new kind of conflict that the imperial military, for all its strength, was unable to win.

NEW ENGLAND

For the first year of the conflict—from the spring of 1775 to the spring of 1776—many British authorities thought their forces were not fighting a real war, but simply quelling pockets of rebellion in the contentious area around Boston. After the redcoats withdrew from Lexington and Concord in April, American forces besieged them in Boston. In the Battle of Bunker Hill (actually fought on Breed's Hill) on June 17, 1775, the Patriots suffered severe casualties and withdrew. But they inflicted even greater losses on the enemy. The siege continued. Early in 1776, finally, the British decided that Boston was a poor place from which to fight. It was in the center of the most anti-British part of America and tactically difficult to defend because it was easily isolated and surrounded. And so, on March 17, 1776, the redcoats evacuated Boston for Halifax, Nova Scotia, with hundreds of **Loyalist**, or **Tory**, refugees (Americans still loyal to Britain and the king).

In the meantime, the Americans began an invasion of Canada. After General **Benedict Arnold** and Ethan Allen seized Fort Ticonderoga in upstate New York in May 1775, Arnold

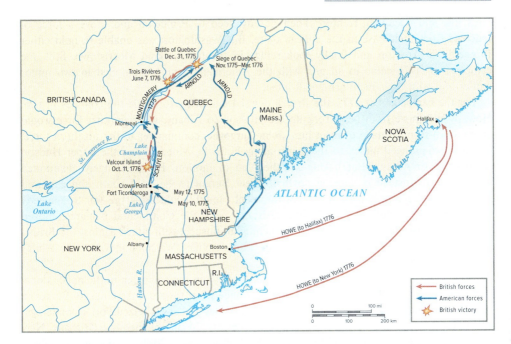

THE REVOLUTION IN THE NORTH, 1775–1776 After initial battles in and around Boston, the British forces left Massachusetts and (after a brief stay in Halifax, Canada) moved south to New York. In the meantime, American forces moved north in an effort to capture British strongholds in Montreal and Quebec, with little success. • *Why might the British have considered New York a better base than Boston?*

and General Richard Montgomery unsuccessfully threatened Quebec in late 1775 and early 1776, in a battle in which Montgomery was killed and Arnold was wounded.

By the spring of 1776, it had become clear to the British that the conflict was not just a local phenomenon. The American campaigns in Canada, along with new agitation in the mid-Atlantic colonies and the South and growing evidence of colonial unity, all suggested that Great Britain must prepare to fight a much larger conflict.

The Mid-Atlantic

During the next phase of the war, which lasted from 1776 until early 1778, the British were in a good position to win. Indeed, only a series of errors and misfortunes prevented them from crushing the rebellion.

The British regrouped quickly after their retreat from Boston. During the summer of 1776, hundreds of British ships and 32,000 British soldiers arrived in New York, under the command of General **William Howe**. He offered Congress a choice: surrender with royal pardon or face a battle against apparently overwhelming odds. To oppose Howe's great force, Washington could muster only about 19,000 soldiers and had no navy at all. Even so, the Americans rejected Howe's offer. The British then pushed the Patriot forces out of Manhattan and off Long Island and drove them in slow retreat over the plains of New Jersey, across the Delaware River, and into Pennsylvania.

The British settled down for the winter in northern and central New Jersey, with an outpost of Hessians at Trenton, on the Delaware River. But Washington did not sit still. On Christmas night 1776, he recrossed the icy Delaware River, surprised and scattered the

Hessians, and occupied Trenton. Then he advanced to Princeton and drove a force of redcoats from their base in the college there. But Washington was unable to hold either Princeton or Trenton and finally took refuge in the hills around Morristown. Still, the campaign of 1776 came to an end with the Americans having triumphed in two minor battles and with their main army still intact.

For the campaigns of 1777, the British devised a strategy to divide the colonies in two. Howe would move from New York up the Hudson to Albany, while another force would come down from Canada to meet him. **John Burgoyne**, commander of the northern force, began a two-pronged attack to the south along both the Mohawk and the upper Hudson approaches to Albany. But having set the plan in motion, Howe strangely abandoned his part of it. Instead of moving north to meet Burgoyne, he went south and captured Philadelphia, hoping that his seizure of the rebel capital would bring the war to a speedy conclusion. Philadelphia fell with little resistance, and the Continental Congress moved into exile in York, Pennsylvania. After launching an unsuccessful attack against the British on October 4 at Germantown (just outside Philadelphia), Washington went into winter quarters at Valley Forge.

Howe's move to Philadelphia left Burgoyne to carry out the campaign in the north alone. He sent Colonel Barry St. Leger up the St. Lawrence River toward Lake Ontario. Burgoyne himself advanced directly down the upper Hudson Valley and easily seized Fort Ticonderoga. But Burgoyne soon experienced two staggering defeats. In one of them—at Oriskany, New York, on August 6—Patriots and the Oneida held off a force of Mohawk, Seneca, Loyalists, and British (all three indigenous nations had once been allied in the Iroquois Confederacy). That development allowed Benedict Arnold to close off the Mohawk Valley to St. Leger's advance. In the other battle—at Bennington, Vermont, on August 16—New England militiamen mauled a detachment that Burgoyne had sent to seek supplies. Short of materials, with all help cut off, Burgoyne fought several costly engagements and then withdrew to **Saratoga**, where General Horatio Gates surrounded him. On October 17, 1777, Burgoyne surrendered.

The campaign in upstate New York was not just a British defeat. It signaled the splintering of the Iroquois Confederacy, which had declared its neutrality in 1776 but now saw members ally themselves with both sides. Among those joining the British were a Mohawk brother and sister, **Joseph and Mary Brant**. This ill-fated alliance further divided the already weakened Iroquois Confederacy, because only three of the Iroquois nations (the Mohawk, Seneca, and Cayuga) followed the Brants in support of the British. A year after the defeat at Oriskany, Iroquois forces joined British troops in a series of raids on white settlements in upstate New York.

In late 1779, Patriots under the command of General John Sullivan harshly retaliated, burning homes and towns, destroying crops, and leaving the land uninhabitable. The harsh winter that followed saw mass Iroquois starvation and disease. The Patriots wreaked such destruction on native settlements that large groups of Iroquois allied with the British fled north into Canada to seek refuge. Many never returned.

Meanwhile the fighting in the North settled into a stalemate. Sir Henry Clinton replaced the unsuccessful William Howe in May 1778 and moved what had been Howe's army from Philadelphia back to New York. The British troops stayed there for more than a year. In the meantime, George Rogers Clark led a Patriot expedition over the Appalachian Mountains and captured settlements in the Illinois country from the British and their indigenous allies. On the whole, however, there was relatively little military activity in the North after 1778. There was, however, considerable intrigue. In the fall of 1780, American forces were shocked by the exposure of treason on the part of General Benedict Arnold. Convinced

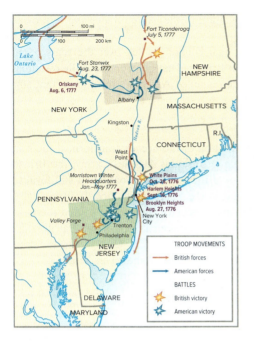
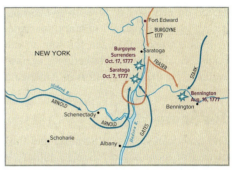

THE REVOLUTION IN THE MIDDLE COLONIES, 1776–1778 These maps illustrate the major campaigns of the Revolution in the middle colonies—New York, New Jersey, and Pennsylvania—between 1776 and 1778. The large map on the left shows the two prongs of the British strategy: first, a movement of British forces south from Canada into the Hudson Valley and, second, a movement of other British forces, under General William Howe, out from New York. The strategy was designed to trap the American army between the two British movements. • *What movements of Howe helped thwart that plan?* The two smaller maps on the right show a detailed picture of some of the major battles. The upper one pictures the surprising American victory at Saratoga. The lower one shows a series of inconclusive battles between New York and Philadelphia in 1777 and 1778.

that the American cause was hopeless, Arnold conspired with British agents to betray the Patriot stronghold at West Point on the Hudson River. When the scheme was exposed and foiled, Arnold fled to the safety of the British camp, where he spent the rest of the war.

SECURING AID FROM ABROAD

The leaders of the American effort knew that victory would not be likely without aid from abroad. Their most promising allies, they realized, were the French, who stood to gain from seeing Britain lose a crucial part of its empire. At first, France provided the United States with badly needed supplies. But France remained reluctant to formally acknowledge the new nation, despite the efforts of Benjamin Franklin in Paris to lobby for aid and diplomatic recognition. France's foreign minister, the Count de Vergennes, wanted evidence that the Americans had a real chance of winning. The British defeat at Saratoga, he believed, offered that evidence.

When the news from Saratoga arrived in London and Paris in early December 1777, a shaken Lord North made a new peace offer: complete home rule within the empire for Americans if they would quit the war. Vergennes feared the Americans might accept the offer and thus destroy France's opportunity to weaken Britain. Encouraged by Franklin, he agreed on February 6, 1778, to give formal recognition to the United States and to provide it with greatly expanded military assistance.

France's decision made the war an international conflict, which over the years pitted France, Spain, and the Netherlands against Great Britain. That helped reduce the resources available for the British effort in America. France remained America's most important ally.

THE SOUTH

The American victory at Saratoga and the intervention of the French transformed the war. Instead of mounting a full-scale military struggle against the American army, the British now tried to enlist the support of those elements of the American population still loyal to the crown. Since Loyalist sentiment was strongest in the South, and since the British also enticed the enslaved to rally to their cause, the main focus of their effort shifted there. In the Carolinas in particular, something like a civil war between Loyalists and Patriots had already been brewing.

Late in 1775, colonials fought the British army at Kemp's Landing in Virginia. Three months later, on February 27, 1776, a band of southern Patriots crushed an uprising of Loyalists at Moore's Creek Bridge in North Carolina. The British tried and failed to take Charleston, South Carolina, in June 1776, but were more successful at a different port, Savannah, Georgia, on December 29, 1778. They captured Savannah despite the opposition of a contingent of Patriots, Frenchmen, and Black soldiers from Saint-Domingue. British forces spent three years (from 1778 to 1781) moving through the South, ultimately taking Charleston in May 1780, and then advancing into the interior.

But the British overestimated the extent of Loyalist sentiment, and underestimated the logistical problems they would face. Patriot forces could move at will throughout the region, blending in with the civilian population. Although the southern Continental army had been weakened by the loss of Charleston, the British faced constant harassment from such Patriots as Thomas Sumter, Andrew Pickens, and Francis Marion, the "Swamp Fox." British attempts to destabilize the South by offering refuge to enslaved people drew tens of thousands of escapees, but also had the effect of galvanizing white southern dedication to the Revolution. With neighbors fighting neighbors in the Carolina backcountry, both regular armies sent commanders to direct the action. **Lord Cornwallis** was named by Clinton to head British forces in the South, Horatio Gates of Saratoga fame to lead the Patriots. Penetrating to Camden, South Carolina, Cornwallis met and crushed a Patriot force under Gates on August 16, 1780. That October, at King's Mountain (near the North Carolina–South Carolina border), a band of Patriot riflemen from the backwoods killed, wounded, or captured every man in a force of 1,100 New York and South Carolina Loyalists upon whom Cornwallis had depended.

Congress recalled Gates, and Washington replaced him with Nathanael Greene, one of the ablest American generals of his time. Once Greene arrived, he confused and exasperated Cornwallis by dividing the American forces into fast-moving contingents while avoiding open, conventional battles. One of the contingents inflicted what Cornwallis admitted was "a very unexpected and severe blow" at Cowpens, near the border between the Carolinas, on January 17, 1781. Finally, after receiving reinforcements, Greene combined all his forces and maneuvered to meet the British at Guilford Court House, North Carolina. After a hard-fought battle there on March 15, 1781, Greene was driven from the field. Cornwallis had lost so many men that he decided to abandon the Carolina campaign. Instead, he moved north, hoping to conduct raids in the interior

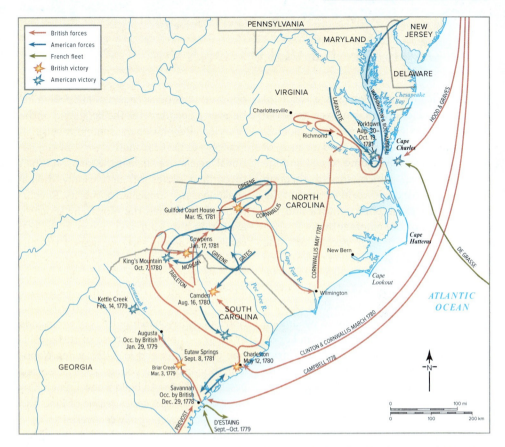

THE REVOLUTION IN THE SOUTH, 1778–1781 The final phase of the American Revolution occurred largely in the South, which the British thought would be a more receptive region for their troops. • *Why did they believe that?* This map reveals the many scattered military efforts of the British and the Americans in those years, none of them conclusive. It also shows the final chapter of the Revolution around Chesapeake Bay and the James River.

of Virginia. Clinton, fearful that the southern army might be destroyed, ordered him to take up a defensive position at **Yorktown**.

American and French forces quickly descended on Yorktown along with the battle-hardened, all-Black First Rhode Island regiment. Washington and the Count de Rochambeau marched a French-American army from New York to join the Marquis de Lafayette in Virginia, while Admiral de Grasse took a French fleet with additional troops up Chesapeake Bay to the York River. These joint operations caught Cornwallis between land and sea. After a few shows of resistance, he gave up on October 17, 1781, the First Rhode Island having taken part in a key assault on Redoubt 10 a few nights earlier. Two days later, as a military band played "The World Turned Upside Down," Cornwallis surrendered his whole army of more than 7,000.

WINNING THE PEACE

Cornwallis's defeat provoked outcries in Britain against continuing the war. Lord North resigned as prime minister, Lord Shelburne emerged from the political wreckage to succeed

AMERICA IN THE WORLD

THE AGE OF REVOLUTIONS

The American Revolution was a result of tensions and conflicts between imperial Britain and its North American colonies. But it was also both a part, and a cause, of what historians have come to call an "age of revolutions," which spread through much of the Western world in the last decades of the eighteenth century and the first decades of the nineteenth.

The modern idea of revolution—the overturning of old systems and regimes and the creation of new ones—was to a large degree a product of the ideas of the Enlightenment. Among those ideas was the notion of popular sovereignty, articulated by, among others, the English philosopher John Locke. Locke argued that political authority did not derive from the divine right of kings or the inherited authority of aristocracies, but from the consent of the governed. A related Enlightenment idea was the concept of individual freedom, which challenged the traditional belief that governments had the right to prescribe the way people act, speak, and even think. Champions of individual freedom in the eighteenth century—among them the French philosopher Voltaire—advocated religious toleration and freedom of thought and expression. The Enlightenment also helped spread the idea of political and legal equality for all people—the end of special privileges for aristocrats and elites and the right of all citizens to participate in the formation of policies and laws. Jean-Jacques Rousseau, a Swiss-French theorist, helped define these new ideas of equality. Together, Enlightenment ideas formed the basis for challenges to existing social orders in many parts of the Western world, and eventually beyond it.

The American Revolution was the first of the Enlightenment-derived uprisings against established orders. It served as an inspiration to people in other lands who opposed unpopular regimes. In 1789, a little over a decade after the beginning of the American Revolution, dissenters rebelled in France—at first through a revolt by the national legislature against the king, and then through a series of increasingly radical challenges to established authority. Revolutionaries abolished the monarchy (publicly executing the king and queen in 1793), challenged and greatly weakened the authority of the Catholic Church, and at the peak of revolutionary chaos during the Jacobin period (1793–1794), killed over 40,000 suspected enemies of the revolution and imprisoned hundreds of thousands of others. The radical phase of the revolution came to an end in 1799, when Napoleon Bonaparte, a young general, seized power and began to build a new French Empire. But France's *ancien regime* of king and aristocracy never wholly recovered.

Together, the French and American Revolutions helped inspire many other Atlantic World uprisings, which in many ways sought to expand Enlightenment promises of liberty and equality to *all* people, not just property-holding whites. In 1791, a major slave revolt began in Saint-Domingue, or Haiti, the greatest sugar-producing colony in the world, and soon attracted over 100,000 rebels. The army of enslaved people defeated both the white settlers of the island and the French colonial armies sent to quell their rebellion, then British and Spanish forces attempting to claim the lucrative colony for themselves. Under the leadership of Toussaint-Louverture, they began to agitate for independence, which they obtained on January 1, 1804, a few months after Toussaint's death in a French prison.

The ideas of these revolutions spread next into Spanish and Portuguese colonies in the Americas, particularly among the

STORMING THE BASTILLE This illustration portrays the storming of the great Parisian fortress and prison, the Bastille, on July 14, 1789. The Bastille was a despised symbol of royal tyranny to many of the French because of the arbitrarily arrested and imprisoned people who were sent there. The July assault was designed to release the prisoners, but in fact the revolutionaries found only seven people in the vast fortress. Even so, the capture of the Bastille—which marked one of the first moments in which ordinary Frenchmen joined the Revolution—became one of the great moments in modern French history. The anniversary of the event, "Bastille Day," remains the French national holiday.

so-called Creoles, people of European ancestry born in the Americas. In the late eighteenth century, they began to resist the continuing authority of colonial officials from Spain and Portugal and to demand a greater say in governing their own lands. When Napoleon's French armies invaded Spain and Portugal in 1807, they weakened the ability of the European regimes to sustain authority over their American colonies. In the years that followed, revolutions swept through much of Latin America. Mexico became an independent nation in 1821, and provinces of Central America that had once been part of Mexico (Guatemala, El Salvador, Honduras, Nicaragua, and Costa Rica) established their independence three years later. Simón Bolívar, modeling his efforts on those of George Washington, led a movement that helped inspire revolutionary campaigns in Venezuela, Ecuador, and Peru, all of which won their independence in the 1820s. At about the same time, Greek patriots, drawing from the examples of other revolutionary nations, launched a movement to win their independence from the Ottoman Empire, which finally succeeded in 1830.

The age of revolutions left many new, independent nations in its wake. It did not, however, succeed in establishing the ideals of popular sovereignty, individual freedom, and political equality in all the nations it affected. Slavery survived in the United States and in many areas of Latin America. New forms of aristocracy and even monarchy emerged in France, Mexico, Brazil, and elsewhere. Women—many of whom had hoped the revolutionary age would win them new rights—made few legal or political gains in this era. But the ideals that the revolutionary era introduced to the Western world continued to shape the histories of nations throughout the nineteenth century and beyond. •

UNDERSTAND, ANALYZE, & EVALUATE

1. How did the American Revolution influence the French Revolution, and how were other nations affected by it?
2. What was the significance of the revolution in Haiti?

him, and British emissaries appeared in France to talk informally with the American diplomats there: Benjamin Franklin, John Adams, and John Jay.

The Americans were under instructions to cooperate with France in their negotiations with Britain. But Vergennes insisted that France could not agree to any settlement with the British until its ally Spain had achieved its principal war aim: winning back Gibraltar from British control. There was no real prospect of that happening soon, and the Americans began to fear that the alliance with France might keep them at war indefinitely. As a result, the Americans began proceeding on their own, without informing Vergennes, and soon drew up a preliminary treaty with Great Britain, which was signed on November 30, 1782. Benjamin Franklin, in the meantime, skillfully pacified Vergennes and avoided an immediate rift in the French–American alliance.

The final treaty, signed September 3, 1783, was, on the whole, remarkably favorable to the United States. It provided a clear-cut recognition of independence and a large, though ambiguous, cession of territory to the new nation—from the southern boundary of Canada to the northern boundary of Florida and from the Atlantic to the Mississippi. The American people had good reason to celebrate as the last of the British occupation forces left New York. Dissenters around the world, too, found inspiration in news of the Revolution. (See "America in the World: The Age of Revolutions.")

WAR AND SOCIETY

Historians have long debated whether the American Revolution was a social as well as a political revolution, whether it was radical or conservative, whether it arose from economic or ideological agendas. But whatever the intention of those who launched and fought the war, the conflict implicated people from every corner of American society, and accelerated, as wars tend to do, certain kinds of social change, even if temporarily.

Loyalists and Religious Groups

Estimates differ as to how many Americans remained loyal to Britain during the Revolution, but it is clear that there were many—at least one-fifth (and some historians estimate as much as one-third) of the white population. Some were officeholders in the imperial government. Others were merchants whose trade was closely tied to the imperial system. Still others were people who lived in relative isolation and had simply retained their traditional loyalties. And there were those who, expecting the British to win the war, were currying favor with the anticipated victors. However they came to their positions, Loyalists held what had been the dominant and respected view until rather recently: that colonists should remain faithful to their sovereign.

Many of these Loyalists were hounded by Patriots in their communities and harassed by legislative and judicial actions. Up to 100,000 fled the country. Those who could afford it moved to Britain. Others moved to Canada, establishing the first English-speaking community in the French-speaking province of Quebec. Some returned to America after the war and gradually reentered the life of the nation.

The war weakened other groups as well. The Anglican Church, headed officially by the king and counting many Loyalist members, lost its status as the official religion of Virginia and Maryland. By the time the fighting ended, many Anglican parishes could no longer even afford clergymen. Also weakened were the Quakers, whose pacifism generated widespread derision.

Other Protestant denominations, however, grew stronger. Presbyterian, Congregational, and Baptist churches successfully tied themselves to the Patriot cause. Most American Catholics also supported the Patriots and won increased popularity as a result. Shortly after the peace treaty was signed, the Vatican provided the United States with its own hierarchy and, in 1789, its first bishop.

THE WAR AND SLAVERY

For some African Americans, the war meant freedom, with the British enabling escapees to leave the country as a way of disrupting the American war effort. The Dunmore Proclamation was aimed at those held by Patriots, not Loyalists, but many enslaved peoples made no such distinction once they learned of the policy. In South Carolina, for example, nearly one-third of the enslaved defected during the war. Some slaveholders in Virginia and Maryland freed or "manumitted" people, and some southern churches flirted briefly with voicing objections to the system. But neither these trends, nor the revolutionary ideals of which they were parts, ever seriously threatened slavery or white supremacy in the South.

In much of the North, the combination of revolutionary sentiment and evangelical Christian fervor helped spread antislavery ideology widely. But even there, white supremacy flourished, and qualifications to the laws limited the character or speed of emancipation, which tended to be either "absolute" or "gradual." Vermont seceded from New York in 1777 and became the first colony to abolish slavery outright (though it was technically an independent republic until admitted to the union in 1791). Soon New Hampshire and Massachusetts joined on the side of absolute emancipation. In Massachusetts, an enslaved woman named Mum Bett, servant to wealthy revolutionaries, heard talk of freedom and equality and understood her condition to violate such precepts. She successfully sued for her freedom in 1781, taking the name Elizabeth Freeman, and others followed suit, leading to abolition in that state in 1783. (Freeman's great-grandson was civil rights activist W. E. B. Du Bois, born in 1868.) Meanwhile Pennsylvania had become the first state to pass gradual emancipation in 1780, and Rhode Island and Connecticut followed suit in 1784. Gradual emancipation generally meant curtailing the importation of new enslaved people and granting freedom to those born after the passage of the laws. The currently enslaved tended to stay enslaved unless states passed full abolition—in the case of Pennsylvania, sixty-seven years later, in 1847.

At war's end, slavery was intact in New York and New Jersey, where the institution played a larger role in the economy than in New England. Both states passed gradual abolition in 1799 and 1804, respectively, but the New York law demonstrated the incomplete and cruel nature of such emancipation. The 1799 measure subjected children born after that date to indentured servitude until young adulthood. Then, the legislature decreed in 1817 that those born before 1799 would be freed—but not until 1827. Across the North, a significant though dwindling number of enslaved persons could be found for several decades after the Revolution, and of course they remained in massive numbers in the South. At some point during the war, almost all states in both regions banned the transatlantic slave trade, but in many cases as part of a broader prohibition of commerce with Britain. It resumed after the war, and southern slaveowners succeeded in pushing back a federal ban on the slave trade until 1808. But the heartbreaking, family-splitting internal traffic in enslaved peoples continued as long as bondage was legal.

Thus the Revolution exposed the continuing contradiction between the nation's commitment to liberty and its simultaneous commitment to slavery, with many colonists even using (without irony) the terminology of "enslavement" to characterize their subjugation by Britain. Many white Southerners and some Northerners believed, in fact, that enslaving Africans—whom they considered inferior and unfit for citizenship—was the best way to ensure liberty for white people. Such white supremacist ideology held that without the enslaved, it would be necessary to recruit a servile white workforce in the South, and that the resulting inequalities would jeopardize the survival of liberty. Even men such as Washington and Jefferson, who had moral misgivings about slavery, struggled to envision any alternative to it. Washington, in fact, insisted his enslaved people be returned to him at war's end by the British, under whom they had served by the terms of Lord Dunmore's Proclamation of 1775, and Jefferson did little to manumit the hundreds of enslaved he regarded as his possessions. If slavery was abolished nationwide, these and other white Americans asked, what would happen to Black people in America? Few whites, North or South, believed freed men and women could be integrated into American society as equals.

NATIVE AMERICANS AND THE REVOLUTION

Indigenous peoples viewed the American Revolution with considerable trepidation, sensing that the battle between white forces was essentially a battle over their own lands. Most nations chose to stay out of the war, but those who fought did so for their own purposes, purposes in fact shared by the colonists: to secure freedom from encroachment and interference. But because many Native Americans feared the Revolution would replace a somewhat trustworthy ruling group (the British, who had tried to limit the expansion of white settlement) with one they considered hostile to them (the Patriots, who had spearheaded the expansion), most of them who chose sides joined the British cause. Thus despite their similar agendas, the colonists developed a searing resentment of indigenous groups during the Revolution. In the Declaration of Independence, Thomas Jefferson even counted the inflaming of "merciless Indian Savages" among the British king's crimes.

In the western Carolinas and Virginia, Cherokee led by Chief Dragging Canoe launched a series of attacks on outlying white settlements in the summer of 1776. Patriot militias responded in great force, ravaging Cherokee lands and forcing the chief and many of his followers to flee west across the Tennessee River. Those Cherokee who remained behind agreed to a new treaty by which they gave up still more land. Some Iroquois, despite the setbacks at Oriskany, continued to wage war against Americans in the West and caused widespread destruction in agricultural areas of New York and Pennsylvania. American armies inflicted heavy retaliatory losses, but the attacks continued.

In the end, the Revolution generally weakened the position of Native Americans in several ways. The Patriot victory increased white demand for western lands. Many whites resented the assistance such nations as the Mohawk had given the British and insisted on treating them as conquered peoples. Others drew from the Revolution a paternalistic view of indigenous groups. Jefferson, for example, came to view native peoples as "noble savages," uncivilized in their present state but redeemable if they were willing to adapt to the norms of white society.

The triumph of the Patriots in the Revolution contributed to the ultimate defeat of the indigenous nations. To white Americans, independence meant, among other things, the right to move aggressively into western lands. The Patriots' victory delivered "the greatest blow that could have been dealt us," one native leader warned.

Women's Rights and Roles

The long Revolutionary War had a profound effect on white American women. The departure of so many men to fight in the Patriot armies left women in charge of farms and businesses. Often, women handled these tasks with great success. But inflation, the unavailability of male labor, and the threat of enemy troops posed significant challenges. Some women whose husbands or fathers were called away to war did not have even a farm or shop to fall back on. Cities and towns had significant populations of impoverished women, who on occasion led protests against price increases, rioted, or looted food. At other times, women launched attacks on occupying British troops, whom they were required to house and feed at considerable expense.

Not all women stayed behind when the men went off to war. Some joined their male relatives in the camps of the Patriot armies. These female "camp followers" increased army morale and provided a ready source of volunteers to cook, launder, nurse, and do other necessary tasks. In the rough environment of the camps, traditional gender distinctions proved difficult to maintain. Considerable numbers of women became involved, at least intermittently, in combat. A few women even disguised themselves as men to be able to fight. The former indentured servant Deborah Sampson of Massachusetts refashioned herself as Robert Shurtleff and served as a scout and combatant in New York and at the final siege of Yorktown. Her sex was later discovered but the state of Massachusetts granted her an army pension.

The emphasis on liberty and the "rights of man" led some women to begin to question their own position in society. **Abigail Adams** wrote to her husband, John Adams, in 1776, "In the new code of laws which I suppose it will be necessary for you to make, I desire you would remember the ladies and be more generous and favorable to them than your ancestors." (See "Consider the Source: The Correspondence of Abigail Adams on Women's Rights.") Adams was simply calling for new protections against abusive and tyrannical men. A few women, however, went further. Judith Sargent Murray, one of the leading essayists of the late eighteenth century, wrote in 1779 that women's minds were as good as those of men and that girls as well as boys deserved access to education.

But little changed as a result. Under English common law, an unmarried woman had some legal rights, but a married woman had virtually none at all. Everything she owned and everything she earned belonged to her husband. Because she had no property rights, she could not engage in any legal transactions on her own. She could not vote. She had no legal authority over her children. Nor could she initiate a divorce; that, too, was a right reserved almost exclusively for men. After the Revolution, it did become easier for women to obtain divorces in a few states. Otherwise, there were few advances and some setbacks—including the loss of widows' rights to regain their dowries from their husbands' estates. The Revolution, in other words, did not really challenge, but actually confirmed and strengthened, the patriarchal legal system.

Still, the Revolution did encourage people of both sexes to reevaluate the contribution of women to the family and society. As the new republic searched for a cultural identity for itself, it attributed a higher value to the role of women as mothers. The new nation was, many Americans liked to believe, producing a new kind of citizen, steeped in the principles of liberty. Mothers had a particularly important task, therefore, in instructing their children in the virtues that the republican citizenry now was expected to possess.

CONSIDER THE SOURCE

THE CORRESPONDENCE OF ABIGAIL ADAMS ON WOMEN'S RIGHTS (1776)

Between March and May 1776, Abigail Adams exchanged letters with her husband John on the matter of women's rights in revolutionary America. She was the daughter of a minister and read widely despite lacking a formal education. The couple kept up a regular correspondence on political and social issues of the day.

Abigail Adams to John Adams, March 31, 1776:
I long to hear that you have declared an independency. And, by the way, in the new code of laws which I suppose it will be necessary for you to make, I desire you would remember the ladies and be more generous and favorable to them than your ancestors. Do not put such unlimited power into the hands of the husbands. Remember, all men would be tyrants if they could. If particular care and attention is not paid to the [150] ladies, we are determined to foment a rebellion, and will not hold ourselves bound by any laws in which we have no voice or representation.

That your sex are naturally tyrannical is a truth so thoroughly established as to admit of no dispute; but such of you as wish to be happy willingly give up the harsh title of master for the more tender and endearing one of friend. Why, then, not put it out of the power of the vicious and the lawless to use us with cruelty and indignity with impunity? Men of sense in all ages abhor those customs which treat us only as the vassals of your sex; regard us then as beings placed by Providence under your protection, and in imitation of the Supreme Being make use of that power only for our happiness.

John Adams to Abigail Adams, April 14, 1776:
As to your extraordinary code of laws, I cannot but laugh. We have been told that our struggle has loosened the bonds of government everywhere; that children and apprentices were disobedient; that schools and colleges were grown turbulent; that Indians slighted their guardians, and negroes grew insolent to their masters. But your letter was the first intimation that another tribe, more numerous and powerful than all the rest, were grown discontented. This is rather too coarse a compliment, but you are so saucy, I won't blot it out. Depend upon it, we know better than to repeal our masculine systems. Although they are in full force, you know they are little more than theory. We dare not exert our power in its full latitude. We are obliged to go fair and softly, and, in practice, you know we are the subjects. We have only the name of masters, and rather than give up this, which would completely subject us to the despotism of the petticoat, I hope General Washington and all our brave heroes would fight; I am sure every good politician would plot, as long as he would against despotism, empire, monarchy, aristocracy, oligarchy, or ochlocracy.

Abigail Adams to John Adams, May 7, 1776:
I cannot say that I think you are very generous to the ladies; for, whilst you are proclaiming peace and good-will to men, emancipating all nations, you insist upon retaining an absolute power over wives. But you must remember that arbitrary power is like most other things which are very hard, very liable to be broken; and, notwithstanding all your wise laws and maxims, we have it in our power, not only to free ourselves, but to subdue our masters, and, without violence, throw both your natural and legal authority at our feet;—
"Charm by accepting, by submitting sway, Yet have our humor most when we obey."

UNDERSTAND, ANALYZE, & EVALUATE

1. What was Abigail Adams's opinion of men in power and what did she request of John Adams as they declared independence?
2. To what other social developments did John Adams compare his wife's request? What did he mean by the "despotism of the petticoat" (a women's undergarment)?
3. What did Abigail Adams predict in her May 7 letter to John Adams? What do you think of her assessment of "arbitrary power"?

Source: Adams, John, Abigail Adams, and Charles Francis Adams, *Familiar Letters of John Adams and His Wife Abigail Adams During the Revolution*. Project Gutenberg, 1776.

THE WAR ECONOMY

The Revolution also produced important changes in the structure of the American economy. After more than a century of dependence on the British imperial system, American commerce suddenly found itself on its own. British ships no longer protected American vessels. In fact, they tried to drive them from the seas. British imperial ports were closed to American trade. But this disruption in traditional economic patterns strengthened the American economy in the long run. Enterprising merchants in New England and elsewhere began to develop new commercial networks in the Caribbean and South America. By the mid-1780s, American merchants were also developing an important trade with Asia.

When British imports to America were cut off, states desperately tried to stimulate domestic manufacturing. No great industrial expansion resulted, but there was a modest increase in production. Trade also increased substantially among the American states.

THE CREATION OF STATE GOVERNMENTS

At the same time as Americans were struggling to win their independence on the battlefield, they were also struggling to create new institutions of government to replace the British system they had repudiated.

THE PRINCIPLES OF REPUBLICANISM

If Americans agreed on nothing else, they agreed that their new governments would be republican. To them, **republicanism** meant a political system in which all power came from the people, rather than from some supreme authority like a king. The success of such a government depended on the character of its citizenry. If the population consisted of sturdy, independent property owners committed to the common good, then the republic could survive. If it consisted of a few powerful aristocrats and a great mass of dependent workers, then it would be in danger. From the beginning, therefore, the ideal of the small freeholder (the independent landowner) was basic to American political ideology. Jefferson, the great champion of the independent **yeoman farmer**, once wrote, "Dependence begets subservience and venality, suffocates the germ of virtue, and prepares fit tools for the designs of ambition."

Another crucial part of republican ideology was the concept of equality. The Declaration of Independence had given voice to the idea in its most ringing phrase: "All men are created equal." This idea would provide a powerful rhetorical framework for claimants to the rights of American citizenship for generations to come.

But for now, those rights went to a limited population of Americans. The United States was not a nation in which all men were independent property holders, and those who were not found their citizenship rights circumscribed. From the beginning, there was a sizable dependent labor force, white and Black. White women remained both politically and economically subordinate. Native Americans were systematically exploited and displaced. Nor was there ever full equality of opportunity. American society was more open and fluid than that of most European nations, but the condition of a person's birth was almost always a crucial determinant of success. And even in northern states that abolished slavery, emancipation could be a slow process.

Important Revolutionary leaders, including John Adams, continued to defend a very narrow vision of citizenship and suffrage beyond the year 1776 and shuddered at the increasingly egalitarian calls for truly democratic government from the people of the young country. One day, he predicted with fear, "new claims will arise; women will demand a vote; lads from twelve to twenty-one will think their claims not closely attended to; and every man who has not a farthing will demand an equal voice with any other, in all acts of state."

The First State Constitutions

Two states—Connecticut and Rhode Island—already had governments that were republican in all but name. They simply deleted references to Britain and the king from their charters and adopted them as constitutions. The other eleven states, however, produced new documents.

The first and perhaps most basic decision was that the constitutions were to be explicitly crafted and recorded, unlike Britain's unwritten constitution. The second decision was that the power of the executive, which Americans believed had grown too great in Britain, must be limited. Pennsylvania eliminated the executive altogether. Most other states inserted provisions limiting the power of governors over appointments, reducing or eliminating their right to veto bills, and preventing them from dismissing the legislature. Most important, every state forbade the governor or any other executive officer from holding a seat in the legislature, thus ensuring that, unlike in Britain, the executive and legislative branches of government would remain separate.

Even so, most new constitutions did not embrace direct popular rule. In Georgia and Pennsylvania, the legislature consisted of one popularly elected house. But in every other state, there were upper and lower chambers, and in most cases the upper chamber was designed to represent the "higher orders" of society. There were property requirements for voters—some modest, others substantial—in all states. All of this roughly mimicked the workings of Parliament in Great Britain.

Revising State Governments

By the late 1770s, Americans were growing concerned about the apparent instability of their new state governments. Many believed, once again, the problem was one of too much democracy. As a result, most of the states began to revise their constitutions to limit popular power. Massachusetts, which ratified its first constitution in 1780, became the first state to act on the new concerns.

Two changes in particular differentiated the Massachusetts and later constitutions from the earlier ones. The first was a change in the process of constitution writing itself. Most of the first documents had been written by state legislatures and thus could easily be amended (or violated) by them. Massachusetts created the constitutional convention: a special assembly of the people that would meet only for the purpose of writing the constitution.

The second change was a significant strengthening of the executive. The 1780 Massachusetts constitution made the governor one of the strongest in any state. He was to be elected directly by the people; he was to have a fixed salary and thus would not be dependent on the legislature each year for his wages; he would have significant appointment powers and a veto over legislation. Other states followed. Those with weak or nonexistent upper houses strengthened or created them. Most increased the powers of the governor. Pennsylvania, which had no executive at all at first, now produced a strong one. By the late 1780s, almost every state had either revised its constitution or drawn up an entirely new one in an effort to produce greater stability in government.

Most Americans continued to believe that religion should play some role in government, but they did not wish to give special privileges to any particular denomination. The privileges that churches had once enjoyed were now largely stripped away. In 1786, Virginia enacted the Statute of Religious Liberty, written by Thomas Jefferson, which called for the separation of church and state. But in Massachusetts, state-sponsored religion and religious constitutional clauses survived well into the nineteenth century.

THE SEARCH FOR A NATIONAL GOVERNMENT

Americans were much quicker to agree on state institutions than they were on the structure of their national government. At first, most believed that the central government should remain relatively weak and that each state would be virtually a sovereign nation. It was in response to such ideas that the Articles of Confederation emerged.

THE CONFEDERATION

The Articles of Confederation, which the Continental Congress had adopted in 1777, provided for a national government much like the one already in place before independence. Congress remained the central—indeed the only—institution of national authority. Its powers expanded to give it authority to conduct wars and foreign relations and to appropriate, borrow, and issue money. But it did not have power to regulate trade, draft troops, or levy taxes directly on the people. For troops and taxes, it had to make formal requests of the state legislatures, which could—and often did—refuse them. There was no separate executive; the "president of the United States" was merely the presiding officer at the sessions of Congress. Each state had a single vote in Congress, and at least nine of the states had to approve any important measure. All thirteen state legislatures had to approve any amendment of the Articles.

During the process of ratifying the Articles of Confederation (which required approval by all thirteen states), broad disagreements over the plan became evident. The small states had insisted on equal state representation, but the larger states wanted representation based on population. The smaller states prevailed on that issue. More important, the states claiming western lands wished to keep them, but the rest of the states demanded that all such territory be turned over to the national government. New York and Virginia had to give up their western claims before the Articles were finally approved. They went into effect in 1781.

The Confederation, which existed from 1781 until 1789, was not a complete failure, but it was far from a success. It lacked adequate powers to deal with interstate issues or to enforce its will on the states.

Diplomatic Failures

In the peace treaty of 1783, the British had promised to evacuate American territory, but British forces continued to occupy a string of frontier posts along the Great Lakes within the United States. Nor did the British honor their agreement to make restitution to slave holders whose enslaved had made their way to British lines. Disputes also erupted over the northeastern boundary of the new nation and over the border between the United States and Florida. Most American trade remained within the British Empire, and Americans wanted full access to British markets. The government in London, however, placed sharp restrictions on that access.

In 1784, Congress sent John Adams as minister to London to resolve these differences, but Adams made no headway with the British, who were never sure whether he represented a single nation or thirteen different ones. Throughout the 1780s, the British government refused even to send a diplomatic minister to the American capital.

Confederation diplomats agreed to a treaty with Spain in 1786. The Spanish accepted the American interpretation of the Florida boundary. In return, the Americans recognized the Spanish possessions in North America and accepted limits on the right of U.S. vessels to navigate the Mississippi for twenty years. But southern states, incensed at the idea of giving up their access to the Mississippi, blocked ratification.

The Confederation and the Northwest

The Confederation's most important accomplishment was its resolution of controversies involving western lands, rapidly being cleared of their indigenous populations by white settlement. The Confederation sought to include these areas in the political structure of the new nation.

The Ordinance of 1784, based on a proposal by Thomas Jefferson, divided the western territory into ten self-governing districts, each of which could petition Congress for statehood when its settler population equaled the number of free inhabitants of the smallest existing state. Then, in the Ordinance of 1785, Congress created a system for surveying and selling western lands. The territory north of the Ohio River was to be surveyed and marked off into neat rectangular townships, each divided into thirty-six identical sections. In every township, four sections were to be reserved by the federal government for future use or sale (a policy that helped establish the idea of "public land"). The revenue from the sale of one of these federally reserved sections was to support creation of a public school.

The precise rectangular pattern imposed on the Northwest Territory—the grid—became a model for all subsequent land policies of the federal government and for many other planning decisions in states and localities. The grid also became characteristic of the layout of many American cities. It had many advantages from the perspective of the national government. It eliminated the uncertainty about property borders that earlier, more informal land systems had produced. At the expense of displaced Native Americans, it sped the development of western lands by making land ownership simple and understandable. But it also encouraged a dispersed form of settlement—with each farm family separated from its neighbors—that made the formation of community more difficult. The 1785 Ordinance made a dramatic and indelible mark on the American landscape.

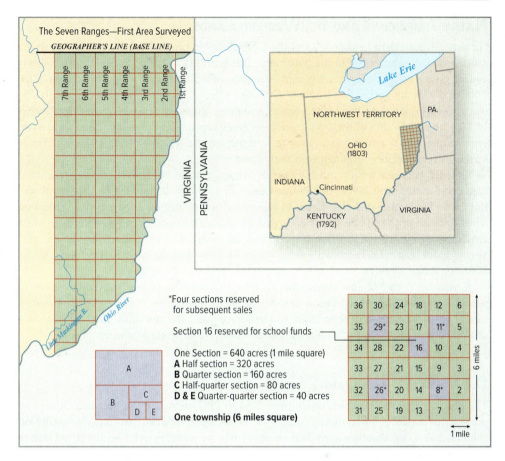

LAND SURVEY: ORDINANCE OF 1785 In the Ordinance of 1785, the Congress established a new system for surveying and selling western lands. These maps illustrate the way in which the lands were divided in an area of Ohio. Note the highly geometrical grid pattern that the ordinance imposed on these lands. Each of the squares in the map on the left was subdivided into 36 sections, as illustrated in the map at the lower right. • *Why was this grid pattern so appealing to the planners of the western lands?*

These ordinances proved highly favorable to land speculators and less so to ordinary settlers, many of whom could not afford the price of the land. Congress compounded the problem by selling much of the best land to the Ohio and Scioto Companies before making it available to anyone else. Criticism of these policies led to the passage in 1787 of another law governing western settlement—legislation that became known as the "**Northwest Ordinance.**" The 1787 Ordinance maintained the grid system, but it abandoned the ten districts established in 1784 and created a single Northwest Territory out of the lands north of the Ohio. The territory could be divided subsequently into three to five territories. It also specified a settler population of 60,000 as a minimum for statehood, guaranteed freedom of religion and the right to trial by jury to residents of the region, and prohibited slavery throughout the territory.

The western lands south of the Ohio River received less attention from Congress. The region that became Kentucky and Tennessee developed rapidly in the late 1770s as slave holding territories, and in the 1780s settlers there began setting up governments and asking for statehood. The Confederation Congress was never able to resolve the conflicting claims in that region successfully.

Native Americans and the Western Lands

On paper, the western land policies of the Confederation brought order and stability to the process of white settlement in the Northwest. But in reality, order and stability came slowly and at great cost to indigenous groups. Congress tried to resolve that problem in 1784, 1785, and 1786 by persuading Iroquois, Choctaw, Chickasaw, and Cherokee leaders to sign treaties ceding lands to the United States. However, those agreements proved ineffective. In 1786, the leadership of the Iroquois Confederacy repudiated the treaty it had signed two years earlier. Other nations had never really accepted the treaties affecting them and continued to resist white movement into their lands.

Violence between whites and indigenous peoples on the Northwest frontier reached a crescendo in the early 1790s. In 1790 and again in 1791, the Miami, led by the famed warrior Little Turtle, defeated U.S. forces in two major battles. Efforts to negotiate a settlement failed because of the Miami's insistence that no treaty was possible unless it forbade white settlement west of the Ohio River. Negotiations did not resume until after General Anthony Wayne led 4,000 soldiers into the Ohio Valley in 1794 and defeated native forces in the **Battle of Fallen Timbers**.

A year later, the Miami signed the Treaty of Greenville, ceding substantial lands in present-day Ohio, Indiana, Illinois, and Michigan to the United States in exchange for a formal acknowledgment of their claim to territory beyond the new lines. But that hard-won assurance proved a frail protection against the pressure of white expansion.

Debts, Taxes, and Daniel Shays

The postwar depression, which lasted from 1784 to 1787, increased the perennial American problem of an inadequate money supply, a burden that weighed particularly heavily on debtors. The Confederation itself had an enormous outstanding debt, accumulated during the Revolutionary War, and few means with which to pay it down. It had sold war bonds that were now due to be repaid, it owed money to its soldiers, and it carried substantial debts abroad. But with no power to tax, it could request money only from the states, and it received only about one-sixth of the money it asked for. The fragile new nation was faced with the grim prospect of defaulting on its obligations.

This alarming possibility brought to prominence a group of leaders who would play a crucial role in the shaping of the republic for several decades. Robert Morris, the head of the Confederation's treasury, Alexander Hamilton, his young protégé, James Madison of Virginia, and others—all called for a "continental impost," a 5 percent duty on imported goods to be levied by Congress and used to fund the debt. Many Americans, however, feared that the impost plan would concentrate too much financial power in the hands of Morris and his allies in Philadelphia. Congress failed to approve the impost in 1781 and again in 1783.

The states had war debts, too, and they generally relied on increased taxation to pay them. But poor farmers, already burdened by debt, considered such policies unfair. They demanded that the state governments issue paper currency to increase the money supply and make it easier for them to meet their obligations. Resentment was especially high among farmers in New England, who felt that the states were squeezing them to enrich already wealthy bondholders in Boston and other towns. Elite merchants and their allies controlled the state legislatures implementing the taxes in places like Massachusetts, further exacerbating class conflict. Here, in microcosm, were some of the tensions that had concerned the revolutionary generation since 1776. In crafting a decentralized government and favoring the states, the framers of the Confederation created a weak Congress with debts and responsibilities but no way to meet them. When the states exercised their power to tax, it fell on common people to fund the public debt, a burden rather resembling those of the colonial period. Farmers may have been technically "represented" in the legislatures, but elites still controlled them. Like before 1776, all of this might erupt into mob rule.

True to those fears, throughout the late 1780s, bands of distressed farmers rioted periodically in various parts of New England. Dissidents in the Connecticut Valley and the Berkshire Hills of Massachusetts rallied behind Daniel Shays, a former captain in the Continental army representing farmers already impoverished by the dislocations of the war. Shays issued a set of demands that included paper money, tax relief, a moratorium on debts, and the abolition of imprisonment for debt. During the summer of 1786, the Shaysites prevented the collection of debts, private or public, and used force to keep courts from convening and sheriffs from selling confiscated property. When winter came, the rebels advanced on Springfield, hoping to seize weapons from the arsenal there. In January 1787, an army of state militiamen set out from Boston, met Shays's band, and dispersed his ragged troops.

As a military enterprise, **Shays's Rebellion** was a failure, although it did produce some concessions to the aggrieved farmers. Shays and his lieutenants, at first sentenced to death, were later pardoned, and Massachusetts offered the protesters some tax relief and a postponement of debt payments. But the rebellion had important consequences for the future of the United States, for it added urgency to the movement to produce a new, national constitution.

(National Portrait Gallery, Smithsonian Institution)

DANIEL SHAYS AND JOB SHATTUCK Shays and Shattuck were the principal leaders of the 1786 uprising of poor Massachusetts farmers demanding relief from their indebtedness. Shattuck led an insurrection in the east, which collapsed when he was captured on November 30. Shays organized the rebellion in the west, which continued until it was finally dispersed by state militia in late February 1787. The following year, state authorities pardoned Shays. Even before that, the legislature responded to the rebellion by providing some relief to the impoverished farmers. This drawing is part of a hostile account of the rebellion published in 1787 in a Boston almanac.

CONCLUSION

Between a small, inconclusive battle on a village green in New England in 1775 and a momentous surrender at Yorktown in 1781, the American people fought a terrible war against the mightiest military nation in the world. Few would have predicted in 1775 that the makeshift armies of the colonies could withstand the forces of the British Empire. But a combination of luck, determination, costly errors by the British, and timely aid from abroad allowed the Patriots, as they called themselves, to make full use of the advantages of fighting on their home soil and to frustrate British designs.

This historic military event also propelled the colonies to unite, to organize, and to declare their independence. Having done so, they fought to defend an actual, fledgling nation. By the end of the war, they had created new governments at both the state and national level and had begun experimenting with new political forms.

Yet although the Revolution hinged rhetorically on the notion of freedom from the British monarchy, many of its architects harbored a narrow vision of that concept when it came to the actual practices of governance and power. Many of them openly believed only the best minds and the economically independent should enjoy the rights of citizenship, and in their view, such rights should be limited to white males and usually just ones with property. But in the republic's first days, the rhetoric of equality was already underwriting murmurings to expand the electorate, frightening some of the founders. And cracks were forming in the edifice of slavery, though it would be many generations before the institution would crumble.

Victory in the American Revolution thus solved many of the problems of the new nation, but it also produced others. What should the United States do about its relations with Native Americans and its neighbors to the north and south? What should it do about the distribution of western lands? What should it do about slavery? How should it balance its commitment to liberty with its need for order? These questions bedeviled the new national government in its first years of existence.

KEY TERMS/PEOPLE/PLACES/EVENTS

Abigail Adams 123
American Patriots 111
Articles of
  Confederation 111
Battle of Fallen Timbers 130
Benedict Arnold 112
Common Sense 110
Declaration of
  Independence 111
George Washington 111
Hessians 107
John Burgoyne 114
Joseph and Mary Brant 114
Lord Cornwallis 116
Lord Dunmore's
  Proclamation 107
Loyalists (Tories) 112
Northwest Ordinance 129
republicanism 125
Saratoga 114
Second Continental
  Congress 107
Shays's Rebellion 131
Thomas Jefferson 111
William Howe 113
yeoman farmers 125
Yorktown 117

RECALL AND REFLECT

1. What questions did the Second Continental Congress debate, and how did it address them?
2. What was the impact of Thomas Paine's *Common Sense* on Americans' view of the war with Britain?
3. What were the ideals of the new state and national governments, and did those ideals match the realities of power and influence in American society?
4. What was the purpose of the Articles of Confederation?

6 THE CONSTITUTION AND THE NEW REPUBLIC

FRAMING A NEW GOVERNMENT
ADOPTION AND ADAPTATION
FEDERALISTS AND REPUBLICANS
ESTABLISHING NATIONAL SOVEREIGNTY
THE DOWNFALL OF THE FEDERALISTS

LOOKING AHEAD

1. What were the most important questions debated at the Constitutional Convention of 1787, and how were they resolved?
2. What were the main tenets of the Federalist and Antifederalist arguments on ratification of the Constitution?
3. What were the origins of the United State's "first party system"?

BY THE LATE 1780S, many Americans had grown dissatisfied with the Confederation. It was, they believed, ridden with factions, unable to deal effectively with economic problems, and frighteningly powerless in the face of Shays's Rebellion. A decade earlier, Americans had deliberately avoided creating a strong national government, fearing it would encroach on the sovereignty of the individual states. Now they reconsidered.

In the summer of 1787, delegates from every state except Rhode Island gathered in Philadelphia to produce a new governing document for the country. Behind closed doors, disagreements flared over how to represent the states in a new Congress, how to treat the matter of slavery, how to balance individual rights and the common good, and perhaps above all, how to share power between the federal government and the states and mitigate against dangerous aggregations of authority.

By September, the delegates had produced a new constitution that created a much more powerful government with three independent branches. The document then came in for intense debate while the states considered ratification. "Federalists" defended the Constitution and thought it properly checked both the power of the masses and the various components of the government it created; "Antifederalists" argued it gave too much power to the federal government and worried about the rights of citizens. In 1788, the last of the nine states necessary for ratification voted to ratify, with assurances that amendments would be added to guarantee individual rights. But the adoption of the Constitution did not complete the creation of the republic, for although most people came to agree that the Constitution should guide American governance, they often disagreed on what that document meant.

FRAMING A NEW GOVERNMENT

The Confederation Congress had become so unpopular and ineffectual by the mid-1780s that it began to lead an almost waif-like existence. In 1783, its members timidly withdrew from Philadelphia to escape army veterans demanding their back pay. They took refuge for a while in Princeton, New Jersey, then moved on to Annapolis, Maryland, and in 1785 settled in New York. Delegates were often scarce. Only with great difficulty could Congress produce a quorum to ratify the treaty with Great Britain, ending the Revolutionary War.

Advocates of Reform

In the 1780s, some of the wealthiest and most powerful groups in the population began to clamor for a stronger national government. By 1786, such demands had grown so intense that even defenders of the existing system reluctantly agreed that the government needed strengthening at its weakest point—its lack of power to tax.

The most effective advocate of a stronger national government was **Alexander Hamilton**, a successful New York lawyer and illegitimate son of a Scottish merchant in the West Indies. Hamilton now called for a national

TIME LINE

1786
Annapolis Conference

1787
Constitutional Convention;
Constitution adopted

1787–1788
Constitution is ratified by the nine states necessary

1789
Washington becomes first president
Bill of Rights
French Revolution
Judiciary Act

1791
First Bank of U.S. chartered

1792
Washington reelected

1794
Whiskey Rebellion
Jay's Treaty

1795
Pinckney's Treaty

1796
John Adams elected president

1797
XYZ Affair
Alien and Sedition Acts

1798–1799
Quasi war with France
Virginia and Kentucky Resolutions

1800
Jefferson elected president

convention to overhaul the Articles of Confederation. He found an important ally in James Madison of Virginia, who persuaded the Virginia legislature to convene an interstate conference on commercial questions. Only five states sent delegates to the meeting, which took place at Annapolis in 1786, but the conference approved a proposal by Hamilton for a convention of special delegates from all the states to meet in Philadelphia the next year.

At first there seemed little reason to believe the Philadelphia convention would attract any more delegates than had the Annapolis meeting. Then, early in 1787, the news of Shays's Rebellion spread throughout the nation, alarming many previously apathetic leaders, including George Washington, who promptly made plans to travel to Philadelphia for the Constitutional Convention. Washington's support gave the meeting wide credibility.

A Divided Convention

Fifty-five men, representing all the states except Rhode Island, attended one or more sessions of the convention that sat in the Philadelphia State House from May to September 1787. These "Founding Fathers," as they became known much later, averaged forty-four years in age and were well educated by the standards of their time. Most were wealthy property owners, and many feared what one of them called the "turbulence and follies" of democracy. Yet all retained the revolutionary suspicion of concentrated power.

The convention unanimously chose Washington to preside over its sessions and closed it to the public and press. It then ruled that each state delegation would have a single vote and that major decisions would require not unanimity, as they did in Congress, but a simple majority. Almost all the delegates agreed that the United States needed a stronger central government. But there agreement ended.

Virginia, the most populous state, sent a well-prepared delegation to Philadelphia led by James Madison, who had devised in some detail a plan for a new "national" government. The Virginia Plan shaped the agenda of the convention from the moment Edmund Randolph of Virginia opened the debate by proposing that "a national government ought to be established, consisting of a supreme Legislative, Executive, and Judiciary." Even that brief description outlined a government very different from the Confederation. But the delegates were so committed to fundamental reform that they approved the resolution after only brief debate.

There was less agreement about the details of Madison's **Virginia Plan**. It called for a national legislature of two houses, with states represented in both bodies in proportion to their population. Smaller states, quite predictably, raised immediate objections. William Paterson of New Jersey offered an alternative (the **New Jersey Plan**) that would retain the essence of the Confederation with its one-house legislature in which all states had equal representation. It would, however, give Congress expanded powers to tax and to regulate commerce. The convention rejected Paterson's proposal, but supporters of the Virginia Plan now realized they would have to make concessions to the smaller states.

Many questions remained unresolved. Among the most important was the question of slavery. There was no serious discussion of abolishing slavery during the convention, but other issues were debated heatedly. Would enslaved people be counted as part of the population in determining representation in Congress, or would they be considered property and therefore not counted? Everyone wanted resolutions to these questions that would benefit their own states. Representatives from slave states argued that the enslaved should be considered persons in determining representation but as property if the new government levied taxes on the states on the basis of population. Representatives from states where slavery had disappeared or was expected to disappear argued the opposite, that the enslaved should be included in calculating taxation but not representation.

Compromise

The delegates bickered for weeks. By the end of June, with both temperature and tempers rising, the convention seemed in danger of collapsing. Finally, on July 2, the convention created a "grand committee," comprising one delegate from each state, which produced a proposal that became the basis of the "Great Compromise." It called for a two-house legislature. In the lower House of Representatives, the states would be represented on the basis of population. Each enslaved person would be counted as three-fifths of a free person in determining the basis for both representation and direct taxation. In the upper Senate, each state, regardless of population, would be represented by two members, with those senators elected by state legislatures, not the general voting public. On July 16, 1787, the convention voted to accept the compromise.

In the next few weeks, the convention agreed to another important compromise. To placate southern delegates, who feared the new government would interfere with slavery, the convention agreed to bar the new government from stopping the importation of enslaved people for twenty years.

Some significant issues remained unaddressed. The **Constitution** provided no definition of citizenship. Nor did it resolve the status of Native American tribes. Most important to many Americans was the absence of a list of individual rights, which would restrain the powers of the national government. Madison opposed the idea, arguing that specifying rights that were reserved to the people would, in effect, limit those rights. Others, however, feared that without such protections the national government might abuse its new authority.

The Constitution of 1787

Many people contributed to the creation of the American Constitution, but the most important person in the process was **James Madison**. Madison had devised the Virginia Plan, and he did most of the drafting of the Constitution itself. Madison's most important achievement, however, was helping to resolve two important philosophical questions: the question of sovereignty and the question of limiting power. (For historians' evolving views on the Constitution's purpose, see "Debating the Past: The Meaning of the Constitution.")

How could a national government exercise sovereignty concurrently with state governments? Where did ultimate sovereignty lie? The answer, Madison and his contemporaries decided, was that all power, at all levels of government, flowed ultimately from the people. Thus neither the federal government nor the state governments were truly sovereign. All of them derived their authority from below. The resolution of the problem of sovereignty made possible one of the distinctive features of the Constitution—its **federalism**, or division of powers between the national and state governments. The Constitution and the government it created were to be the "supreme law" of the land. At the same time, however, the Constitution left important powers in the hands of the states.

In addition to addressing the question of sovereignty, the writers of the Constitution resolved to spread authority over several centers of power. Drawing from the ideas of the French philosopher Baron de Montesquieu, the framers endeavored to prevent any single group, or tyrannical individual, from dominating the government. The Constitution provided for a **separation of powers** within the government, managed by a system of **checks and balances** among the legislative, executive, and judicial branches. The forces within the government would constantly check one another. Congress would have two chambers, each constraining the other, since both would have to agree before any law could be passed. The president would have the power to veto acts of Congress, but Congress could override those vetoes and also check the

DEBATING THE PAST

The Meaning of the Constitution

The Constitution of the United States inspired debate from the moment it was drafted. Some argue that the Constitution is a flexible document intended to evolve in response to society's evolution. Others counter that it has a fixed meaning, rooted in the "original intent" of the framers, and that to move beyond that is to deny its value.

Historians, too, disagree about why the Constitution was written and what it meant. To some scholars, the creation of the federal system was an effort to preserve the ideals of the Revolution and to create a strong national government capable of exercising real authority. To others, the Constitution was an effort to protect the economic interests of existing elites, even at the cost of betraying the principles of the Revolution. And to still others, the Constitution was designed to protect individual freedom and to limit the power of the federal government.

The first influential exponent of the heroic view of the Constitution as the culmination of the Revolution was John Fiske, whose book *The Critical Period of American History* (1888) painted a grim picture of political life under the Articles of Confederation. Many problems, including economic difficulties, the weakness and ineptitude of the national government, threats from abroad, interstate jealousies, and widespread lawlessness, beset the new nation. Fiske argued that only the timely adoption of the Constitution saved the young republic from disaster.

In *An Economic Interpretation of the Constitution of the United States* (1913), Charles A. Beard presented a powerful challenge to Fiske's view. According to Beard, the 1780s had been a "critical period" primarily for conservative business interests who feared that the decentralized political structure of the republic imperiled their financial position. Such men, he claimed, wanted a government able to promote industry and trade, protect private property, and perhaps, most of all, make good the public debt—much of which was owed to them. The Constitution was, Beard claimed, "an economic document drawn with superb skill by men whose property interests were immediately at stake" and who won its ratification over the opposition of a majority of the people.

A series of powerful challenges to Beard's thesis emerged in the 1950s. The Constitution, many scholars now began to argue, was not an effort to preserve property but an enlightened effort to ensure stability and order. Robert E. Brown, for example, argued in 1956 that "absolutely no correlation" could be shown between the wealth of the delegates to the Constitutional Convention and their position on the Constitution. Examining the debate between the Federalists and the Antifederalists, Forrest McDonald, in *We the People* (1958), also concluded that there was no consistent relationship between wealth and property and support for the Constitution. Instead, opinion on the new system was far more likely to reflect local and regional interests. These challenges greatly weakened Beard's argument. Few historians any longer accept his thesis without reservation.

In the 1960s, scholars began again to revive an economic interpretation of the Constitution—one that differed from Beard's but nevertheless emphasized social and economic factors as motives for supporting the federal system. Jackson Turner Main argued in *The Anti-federalists* (1961) that supporters of the Constitution were "cosmopolitan commercialists," eager to advance the economic development of the

138

nation; the Antifederalists, by contrast, were "agrarian localists," fearful of centralization. Gordon Wood, in *The Creation of the American Republic* (1969), suggested that the debate over the state constitutions in the 1770s and 1780s reflected profound social divisions and that those same divisions helped shape the argument over the federal Constitution. The Federalists, Wood suggested, were largely traditional aristocrats who had become deeply concerned by the instability of life under the Articles of Confederation and were particularly alarmed by the decline in popular deference toward social elites. The creation of the Constitution was part of a larger search to create a legitimate political leadership based on the existing social hierarchy. It reflected the efforts of elites to contain what they considered the excesses of democracy.

More recently, historians have continued to examine the question of "intent." Did the framers intend a strong, centralized political system; or did they intend to create a decentralized system with a heavy emphasis on individual rights? The answer, according to Jack Rakove in *Original Meanings* (1996), and *Revolutionaries* (2010), is both—and many other things as well. The Constitution, he argues, was the result of a long and vigorous debate through which the views of many different groups found their way into the document. James Madison, generally known as the father of the Constitution, was a strong nationalist, as was Alexander Hamilton. They believed that only a powerful central government could preserve stability in a large nation, and they saw the Constitution as a way to protect order and property and defend the nation against the dangers of too much liberty. But if Madison and Hamilton feared too much liberty, they also feared too little. And that made them receptive to the demands of the Antifederalists for protections of individual rights, which culminated in the Bill of Rights. The very "middling sorts" who had exercised more and more power since 1776, scaring many conservative founders, also helped push for such citizen rights, Woody Holton argues in *Unruly Americans and the Origins of the Constitution* (2007).

Just as crucially, what did the framers intend regarding slavery? On one side of a debate that rather parallels the one over the meaning of the Revolution, some scholars maintain that the Constitution may not have called for an immediate end to human bondage, but that it ultimately laid the groundwork for slavery's destruction. They point, for instance, to the failure of the Constitution to mention slavery, and to Frederick Douglass's speech in 1852 declaring the Constitution a "glorious liberty document," that, if properly understood, offered no explicit succor to slavery. Sean Wilentz argues in *No Property in Man: Slavery and Antislavery at the Nation's Founding* (2018) that the Constitution contained both pro- and antislavery components, but that ultimately the framers refused to sanction the institution with explicit protections.

Yet other scholars find this picture too celebratory, arguing that it downplays the pro-slavery character of the Constitution. While granting that concessions were made to antislavery voices at the convention, many historians agree with the views of David Waldstreicher in *Slavery's Constitution: From Revolution to Ratification* (2009) and George

(National Archives and Records Administration)

William Van Cleve in *A Slaveholder's Union: Slavery, Politics, and the Constitution in the Early American Republic* (2010). Even if the Constitution withheld specific rhetorical sanction for the idea of "property in man," it rewarded slaveholders with potentially escalating political representation for holding human beings in bondage, allowed the slave trade to continue and in fact flourish by pledging not to terminate it for at least twenty years, and extended various other economic and legal protections for the institution, such as mandating the return of fugitive enslaved people to slaveholders. In Waldstreicher's words, "In growing their government, the framers and their constituents created fundamental laws that sustained human bondage."

UNDERSTAND, ANALYZE, & EVALUATE

1. Is the Constitution a conservative, liberal, or radical document?
2. Did the framers consider the Constitution something "finished" (with the exception of constitutional amendments), or did they consider it a document that would evolve in response to changes in society over time?
3. Which parts of the Constitution suggest that the framers' intent was to create a strong, centralized political system? Which parts suggest that the framers' intent was to create a decentralized system with heavy emphasis on individual rights?

executive through impeachment. The judiciary would monitor the constitutionality of laws and orders coming from the executive and legislative branches. Federal courts would be protected from both of those entities, because judges would serve for life, but the executive would appoint federal judges in the first place and the legislature would confirm them.

The "federal" structure of the government was designed to protect the United States from the kind of despotism that Americans believed had emerged in Britain. Framers of the Constitution wanted a stronger central government, but one not *too* strong. Likewise, they wanted a government representative of and answerable to the popular will, but not *too* much so. Many of them harbored limited trust in the abilities of citizens to put the common good before their individual needs, and pointed to Shays's Rebellion as recent evidence of popular power run amok. Thus in the new government, only the members of the House of Representatives would be elected directly by the people. Senators would be chosen by state legislatures. The president would be chosen by an electoral college, with each state promoting electors to that body however it saw fit but equal to the total number of the state's members of Congress (Senate plus House). No requirement was written into the Constitution that these electors cast their ballots for president and vice president according to the popular will in their states, though that later became the accepted practice when states began recording a popular vote.

On September 17, 1787, thirty-nine delegates signed the Constitution. The document established a democratic republic that would be governed by white men. The framers did not explicitly define **citizenship**—the legal recognition of a person's inclusion in a body politic through the granting of rights and privileges—but common wisdom and jurisprudence held that birth in the United States and whiteness made one a citizen. Congress made this explicit for immigrants with the Naturalization Act of 1790, which helped legalize the stream of newcomers and allowed them to become citizens—provided they were "free white person[s]."

States were left to adjudicate the particulars of citizenship and suffrage rights, but in general they reserved that status for white people and those privileges for white male property owners. New Jersey allowed propertied white women to vote, but brought its suffrage laws into line with those of the other (male suffrage only) states in 1807. A few states would extend citizenship and suffrage rights to free Blacks, and free Blacks in South Carolina and North Carolina petitioned their state legislature and the U.S. Congress in 1791 and 1797, respectively, for some of the protections afforded white people by the Constitution. Both attempts met rejection, and indeed, Black citizenship at the state level was not the norm. They were not, one southern official noted, "constituent members of our society."

Thomas Jefferson worried about excluding "a whole race of men" from the natural rights he had done much to promote. But he could never accept the idea that Black men and women could attain the level of knowledge and intelligence of white people, despite an intimate relationship with a Black woman, Sally Hemings, whom Jefferson enslaved on his Virginia plantation. He fathered children with her, yet he did not change his position on slavery. Unlike George Washington, who freed his enslaved persons after his death, Jefferson (deeply in debt) required his heirs to sell his enslaved workers upon his death, after liberating a few members of the Hemings family.

Jefferson did profess to believe Native Americans could be taught the ways of "civilization" and live as white Americans did. And indigenous groups had at least the semblance of a legal status within the nation, through treaties that assured them of land possession. But most of these treaties did not survive for long, and native groups found themselves driven farther and farther west without very much of the protection the government had promised. Efforts to teach Anglo farming methods, whereby men did the farming and women cared for the home, clashed with Native American practices and traditions.

Thus indigenous groups, African Americans, and women enjoyed virtually none of the citizenship rights offered to the white male population. It was not until 1868 that the Fourteenth Amendment guaranteed people of color born in the United States the status, if not yet the privileges, of citizenship. Native Americans were not granted birthright citizenship in the United States until the 1920s. And though some states passed woman suffrage laws in the late nineteenth century, it wasn't until 1920 that women secured ratification of the Nineteenth Amendment, giving them the ballot throughout the nation.

ADOPTION AND ADAPTATION

The delegates at Philadelphia had greatly exceeded their instructions from Congress and the states. Instead of making simple revisions to the Articles of Confederation, they had produced a plan for a completely different form of government. They feared that the Constitution would not be ratified under the rules of the Articles of Confederation, which required unanimous approval by the state legislatures. So the convention changed the rules, proposing that the new government would come into being when nine of the thirteen states ratified the Constitution and recommending that state conventions, not state legislatures, be called to ratify it.

FEDERALISTS AND ANTIFEDERALISTS

The Congress in New York accepted the convention's work and submitted it to the states for approval. All the state legislatures except Rhode Island elected delegates to ratifying conventions, most of which began meeting in early 1788. Even before the ratifying conventions convened, however, a great national debate on the new Constitution had begun.

Supporters of the Constitution had a number of advantages. Better organized than their opponents, they seized an appealing label for themselves: **Federalists**—a term that opponents of centralization had once used to describe themselves—thus implying that they were less committed to a "nationalist" government than in fact they were. In addition, the Federalists had the support of not only the two most eminent men in the United States, Ben Franklin and George Washington, but also the ablest political philosophers of their time: Alexander Hamilton, James Madison, and John Jay. Under the joint pseudonym *Publius*, these three men wrote a series of essays, widely published in newspapers throughout the nation (Federalists largely controlled the press), explaining the meaning and virtues of the Constitution. The essays were later gathered together and published as a book known today as ***The Federalist Papers***. In just one example, the papers defended the controversial "necessary and proper" clause of the Constitution, an important measure that gave Congress sweeping authority to make laws "necessary and proper" to executing the federal government's authority. The *states*, not Congress, had been given that sort of incidental power by the outgoing Articles of Confederation.

The Federalists called their critics "**Antifederalists**," suggesting that their rivals had nothing to offer except opposition. But the Antifederalists, led by such distinguished revolutionary leaders as Patrick Henry and Samuel Adams, believed themselves to be defenders of the Revolution's true principles. They argued that the Constitution would increase taxes, weaken the states, grant the central government dictatorial powers, favor the "well-born" over the common people, and abolish individual liberty. Antifederalists found the necessary and proper clause a particularly frightening transfer of powers not expressly "delegated" by the Constitution from the states to the federal government. But their biggest complaint was that the Constitution lacked a bill of rights. Only by enumerating the natural rights of the people, they argued, could there be any certainty that those rights would be protected.

Despite the efforts of the Antifederalists, ratification proceeded quickly during the winter of 1787–1788. The Delaware convention, the first to act, ratified the Constitution unanimously, as did conventions in New Jersey and Georgia. And in June 1788, New Hampshire, the critical ninth state, ratified the document. It was now theoretically possible for the Constitution to go into effect. But a new government could not hope to succeed without Virginia and New York, the largest states, whose conventions remained closely divided. By the end of June, first Virginia and then New York consented to the Constitution by narrow margins and on the assumption that a bill of rights would be added in the form of amendments to the Constitution. North Carolina's convention adjourned without taking action, waiting to see what happened with the amendments. Rhode Island, controlled by staunch opponents of centralized government, did not even consider ratification.

COMPLETING THE STRUCTURE

The first elections under the Constitution were held in the early months of 1789. There was never any doubt about who would be the first president. George Washington had presided at the Constitutional Convention, and many who had favored ratification did so only because they expected him to preside over the new government as well. Washington received the votes of all the presidential electors, and almost all of the "popular vote," which in this first presidential election incorporated only a tiny percentage even of adult white males. **John Adams**, a leading Federalist, came in second to Washington and thereby became vice president. After a journey from his estate at Mount Vernon, Virginia, marked by elaborate celebrations along the way, Washington was inaugurated in New York on April 30, 1789.

The first Congress served in many ways as a continuation of the Constitutional Convention. Its most important task was drafting a bill of rights. By early 1789, even Madison had come to agree that some such bill would be essential to legitimizing the new government. On September 25, 1789, Congress approved twelve amendments, ten of which were ratified by the states by the end of 1791. These first ten amendments to the Constitution comprise what we know as the **Bill of Rights**. Nine of them placed limitations on the new government by forbidding it to infringe on certain fundamental rights: freedom of religion, speech,

(Image courtesy National Gallery of Art)

GEORGE WASHINGTON AT MOUNT VERNON Washington was in his first term as president in 1790 when an anonymous folk artist painted this view of his home at Mount Vernon, Virginia. Washington appears in uniform, along with members of his family, on the lawn. After he retired from office in 1797, Washington returned happily to his plantation and spent the two years before his death in 1799 "amusing myself in agricultural and rural pursuits." He also played host to an endless stream of visitors from throughout the country and Europe.

and the press; immunity from arbitrary arrest; trial by jury, and others. The amendments contained language, such as the prohibition of "cruel and unusual punishment," that would remain open to interpretation and jurisprudence for generations.

Provisions for the judiciary branch were even more vague. The Constitution said only: "The judicial power of the United States shall be vested in one Supreme Court, and in such inferior courts as the Congress may from time to time ordain and establish." It was left to Congress to determine the number of Supreme Court judges to be appointed and the kinds of lower courts to be organized. In the Judiciary Act of 1789, Congress provided for a Supreme Court of six members and a system of lower district courts and courts of appeal. It also gave the Supreme Court the power to make the final decision in cases involving the constitutionality of state laws.

The Constitution also said little about the organization of the executive branch. It referred indirectly to executive departments but did not specify which ones or how many there should be. The first Congress created three such departments—state, treasury, and war—and also established the offices of the attorney general and postmaster general. To the office of secretary of the treasury, Washington appointed Alexander Hamilton of New York. For secretary of war, he chose a Massachusetts Federalist, General Henry Knox. He named Edmund Randolph of Virginia as attorney general and chose another Virginian, Thomas Jefferson, for secretary of state.

FEDERALISTS AND REPUBLICANS

The framers of the Constitution had dealt with many controversies by papering them over with a series of vague compromises. As a result, the disagreements survived to plague the new government.

At the heart of the controversies of the 1790s was the same basic difference in philosophy that had fueled the debate over the Constitution. On one side stood a powerful group who envisioned the United States as a genuine nation-state, with centralized authority and a complex commercial economy. On the other side stood thinkers who envisioned a more modest national government. Rather than aspire to be a highly commercial or urban nation, they believed the new nation should remain predominantly rural and agrarian. The centralizers became known as the Federalists and gravitated to the leadership of Alexander Hamilton. Their opponents acquired the name **Republicans** and admired the views of Thomas Jefferson as well as James Madison, who grew skeptical of Federalist rule.

Hamilton and the Federalists

For twelve years, the Federalists retained firm control over the new government. That was in part because George Washington had always envisioned a strong national government. But the president, Washington believed, should stand above political controversies, and so he avoided personal involvement in the deliberations of Congress. As a result, the dominant figure in his administration became Alexander Hamilton. Of all the national leaders of his time, Hamilton was one of the most aristocratic in his political philosophy. He believed a stable and effective government required an elite ruling class. Thus the new government needed the support of the wealthy and powerful, and to get that, it needed to give elites a stake in its success.

Hamilton proposed, therefore, that the existing public debt be "funded." This meant that the various certificates of indebtedness that the old Congress had issued during and after the Revolution—many of them now in the possession of wealthy speculators—be called in and exchanged for interest-bearing bonds. He also recommended that the states' revolutionary debts be "assumed" (taken over) to cause state bondholders to look to the central government for eventual payment. Hamilton wanted to create a permanent national debt, with new bonds being issued as old ones were paid off. He hoped to motivate the wealthy classes, who were the most likely to lend money to, and therefore perpetually support, the survival of the federal state. But Hamilton additionally believed that centralizing and assuming debt in this way would avoid defaulting on old debts and therefore lend the United States credibility on the global economic stage.

Hamilton also wanted to create a national bank. It would provide loans and currency to businesses, give the government a safe place for the deposit of federal funds, facilitate the collection of taxes and the disbursement of the government's expenditures, and provide a stable center to the nation's small and feeble banking system. The bank would be chartered by the federal government, but much of its capital would come from private investors.

The funding and assumption of debts would require new sources of revenue. Hamilton recommended two kinds of taxes to complement the receipts anticipated from the sales of public land. One was an excise tax on alcoholic beverages, a tax that would be most burdensome to the whiskey distillers of the backcountry, small farmers who converted part of their corn and rye crops into whiskey. The other was a tariff on imports, which Hamilton saw not only as a source of revenue but also as a way to protect domestic industries from foreign competition. In his famous "Report on Manufactures" of 1791, he outlined a plan for stimulating the growth of industry and spoke glowingly of the advantages to society of a healthy manufacturing sector.

The Federalists, in short, offered more than a plan for a stable new government. They offered a vision of the sort of nation the United Nations should become—a nation with a wealthy, enlightened ruling class; a vigorous, independent commercial economy; and a thriving manufacturing sector.

Enacting the Federalist Program

Few members of Congress objected to Hamilton's plan for funding the national debt, but many did oppose his proposal to exchange new bonds for old certificates of indebtedness on a dollar-for-dollar basis. Many of the original holders had been forced to sell during the hard times of the 1780s to speculators, who had bought them at a fraction of their face value. James Madison, now a House representative from Virginia, argued for a plan by which the new bonds would be divided between the original purchasers and the speculators. But Hamilton's allies insisted that the honor of the government required a literal fulfillment of its earlier promises to pay whoever held the bonds. Congress finally passed the funding bill Hamilton wanted.

Hamilton's proposal that the federal government assume state debts encountered greater difficulty. Its opponents argued that if the federal government took over the state debts, the states with small debts would have to pay taxes to service the states with large ones. Massachusetts, for example, owed much more money than did Virginia. Only by striking a bargain with the Virginians were Hamilton and his supporters able to win passage of the assumption bill.

The deal involved the location of the national capital, which the Virginians wanted near them in the South. Hamilton and Jefferson met and agreed to exchange northern support for placing the capital in the South for Virginia's votes for the assumption bill. The bargain called for the construction of a new capital city on the banks of the Potomac River, which divided Maryland and Virginia, on land to be selected by George Washington.

As for Hamilton's bank bill, Madison, Jefferson, Randolph, and others argued that because the Constitution made no provision for a national bank, Congress had no authority to create one. But Congress agreed to Hamilton's bill despite these objections, and Washington signed it. The Bank of the United States began operations in 1791.

Hamilton also had his way with the excise tax, although protests from farmers later forced revisions to reduce the burden on smaller distillers. He failed to win passage of a tariff as highly protective as he had hoped for, but the tariff law of 1792 did raise the rates somewhat.

Once enacted, Hamilton's program won the support of manufacturers, creditors, and other influential segments of the population. But others found it less appealing. Small farmers complained they were being taxed excessively. They and others began to argue that the Federalist program served the interests of a small number of wealthy elites rather than the people at large. From these sentiments, an organized political opposition arose.

THE REPUBLICAN OPPOSITION

The Constitution made no reference to political parties. Most of the framers believed that organized parties were dangerous "factions" to be avoided. Disagreement was inevitable on particular issues, but they believed that such disagreements need not and should not lead to the formation of permanent factions.

Yet not many years had passed after the ratification of the Constitution before Madison and others became convinced that Hamilton and his followers had become dangerous and self-interested. The Federalists had used the powers of their offices to reward their supporters and win additional allies. They were doing many of the same things, their opponents believed, that British governments of the early eighteenth century had done.

Because the Federalists appeared to their critics so menacing and tyrannical, there was no alternative but to organize a vigorous opposition. The result was the emergence of an alternative political organization, whose members ultimately called themselves "Democratic-Republicans" or just Republicans. (These first Republicans are not institutionally related to the modern Republican Party, which was created in the 1850s.) By the late 1790s, Republicans were going to even greater lengths than Federalists to create vehicles of partisan influence. In every state they formed committees, societies, and caucuses. Republican groups banded together to influence state and local elections. Neither side was willing to admit that it was acting as a party, nor would either concede the right of the other to exist. This institutionalized factionalism is known to historians as the "first party system."

(Image courtesy National Gallery of Art)

THE JEFFERSONIAN IDYLL American artists in the early nineteenth century were drawn to tranquil rural scenes, symbolic of the Jeffersonian vision of a nation of small, independent farmers. By 1822, when Francis Alexander painted *Ralph Wheelock's Farm*, the simple agrarian republic was already being transformed by rapid economic growth.

From the beginning, the preeminent figures among the Republicans were Thomas Jefferson and James Madison. Jefferson, the most prominent spokesman for the cause, promoted a vision of an agrarian republic in which most citizens would farm their own land. Jefferson did not scorn commercial or industrial activity. But he believed that the nation should be wary of too much urbanization and industrialization.

Although both parties had supporters across the country and among all classes, there were regional and economic differences. Federalists were most numerous in the commercial centers of the Northeast and in such southern seaports as Charleston. Republicans were stronger in rural areas of the South and West. The difference in their philosophies was visible in, among other things, their reactions to the progress of the French Revolution. As that revolution grew increasingly radical in the 1790s, the Federalists expressed horror. But the Republicans applauded the democratic, anti-aristocratic spirit they believed the French Revolution embodied.

When the time came for the nation's second presidential election, in 1792, both Jefferson and Hamilton urged Washington to run for a second term. The president reluctantly agreed. But while Washington had the respect of both factions, he was, in reality, more sympathetic to the Federalists than the Republicans.

ESTABLISHING NATIONAL SOVEREIGNTY

The Federalists consolidated their position by exerting effective and sometimes brutal control over the western territories and by their management of international diplomacy.

Securing the West

Despite the Northwest Ordinance, the old Congress had largely failed to tie the outlying western areas of the country firmly to the national government. Farmers in western Massachusetts had rebelled. Settlers in Vermont, Kentucky, and Tennessee had flirted with seceding from the Union. At first, the new government under the Constitution faced similar problems.

In 1794, farmers in western Pennsylvania raised a major challenge to federal authority when they refused to pay the new whiskey excise tax and began terrorizing tax collectors in the region. But the federal government did not leave settlement of the so-called **Whiskey Rebellion** to the authorities of Pennsylvania. At Hamilton's urging, Washington called out the militias of three states and assembled an army of nearly 15,000, and he personally led the troops into Pennsylvania. At the approach of the militiamen, the rebellion quickly collapsed.

The federal government won the allegiance of the whiskey rebels through intimidation. It secured the loyalties of other western people by accepting new states as members of the Union. The last two of the original thirteen colonies joined the Union once the Bill of Rights had been appended to the Constitution—North Carolina in 1789 and Rhode Island in 1790. Vermont became the fourteenth state in 1791, after New York and New Hampshire agreed to give up their rights to it. Next came Kentucky, in 1792, when Virginia relinquished its claim to that region. After North Carolina ceded its western lands to the Union, Tennessee became a state in 1796.

The new government faced a greater challenge in more distant areas of the Northwest and the Southwest. The ordinances of 1784-1787, establishing the terms of white settlement in the West, had produced a series of border conflicts with indigenous peoples. The new government inherited these tensions, which continued with few interruptions for nearly a decade.

Such clashes revealed another issue the Constitution had done little to resolve: the place of Native Americans within the new federal structure. The Constitution gave Congress power to "regulate Commerce . . . with the Indian tribes." And it bound the new government to respect treaty agreements negotiated by the Confederation, most of which had been struck with indigenous representatives. But none of this did very much to clarify the precise legal standing of Native Americans within the United States. The nations received no direct representation in the new government. Above all, the Constitution did not address the major issue of land. Native Americans lived within the boundaries of the United States, yet they claimed (and the white government at times agreed) that they had some measure of sovereignty over their own land. But neither the Constitution nor common law offered any clear guide to the rights of a "nation within a nation" or to the precise nature of indigenous sovereignty.

Maintaining Neutrality

A crisis in Anglo-American relations emerged in 1793, when the revolutionary French government went to war with Great Britain. Both the president and Congress took steps to establish American neutrality in the conflict, but that neutrality was severely tested.

Early in 1794, the Royal Navy began seizing hundreds of American ships engaged in trade in the French West Indies. Hamilton was deeply concerned. War would mean an end to imports from Britain, and most of the revenue for maintaining his financial system came from duties on those imports. Hamilton and the Federalists did not trust the State Department, now in the hands of the ardently pro-French Edmund Randolph, to find a solution to the crisis. So they persuaded Washington to name a special commissioner—the chief justice of the Supreme Court, John Jay—to go to England and negotiate a solution. Jay was instructed to secure compensation for the recent British assaults on American shipping, to demand withdrawal of British forces from their posts on the frontier of the United States, and to negotiate a commercial treaty with Britain.

The long and complex treaty Jay negotiated in 1794 failed to achieve all these goals. But it settled the conflict with Britain, avoiding a likely war. It provided for undisputed

CONSIDER THE SOURCE

WASHINGTON'S FAREWELL ADDRESS, *AMERICAN DAILY ADVERTISER*, SEPTEMBER 19, 1796

In this open letter to the American people, drafted by James Madison in 1792 and later revised with the aid of Alexander Hamilton, President Washington defended the young Constitution and warned against disunity among the nation's various states and political factions. Here he cautions citizens about another threat to the republic—entangling engagements abroad.

Observe good faith and justice toward all nations. Cultivate peace and harmony with all. Religion and morality enjoin this conduct. And can it be that good policy does not equally enjoin it? It will be worthy of a free, enlightened, and, at no distant period, a great nation to give to mankind the magnanimous and too novel example of a people always guided by an exalted justice and benevolence....

In the execution of such a plan nothing is more essential than that permanent, inveterate antipathies against particular nations and passionate attachments for others should be excluded, and that, in place of them just and amicable feelings toward all should be cultivated. The nation which indulges toward another an habitual hatred or an habitual fondness is in some degree a slave. It is a slave to its animosity or to its affection either of which is sufficient to lead it astray from its duty and its interest. The nation prompted by ill will and resentment sometimes impels to war the government, contrary to the best calculations of policy. The government sometimes participates in the national propensity, and adopts through passion what reason would reject....

So, likewise, a passionate attachment of one nation for another produces a variety of evils. Sympathy for the favorite nation, facilitating the illusion of an imaginary common interest in cases where no real common interest exists, and infusing into one the enmities of the other, betrays the former into a participation in the quarrels and wars of the latter without adequate inducement or justification....

As avenues to foreign influence in innumerable ways, such attachments are particularly alarming to the truly enlightened and independent patriot. How many opportunities do they afford to tamper with domestic factions to practice the arts of seduction, to mislead public opinion, to influence or awe the public councils! Such an attachment of a small or weak toward a great and powerful nation dooms the former to be the satellite of the latter.

Against the insidious wiles of foreign influence (I conjure you to believe me, fellow citizens) the jealousy of a free people ought to be constantly awake, since history and experience prove that foreign influence is one of the most baneful foes of republican government....

The great rule of conduct for us in regard to foreign nations is, in extending our commercial relation to have with them as little political connection as possible. So far as we have already formed engagements, let them be fulfilled with perfect good faith. Here let us stop.

Europe has a set of primary interests which to us have no, or a very remote, relation. Hence she must be engaged in frequent controversies, the causes of which are essentially foreign to our concerns. Hence, therefore, it must be unwise in us to implicate ourselves by artificial ties in the ordinary vicissitudes of her politics, or the ordinary combinations and collisions of her friendships or enmities.

Our detached and distant situation invites and enables us to pursue a different course. If we remain one people, under an efficient government, the period is not far off when we may defy material injury from external annoyance; when we may take such an attitude as will cause the neutrality we may at any time resolve upon to be scrupulously respected; when belligerent nations, under the impossibility of making acquisitions upon us, will not lightly hazard giving us provocation; when we may choose peace or war, as our interest, guided by justice, shall counsel.

Why forego the advantages of so peculiar a situation? Why quit our own to stand upon foreign ground? Why, by interweaving our destiny with that of any part of Europe, entangle our peace and prosperity in the toils of European ambition, rivalship, interest, humor, or caprice?

It is our true policy to steer clear of permanent alliances with any portion of the foreign world, so far, I mean, as we are now at liberty to do it. For let me not be understood as capable of patronizing infidelity to existing engagements. I hold the maxim of less applicable to public than to private affairs that honesty is always the best policy. I repeat therefore, let those engagements be observed in their genuine sense. But in my opinion it is unnecessary and would be unwise to extend them.

Taking care always to keep ourselves by suitable establishments on a respectable defensive posture, we may safely trust to temporary alliances for extraordinary emergencies.

Harmony, liberal intercourse with all nations, are recommended by policy, humanity, and interest. But even our commercial policy should hold an equal and impartial hand, neither seeking nor granting exclusive favors or preferences; . . . constantly keeping in view that it is folly in one nation to look for disinterested favors from another; that it must pay with a portion of its independence for whatever it may accept under that character; that by such acceptance it may place itself in the condition of having given equivalents for nominal favors, and yet of being reproached with ingratitude for not giving more. There can be no greater error than to expect or calculate upon real favors from nation to nation. It is an illusion which experience must cure, which a just pride ought to discard.

UNDERSTAND, ANALYZE, & EVALUATE

1. What advice did George Washington offer on foreign policy?
2. Did Washington advocate the complete isolation of the United States from Europe? Explain.
3. How did Washington characterize Europe? What circumstances of the 1790s may have inspired this assessment?

Source: www.senate.gov/artandhistory/history/minute/Washingtons_Farewell_Address.htm.

American sovereignty over the entire Northwest and produced a reasonably satisfactory commercial relationship. Nevertheless, when the terms became known in the United States, criticism was intense. Opponents of the treaty—Jeffersonian Republicans fearful that economic ties to Britain would fund and strengthen the Federalist agenda—went to great lengths to defeat it. But in the end the Senate ratified what was by then known as **Jay's Treaty**.

Jay's Treaty paved the way for a settlement of important American disputes with Spain. Under **Pinckney's Treaty** (negotiated by Thomas Pinckney and signed in 1795), Spain recognized the right of Americans to navigate the Mississippi to its mouth and to deposit goods at New Orleans for reloading on oceangoing ships; agreed to fix the northern boundary of Florida along the 31st parallel; and commanded its authorities to prevent Native Americans in Florida from launching raids north across that border. (For President Washington's views on such matters of foreign policy, see "Consider the Source: Washington's Farewell Address.")

THE DOWNFALL OF THE FEDERALISTS

Almost everyone in the 1790s agreed there was no place in a stable republic for organized parties. So to the Federalists, the emergence of the Republicans as a powerful and apparently permanent opposition seemed a grave threat to national stability and to what they considered the ideals of the Revolution. The Republicans, in turn, viewed the Federalists as pro-British monarchists who would undo their vision of what the Revolution had meant. It was through this prism of hostility and suspicion that both sides regarded the international perils that confronted the government in the 1790s. The success and sustainability of the Revolution, and even the survival of the new nation, seemed at stake to Republicans and Federalists alike. And so it was that Hamilton and his followers moved forcefully against what they considered dangerous and illegitimate dissent.

The Election of 1796

George Washington refused to run for a third term as president in 1796. Jefferson was the obvious presidential candidate of the Republicans that year, but the Federalists faced a more difficult choice. Hamilton had created too many enemies to be a credible candidate. Vice President John Adams, who was not directly associated with any of the controversial Federalist achievements, received the party's nomination for president at a caucus of Federalists in Congress.

The Federalists were still clearly the dominant party. But without Washington to mediate, they fell victim to fierce factional rivalries. Adams defeated Jefferson by only three electoral votes and assumed the presidency as head of a divided party facing a powerful opposition. Jefferson became vice president by finishing second. (Not until the adoption of the Twelfth Amendment in 1804 did electors vote separately for president and vice president.)

The Quasi War with France

American relations with Great Britain and Spain improved as a result of Jay's and Pinckney's treaties. But the nation's relations with revolutionary France quickly deteriorated. French vessels captured American ships on the high seas. The French government refused to receive Charles Cotesworth Pinckney when he arrived in Paris as the new American minister. In an effort to stabilize relations, Adams appointed a bipartisan commission to negotiate with France. When the Americans arrived in Paris in 1797, three agents of the French foreign minister, Prince Talleyrand, demanded a loan for France and a bribe for French officials before any negotiations could begin. Pinckney, a member of the commission, responded, "No! No! Not a sixpence!"

Even when Adams heard of the failure of diplomacy, he remained reluctant to go to war. Under pressure from Congress, including Federalists eager to fight the French, he sent that body the commissioners' report, though not before deleting the names of the three French agents and designated them only as Messrs. X, Y, and Z. When the report was published, the "**XYZ Affair**," as it quickly became known, provoked widespread popular outrage at France's actions and strong popular support for the Federalists' response. For nearly two years, 1798 and 1799, the United States found itself engaged in a **quasi war** with France.

Adams never asked for a declaration of war, but Congress passed measures cutting off all trade with France, nullifying the Treaty of Alliance of 1778, authorizing American vessels to capture French armed ships, and creating the Department of the Navy. The new maritime force soon won a number of battles and captured a total of eighty-five French ships. The United States also began cooperating closely with the British. At last, the French began trying to conciliate the United States. Adams sent another commission to Paris in 1800, and the new French government (headed now by "First Consul" Napoleon Bonaparte) agreed to a treaty with the United States that canceled the old agreements of 1778 and established new commercial arrangements. As a result, the "war" came to a reasonably peaceful end.

Repression and Protest

The conflict with France helped the Federalists increase their majorities in Congress in 1798. They now began to consider ways to silence the Republican opposition. The result was some of the most controversial legislation in American history: a series of measures known collectively as the **Alien and Sedition Acts**.

The Naturalization Act placed new obstacles in the way of foreigners who wished to become American citizens, a move designed by the Federalists to deprive the Republicans of voters. The Alien Friends Act and Alien Enemies Act empowered the president to imprison and deport immigrants considered to be plotting against the government. The Sedition Act allowed the government to prosecute citizens who engaged in "sedition" against the government. In theory, only libelous or treasonous activities were subject to prosecution, but since such activities had no clear definition, the law, in effect, criminalized opposition against the state. The Republicans interpreted the new laws, quite reasonably, as part of a Federalist campaign to destroy them.

President Adams signed the new laws. He may have been cautious in implementing them, declining to deport any aliens, for instance, but the laws nonetheless collectively imagined the Republicans as internal, disloyal enemies of the state. The Alien and Sedition Acts discouraged immigration, encouraged some foreigners already in the country to leave, and chilled dissent. And the administration used them to arrest and convict ten men, most of them Republican newspaper editors whose only crime had been criticism of Federalists in government.

Republican leaders contemplated ways to reverse the Alien and Sedition Acts. Some looked to the state legislatures for help. They solidified a theory to justify action by the states against the federal government in two sets of resolutions of 1798–1799, one written (anonymously) by Jefferson and adopted by the Kentucky legislature, and the other drafted by Madison and approved by the Virginia legislature. The **Virginia and Kentucky Resolutions**, as they were known, relied on the ideas of John Locke and the Tenth Amendment to the Constitution, which gave to the states powers not explicitly granted to the federal government. They argued that the federal government had been formed by a "compact," or contract, among the states and possessed only certain delegated powers. Whenever a state decided that the central government had exceeded those powers, it had the right to "nullify" the laws in question.

The Republicans did not win wide support for the nullification idea. They did, however, succeed in elevating their dispute with the Federalists to the level of a national crisis. By the late 1790s, the entire nation was deeply and bitterly politicized. State legislatures at times resembled battlegrounds. Even the U.S. Congress was plagued with violent

CONGRESSIONAL BRAWLERS, 1798 This cartoon was inspired by the celebrated fight on the floor of the House of Representatives between Matthew Lyon, a Republican representative from Vermont, and Roger Griswold, a Federalist from Connecticut. Griswold (at right) attacks Lyon with his cane, and Lyon retaliates with fire tongs. Other members of Congress are portrayed enjoying the battle.

disagreements. In one celebrated incident in the chamber of the House of Representatives, Matthew Lyon, a Republican from Vermont, responded to an insult from Roger Griswold, a Federalist from Connecticut, by spitting on Griswold. When Griswold later attacked Lyon with his cane, Lyon fought back with a pair of fire tongs, and soon the two men were wrestling on the floor. Several months later, Lyon was found guilty of violating the Sedition Act, but still managed re-election from his prison cell.

The "Revolution" of 1800

These bitter controversies shaped the presidential election of 1800. The presidential candidates were the same as four years earlier: Adams for the Federalists, Jefferson for the Republicans. But the campaign of 1800 was very different from the prior one. Adams and Jefferson themselves displayed reasonable dignity, but their supporters showed no such restraint. The Federalists accused Jefferson and his followers of dangerous radicalism, men whom if so empowered would bring on a reign of terror comparable to that of the French Revolution. The Republicans portrayed Adams as a tyrant conspiring to become king, accusing the Federalists of plotting to impose a kind of monarchical authoritarianism. The election was close, and the crucial contest was in New York. There, Aaron Burr mobilized an organization of Revolutionary War veterans, the Tammany Society, to serve as a Republican political machine. Through Tammany's efforts, the party carried the city by a large majority, and with it the state. Jefferson, it seemed, had won.

But an unexpected complication soon jeopardized the Republican victory. The Constitution called for each elector to "vote by ballot for two persons." The expectation was that an elector would cast one vote for his party's presidential candidate and the other for his party's vice presidential candidate. To avoid a tie, the Republicans had intended that one elector would refrain from voting for the party's vice presidential candidate, Aaron Burr. But when the votes were counted, Jefferson and Burr each had 73. No candidate had a majority, and the House of Representatives had to choose between the two top candidates, Jefferson and Burr. Each state delegation would cast a single vote.

The new Congress, elected in 1800 with a Republican majority, was not to convene until after the inauguration of the president, so it was the Federalist Congress that had to decide the question. After a long deadlock, several leading Federalists concluded, following Hamilton's advice, that Burr was too unreliable to trust with the presidency. On the thirty-sixth ballot, Jefferson was elected.

After the election of 1800, the only branch of the federal government left in Federalist hands was the judiciary. The Adams administration spent its last months in office taking steps to make the party's hold on the courts secure. With the Judiciary Act of 1801, the Federalists reduced the number of Supreme Court justiceships by one but greatly increased the number of federal judgeships as a whole. Adams quickly appointed Federalists to the newly created positions. He also appointed a leading Federalist, John Marshall, to be chief justice of the Supreme Court, a position Marshall held for thirty-four years. Indeed, there were charges that Adams stayed up until midnight on his last day in office to finish signing the new judges' commissions. These officeholders became known as the "midnight appointments."

Even so, the Republicans viewed their victory as almost complete. The nation had, they believed, been saved from tyranny. The exuberance with which the victors viewed the future—and the importance they ascribed to the defeat of the Federalists—was evident in the phrase Jefferson himself later used to describe his election. He called it the "**Revolution of 1800**."

CONCLUSION

The Constitution of 1787 created a federal system of dispersed authority, divided among national and state governments and among an executive, a legislature, and a judiciary. The young nation thus sought to balance its need for an effective central government against its fear of concentrated and despotic power. The ability of the delegates to the Constitutional Convention to compromise revealed their yearning for a stable political system. The same willingness to compromise allowed the greatest challenge to the ideals of the new democracy—slavery—to survive intact.

The writing and ratifying of the Constitution settled some questions about the shape of the new nation. The first twelve years under the government created by the Constitution solved others. And yet by the year 1800, a basic disagreement about the future of the nation remained unresolved, creating bitter divisions and conflicts on the political scene. The election of Thomas Jefferson to the presidency that year opened a new chapter in the nation's history, solidly repudiating the Federalist grip on power. It also brought to a close, at least temporarily and surprisingly peacefully, savage political conflicts that had seemed to threaten the nation's future.

KEY TERMS/PEOPLE/PLACES/EVENTS

Alexander Hamilton 135
Alien and Sedition Acts 151
Antifederalists 142
Bill of Rights 142
checks and balances 137
citizenship 140
Constitution 137
federalism 137
Federalists 141
James Madison 137
Jay's Treaty 149
John Adams 142
New Jersey Plan 136
Pinckney's Treaty 149
quasi war 150
Republicans 144
Revolution of 1800 153
separation of powers 137
The Federalist Papers 142
Virginia and Kentucky
 Resolutions 151
Virginia Plan 136
Whiskey Rebellion 147
XYZ Affair 150

RECALL AND REFLECT

1. How did the Constitution of 1787 attempt to resolve the weaknesses of the Articles of Confederation?
2. What role did *The Federalist Papers* play in the battle over ratification of the Constitution?
3. What were the main tenets of Alexander Hamilton's financial program?
4. What diplomatic crises did the United States face in the first decade of its existence, and how did the new government respond to these crises?
5. What was the "Revolution of 1800" and in what way was it a revolution? What were the bitter disagreements of the country's first party system?

7 THE JEFFERSONIAN ERA

THE RISE OF CULTURAL NATIONALISM
STIRRINGS OF INDUSTRIALISM
JEFFERSON THE PRESIDENT
DOUBLING THE NATIONAL DOMAIN
EXPANSION AND WAR
THE WAR OF 1812

LOOKING AHEAD

1. How successful was Jefferson's effort to create a "republican" society dominated by sturdy, independent farmers?
2. How did the Napoleonic Wars affect the United States?
3. What events and issues led to the War of 1812?

THOMAS JEFFERSON AND HIS FOLLOWERS assumed control of the national government in 1801 as the champions of a distinctive vision of the United States. They favored a society of sturdy, independent farmers, happily free from the workshops, industrial towns, and urban mobs of Europe. They celebrated localism and simplicity. Above all, they proposed a federal government of sharply limited power.

Almost nothing worked out as they had planned, for during the Republican years in power the young republic developed in ways that exposed the impracticalities of their vision. The American economy became steadily more diversified and complex, making the ideal of a simple, agrarian society difficult to maintain. American cultural life was dominated by a vigorous and ambitious nationalism. Jefferson himself contributed to the changes by exercising strong national authority at times and by arranging the greatest single increase in the size of the United States in its history.

THE RISE OF CULTURAL NATIONALISM

In many respects, American cultural life in the early nineteenth century reflected the Republican vision of the nation's future. Opportunities for education increased, the nation's literary and artistic life began to free itself from European influences, and American religion began to adjust to the spread of Enlightenment rationalism. In other ways, however, the new culture was posing a serious challenge to Republican ideals.

EDUCATIONAL AND LITERARY NATIONALISM

Central to the Republican vision of the United States was the concept of a virtuous and enlightened citizenry. Republicans believed, therefore, in the creation of a nationwide system of public schools in which all male citizens would receive free education. Such hopes were not fulfilled, as it would be many decades before some states, particularly in the South, established viable public education systems. Schooling remained primarily the responsibility of private institutions, most of which were open only to those who could afford to pay for them. In southern and mid-Atlantic states, most schools were run by religious groups. In New England, private academies were often more secular, many of them modeled on those founded by the Phillips family at Andover, Massachusetts, in 1778, and at Exeter, New Hampshire, three years later. Many were frankly aristocratic in outlook. Some educational institutions were open to the poor, but not nearly enough to accommodate everyone, and the education they offered was usually clearly inferior to that provided for more prosperous students.

Private secondary schools such as those in New England as well as many public schools generally excluded girls and young

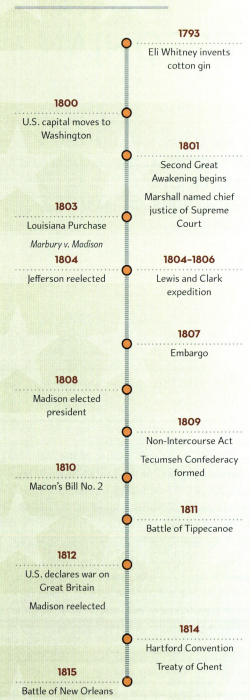

TIME LINE

1793 — Eli Whitney invents cotton gin

1800 — U.S. capital moves to Washington

1801 — Second Great Awakening begins; Marshall named chief justice of Supreme Court

1803 — Louisiana Purchase; *Marbury v. Madison*

1804 — Jefferson reelected

1804–1806 — Lewis and Clark expedition

1807 — Embargo

1808 — Madison elected president

1809 — Non-Intercourse Act; Tecumseh Confederacy formed

1810 — Macon's Bill No. 2

1811 — Battle of Tippecanoe

1812 — U.S. declares war on Great Britain; Madison reelected

1814 — Hartford Convention; Treaty of Ghent

1815 — Battle of New Orleans

women from the classroom. Yet the late eighteenth and early nineteenth centuries did see some important advances in education for women. As Americans began to place a higher value on the importance of the "republican mother" who would help train the new generation for citizenship, people began to ask how mothers could raise their children to be enlightened if the mothers themselves were uneducated. Such concerns helped speed the creation of female academies throughout the nation, usually for the daughters of affluent families. In 1789, Massachusetts required that its public schools serve all children regardless of gender. Other states, although not all, soon followed.

Some women aspired to more. In 1784, **Judith Sargent Murray** published an essay defending the right of women to education. Women and men were equal in intellect and potential, Murray argued, so their educational opportunities should be equivalent. And they should have opportunities to earn their own livings and establish roles in society apart from their husbands and families. Murray's ideas attracted relatively little support.

Because Jefferson and his followers characterized Native Americans as "noble savages" (uncivilized but not necessarily uncivilizable), they hoped that schooling indigenous children in white culture would "uplift" the nations. Missionaries and mission schools proliferated among Native Americans. There were no comparable efforts to educate enslaved African Americans.

Higher education similarly diverged from Republican ideals. The number of colleges and universities in the United States grew substantially, from nine at the time of the Revolution to twenty-two in 1800. None of the new schools, however, were truly public. Even universities established by state legislatures relied on private contributions and tuition fees to survive. Scarcely more than one white man in a thousand (and virtually no women, Blacks, or indigenous people) had access to any college education, and those few who did attend universities were, almost without exception, members of prosperous, propertied families.

MEDICINE AND SCIENCE

The modest expansion of education accompanied changes in the areas of medicine and science. The two were not always closely connected to each other in the early nineteenth century, but many physicians were working hard to strengthen the link. The University of Pennsylvania created the first American medical school in 1765. Most doctors, however, studied medicine by working with an established practitioner. Some American physicians believed in applying new scientific methods to medicine, but they had to struggle against age-old prejudices and superstitions. Efforts to teach anatomy, for example, encountered strong public hostility because of the dissection of cadavers (often stolen from graveyards) that the study required. Municipal authorities had virtually no understanding of medical science and almost no idea of what to do in the face of the severe epidemics that often swept their populations. Only slowly did they respond to warnings that the lack of adequate sanitation programs was to blame for much disease.

Individual patients often had more to fear from their doctors than from their illnesses. Even the leading advocates of scientific medicine often embraced ineffective or dangerous treatments. George Washington's death in 1799 was probably less a result of the minor throat infection that had afflicted him than of his physicians' efforts to cure him by bleeding and purging.

The medical profession also used its newfound commitment to the scientific method to justify expanding its control over kinds of care traditionally outside its domain. Most childbirths, for example, had been attended by female midwives in the eighteenth century. In the early nineteenth century, physicians began to handle deliveries themselves. Among the results of that change were a narrowing of opportunities for women and a restriction of access to childbirth care for poor mothers, who could afford midwives but not physicians.

Cultural Aspirations of the New Nation

Many Americans dreamed of an American literary and artistic life that would rival the greatest achievements of Europe. The 1772 "Poem on the Rising Glory of America" predicted that America was destined to become the "seat of empire" and the "final stage" of civilization. **Noah Webster**, the Connecticut schoolmaster, lawyer, and author of widely used American spellers and dictionaries, echoed such sentiments, arguing that the American schoolboy should be educated as a nationalist. "As soon as he opens his lips," Webster wrote, "he should rehearse the history of his own country; he should lisp the praise of liberty, and of those illustrious heroes and statesmen who have wrought a revolution in her favor."

A growing number of writers began working to create a strong American literature. Among the most popular was **Washington Irving** of New York, who won popular acclaim for his satirical histories of early American life and his powerful fables of society in the New World. His folktales, recounting the adventures of American fictional literary heroes such as Ichabod Crane and Rip Van Winkle, made him the widely acknowledged leader of American literary life in the early nineteenth century.

Religion and Revivalism

By elevating ideas of individual liberty and reason, the American Revolution had weakened traditional forms of religious practice and challenged many ecclesiastical traditions. By the 1790s, only a small proportion of white Americans were members of formal churches, and ministers were complaining about the "decay of vital piety."

Religious traditionalists were particularly alarmed about the emergence of new, "rational" religious doctrines—theologies that reflected modern, scientific attitudes. **Deism**, which had originated among Enlightenment philosophers in France, attracted such educated Americans as Jefferson and Benjamin Franklin and by 1800 was reaching a moderately broad popular audience. Deists accepted the existence of God, but they considered God a remote "watchmaker" who, after having created the universe, had withdrawn from direct involvement with the human race and its sins. Religious skepticism also produced the philosophies of "universalism" and "unitarianism." Disciples of these new ideas rejected the traditional Calvinist belief in predestination and the idea of the Trinity. Jesus was only a great religious teacher, they claimed, not the son of God.

But religious skepticism attracted relatively few people. Most Americans who remained religious clung to more traditional faiths. And beginning in 1801, those traditions staged a dramatic comeback in a series of revivalist waves known as the **Second Great Awakening**. The origins of the Awakening lay in the efforts of conservative theologians to fight the spread of religious rationalism. Presbyterians expanded their efforts on the western fringes of white settlement. Itinerant Methodist preachers traveled throughout the nation to win recruits for their new church, which soon became the fastest-growing denomination in the United States. Almost as successful were the Baptists, who found an especially fervent following in the South.

By the early nineteenth century, the revivalist energies of all these Protestant denominations were combining to create the greatest surge of evangelical fervor since the first Great Awakening sixty years before. In only a few years, membership in churches embracing revivalism was mushrooming. At **Cane Ridge**, Kentucky, in the summer of 1801, a group of evangelical ministers presided over the nation's first "camp meeting"—an extraordinary revival that lasted several days and impressed all who saw it with its fervor and its size (some estimated that 25,000 people attended). Such events became common in subsequent years.

(Library of Congress Prints and Photographs Division [LC-USZC4-772])

METHODIST CAMP MEETING, c. 1819 Camp (or revival) meetings were popular among some evangelical Christians in the United States as early as 1800. By the 1820s, there were approximately 1,000 meetings a year, most of them in the South and the West. After one such meeting in 1806, a participant wrote: "Will I ever see anything more like the day of Judgement on the side of eternity—to see the people running, yes, running from every direction to the stand, weeping, shouting, and shouting for joy. . . . O! Glorious day they went home singing shouting." This image from the 1810s suggests the degree to which women participated in many revivals.

The basic message of the Second Great Awakening was that individuals must readmit God and Christ into their daily lives. They must embrace a fervent, active piety, and they must reject the skeptical rationalism that threatened traditional beliefs. Yet the wave of revivalism did not restore the religion of the past whole cloth. Few denominations any longer accepted the idea of predestination, and the belief that people could affect their own destinies added intensity to the individual's search for salvation. The Awakening, in short, combined a more active piety with a belief in a God whose grace could be attained through faith and good works.

One of the striking features of the Awakening was the preponderance of women within it. Female converts far outnumbered males. This may have been due in part to the movement of certain kinds of work out of the home and into the factory that ultimately developed over the first half of the nineteenth century. That process ultimately robbed women of a key role as part of a household-based economy and left many feeling isolated. Religious enthusiasm provided, among other things, access to a new range of activities—charitable societies ministering to orphans and the poor, missionary organizations, and others—in which women came to play important parts over the coming decades.

In some areas of the country, revival meetings were open to people of all races. From these revivals emerged a group of Black preachers who became important figures within the enslaved community. Some of them translated the apparently egalitarian religious message of the Awakening—that salvation was available to all—into a similarly liberating message for people of color in the present world. In part out of Black revival meetings in Virginia, an elaborate plan arose in 1800 (devised by Gabriel Prosser, the brother of a Black preacher) for a slave rebellion and an attack on Richmond, inspired as well by the French and Haitian Revolutions. The plan was discovered and foiled in advance by whites, but revivalism continued in subsequent years to inflame occasional racial unrest in the South.

The spirit of revivalism was particularly strong among Native Americans. Presbyterian and Baptist missionaries were active among the southern nations and sparked a wave of conversions. But the most important revivalist was **Handsome Lake**, a Seneca whose seemingly miraculous "rebirth" after years of alcoholism helped give him a special stature within his nation. Handsome Lake blended traditionalism and change, calling for a revival of indigenous ways and a repudiation of the individualism of white society, but also encouraging Christian missionaries to become active within the nations and Iroquois men to abandon their roles as hunters and become sedentary farmers. As in much of white society, Iroquois women, who had traditionally done the farming, were to move into more domestic roles. His mixture of messages spread through the scattered Iroquois communities that had survived the military and political devastation of previous decades and inspired many indigenous people to give up whiskey, gambling, and other destructive customs derived from white society.

STIRRINGS OF INDUSTRIALISM

While Americans had been engaged in a revolution to win their independence, a momentous economic transformation had been in progress in Great Britain: the **Industrial Revolution**. Power-driven machines were making manufacturing more rapid and extensive, with profound social and economic consequences. (See "America in the World: The Global Industrial Revolution.")

Technology in the United States

Nothing comparable to the European Industrial Revolution occurred in the United States in the first two decades of the nineteenth century. Yet even while Jeffersonians warned of the dangers of rapid economic change, Americans witnessed a series of technological advances that would ultimately transform their nation. Some of these innovations were British imports. Despite efforts by the London government to prevent the export of textile machinery or the emigration of skilled mechanics, a number of immigrants with advanced technological knowledge arrived in the United States, eager to introduce the new machines in their new home. Samuel Slater, for example, used knowledge he had acquired before leaving England to build a spinning mill in Pawtucket, Rhode Island, for the Quaker merchant Moses Brown in 1790.

The United States also produced notable inventors of its own. In 1793, **Eli Whitney** developed a machine that performed the arduous task of removing the seeds from short-staple cotton quickly and efficiently. It was dubbed the cotton "gin," a derivative of "engine," a device whose impact fell almost entirely on the enslaved. Now a single operator could clean as much cotton in a few hours as it once took a group of enslaved people to do in a day. The results were profound. Previously cotton cultivation had been restricted largely to the coast and the Sea Islands, the only places where long-staple cotton—easily cleaned by hand—could be grown. With the invention of an efficient mechanical means for cleaning short-staple cotton, it too became a profitable crop, and one that could be grown throughout the South. Cotton cultivation spread, and within a decade, the total crop increased eightfold. Whatever wavering the institution of slavery had experienced after the Revolution,

(Bettmann/Getty Images)

PAWTUCKET BRIDGE AND FALLS One reason for the growth of the textile industry in New England in the early nineteenth century was ready access to water power to run machinery in the factories. That was certainly the case with Slater's Mill, one of the first American textile factories. It was located in Pawtucket, Rhode Island, alongside a powerful waterfall, demonstrating the critical importance of water power to early American industry.

as tobacco production declined, it expanded once again amidst the economic incentives of the cotton trade. The large supply of domestically produced fiber also encouraged entrepreneurs in New England and elsewhere to develop a native textile industry.

In his development of the cotton gin, Whitney helped introduce the concept of interchangeable parts to the United States. As owners expanded their use of machines, they needed access to spare parts engineered to fit the machines properly. Whitney designed not only the cotton gin, but also machine tools that could manufacture its component parts to exact specifications. The U.S. government later commissioned Whitney to manufacture 1,000 muskets for the army. Each part of the gun had to be interchangeable with the equivalent part in every other gun.

Interchangeability was of great importance in the United States because of the great distances many people had to travel to reach towns or cities and the relatively limited transportation systems available to them. Interchangeable parts meant a farmer could repair a machine himself. But interchangeability was not easy to achieve. In theory, many parts were designed to be interchangeable. In reality, the manufacturing of such parts was for many years not nearly precise enough. Farmers and others often had to do considerable fitting before the parts would work in their equipment. Not until later in the century would machine tools be developed to the point that they could make truly interchangeable parts.

AMERICA IN THE WORLD

THE GLOBAL INDUSTRIAL REVOLUTION

While Americans were engaged in a revolution to win their independence, they were also taking the first steps toward another great revolution—one that was already in progress in Europe. It was the emergence of modern industrialism. Historians differ over precisely when the Industrial Revolution began, but it is clear that by the end of the eighteenth century it was well under way in many parts of the world. A hundred years later, the global process of industrialization had transformed the societies of Britain, most of continental Europe, Japan, and the United States. Its social and economic consequences were complex and profound and continue to shape the nature of global society.

For Americans, the Industrial Revolution largely resulted from rapid changes in Great Britain, the nation with which they had the closest relations. Britain was the first nation to develop significant industrial capacity. The factory system took root in Britain in the late eighteenth century, revolutionizing the manufacture of cotton thread and cloth. One invention followed another in quick succession. Improvements in weaving drove improvements in spinning, and these changes created a demand for new devices for carding (combing and straightening the fibers for the spinner). Water, wind, and animal power continued to be important in the textile industry. But more important was the emergence of steam power, which began to proliferate after the appearance of James Watt's advanced steam engine (patented in 1769). Cumbersome and inefficient by modern standards, Watt's engine was nevertheless a major improvement over earlier "atmospheric" engines. Britain's textile industry quickly became the most profitable in the world, and it helped encourage comparable advances in other fields of manufacturing as well.

Despite the efforts of the British government to prevent the export of industrial technology, knowledge of the new machines reached other nations quickly, usually through the emigration of people with knowledge of British factories. The United States benefited the most because it received more immigrants from Great Britain than from any other country, but technology spread quickly to the nations of continental Europe as well. Belgium was the first, developing a significant coal, iron, and armaments industry in the early nineteenth century. France, profiting from the immigration of approximately fifteen thousand British workers with advanced technological skills, had created a substantial industrial capacity in textiles and metals by the end of the 1820s, which in turn contributed to a great boom in railroad construction later in the century. German industrialization progressed rapidly after 1840, beginning with coal and iron production and then, in the 1850s, moving into large-scale railroad construction. By the late nineteenth century, Germany had created some of the world's largest industrial corporations. In Japan, the sudden intrusion of American and European traders helped spur the so-called Meiji reforms of the 1880s and 1890s, which launched a period of rapid industrialization there as well.

Industrialization changed not just the world's economies but also its societies. First in England and then in continental Europe, the United States, and Japan, social systems underwent wrenching changes. Hundreds of thousands of men and women moved from rural areas into cities to work in factories, where they experienced both the benefits and the costs of industrialization. The standard of living of the new working class, when objectively quantified, was usually significantly higher than that of the rural poor, and factory laborers experienced some improvement in nutrition and other material circumstances. But the psychological costs of moving from one way of life into a fundamentally different one could outweigh the economic gains. There was little in most workers' prior experience to prepare them for the nature of industrial labor. It was disciplined, routinized work with a fixed and rigid schedule, a sharp contrast to the varying, seasonal work pattern of the rural economy. Nor were many factory workers prepared for life in the new industrial towns and expanding cities.

Industrial workers experienced, too, a fundamental change in their relationship with their employers. Unlike rural landlords and local aristocrats, factory owners and managers were usually remote and inaccessible figures. The new class of industrial **capitalists**, many of them accumulating unprecedented wealth, dealt with their workers impersonally, and the result was a growing schism between the two classes, each lacking access to or understanding of the other. Working men and women throughout the globe began thinking of themselves as a distinct class, with common goals and interests. And their efforts simultaneously to adjust to their new way of life and to resist its most damaging aspects sometimes created great social turbulence. Battles between workers and employers became a characteristic feature of industrial life throughout the world.

Life in industrial nations changed at every level. Populations grew rapidly, and people began to live longer. At the same time, pollution, crime, and infectious disease (until modern sanitation systems emerged) increased greatly in industrialized cities. Around the industrial world, middle classes expanded and came, in varying degrees, to dominate the economy, although not always the culture or the politics, of their nations.

Not since the agrarian revolution thousands of years earlier, when many humans had turned from hunting to farming for sustenance, had there been an economic change comparable to the Industrial Revolution. Centuries of traditions, social patterns, and cultural and religious assumptions were challenged and often shattered. The tentative stirrings of industrialism in the United States in the early nineteenth century were part of a vast movement that over the course of the next century transformed much of the globe. •

UNDERSTAND, ANALYZE, & EVALUATE

1. Why did the British government attempt to prevent the export of Britain's industrial technology?
2. What did the Industrial Revolution mean for ordinary people around the world?

Transportation Innovations

One of the prerequisites for industrialization is a transportation system that allows the efficient movement of raw materials to factories and of finished goods to markets. The United States had no such system in the early years of the republic, and thus it had no domestic market extensive enough to justify large-scale production. But projects were under way that would ultimately expand the transportation network.

One such project was the development of the steamboat. Britain had pioneered steam power, and even steam navigation, in the eighteenth century, and there had been experiments in the United States in the 1780s and 1790s in various forms of steam-powered

PATTERNS OF POPULAR CULTURE

Horse Racing

Informal horse racing began in North America almost as soon as Europeans settled the English colonies. Formal racing followed quickly. The first racetrack in North America—New Market (named for a popular racecourse in England)—was established in 1665 on Long Island, near present-day Garden City, New York. Tracks quickly developed wide appeal, and soon horse racing had spread up and down the Atlantic Coast. By the time of the American Revolution, it was popular in almost every colony and was moving as well into the newly settled areas of the Southwest. Andrew Jackson was a founder of the first racetrack in Nashville, Tennessee, in the early nineteenth century. Kentucky—whose native bluegrass was early recognized as ideal for grazing horses—had eight tracks by 1800.

Like almost everything else in the life of the early United States, the world of horse racing was bounded by lines of class and race. For many years, it was considered the exclusive preserve of "gentlemen," so much so that in 1674 a Virginia court fined James Bullocke, a tailor, for proposing a race, "it being contrary to Law for a Labourer to make a race, being a sport only for Gentlemen." But while white aristocrats retained control of racing, they were not the only people who participated in it. Southern planters often trained young enslaved men as jockeys for their horses, just as northern horse owners employed the services of free Black people as riders. In the North and the South, African Americans eventually emerged as some of the most talented and experienced trainers of racing horses. And despite social and legal pressures, free Blacks and poor whites often staged their own informal races.

Racing also began early to reflect the growing sectional rivalry between North and South. In 1824, the Union Race Course on Long Island established an astounding $24,000 purse for a race between two famous thoroughbreds: American Eclipse (from the North) and Sir Henry (from the South). American Eclipse won two of the three heats. A southern racehorse prevailed in another such celebrated contest in 1836. These intersectional races, which drew enormous crowds and created tremendous publicity, continued into the 1850s, until the North–South rivalry began to take a more deadly form.

Horse racing remained popular after the Civil War, but two developments changed its character considerably. One was the successful effort to drive African Americans out of the sport. At least until the 1890s, Black jockeys and trainers remained central to racing. At the first Kentucky Derby, in 1875, fourteen of the fifteen horses had African American riders. One Black man, Isaac Murphy, won a remarkable 44 percent of all races in which he rode, including three Kentucky Derbys. Gradually, however, the enforced racial segregation of so many other areas of American life penetrated racing as well. By the beginning of the twentieth century, white jockeys and organized jockey clubs had driven almost all Black riders and many Black trainers out of the sport.

The second change was the introduction of formalized betting. In the late nineteenth century, racetracks created betting systems to lure customers to the races. At the same time that the breeding of racehorses was moving into the hands of enormously wealthy families,

(Yale University Art Gallery)

"TROTTING CRACKS" ON THE SNOW, 1858 This lithograph by Louis Maurer portrays trotting racehorses hitched to sleighs. The publishing duo Currier and Ives circulated this and many other images of trotters, reflecting and contributing to the popularity of the sport in the nineteenth century.

the audience for racing was becoming increasingly working class and lower middle class. The people who now came to racetracks were mostly white men, and some white women, who were lured not by a love of horses but by the usually futile hope of quick and easy riches through gambling. •

UNDERSTAND, ANALYZE, & EVALUATE

1. Why do you think horse racing was such a popular spectator sport in the early United States? Why has it continued to be popular?
2. How did changes in the sport of horse racing reflect similar changes in American society at large?

transportation. A major advance emerged out of the efforts of the inventor **Robert Fulton** and the promoter Robert R. Livingston, who made possible the launching of a steamboat large enough to carry passengers. Their *Clermont,* equipped with paddle wheels and a British-built engine, sailed up the Hudson River in the summer of 1807.

Meanwhile, travelers witnessed the beginnings of what became known as the "turnpike era." In 1794, a corporation built a toll road running the sixty miles from Philadelphia to Lancaster, Pennsylvania, with a hard-packed surface of crushed stone that provided a good year-round surface with effective drainage, but was very expensive to construct. The Pennsylvania venture proved so successful that similar turnpikes, so named from the kind of tollgate frequently used, were laid out from other cities to neighboring towns. Like they do today, travelers on these roads paid to use them.

The process of building the turnpikes was difficult. Companies had to survey routes with many things in mind, particularly elevation. Horse-drawn vehicles had great difficulty traveling along roads with more than a five-degree incline, which required many roads to take very circuitous routes to avoid steep hills. Building pathways over mountains was an almost

COUNTRY AND CITY

Despite all these changes, the United States remained an overwhelmingly rural and agrarian nation. Only 3 percent of the population lived in towns of more than 8,000 in 1800. Even the nation's largest cities could not begin to compare with such European capitals as London and Paris, though Philadelphia, with 70,000 residents, New York, with 60,000, and others were becoming centers of commerce, learning, and urban culture comparable to many of the secondary cities of Europe.

People in cities and towns lived differently from the vast majority of Americans who continued to work as farmers. Amidst the abject poverty among many urban residents, affluent city people nonetheless sought increasing elegance and refinement in their homes, their grounds, and their dress. They also looked for diversions—music, theater, dancing, and, for many people, horse racing. Informal horse racing had begun as early as the 1620s, and the first formal race course in North America opened near New York City in 1665. By the early nineteenth century, it was a popular activity in most areas of the country, urban and more rural alike. The crowds that gathered at horse races were an early sign of the vast appetite for popular, public entertainments that would be an enduring part of American culture. (See "Patterns of Popular Culture: Horse Racing.")

It was still possible for some to believe that this small nation might not become a complex modern society. But the forces pushing such a transformation were already at work. And Thomas Jefferson, for all his commitment to the agrarian ideal, found himself as president obliged to confront and accommodate them.

JEFFERSON THE PRESIDENT

Privately, Thomas Jefferson may well have considered his victory over John Adams in 1800 to be what he later termed it: a revolution "as real . . . as that of 1776." Publicly, however, he was restrained and conciliatory, attempting to minimize the differences between the two parties and calm the passions that the bitter campaign had aroused. There was no public repudiation of Federalist policies, no true "revolution." Indeed, at times Jefferson seemed to outdo the Federalists at their own work.

THE FEDERAL CITY AND THE "PEOPLE'S PRESIDENT"

The modest character of the federal government during the Jeffersonian era was symbolized by the newly founded national capital, the city of **Washington, D.C.** There were many who envisioned that the uncompleted town, designed by the French architect Pierre L'Enfant, would soon emerge as the Paris of the United States.

In reality, throughout most of the nineteenth century Washington remained little more than a straggling, provincial village. Although the population increased steadily from the 3,200 counted in the 1800 census, it never rivaled that of New York, Philadelphia, or the other major cities of the nation and remained a raw, inhospitable community. Members of Congress viewed Washington not as a home but as a place to visit briefly during sessions of the legislature. Most lived in a cluster of simple boardinghouses in the vicinity of the Capitol. It was not unusual for a member of Congress to resign his seat in the midst of a session to return home if he had an opportunity to accept the more prestigious post of member of his state legislature.

Jefferson was a wealthy planter and lawyer by background, but as president he conveyed to the public an image of plain, almost crude disdain for pretension. Like an ordinary citizen, he walked to and from his inauguration at the Capitol. In the presidential mansion, which had not yet acquired the name "White House," he disregarded the courtly etiquette of his predecessors. He did not always bother to dress up, prompting the British ambassador to complain of being received by the president in clothes that were "indicative of utter slovenliness and indifference to appearances."

Yet Jefferson managed to impress most of those who knew him. He probably had a wider range of interests and accomplishments than any other major political figure in American history, with the possible exception of Benjamin Franklin. In addition to politics, law, and diplomacy, he was an active architect, educator, inventor, farmer, and philosopher-scientist.

Jefferson was a shrewd and practical politician. He worked hard to exert influence as the leader of his party, giving direction to Republicans in Congress by quiet and sometimes even devious means. Although the Republicans had objected strenuously to the efforts of their Federalist predecessors to build a network of influence through patronage, Jefferson

(Bettmann/Getty Images)

THOMAS JEFFERSON This 1805 portrait by the noted American painter Rembrandt Peale shows Jefferson at the beginning of his second term as president. It also conveys (through the simplicity of dress and the slightly unkempt hair) the image of democratic simplicity that Jefferson liked to project as the champion of the "common man."

used his powers of appointment as an effective political weapon. Like Washington before him, he believed that federal offices should be filled with men loyal to the principles and policies of the administration. By the end of his second term, practically all federal jobs were held by loyal Republicans. Jefferson was a popular president and had little difficulty winning reelection against Federalist Charles C. Pinckney in 1804. Jefferson won by an overwhelming electoral college majority of 162 to 14, and Republican membership of both houses of Congress increased.

Dollars and Ships

Under Washington and Adams, the Republicans believed, the government had been needlessly extravagant. Yearly federal expenditures had almost tripled between 1793 and 1800, as Hamilton had hoped. The public debt had also risen, and an extensive system of internal taxation had been erected.

The Jefferson administration moved deliberately to reverse these trends. In 1802, the president persuaded Congress to abolish all internal taxes, leaving customs duties and the sale of western lands (after expropriating them from indigenous peoples) as the only sources of revenue for the government. Meanwhile, Secretary of the Treasury Albert Gallatin drastically reduced government spending. Although Jefferson was unable to retire the national debt as he had hoped, he did cut it almost in half (from $83 million to $45 million).

For a time Jefferson also scaled down the armed forces. He reduced the already tiny army of 4,000 men to 2,500 and pared down the navy from twenty-five ships in commission to seven. Anything but the smallest of standing armies, he argued, might menace civil liberties and civilian control of government. Yet Jefferson was not a pacifist. At the same time that he was reducing the size of the army and navy, he helped establish the U.S. Military Academy at West Point, founded in 1802. And when trouble started brewing overseas, he began again to build up the fleet. Such trouble appeared first in the Mediterranean, off the coast of northern Africa.

For years the Barbary states of North Africa—Morocco, Algiers, Tunis, and Tripoli—had been demanding protection money from all nations whose ships sailed the Mediterranean. Even Great Britain regularly paid off the Barbary pirates to ensure safe passage. During the 1780s and 1790s the United States, too, had agreed to treaties providing for annual tribute to the Barbary states to protect American vessels trading in the region. But Jefferson showed reluctance to continue this policy of appeasement. "Tribute or war is the usual alternative of these Barbary pirates," he said. "Why not build a navy and decide on war?"

He got it. In 1801, the pasha (leader) of Tripoli forced Jefferson's hand. Unhappy with American responses to his demands for tribute, he ordered the flagpole of the American consulate chopped down and declared war. Jefferson built up American naval forces in the area and the Marines defeated a contingent of the pasha's forces. In 1805, Jefferson agreed to terms by which the United States ended the payment of tribute to Tripoli but paid a substantial ransom for the release of American prisoners seized by Barbary pirates.

Conflict with the Courts

Having won control of the executive and legislative branches of government, the Republicans looked with suspicion on the judiciary, which remained largely in the hands of Federalist judges. Soon after Jefferson's first inauguration, his followers in Congress launched an attack on this last preserve of the opposition. Their first step was the repeal of the Judiciary Act of 1801, thus eliminating the judgeships to which Adams had made his "midnight appointments."

The debate over the courts led to one of the most important judicial decisions in the history of the nation. Federalists had long maintained that the Supreme Court had the authority to nullify acts of Congress, and the Court itself had actually exercised the power of judicial review in 1796 when it upheld the validity of a law passed by Congress. But the Court's authority in this area would not be secure, it was clear, until it actually declared a congressional act unconstitutional. In 1803, in the case of *Marbury v. Madison*, it did so. William Marbury, one of Adams's midnight appointments, had been named a justice of the peace in the District of Columbia. But his commission, although signed and sealed, had not been delivered to him before Adams left office. When Jefferson took office, his secretary of state, James Madison, refused to hand over the commission. Marbury asked the Supreme Court to direct Madison to perform his official duty. But the Court ruled that while Marbury had a right to his commission, the Court had no authority to order Madison to deliver it. On the surface, therefore, the decision was a victory for the administration. But of much greater importance than the relatively insignificant matter of Marbury's commission was the Court's reasoning in the decision.

The original Judiciary Act of 1789 had given the Court the power to compel executive officials to act in such matters as the delivery of commissions, and it was on that basis that Marbury had filed his suit. But the Court ruled that Congress had exceeded its authority, that the Constitution defined the powers of the judiciary, and that the legislature had no right to expand them. The relevant section of the 1789 act was, therefore, unconstitutional and void. In seeming to deny its own authority, the Court was in fact radically enlarging it. The justices had repudiated a relatively minor power (the power to force the delivery of a commission) by asserting a vastly greater one (the power to nullify an act of Congress).

The chief justice of the United States at the time of the ruling (and until 1835) was **John Marshall**. A leading Federalist and prominent Virginia lawyer, he had served John Adams as secretary of state. Ironically, it was Marshall who had failed to deliver Marbury's commission. In 1801, just before leaving office, Adams had appointed him chief justice, and almost immediately Marshall established himself as the dominant figure of the Court, shaping virtually all its most important rulings, including *Marbury v. Madison*. Through a succession of Republican presidents, he battled to give the federal government unity and strength. And in so doing, he established the judiciary as a coequal branch of government with the executive and the legislature.

DOUBLING THE NATIONAL DOMAIN

In the same year Jefferson was elected president of the United States, Napoleon Bonaparte made himself ruler of France with the title of first consul. In the year Jefferson was reelected, Napoleon named himself emperor. The two men had little in common, yet for a time they were of great assistance to each other in international politics.

JEFFERSON AND NAPOLEON

Having failed in a grandiose plan to seize India from the British Empire, Napoleon began to dream of restoring French power in the New World. The territory east of the Mississippi, which France had ceded to Great Britain in 1763, was now part of the United States, but Napoleon hoped to regain the lands west of the Mississippi, which had belonged to Spain since the end of the Seven Years' War. In 1800, under the secret Treaty of San Ildefonso,

France regained title to Louisiana, which included almost the whole of the Mississippi Valley to the west of the river. The Louisiana Territory would, Napoleon hoped, become the heart of a great French empire in North America.

Jefferson was unaware at first of Napoleon's imperial ambitions on the continent. For a time he pursued a foreign policy that reflected his well-known admiration for France. But he began to reassess American relations with the French when he heard rumors of the secret transfer of Louisiana. Particularly troubling to Jefferson was French control of New Orleans, the outlet through which the produce of the fast-growing western regions of the United States was shipped to the markets of the world.

Jefferson was even more alarmed when, in the fall of 1802, he learned that Spanish officials at New Orleans (the French had not yet taken formal possession) had announced a disturbing new regulation. American ships sailing the Mississippi River had for many years been accustomed to depositing their cargoes in New Orleans for transfer to oceangoing vessels. The Spanish now forbade the practice, even though they had guaranteed Americans that right in the Pinckney Treaty of 1795.

Westerners demanded that the federal government do something to reopen the river, and the president faced a dilemma. If he yielded to the frontier clamor and tried to change the policy by force, he would run the risk of a major war with France. If he ignored the westerners' demands, he would lose political support. But Jefferson envisioned another solution. He instructed Robert Livingston, the American ambassador in Paris, to negotiate for the purchase of New Orleans. Livingston, on his own authority, proposed that the French sell the United States the rest of Louisiana as well.

(Library of Congress Prints and Photographs Division [LC-DIG-pga-00809])

THE LEVEE IN NEW ORLEANS Because of its location near the mouth of the Mississippi River, New Orleans was the principal port of western North America in the early nineteenth century. Through it, western farmers shipped their produce to markets in the East and Europe. This 1884 lithograph shows a busy traffic in goods through the port.

In the meantime, Jefferson persuaded Congress to appropriate funds for an expansion of the army and the construction of a river fleet, and he hinted that American forces might descend on New Orleans and that the United States might form an alliance with Great Britain if the problems with France were not resolved. But Napoleon required no threat to change his mind, as his plans for an empire had gone seriously awry. A yellow fever epidemic had wiped out much of the French army sent to quell the rebellion in Saint-Domingue (present-day Haiti), and the expeditionary force he wished to send to reinforce the troops had been icebound in a Dutch harbor through the winter of 1802–1803. By the time the harbor thawed in the spring of 1803, Napoleon was preparing for a renewed war in Europe. He would not, he realized, have the resources to secure an empire in North America. Napoleon thus decided to offer the United States the entire Louisiana Territory.

The Louisiana Purchase

Faced with Napoleon's sudden proposal, Livingston and James Monroe, whom Jefferson had sent to Paris to assist in the negotiations, had to decide whether they should accept it even though they had no authorization to do so. Fearful that Napoleon might withdraw the offer, they decided to proceed. After some haggling over the price, Livingston and Monroe signed an agreement with Napoleon on April 30, 1803.

By the terms of the treaty, the United States was to pay a total of $15 million to the French government, a bargain at under three cents an acre. The Americans would also grant certain exclusive commercial privileges to France in the port of New Orleans and incorporate the white residents of Louisiana into the Union with the same rights and privileges as other citizens. The boundaries of the purchase were not clearly defined. What was clear, however, was that the lands were far from uninhabited. The Osage, Kiowa, Mandan, Pawnee, Arapaho, Cheyenne, Shoshone, Omaha, Arikara, Sioux, Oto, Crow, and other indigenous groups lived within the Louisiana Purchase. Terrible smallpox epidemics had ravaged the West during the Revolution and then again in 1801. The decimation, though not destruction, of these native populations allowed white people in the east to tell themselves stories of an empty West awaiting divinely ordained American expansion.

In Washington, the president was both pleased and embarrassed when he received the treaty. He was pleased with the terms of the bargain, but he was uncertain about his authority to accept it, since the Constitution said nothing about the acquisition of new territory. Jefferson's advisers persuaded him that his treaty-making power under the Constitution would justify the purchase. Presented to Congress as a treaty, the deal was approved despite Federalist opposition. Finally, late in 1803, General James Wilkinson, a commissioner of the United States, took formal control of the territory. Before long, the Louisiana Territory would be organized into states on the general pattern of the Northwest Territory, except that for now slavery was sanctioned. The first state from the purchase admitted to the Union was Louisiana in 1812.

Exploring the West

Meanwhile, a series of explorations revealed the geography of the far-flung new territory to white Americans. In 1803, Jefferson helped plan an expedition that was to cross the continent to the Pacific Ocean, gather geographical information, and investigate prospects for trade with indigenous peoples. (See "Consider the Source: Thomas Jefferson to Meriwether Lewis.") The expedition began in May 1804. Jefferson named as its leader twenty-nine-year-old

CONSIDER THE SOURCE

THOMAS JEFFERSON TO MERIWETHER LEWIS (1803)

In the summer of 1803, between the purchase and the incorporation of the Louisiana Territory, President Jefferson sent the following instructions to the explorer Meriwether Lewis. Here Jefferson reveals not only his own expansive curiosity, but also his administration's plans for the newly acquired lands.

To Meriwether Lewis, esquire, Captain of the 1st regiment of infantry of the United States of America: Your situation as Secretary of the President of the United States has made you acquainted with the objects of my confidential message of Jan. 18, 1803, to the legislature . . . you are appointed to carry them into execution.

Instruments for ascertaining by celestial observations the geography of the country thro' which you will pass, have already been provided. Light articles for barter, & presents among the Indians, arms for your attendants, say for from 10 to 12 men, boats, tents, & other travelling apparatus, with ammunition, medicine, surgical instruments & provisions you will have prepared with such aids as the Secretary at War can yield in his department; & from him also you will receive authority to engage among our troops, by voluntary agreement, the number of attendants above mentioned, over whom you, as their commanding officer are invested with all the powers the laws give in such a case. . . .

Your mission has been communicated to the Ministers here from France, Spain & Great Britain, and through them to their governments: and such assurances given them as to it's objects as we trust will satisfy them. The country of Louisiana having been ceded by Spain to France, the passport you have from the Minister of France, the representative of the present sovereign of the country, will be a protection with all its subjects: And that from the Minister of England will entitle you to the friendly aid of any traders of that allegiance with whom you may happen to meet.

The object of your mission is to explore the Missouri river, & such principal stream of it, as, by it's course & communication with the waters of the Pacific Ocean, may offer the most direct & practicable water communication across this continent, for the purposes of commerce.

Beginning at the mouth of the Missouri, you will take observations of latitude & longitude, at all remarkable points on the river, & especially at the mouths of rivers, at rapids, at islands & other places & objects distinguished by such natural marks & characters of a durable kind, as that they may with certainty be recognized hereafter. . . . The interesting points of the portage between the heads of the Missouri & the water offering the best communication with the Pacific Ocean should also be fixed by observation, & the course of that water to the ocean, in the same manner as that of the Missouri.

Your observations are to be taken with great pains & accuracy, to be entered distinctly, & intelligibly for others as well as yourself, to comprehend all the elements necessary, with the aid of the usual tables, to fix the latitude and longitude of the places at which they were taken, & are to be rendered to the war office, for the purpose of having the calculations made concurrently by proper persons within the U.S. Several copies of these, as well as your other notes, should be made at leisure times & put into the care of the most trustworthy of your attendants, to guard by multiplying them, against the accidental losses to which they will be exposed. A further guard would be that one of these copies be written on the paper of the birch, as less liable to injury from damp than common paper.

The commerce which may be carried on with the people inhabiting the line you will pursue, renders a knowledge of these people important. You will therefore endeavor to make yourself acquainted, as far as a diligent pursuit of your journey shall admit, with the names of the nations & their numbers; the extent & limits of their possessions; their relations with other tribes or nations; their language, traditions, monuments; their ordinary occupations in agriculture, fishing, hunting, war, arts, & the implements for these; their food, clothing, & domestic accomodations; the diseases prevalent among them, & the remedies they use; moral & physical circumstances which distinguish them from the tribes we know; peculiarities in their laws, customs & dispositions; and articles of commerce they may need or furnish, & to what extent.

And considering the interest which every nation has in extending & strengthening the authority of reason & justice among the people around them, it will be useful to acquire what knowledge you can of the state of morality, religion & information among them, as it may better enable those who endeavor to civilize & instruct them, to adapt their measures to the existing notions & practises of those on whom they are to operate.

Other objects worthy of notice will be the soil & face of the country, it's growth & vegetable productions; especially those not of the U.S.; the animals of the country generally, & especially those not known in the U.S., the remains and accounts of any which may be deemed rare or extinct; the mineral productions of every kind; but more particularly metals, limestone, pit coal & saltpetre; salines & mineral waters, noting the temperature of the last, & such circumstances as may indicate their character. Volcanic appearances. Climate as characterized by the thermometer, by the proportion of rainy, cloudy & clear days, by lightening, hail, snow, ice, by the access & recess of frost, by the winds prevailing at different seasons, the dates at which particular plants put forth or lose their flowers, or leaf, times of appearance of particular birds, reptiles or insects.

[. . .]

In all your intercourse with the natives treat them in the most friendly & conciliatory manner which their own conduct will admit; allay all jealousies as to the object of your journey, satisfy them of it's innocence, make them acquainted with the position, extent, character, peaceable & commercial dispositions of the U.S., of our wish to be neighborly, friendly & useful to them, & of our dispositions to a commercial intercourse with them; confer with them on the points most convenient as mutual emporiums, & the articles of most desireable interchange for them & us. If a few of their influential chiefs, within practicable distance, wish to visit us, arrange such a visit with them, and furnish them with authority to call on our officers, on their entering the U.S. to have them conveyed to this place at public expence. If any of them should wish to have some of their young people brought up with us, & taught such arts as may be useful to them, we will receive, instruct & take care of them, such a mission.

UNDERSTAND, ANALYZE, & EVALUATE

1. At the time that Jefferson wrote this letter, who held official possession of Louisiana? What European nations were present in the Louisiana Territory?
2. What do the details of this letter reveal about Jefferson's own interest in nature and science?
3. What guidance did Jefferson offer Lewis in regard to native peoples? What policy toward Native Americans did Jefferson seem to have in mind for the future?

Source: Barth, Gunther (ed.), *The Lewis and Clark Expedition: Selections from the Journals, Arranged by Topic.* New York: Bedford St. Martin's, 1998, 18–22. Original manuscript in *Bureau of Rolls, Jefferson Papers,* ser. 1, vol. 9, doc. 269, reprinted in Thwaites, *Original Journals of the Lewis and Clark Expedition,* 7:247–252.

Meriwether Lewis, a veteran of wars against Native Americans and his personal secretary. Lewis chose as a colleague the thirty-four-year-old William Clark, an experienced frontiersman and soldier. **Lewis and Clark**, with a company of four dozen men, started up the Missouri River from St. Louis. With the Shoshone woman Sacajawea as their interpreter, they eventually crossed the Rocky Mountains, descended along the Snake and Columbia Rivers, and in the late autumn of 1805 camped on the Pacific Coast. In September 1806, they were back in St. Louis with elaborate records of the geography and the native civilizations they had observed along the way.

While Lewis and Clark explored, Jefferson dispatched groups to other parts of the Louisiana Territory. Lieutenant Zebulon Montgomery Pike, twenty-six years old, led an expedition in the fall of 1805 from St. Louis into the upper Mississippi Valley. In the summer of 1806, he set out again, proceeding up the valley of the Arkansas River and into what later became Colorado. His account of his western travels helped create an enduring (and inaccurate) impression among most Americans that the land between the Missouri River and the Rockies was an uncultivable desert.

EXPLORING THE LOUISIANA PURCHASE, 1804–1807 When Jefferson purchased the Louisiana Territory from France in 1803, he doubled the size of the nation. But few Americans knew what they had bought. The Lewis and Clark expedition set out in 1804 to investigate the new territory, and this map shows their route, along with that of another explorer, Zebulon Pike. Note the vast distances the two parties covered (including, in both cases, a great deal of land outside the Louisiana Purchase), as well as the fact these lands were already inhabited by indigenous groups. Note, too, how much of this enormous territory lay outside the orbit of even these ambitious explorations. • *What might explain why the explorers took such winding routes through the territories?*

THE BURR CONSPIRACY

Jefferson's triumphant reelection in 1804 suggested that most of the nation approved of the new territorial acquisition. But some New England Federalists raged against it. They realized that the more new states that joined the Union, the less power their region and party would retain. In Massachusetts, a group of the most extreme Federalists, known as the Essex Junto, concluded that the only recourse for New England was to secede from the Union and form a separate "northern confederacy." If such a breakaway state were to have any hope for survival, the Federalists believed, it would have to include New York and New Jersey as well as New England. But the leading Federalist in New York, Alexander Hamilton, refused to support the secessionist scheme.

Federalists in New York then turned to Hamilton's greatest political rival, Vice President Aaron Burr. Burr accepted a Federalist proposal that he become their candidate for governor of New York in 1804, and there were rumors he had also agreed to support the Federalist plans for **secession**. Hamilton accused Burr of plotting treason and made numerous private remarks, widely reported in the press, about Burr's "despicable" character. When Burr lost the election, he blamed his defeat on Hamilton's malevolence and challenged him to a duel. Hamilton feared that refusing Burr's challenge would brand him a coward. And so, on a July morning in 1804, the two men met at Weehawken, New Jersey. Hamilton was wounded and died the next day.

Burr now had to flee New York to avoid an indictment for murder. He found new outlets for his ambitions in the West. Even before the duel, he had begun corresponding with General James Wilkinson, now governor of the Louisiana Territory. Burr and Wilkinson, it seems clear, hoped to lead an expedition that would capture Mexico from the Spanish. But there were also rumors they wanted to separate the Southwest from the Union and create a western empire that Burr would rule. There is little evidence that these rumors were true.

Whether true or not, many of Burr's opponents chose to believe the rumors, including, ultimately, Jefferson himself. When Burr led a group of armed followers down the Ohio River by boat in 1806, disturbing reports flowed into Washington (the most alarming from Wilkinson, who had suddenly turned against Burr) that an attack on New Orleans was imminent. Jefferson ordered the arrest of Burr and his men as traitors. Burr was brought to Richmond for trial. But to Jefferson's chagrin, Chief Justice Marshall limited the evidence the government could present and defined the charge in such a way the jury had little choice but to acquit. Burr soon faded from the public eye. But when he learned of the Texas revolution against Mexico years later, he said, "What was treason in me thirty years ago is patriotic now."

The Burr conspiracy was in part the story of a single man's soaring ambitions and flamboyant personality. But it also exposed the larger perils still facing the new nation. With a central government that remained deliberately weak, with ambitious political leaders willing, if necessary, to circumvent normal channels in their search for power, the legitimacy of the federal government—and indeed the existence of the United States as a stable and united nation—remained tenuous.

EXPANSION AND WAR

Two very different conflicts were taking shape in the last years of Jefferson's presidency. One was the continuing tension in Europe, which in 1803 escalated once again into a full-scale conflict (the Napoleonic Wars). As fighting between the British and the French increased, each side took steps to prevent the United States from trading with the other. The other conflict occurred in North America itself, a result of the ceaseless westward expansion of white settlement, which was colliding with a native population committed to protecting its lands from intruders. In both the North and the South, the threatened nations mobilized to resist white encroachments. They began as well to forge connections with British forces in Canada and Spanish forces in Florida. The indigenous conflict on land became intertwined with the European conflict on the seas, and ultimately helped cause the War of 1812.

Conflict on the Seas

In 1805, at the Battle of Trafalgar, a British fleet virtually destroyed what was left of the French navy. Because France could no longer challenge the British at sea, Napoleon now chose to pressure Britain in other ways. The result was what he called the Continental System, designed to close the European continent to British trade. Napoleon issued a series of decrees barring British and neutral ships touching at British ports from landing their cargoes at any European port controlled by France or its allies. The British government replied to Napoleon's decrees by establishing a blockade of the European coast. The blockade required that any goods being shipped to Napoleon's Europe be carried either in British vessels or in neutral vessels stopping at British ports—precisely what Napoleon's policies forbade.

In the early nineteenth century, the United States had developed one of the most important merchant marines in the world, one that soon controlled a large proportion of the trade between Europe and the West Indies. But the events in Europe now challenged that control, because American ships were caught between Napoleon's decrees and Britain's blockade. If they sailed directly for the European continent, they risked being captured by the British navy. If they sailed by way of a British port, they ran the risk of seizure by the French. Both of the warring powers were violating the United States' rights as a neutral nation. But most Americans considered the British, with their greater sea power, the worse offender—especially since British vessels frequently stopped American ships on the high seas and seized sailors off the decks, making them victims of **impressment**.

Impressment

Many British sailors called their navy—with its floggings, low pay, and terrible shipboard conditions—a "floating hell." Few volunteered. Most had to be "impressed" (forced) into service, and at every opportunity they deserted. By 1807, many of these deserters had emigrated to the United States and joined the American merchant marine or navy. To check this loss of manpower, the British claimed the right to stop and search American merchantmen and re-impress deserters. They did not claim the right to take native-born Americans, but they did insist on the right to seize naturalized Americans born on British soil. In practice, the British navy often made no careful distinctions, impressing British deserters and native-born Americans alike.

In the summer of 1807, the British went to more provocative extremes. Sailing from Norfolk, with several alleged deserters from the British navy among the crew, the American naval frigate *Chesapeake* was hailed by the British ship *Leopard*. When the American commander, James Barron, refused to allow the British to search the *Chesapeake*, the *Leopard* opened fire. Barron had no choice but to surrender, and a boarding party from the *Leopard* dragged four men off the American frigate.

When news of the *Chesapeake-Leopard* incident reached the United States, there was a great popular clamor for revenge. Jefferson and his secretary of state James Madison tried to maintain the peace. Jefferson expelled all British warships from American waters to lessen the likelihood of future incidents. Then he sent instructions to his minister in London, James Monroe, to demand from the British government an end to impressment. Britain disavowed the actions of the *Leopard*'s commanding officer and recalled him, offered compensation for those killed and wounded in the incident, and promised to return three of the captured sailors (the fourth had been hanged). But the British cabinet refused to renounce impressment and instead reasserted its right to recover deserting seamen.

"Peaceable Coercion"

To prevent future incidents that might bring the nation again to the brink of war, Jefferson persuaded Congress to pass a drastic measure late in 1807. Known as the Embargo Act, it prohibited all foreign trade. The **embargo** was widely evaded, but it was effective enough to create a serious depression throughout most of the nation. Hardest hit were the merchants and shipowners of the Northeast, most of them Federalists.

The presidential election of 1808 came in the midst of this embargo-induced depression. James Madison was elected president, but the Federalist candidate, Charles Pinckney again, ran much more strongly than he had in 1804. The Embargo Act was clearly a growing political liability, and Jefferson decided to back down. A few days before leaving office, he approved a bill ending his experiment with what he called "peaceable coercion."

(North Wind Picture Archives/Alamy Stock Photo)

STRUGGLING WITH THE EMBARGO This cartoon shows a merchant being injured by the terms of the U.S. embargo, which is personified by the snapping turtle. The word *Ograbme* is "embargo" spelled backwards. The embargo not only enraged American merchants but also failed to resolve the maritime tensions with the British that ultimately helped lead to war in 1812.

To replace the embargo, Congress passed the Non-Intercourse Act just before Madison took office. It reopened trade with all nations but Great Britain and France. A year later, in 1810, the Non-Intercourse Act expired and was replaced by Macon's Bill No. 2, which reopened free commercial relations with those two powers but authorized the president to prohibit commerce with either belligerent if it should continue violating neutral shipping after the other had stopped. Napoleon, in an effort to induce the United States to reimpose the embargo against Britain, announced that France would no longer interfere with American shipping. Madison announced that an embargo against Great Britain alone would automatically go into effect early in 1811 unless Britain renounced its restrictions on American shipping.

In time, this new, limited embargo persuaded London to repeal its blockade of Europe. But the repeal came too late to prevent war. In any case, naval policies were only part of growing tensions between Britain and the United States.

Native Americans and the British

Given the ruthlessness with which white settlers in North America had continued to dislodge native groups, it was hardly surprising that indigenous peoples continued to look to England for protection. The British in Canada, for their part, had relied on Native Americans as partners in the lucrative fur trade. There had been relative peace in the Northwest for over a decade after Jay's Treaty and a victory over the nations for Anthony Wayne's forces at the Battle of Fallen Timbers in 1794. But the 1807 war crisis following the *Chesapeake-Leopard* incident revived the conflict between Native Americans and white settlers.

The Virginia-born **William Henry Harrison**, already a veteran of combat against Native Americans at age twenty-six, went to Washington as the congressional delegate from the Northwest Territory in 1799. An advocate of development in the western lands, he was largely responsible for the passage in 1800 of the so-called Harrison Land Law, which enabled white settlers to acquire farms from the public domain on much easier terms than before.

In 1801, Jefferson appointed Harrison governor of the Indiana Territory. Jefferson offered native groups a choice: they could convert themselves into settled farmers and become part of white society, or they could migrate west of the Mississippi. In either case, they would have to agree by treaty to give up claims to their lands in the Northwest.

Jefferson considered the assimilation policy a benign alternative to continuing conflict between white settlers and the Sauk, Shawnee, Kickapoo, Meskwaki, Wea, Miami, and other Native Americans in the territory. But to nations subject to the treaties, the new policy seemed terribly harsh, especially given the cruel efficiency with which Harrison set out to implement it. He used threats, bribes, trickery, and whatever other tactics he felt would help him conclude treaties. In the first decade of the nineteenth century, the number of white Americans who had settled west of the Appalachians had grown to more than 500,000—a population far larger than that of the Native Americans. The nations would face ever-growing pressure to move out of the way of the rapidly growing white settlements. By 1807 the United States had extracted treaty rights to eastern Michigan, southern Indiana, and most of Illinois from reluctant native leaders.

Meanwhile, in the Southwest, white Americans were taking millions of acres from Native Americans in Georgia, Tennessee, and Mississippi. Though overwhelmed as individual nations by the power of the United States, two new factors emboldened them. One was the policy of British authorities in Canada. After the *Chesapeake* incident, they began to expect an American invasion of Canada and therefore renewed efforts to forge alliances with indigenous peoples. A second and more important factor was the rise of two remarkable native leaders, Tenskwatawa and **Tecumseh**.

Tecumseh and the Prophet

Tenskwatawa was a charismatic religious leader and orator known as "**the Prophet**." Like Handsome Lake, he had experienced a mystical awakening in the process of recovering from alcoholism. Having freed himself from what he considered the evil effects of white culture, he began to speak to his people of the superior virtues of native civilization and the sinfulness and corruption of the white world. In the process, he inspired a religious revival that spread through numerous nations and helped unite them. The Prophet's headquarters in present-day Indiana, at the meeting of Tippecanoe Creek and the Wabash River (known as Prophetstown), became a sacred place for native peoples of the region. Out of their common religious experiences, they began to consider joint military efforts as well. Tenskwatawa advocated total separation from white Americans and a culture rooted in tradition. The effort to trade with the Anglos and to borrow from their culture would, he argued, lead to the death of native ways.

Tecumseh—the chief of the Shawnee known as "the Shooting Star"—was in many ways more militant than his brother Tenskwatawa. "Where today are the Pequot," he thundered. "Where are . . . the other powerful tribes of our people? They have vanished before the avarice and oppression of the white man." He warned of his people's extermination if they did not take action against the white Americans moving into their lands.

Tecumseh understood that only through united action could he hope to resist the steady advance of white civilization. Beginning in 1809, he set out to unite all the nations of the Mississippi Valley into what became known as the Tecumseh Confederacy. Together, he promised, they would halt white expansion, recover the whole Northwest, and make the Ohio River the boundary between the United States and native country. He maintained that Harrison and others, by negotiating treaties with individual nations, had obtained no real title to land. The land belonged to all the indigenous peoples; none of them could rightfully cede any of it without the consent of the others. In 1811, Tecumseh left Prophetstown and traveled down the Mississippi to visit the nations of the South and persuade them to join the alliance.

During Tecumseh's absence, Governor Harrison saw a chance to destroy the growing influence of the two leaders. With 1,000 soldiers, he camped near Prophetstown, and on November 7, 1811, he provoked an armed conflict. Although the white forces suffered losses as heavy as those of the Native Americans, Harrison drove off the indigenous people and burned the town. The Battle of Tippecanoe, named for the creek, disillusioned many of the Prophet's followers, and Tecumseh returned to find the confederacy in disarray. But there were still warriors eager for combat, and by the spring of 1812 they were raiding white settlements along the frontier.

The mobilization of the nations resulted largely from indigenous initiative, but Britain's agents in Canada had encouraged and helped supply the uprising. To Harrison and most white residents of the regions, there seemed only one way to make the West safe for Americans: drive the British out of Canada and annex that province to the United States.

Florida and War Fever

While white frontiersmen in the North demanded the conquest of Canada, those in the South looked to the acquisition of Spanish Florida. The territory was a continuing threat to whites in the southern United States. Enslaved people escaped across the Florida border, and Native Americans there launched frequent raids north. But white Southerners also coveted the Florida panhandle's network of rivers that could provide residents of the Southwest with access to valuable ports on the Gulf of Mexico.

In 1810, American settlers in West Florida (presently part of Mississippi, Alabama, and Louisiana) seized the Spanish fort at Baton Rouge and asked the federal government to annex the territory to the United States. President Madison happily agreed and then began planning to get the rest of Florida, too. The desire for Florida became yet another motivation for war with Britain. Spain was Britain's ally, and a war might provide an excuse for taking Spanish as well as British territory.

By 1812, war fever was raging on both the northern and southern borders of the United States. The demands of the residents of these areas found substantial support in Washington among a group of determined young congressmen who earned the name **War Hawks**.

In the congressional elections of 1810, voters elected a large number of representatives of both parties eager for war with Britain. The most influential of them came from the new states in the West or from the backcountry of the old states in the South. Two of their leaders, both recently elected to the House of Representatives, were Henry Clay of Kentucky and John C. Calhoun of South Carolina, men of great intellect, magnetism, and ambition. Both were supporters of war with Great Britain. Clay was elected Speaker of the House in 1811, and he appointed Calhoun to the crucial Committee on Foreign Affairs. Both men began agitating for the conquest of Canada. Madison still preferred peace but was losing control of Congress. On June 18, 1812, he approved a declaration of war against Britain.

THE WAR OF 1812

The British were not eager for conflict with the United States. Even after the Americans declared war, Britain largely ignored them for a time, occupied as they were with fighting the French in the Napoleonic Wars. But in the fall of 1812, Napoleon launched a catastrophic campaign against Russia that left his army in disarray. By late 1813, with the French Empire on its way to final defeat, Britain was able to turn its military attention to the United States.

BATTLES WITH THE NATIONS

In the summer of 1812, American forces invaded Canada through Detroit. They soon had to retreat back to Detroit and in August surrendered the fort there. Other invasion efforts also failed. In the meantime, Fort Dearborn (later Chicago) fell before an indigenous attack.

Things went only slightly better for the United States on the seas. At first, American frigates won some spectacular victories over British warships. But by 1813, the British navy was counterattacking effectively, driving the American frigates to cover and imposing a blockade on the United States.

The United States did, however, achieve significant early military successes on the Great Lakes. First, the Americans took command of Lake Ontario, permitting them to raid and burn York (now Toronto), the capital of Canada. American forces then seized control of Lake Erie, mainly through the work of the young Oliver Hazard Perry, who engaged and dispersed a British fleet at Put-in-Bay on September 10, 1813. This made possible, at last, a more successful invasion of Canada by way of Detroit. William Henry Harrison pushed up the Thames River into upper Canada and on October 5, 1813, won a victory notable for the death of Tecumseh, who was serving as a brigadier general in the British army. The Battle of the Thames resulted in no lasting occupation of Canada, but it weakened and disheartened the Native Americans of the Northwest.

In the meantime, another white military leader was striking an even harder blow at native peoples in the Southwest. Some members of the internally divided Creek, supplied by the Spaniards in Florida, had been attacking white settlers near the Florida border. Andrew Jackson, a wealthy Tennessee planter and a general in the state militia, set off with a small number of Creek allies in pursuit of the Red Stick Creek. On March 27, 1814, in a massacre at Horseshoe Bend, Jackson's men took terrible revenge, slaughtering women and children along with warriors. The nation agreed to cede most of its lands to the United States and would eventually retreat westward. The vicious "battle" also won Jackson a commission as major general in the U.S. Army, and in that capacity he led his men south into Florida. On November 7, 1814, he seized the Spanish fort at Pensacola.

BATTLES WITH THE BRITISH

But these victories did not end the war. After the surrender of Napoleon in 1814, Britain decided to invade the United States. A British armada sailed up the Patuxent River from

THE WAR OF 1812 This map illustrates the military maneuvers of the British and the Americans during the War of 1812. It shows all the theaters of the war, from New Orleans to southern Canada, the extended land and water battle along the Canadian border and in the Great Lakes, and the fighting around Washington and Baltimore. Note how in all these theaters there are about the same number of British and American victories. • *What finally brought this inconclusive war to an end?*

Chesapeake Bay and landed an army that marched to nearby Bladensburg, on the outskirts of Washington, where it dispersed a poorly trained force of American militiamen. On August 24, 1814, British troops entered Washington and put the government to flight. Then they set fire to several public buildings, including the White House, in retaliation for the earlier American burning of the Canadian capital at York.

Leaving Washington in partial ruins, the invading army proceeded up the bay toward Baltimore. But that city, guarded by Fort McHenry, was prepared. To block the approaching fleet, the American garrison had sunk several ships in the Patapsco River (the entry to Baltimore's harbor), thus forcing the British to bombard the fort from a distance. Through the night of September 13, Francis Scott Key, a Washington lawyer on board one of the British ships to negotiate the return of prisoners, watched the bombardment. The next morning, "by the dawn's early light," he could see the flag on the fort still flying. He recorded his pride in the moment by writing a poem, "The Star-Spangled Banner." The British withdrew from Baltimore, and Key's words were soon set to the tune of an old English drinking song. (In 1931 "The Star-Spangled Banner" became the official national anthem.)

Meanwhile, American forces repelled another British invasion in northern New York. At the Battle of Plattsburgh, on September 11, 1814, they turned back a much larger British naval and land force. In the South, a formidable array of battle-hardened British veterans landed below New Orleans and prepared to advance north up the Mississippi. Awaiting the British was Andrew Jackson with a multiracial contingent of Tennesseans, Kentuckians, Creoles, pirates, and regular army troops drawn up behind earthen breastworks. On January 8, 1815, the redcoats advanced on the American fortifications, but the exposed British forces were no match for Jackson's well-protected men. After the Americans had repulsed several waves of attackers, the British finally retreated, leaving behind 700 dead, 1,400 wounded, and 500 prisoners. Jackson's losses were 8 killed and 13 wounded. Only later did news reach North America that the United States and Britain had signed a peace treaty several weeks before the Battle of New Orleans.

The Revolt of New England

With a few notable exceptions, the military efforts of the United States between 1812 and 1815 had failed. As a result, the Republican government became increasingly unpopular. In New England, opposition both to the war and to the Republicans was so extreme that some Federalists celebrated British victories. In Congress, in the meantime, the Republicans had continual trouble with the Federalist opposition, led by the young New Hampshire congressman Daniel Webster.

By now the Federalists were in the minority in the country, but they were still the majority party in New England. Some of them began to dream once again of creating a separate nation. Talk of secession reached a climax in the winter of 1814–1815.

On December 15, 1814, delegates from the New England states met in Hartford, Connecticut, to discuss their grievances against the Madison administration. The would-be seceders at the **Hartford Convention** were outnumbered by a comparatively moderate majority. But while the convention's report only hinted at secession, it reasserted the right of nullification and proposed seven amendments to the Constitution designed to protect New England from the growing influence of the South and the West.

Because the war was going so badly, the New Englanders assumed that the Republicans would have to agree to their demands. Soon after the convention adjourned, however, the news of Jackson's victory at New Orleans reached the cities of the Northeast. A day or two later, reports of a peace treaty arrived from abroad. In the changed atmosphere, the aims of the Hartford Convention and the Federalist Party came to seem futile, irrelevant, even treasonable. The party would never recover from those associations with disloyalty.

THE PEACE SETTLEMENT

Negotiations between the United States and Britain began in August 1814, when American and British diplomats met in Ghent, Belgium. John Quincy Adams, Henry Clay, and Albert Gallatin led the American delegation. Although both sides began with extravagant demands, the final treaty did little except end the fighting itself. The Americans gave up their demand for a British renunciation of impressment and for the cession of Canada to the United States. The British abandoned their call for the creation of an indigenous buffer state in the Northwest and made other, minor territorial concessions. The treaty was signed on Christmas Eve 1814.

Both sides had reason to accept this skimpy agreement. The British, exhausted and in debt from their prolonged conflict with Napoleon, were eager to settle the lesser dispute in North America. The Americans realized that with the defeat of Napoleon in Europe, the British would no longer have much incentive to interfere with American commerce.

Other settlements followed the Treaty of Ghent. A commercial treaty in 1815 gave Americans the right to trade freely with England and much of the British Empire. The Rush-Bagot agreement of 1817 provided for mutual disarmament on the Great Lakes.

For Native Americans east of the Mississippi, the conflict dealt another disastrous blow to their ability to resist white expansion. Tecumseh was dead. The British were gone from the Northwest. And the alliance built by Tecumseh and the Prophet had collapsed. As the end of the war spurred a new white movement westward, indigenous peoples were less able than ever to defend their land.

CONCLUSION

Thomas Jefferson called his election to the presidency the "Revolution of 1800," and his supporters believed that his victory would bring a dramatic change in the character of the nation—a retreat from Hamilton's dreams of a powerful, developing nation and a return to an ideal of a simple agrarian republic.

The changes to American society in this period in some ways undermined that Jeffersonian dream. The nation's population was expanding and diversifying. Its cities were growing, and its commercial life was becoming ever more important. Yet in 1803, the Jefferson administration made one of the most important contributions to the growth of the United States: the Louisiana Purchase, which dramatically expanded the physical boundaries of the nation and extended white settlement, and thus the agrarian republic, deeper into the continent. In the process, it greatly widened the battles between Europeans and Native Americans.

The growing national pride and commercial ambitions of the United States gradually created another serious conflict with Great Britain: the War of 1812, a war that was settled finally in 1814 on terms at least mildly favorable to the United States. By then, the bitter party rivalries that had characterized the first years of the republic had to some degree subsided, and the nation was poised to enter what became known, quite inaccurately, as the "era of good feelings."

KEY TERMS/PEOPLE/PLACES/EVENTS

Cane Ridge 158
capitalists 163
deism 158
Eli Whitney 160
embargo 177
Handsome Lake 160
Hartford Convention 182
impressment 176

Industrial Revolution 160
John Marshall 169
Judith Sargent Murray 157
Lewis and Clark 174
Marbury v. Madison 169
Noah Webster 158
Robert Fulton 165
secession 175

Second Great Awakening 158
Tecumseh 178
the Prophet
 (Tenskwatawa) 179
War Hawks 180
Washington Irving 158
Washington, D.C. 166
William Henry Harrison 178

RECALL AND REFLECT

1. What was the impact of the Second Great Awakening on women, African Americans, and Native Americans?
2. What was the long-term significance of the *Marbury v. Madison* ruling?
3. What were the implications of the Louisiana Purchase?
4. What foreign entanglements and questions of foreign policy did Jefferson have to deal with during his presidency? How did these affect his political philosophy?
5. What were the consequences of the War of 1812?

Design element: Stars and Stripes: McGraw Hill Education.

8 | EXPANSION AND DIVISION IN THE EARLY REPUBLIC

STABILIZING ECONOMIC GROWTH
EXPANDING WESTWARD
THE "ERA OF GOOD FEELINGS"
SECTIONALISM AND NATIONALISM
THE REVIVAL OF OPPOSITION

LOOKING AHEAD

1. How did the economic developments and territorial expansion of this era affect American nationalism?
2. What was the "era of good feelings," and why was it given that name?
3. How did the Marshall Court seek to establish a strong national government?

LIKE A "FIRE BELL IN THE NIGHT," as Thomas Jefferson described it, the issue of slavery arose after the War of 1812 to threaten the unity of the nation. The debate began when the territory of Missouri applied for admission to the Union, raising the question of whether it would be a free or a slaveholding state. But the larger issue, one that would rise again and again to plague the republic, was whether the vast new western regions of the United States would ultimately align politically with the North or the South.

The Missouri crisis, settled in 1820, was a sign of sectional crises to come. But at the time it stood in sharp contrast to the rising spirit of American nationalism that emerged after the War of 1812. Even as slavery was slowly pulling the nation apart, other social forces drew it together and promoted a sense of unity among many white Americans. A set of widely shared sentiments and ideals served to unite them during what historians sometimes dub the "Era of Good Feelings": the memory of the Revolution, the veneration of the Constitution, and the belief that the United States had a special mission in the world

TIME LINE

1815 Treaties of Portage des Sioux

1816 Second Bank of U.S.; Monroe elected president

1818 Seminole War ends

1819 Panic and depression; Dartmouth College v. Woodward; McCulloch v. Maryland

1820 Missouri Compromise; Monroe reelected

1823 Monroe Doctrine

1824 John Quincy Adams elected president

1828 Tariff of abominations; Jackson elected president

STABILIZING ECONOMIC GROWTH

After the War of 1812, the United States continued its economic growth and territorial expansion. But a vigorous postwar boom eventually led to a disastrous bust in 1819. The end of the Napoleonic Wars sparked an artificial bump in demand for American crop exports, which led to a speculative boom fueled by greedy banks. When the Second Bank of the United States grew fearful that financial institutions had overextended credit and made poor investments, it yanked its credit chain and caused a financial panic.

THE GOVERNMENT AND ECONOMIC GROWTH

The War of 1812 produced chaos in shipping and banking and dramatically exposed the inadequacy of the nation's transportation and financial systems. Following the war, new efforts arose to strengthen national economic development.

The wartime experience underlined the need for another national bank. After the expiration of the first bank's charter, a large number of state banks had issued vast quantities of banknotes, creating a confusing variety of currency of widely differing value. It was difficult to tell what any banknote was really worth, and counterfeiting was easy. In response to these problems, Congress chartered a second Bank of the United States in 1816, much like its predecessor of 1791 but with more capital. The national bank could not forbid state banks from issuing notes, but its size and power enabled it to compel the state banks to issue only sound notes or risk being forced out of business.

Congress also acted to promote manufacturing, which the war (by cutting off imports) had already greatly stimulated. The American textile industry, in particular, had grown dramatically. Between 1807 and 1815, the total number of cotton spindles in the country increased more than fifteenfold, from 8,000 to 130,000. Before the war, the textile factories clustered in New England produced only yarn and thread while families operating hand looms did the actual weaving of cloth at home. This began to change when the Boston merchant **Francis Cabot Lowell**, after examining British textile machinery, developed a power loom that improved upon it. In 1813, in Waltham, Massachusetts, Lowell founded the first mill in the United States to commercially spin and weave under a single roof.

The end of the war dimmed the prospects for American industry. British ships now swarmed into American ports and unloaded cargoes of manufactured goods, many priced below cost. In response, in 1816, protectionists in Congress passed a tariff law that effectively limited competition from abroad on a wide range of items, including cotton cloth, despite objections from agricultural interests, who objected to paying higher prices for manufactured goods.

Transportation

A pressing economic need of the early republican period was a better transportation system to link the vast territories of the growing United States. Questions arose as to whether the federal government should help finance roads and other "internal improvements." The idea of using government funds to finance road building was not a new one. When Ohio entered the Union in 1803, the federal government agreed that part of the proceeds from the sale of public lands there should finance road construction. And in 1807, Congress enacted a law proposed by the Jefferson administration that permitted using revenues from Ohio land sales to finance a National Road from the Potomac River to the Ohio. By 1818, the National Road ran as far as Wheeling, Virginia, on the Ohio River; and the Lancaster Pike, financed in part by the state of Pennsylvania, extended westward to Pittsburgh.

At the same time, steam-powered shipping was expanding rapidly. By 1816, river steamers were sailing up the Mississippi to the Ohio River and up the Ohio as far as Pittsburgh. Steamboats were soon carrying more cargo on the Mississippi than all the earlier forms of river transport combined—flatboats, barges, and others. They stimulated the agricultural

(Courtesy National Gallery of Art, Washington)

STEAMBOATS ON THE HUDSON In 1807, inventor Robert Fulton introduced an engine that could propel a boat from Manhattan to Albany, a distance of about 150 miles, in 32 hours. His steam-powered vessels were the first to be large and reliable enough for commercial use. This painting from 1854 by James Bard depicts the towboat "John Birkbeck."

economy of the West and the South by providing cheaper access to markets and enabling eastern manufacturers to send their finished goods west much more readily.

Nevertheless, serious gaps in the nation's transportation network remained, as the War of 1812 exposed. Once the British blockade had cut off Atlantic shipping, the coastal roads had become choked by an unaccustomed volume of north-south traffic. Congress passed a bill, introduced by Representative John C. Calhoun, that would make use of government funds to finance internal improvements. But President James Madison, on his last day in office, vetoed it. He believed that Congress lacked authority to fund such improvements without a constitutional amendment. For a time, state governments and private enterprise were left on their own to build the transportation network necessary for the growing American economy.

EXPANDING WESTWARD

Another reason for the rising interest in internal improvements was the dramatic westward surge of white Americans. By 1820, white settlers had pushed well beyond the Mississippi River, and the western population of citizens was increasing more rapidly than the rest of the nation.

Westward Migration

The westward movement of Euro-Americans was one of the most important developments of the early republican period and the nineteenth century more broadly. "Going West" at this time meant moving to territory east of the Mississippi River, along with Missouri and parts of Louisiana. Many flooded into what they specifically called the "Old Northwest," which is part of present-day Missouri.

People moved for several reasons. One was simply population growth, which drove many white Americans out of the crowded East in search of new opportunities. Between 1800 and 1820, the American population nearly doubled—from 5.3 million to 9.6 million. Most Americans were still farmers, and the agricultural lands of the East were by now largely occupied or exhausted. In the South, landless whites confronted a growing plantation system that made it difficult to purchase land. Land in the West was more plentiful than in the East or South. In the aftermath of the War of 1812, the federal government continued its policy of pushing Native Americans westward, signing the Treaties of Portage des Sioux in 1815 that took more land from indigenous inhabitants.

White Settlers in the Old Northwest

Having arrived at their new lands, most settlers built lean-tos or cabins, hewed clearings out of the forest, and planted crops of corn to supplement wild game and domestic animals. It was a rough and lonely existence. Men, women, and children worked side by side in the fields and had little outside contact for weeks or months.

Life in the western territories was not, however, entirely solitary or individualistic. Migrants often journeyed westward in groups and built communities with schools, churches, and stores. The labor shortage in the interior led neighbors to develop systems of mutual aid. They gathered periodically to raise a barn, clear land, or harvest crops.

Another common feature of life in the Old Northwest was mobility. Individuals and families were constantly on the move, settling for a few years in one place and then selling their land (often at a significant profit) and resettling somewhere else. When new areas for settlement opened farther to the west, it was often the people already on the western edges of white settlement—rather than those coming from the East—who flocked to them first.

The Plantation System in the Old Southwest

In the Old Southwest (later known as the Deep South), the new agricultural economy emerged along different lines. The market for cotton continued to grow, and the Old Southwest contained a broad zone where cotton could thrive. That zone became known as the Black Belt, a region of dark, productive soil in Alabama, Mississippi, and Georgia. It became a center of slavery, where enslaved laborers provided the agricultural know-how and muscle to plant and harvest tens of thousands of acres of cotton and soy that made slaveholders and planters fabulously wealthy.

The first whites to arrive in the Old Southwest were usually small farmers who made rough clearings in the forest. Success in the wilderness was by no means assured, even for the wealthiest settlers. Many managed to do little more than subsist in their new environment, while others experienced utter ruin. But some planters did very well, and bought up cleared land from others as the original settlers moved farther west. They also built up large enslaved workforces as they expanded small clearings into vast cotton fields. Their success inspired wealthier planters from other territories and states to join them. Eventually they replaced the cabins of the early pioneers with more sumptuous log dwellings and ultimately imposing mansions to signify their wealth and their ownership of enslaved people.

The rapid seizure and settlement of the Old Northwest and Southwest resulted in the admission of four new states to the Union: Indiana in 1816, Mississippi in 1817, Illinois in 1818, and Alabama in 1819.

Trade and Trapping in the Far West

In the early decades of the nineteenth century, few Euro-Americans ventured into the far western areas of the continent. The lands comprising what is now Texas, California, and much of the rest of the far Southwest belonged to the Spanish colony of New Spain. But the revolutionary fervor of the age stimulated an independence movement, and in 1821 insurgents declared victory, replacing New Spain with the independent Mexican Empire. Several years later the Mexican Empire became a republic.

After independence, Mexico almost immediately opened its northern territories to trade with and settlement by Americans. The new government hoped that settlers, who were to become Mexican citizens, would help secure their northern border, and that traders would strengthen their connection to the continental economy. Instead, American traders quickly displaced Native American and Mexican traders. In New Mexico, for example, the Missouri merchant William Becknell began to offer American manufactured goods for sale in 1821, priced considerably below the Mexican goods that had dominated the market in the past. Mexico effectively lost its markets in its own territory and a steady traffic of commercial wagon trains began moving back and forth along the Santa Fe Trail between Missouri and New Mexico. Over in Texas, American land speculators like Moses Austin and his son Stephen sold off parcels of their huge land grants from Mexico to small farmers. Rather than assimilating into the Mexican state, the Texas settlers maintained a separate national identity.

Fur trading became more profitable and organized. After the War of 1812, John Jacob Astor's American Fur Company and other firms extended their operations from the Great Lakes area westward to the Rockies. At first, fur traders did most of their business by purchasing pelts from indigenous peoples. But increasingly, white trappers entered the region and joined the Iroquois and other native groups in pursuit of beaver and other furs.

(Yale University Art Gallery)

THE TRAPPERS' CAMP-FIRE This illustration by British artist F. F. Palmer imagines a moment of camaraderie among trappers, with the Rocky Mountains in the background.

The trappers, or "mountain men," who began trading in the Far West were small in number and mostly young, single men. But they developed important commercial relationships with the Native American and Mexican residents of the West. Some entered into intimate relationships with native and Mexican women. They also recruited women as helpers in the difficult work of preparing furs and skins for trading. In some cases, though, white trappers clashed violently with native inhabitants.

In 1822, Andrew and William Ashley founded the Rocky Mountain Fur Company and recruited white trappers to move permanently into the Rockies. The Ashleys dispatched supplies annually to their trappers in exchange for furs and skins. The arrival of the supply train became the occasion for a gathering of scores of mountain men, some of whom lived much of the year in considerable isolation. However isolated their daily lives, these mountain men were closely bound up with the expanding market economy, an economy in which the bulk of profits from the trade flowed to the merchants, not the trappers.

EASTERN IMAGES OF THE WEST

Americans in the East were only dimly aware of the world of the trappers. They were more aware of the explorers, many of them dispatched by the U.S. government. In 1819 and 1820, the War Department ordered **Stephen H. Long** to journey up the Platte and South Platte Rivers through what is now Nebraska and eastern Colorado (where he discovered the peak that would be named for him). He then returned eastward along the Arkansas River, through what is now Kansas. Long wrote an influential report on his trip, which echoed the dismissive conclusions of Zebulon Pike fifteen years before. The region "between the Missouri River and the Rocky Mountains," Long wrote, "is almost wholly unfit for cultivation, and of course uninhabitable by a people depending upon agriculture for their subsistence." On the published map of his expedition, he labeled the Great Plains the "Great American Desert."

THE "ERA OF GOOD FEELINGS"

The expansion of the economy, the growth of white settlement and trade in the West, the creation of new states—all reflected the rising spirit of nationalism that was spreading through the United States in the years following the War of 1812. That spirit found reflection for a time in the character of early republican national politics.

The End of the First Party System

Ever since 1800, the presidency seemed to have been the possession of Virginians. After two terms in office, Jefferson secured the presidential nomination for his secretary of state, James Madison, and after two more terms, Madison did the same for *his* secretary of state, James Monroe, in 1816. Many in the North resented the so-called Virginia Dynasty, but the Republicans had no difficulty electing their candidate that year. Monroe received 183 ballots in the electoral college. His Federalist opponent, Rufus King of New York, received only 34.

Monroe entered office under what seemed to be favorable circumstances. With the decline of the Federalists amid their disloyal talk during the War of 1812, his party faced no serious opposition. And with the conclusion of that conflict, the nation faced no important international threats. Some American politicians had dreamed since the first days of the republic of a time in which partisan divisions and factional disputes might come to an end. In the prosperous postwar years, Monroe attempted to use his office to realize that dream.

He made this aim clear, above all, in the selection of his cabinet. For secretary of state, he chose former Federalist **John Quincy Adams** of Massachusetts, son of the second president. Jefferson, Madison, and Monroe had all served as secretary of state before becoming president. Adams thus became the heir apparent, suggesting the Virginia Dynasty would soon come to an end. Speaker of the House Henry Clay declined an offer to be secretary of war, so Monroe named John C. Calhoun instead.

Soon after his inauguration, Monroe made a goodwill tour through the country. In New England, so recently the scene of rabid Federalist discontent, he was greeted everywhere with enthusiastic demonstrations. The *Columbian Centinel,* a Federalist newspaper in Boston, observed that an "era of good feelings" had arrived. And on the political surface, at least, that seemed to be the case. In 1820, Monroe was reelected without opposition. For all practical purposes, the Federalist Party had ceased to exist. To be sure, the notion of living in an "era of good feelings" was a slight of hand, as the period did not lack social or economic turmoil. Nor was it a time when the issue of slavery was settled.

John Quincy Adams and Florida

John Quincy Adams had spent much of his life in diplomatic service before becoming secretary of state. He was a committed nationalist, and he considered his most important task to be the promotion of American expansion.

His first challenge was Florida. The United States had already annexed West Florida, but that claim was in dispute. Most Americans, moreover, still believed the nation should gain possession of the entire peninsula. In 1817, Adams began negotiations with the Spanish minister, Luis de Onís, over the territory.

In the meantime, however, events in Florida were taking their own dangerous course. Andrew Jackson, now in command of American troops along the Florida frontier, had orders from Secretary of War Calhoun to "adopt the necessary measures" to stop continuing raids on American territory by the Seminole Tribe south of the border. Jackson used those

(De Agostini/DEA Picture Library/Getty Images)

This image of Jackson reflects his rising popularity as an American military hero who violently conquers Native Americans. It also portrays Native Americans as savages with no claim to land or dignity, a view shared by many white Americans.

orders as an excuse to act aggressively: he invaded Florida and seized the Spanish forts at St. Marks and Pensacola. This became the first of several operations known as the **Seminole Wars**. During the war, Jackson ordered the execution of two British citizens accused of supporting the Seminole, which outraged the British government and led it to demand a formal investigation. Cabinet members called for a court martial and Congress nearly censured Jackson.

Yet instead of condemning Jackson's raid, Adams urged the government to assume responsibility for it. The United States, he said, had the right under international law to defend itself against threats from across its borders. Jackson's invasion demonstrated to the Spanish that the United States could easily take Florida by force. Adams implied that the nation might consider doing so.

Onís now realized that he had little choice but to negotiate a settlement. Under the provisions of the **Adams-Onís Treaty** of 1819, Spain ceded all of Florida to the United States and gave up its claim to territory north of the 42nd parallel in the Pacific Northwest. In return, the American government gave up its claims to Texas—for a time.

THE PANIC OF 1819

The Monroe administration had little time to revel in its diplomatic successes, for the nation was facing a serious economic crisis: the **Panic of 1819**. The panic followed a period of high foreign demand for American farm goods, caused by disruptions related to the end of the Napoleonic Wars, and thus of exceptionally high prices for American farmers.

The rising prices for farm goods stimulated a land boom in the western United States. Fueled by speculative investments, land prices soared.

The availability of easy credit to settlers and speculators—from the government (under the land acts of 1800 and 1804), from state banks and wildcat banks, even for a time from the rechartered Bank of the United States—fueled the land boom. Beginning in 1819, however, new management at the national bank began tightening credit, calling in loans, and foreclosing mortgages. This precipitated a series of failures by state banks. The result was a financial panic. Six years of depression followed.

Some Americans saw the Panic of 1819 and the widespread distress that followed as a warning that rapid economic growth and territorial expansion would destabilize the nation. But by 1820, most Americans were irrevocably committed to the idea of growth and expansion.

SECTIONALISM AND NATIONALISM

For a brief but alarming moment in 1819–1820, the increasing differences between the North and the South threatened the unity of the United States. The Missouri Compromise was at the heart of this sectional crisis.

THE MISSOURI COMPROMISE

When Missouri applied for admission to the Union as a state in 1819, slavery was already well established there. Even so, Representative James Tallmadge Jr., of New York proposed an amendment to the Missouri statehood bill that would prohibit the further introduction of enslaved labor

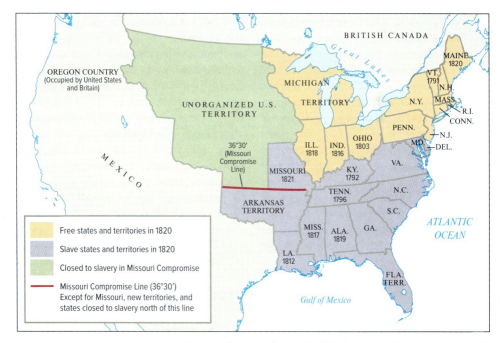

THE MISSOURI COMPROMISE, 1820 This map illustrates the way in which the Missouri Compromise proposed to settle the controversy over slavery in the new western territories of the United States. The compromise rested on the virtually simultaneous admission of Missouri and Maine to the Union, one a slave state and the other a free one. Note the red line extending beyond the southern border of Missouri, which in theory established a permanent boundary between areas in which slavery could be established and areas where it could not be. • *What caused the Missouri Compromise?*

CONSIDER THE SOURCE

THOMAS JEFFERSON REACTS TO THE MISSOURI COMPROMISE (1820)

In this letter to Massachusetts congressman John Holmes, the former president writes of the sectional divisions supposedly resolved by the recent Missouri Compromise. Jefferson wonders how the Union will hold together amid sharp disagreements over slavery and westward expansion.

Monticello, April 22, 1820 I thank you, Dear Sir, for the copy you have been so kind as to send me of the letter to your constituents on the Missouri question. It is a perfect justification to them. I had for a long time ceased to read newspapers, or pay any attention to public affairs, confident they were in good hands, and content to be a passenger in our bark to the shore from which I am not distant. But this momentous question, like a firebell in the night, awakened and filled me with terror. I considered it at once as the knell of the Union. It is hushed, indeed, for the moment. But this is a reprieve only, not a final sentence. A geographical line, coinciding with a marked principle, moral and political, once conceived and held up to the angry passions of men, will never be obliterated; and every new irritation will mark it deeper and deeper. I can say, with conscious truth, that there is not a man on earth who would sacrifice more than I would to relieve us from this heavy reproach, in any *practicable* way.

The cession of that kind of property, for so it is misnamed, is a bagatelle which would not cost me a second thought, if, in that way, a general emancipation and *expatriation* could be effected; and gradually, and with due sacrifices, I think it might be. But as it is, we have the wolf by the ears, and we can neither hold him, nor safely let him go. Justice is in one scale, and self-preservation in the other. Of one thing I am certain, that as the passage of slaves from one state to another would not make a slave of a single human being who would not be so without it, so their diffusion over a greater surface would make them individually happier, and proportionally facilitate the accomplishment of their emancipation, by dividing the burden on a greater number of coadjutors. An abstinence too, from this act of power, would remove the jealousy excited by the undertaking of Congress to regulate the condition of the different descriptions of men composing a state. This certainly is the exclusive right of every state, which nothing in the Constitution has taken from them and given to the general government. Could Congress, for example, say that the non-freemen of Connecticut shall be freemen, or that they shall not emigrate into any other state?

I regret that I am now to die in the belief, that the useless sacrifice of themselves by the generation of '76, to acquire self-government and happiness to their country, is to be thrown away by the unwise and unworthy passions of their sons, and that my only consolation is to be, that I live not to weep over it. If they would but dispassionately weigh the blessings they will throw away against an abstract principle more likely to be effected by union than by scission, they would pause before they would perpetrate this act of suicide on themselves, and of treason against the hopes of the world. To yourself, as the faithful advocate of the Union, I tender the offering of my high esteem and respect.

UNDERSTAND, ANALYZE, & EVALUATE

1. What does Jefferson's metaphor of "a firebell in the night" suggest about his own feelings about the Missouri Compromise and its geographical line?

2. What was Jefferson referring to when he wrote that Americans had "the wolf by the ears"? How appropriate is this metaphor in your assessment?

3. What seemed to be Jefferson's position on the powers of states and the federal government with respect to slavery?

Source: Library of Congress, *Thomas Jefferson Papers, Series 1. General Correspondence. 1651–1827, Thomas Jefferson to John Holmes*, April 22, 1820, http://memory.loc.gov; reproduced in Wayne Franklin (ed.), *The Selected Writings of Thomas Jefferson, A Norton Critical Edition*. New York: W.W. Norton & Company, 2010, 361–362.

into Missouri and provide for the gradual emancipation of the enslaved people already there. The **Tallmadge Amendment** provoked a controversy that raged for the next two years.

Since the beginning of the republic, new states had come into the Union mostly in pairs, one from the North and one from the South. In 1819, there were eleven free states and eleven slave states. The admission of Missouri would upset that balance.

Complicating the Missouri question was the admission of Maine as a new (and free) state. Speaker of the House **Henry Clay** informed northern members that if they blocked Missouri from entering the Union as a slave state, Southerners would block the admission of Maine. Senators wanted to keep the delicate political balance in their chamber between slave and free states. Ultimately the Senate agreed to combine the Maine and Missouri proposals into a single bill. Maine would be admitted as a free state, Missouri as a slave state. Then Senator Jesse B. Thomas of Illinois proposed an amendment prohibiting slavery in the rest of the Louisiana Purchase territory north of the southern boundary of Missouri (the 36°30′ parallel). The Senate adopted the Thomas Amendment, and Speaker Clay, with great difficulty, guided the amended Maine-Missouri bill through the House.

Nationalists in both the North and South welcomed this settlement—which became known as the **Missouri Compromise**—as a timely resolution that would hopefully end growing mistrust between the North and South. Former president Thomas Jefferson, however, was not one of the optimistic voices. He feared a future in which the nation couldn't contain the slavery issue. (See "Consider the Source: Thomas Jefferson Reacts to the Missouri Compromise.")

Marshall and the Court

John Marshall served as chief justice of the U.S. Supreme Court from 1801 to 1835. More than anyone but the framers themselves, he molded the development of the Constitution: strengthening the Supreme Court, increasing the power of the federal government, and advancing the interests of the propertied and commercial classes.

Committed to promoting commerce, the Marshall Court strongly defended the inviolability of contracts. It broke ranks with decades of popular and lower court opinion that had easily and routinely permitted the reversal or modification of contracts. In *Fletcher v. Peck* (1810), which arose out of a series of notorious land frauds in Georgia, the Court had to decide whether the Georgia legislature of 1796 could repeal the act of the previous legislature granting lands under shady circumstances to the Yazoo Land Companies. In a unanimous decision, Marshall held that a land grant was a valid contract and could not be repealed even if corruption was involved.

Dartmouth College v. Woodward (1819) further expanded the meaning of the contract clause of the Constitution. Having gained control of the New Hampshire state government, Republicans tried to revise Dartmouth College's charter to convert the private college into a state university. Daniel Webster argued the college's case. The Dartmouth charter, he insisted, was a contract, protected by the same doctrine that the Court had already upheld in *Fletcher v. Peck*.

(Stock Montage/Archive Photos/Getty Images)

JOHN MARSHALL A former secretary of state, Marshall served as chief justice from 1801 until his death in 1835 at the age of eighty. Such was the power of his intellect and personality that he dominated his fellow justices throughout that period, regardless of their previous party affiliations or legal ideologies. Marshall established the independence of the Court, gave it a reputation for nonpartisan integrity, and established its powers, which were only vaguely described by the Constitution.

The Court ruled for Dartmouth, proclaiming that corporation charters such as the one the colonial legislature had granted the college were contracts and thus inviolable. The decision placed important restrictions on the ability of state governments to control corporations.

In overturning the act of the legislature and the decisions of the New Hampshire courts, the justices also implicitly claimed for themselves the right to override the decisions of state courts. But advocates of states' rights, especially in the South, continued to challenge this right. In *Cohens v. Virginia* (1821), Marshall explicitly affirmed the constitutionality of federal review of state court decisions. The states had given up part of their sovereignty in ratifying the Constitution, he explained, and their courts must submit to federal jurisdiction.

Meanwhile, in *McCulloch v. Maryland* (1819), Marshall confirmed the "implied powers" of Congress by upholding the constitutionality of the Bank of the United States. The Bank had become so unpopular in the South and the West that several states tried to drive branches out of business. This case presented two constitutional questions to the Supreme Court: Could Congress charter a bank? And if so, could individual states ban it or tax it? Daniel Webster, one of the Bank's attorneys, argued that establishing such an institution came within the "necessary and proper" clause of the Constitution and that the power to tax involved a "power to destroy." If the states could tax the Bank at all, they could tax it to death. Marshall adopted Webster's words in deciding for the Bank.

In the case of **Gibbons v. Ogden** (1824), the Court strengthened Congress's power to regulate interstate commerce. The State of New York had granted the steamboat company of Robert Fulton and Robert Livingston the exclusive right to carry passengers on the Hudson River to New York City. Fulton and Livingston then gave Aaron Ogden the business of carrying passengers across the river between New York and New Jersey. But Thomas Gibbons, who had a license granted by Congress, began competing with Ogden for the ferry traffic. Ogden brought suit against him and won in the New York courts. Gibbons appealed to the Supreme Court. The most important question facing the justices was whether Congress's power to give Gibbons a license superseded the State of New York's power to grant Ogden a monopoly. Marshall claimed that the power of Congress to regulate interstate commerce (which, he said, included navigation) was "complete in itself" and might be "exercised to its utmost extent." Ogden's state-granted monopoly, therefore, was void.

The decisions of the Marshall Court established the primacy of the federal government over the states in regulating the economy and opened the way for an increased federal role in promoting economic growth. They protected corporations and other private economic institutions from local government interference.

THE COURT AND NATIVE PEOPLES

The nationalist leanings of the Marshall Court were visible as well in a series of decisions concerning the legal status of indigenous peoples. But these rulings did not simply affirm the supremacy of the United States. They also carved out a distinctive position for Native Americans within the constitutional structure.

The first of the crucial decisions was *Johnson v. McIntosh* (1823). Leaders of the Illinois and Pinakeshaw nations had sold parcels of their land to a group of white settlers (including Johnson) but had later signed a treaty with the federal government ceding territory that included those same parcels to the United States. The government proceeded to grant homestead rights to new white settlers (among them McIntosh) on the land claimed by Johnson. The Court was asked to decide which claim had precedence. Marshall's ruling, not surprisingly, favored the United States. But in explaining it, he offered a preliminary definition of the place of indigenous groups within the nation. Native Americans had a basic right to their tribal lands, he said, that preceded all other American law. Individual American citizens could not buy or take land from native nations. Only the federal government—the supreme authority—could do that.

Even more important was the Court's 1832 decision in **Worcester v. Georgia**, in which the Court invalidated a Georgia law that attempted to regulate access by U.S. citizens to Cherokee country. Only the federal government could do so, Marshall claimed. Indigenous nations, he explained, were sovereign entities in much the same way Georgia was a sovereign entity—"distinct political communities, having territorial boundaries within which their authority is exclusive." In defending the power of the federal government, he affirmed, and indeed expanded, the rights of indigenous peoples to remain free from the authority of *state* governments. Less beneficially, he also made them even more vulnerable to almost whatever the federal government wanted to do.

The Marshall decisions did what the Constitution itself had not: define a place for Native Americans within the American political system. Native nations had basic property rights. They were sovereign entities not subject to the authority of state governments. But the federal government, like a "guardian" governing its "ward," had ultimate authority over indigenous affairs.

(Culture Club/Hulton Archive/Getty Images)

CHEROKEE LEADER SEQUOYAH Sequoyah (who also used the name George Guess) was a Cherokee who translated the Cherokee language into writing through an elaborate syllabary (equivalent to an alphabet) of his own invention, pictured here. He opposed indigenous assimilation into white society and saw the preservation of the Cherokee language as a way to protect his native culture. He moved to Arkansas in the 1820s and became a chief of the western Cherokee Nation.

THE LATIN AMERICAN REVOLUTION AND THE MONROE DOCTRINE

Just as the Supreme Court was asserting American nationalism in shaping the country's economic life, so the Monroe administration was asserting nationalism in formulating foreign policy. Thus far, American diplomacy had been principally concerned with Europe. But in the 1820s, dealing with Europe forced the United States to develop a policy toward Latin America.

Americans looking southward in the years following the War of 1812 beheld a tremendous spectacle: the Spanish Empire in its death throes and a whole continent in revolt. Already the United States had developed a profitable trade with Latin America. Many believed the success of the anti-Spanish revolutions would further strengthen the United State's position in the region.

In 1815, the United States proclaimed neutrality in the wars between Spain and its rebellious colonies. But the United States sold ships and supplies to the revolutionaries, a clear indication that it was trying to help the revolutions. Finally, in 1822, President Monroe established diplomatic relations with five new nations—La Plata (later Argentina), Chile, Peru, Colombia, and Mexico—making the United States the first country to recognize them.

In 1823, Monroe went further and announced a policy that would ultimately be known (beginning some thirty years later) as the **Monroe Doctrine**, even though it was primarily the work of John Quincy Adams. "The American continents," Monroe declared, "are henceforth not to be considered as subjects for future colonization by any European powers." The United States would consider any foreign challenge to the sovereignty of existing

American nations as an unfriendly act. At the same time, he proclaimed, "Our policy in regard to Europe . . . is not to interfere in the internal concerns of any of its powers."

But Monroe also vowed not to interfere with *current* European powers operating in the Americas, and indeed, without a viable seafaring force, the United States relied on the British Royal Navy to make the implicit threats in the statement credible. The intended targets here were the French, Spain's allies, who Americans feared might help Spain retake its lost empire, and the Russians, encroaching on the northern Pacific coastline.

The Monroe Doctrine had few immediate effects, but it was important as an expression of the growing spirit of nationalism in the United States in the 1820s. And it established the idea of the United States as the dominant power in the Western Hemisphere.

THE REVIVAL OF OPPOSITION

After 1816, the Federalist Party ceased to exist, discredited by its seemingly treasonous behavior during the War of 1812 and outmatched in several consecutive presidential races by the party of Jefferson. The Republican Party became the only national political organization in the United States for a short time. In many ways, it now resembled the defunct Federalist Party in its commitment to economic growth and centralized government.

But deep splits were growing, just as they had in the late eighteenth century. By the 1820s, a two-party system was emerging once again. The full name of the mighty Republican Party had always been the Democratic-Republican Party. It now cleaved along the lines its name suggested, with the divisions visible in 1824 but explicit in 1828. Those divisions centered on questions of how much money the federal government should commit to economic development, whether it should continue to raise money through new tariffs, and how powerful it should become in individuals' everyday lives. By the 1828 election, there would be a Democratic Party, which leaned toward the old Jeffersonian vision of a decentralized nation and questioned the federal government's growing role in the economy. The other party was the National Republican Party (later the Whigs and unrelated to the modern Republican Party), which leaned toward the old Federalist belief in a powerful central government. The Whigs believed in a strong national bank and a centralized economy. While both parties championed economic growth and expansion, they disagreed on how strong a role the federal government should play in it.

The "Corrupt Bargain"

Until 1820, presidential candidates were nominated by party caucuses in Congress. But in 1824, "King Caucus" was overthrown. The Republican caucus nominated William H. Crawford of Georgia, the favorite of the extreme states' rights faction of the party. But other candidates received nominations from state legislatures and won endorsements from irregular mass meetings throughout the country.

One of them was Secretary of State John Quincy Adams. He was a man of cold and forbidding manners, with little popular appeal. Another contender was Henry Clay, the Speaker of the House. He had a devoted personal following and a definite and coherent program: the **American System**, which proposed creating a great home market for factory and farm producers by raising the protective tariff, strengthening the national bank, and financing internal improvements. Andrew Jackson, the fourth major candidate and a new member of the U.S. Senate, had no significant political record. But he was a military hero and had the help of shrewd political allies from his home state of Tennessee. All four of these candidates technically ran as Democratic-Republicans; the splintering of the party was obvious.

Jackson received electoral votes, 99, more than any other candidate, but not a majority. Adams was a close second with 84 electoral votes. The Twelfth Amendment to the Constitution (passed in the aftermath of the contested 1800 election) required the House of Representatives, with one vote per state delegation regardless of population, to choose among the three candidates with the largest numbers of electoral votes.

The contest now came down to Jackson and Adams. Crawford was seriously ill and had only won 41 electoral votes. Clay was out of the running, too, with four less electoral votes than Crawford. But as Speaker of the House he could play a decisive role in the outcome. And he did. Jackson was Clay's most dangerous political rival in the West, so Clay threw his support behind Adams, in part because Adams was an ardent nationalist and a likely supporter of Clay's American System. Clay brokered a regional alliance between congressmen from New England and the Ohio Valley who voted for Adams, giving him the presidency. President Adams then named Clay as his Secretary of State. The State Department was the well-established route to the presidency and Adams thus appeared to be naming Clay as his own successor. The Jacksonians were outraged by the back-room dealings that had cost their candidate the election. Jackson had won, after all, the most electoral votes. The anger the Jacksonians expressed at what they called a "corrupt bargain" haunted Adams throughout his presidency.

THE SECOND PRESIDENT ADAMS

Adams proposed an ambitiously nationalist program reminiscent of Clay's American System, but Jacksonians in Congress blocked most of it. Adams also experienced diplomatic frustrations. He appointed delegates to an international conference that the Venezuelan liberator Simón Bolívar had called in Panama in 1826. But Haiti was one of the participating nations, and Southerners in Congress opposed the idea of white Americans mingling with the Black delegates. Congress delayed approving the Panama mission so long that the American delegation did not arrive until after the conference was over.

Even more damaging to the administration was its backing of a new tariff on imported goods in 1828. Tariffs had long been a way that the federal government had raised money, but this particular tariff exposed growing sectional conflict. It originated with the demands of New England and Midwestern farmers and manufactures seeking to to shield themselves from cheaper goods and raw materials imported from Europe. The bill raised the average tariff to 45%, up from 35% in 1824 and 25% in 1816. When Adams signed it into law he antagonized Southerners who depended heavily on foreign goods and industrial products and cursed it as the "tariff of abominations."

JACKSON TRIUMPHANT

By the time of the 1828 presidential election, the new two-party system was now in place. On one side stood the supporters of John Quincy Adams and the National Republicans. Opposing them were the followers of Andrew Jackson, the Democrats. Adams attracted the support of most remaining Federalists. Jackson appealed to a broad coalition that opposed the "economic aristocracy."

But issues seemed to count for little in the end, as the campaign degenerated into a war of personal invective. The Jacksonians charged that Adams had been guilty of gross waste and extravagance. Adams's supporters hurled even worse accusations at Jackson. They called him a murderer and distributed a "coffin handbill," which listed, within coffin-shaped outlines, the names of militiamen whom Jackson was said to have shot in cold blood during the War of 1812. (The men had been deserters who were legally executed after sentence by a court-martial.) And they called his wife a bigamist. Jackson had married his beloved Rachel at a time when the pair incorrectly believed her first husband had divorced her. (When Jackson's wife read of the accusations against her, she collapsed and, a few weeks later, died.)

Jackson's victory was decisive, but sectional. Adams swept virtually all of New England and showed significant strength in the mid-Atlantic region. Nevertheless, the Jacksonians considered their victory as complete and as important as Jefferson's in 1800. Once again, the forces of privilege had been driven from Washington. Once again, a champion of democracy would occupy the White House. The United States had entered, some Jacksonians claimed, a new era of democracy, the "era of the common man."

CONCLUSION

In the aftermath of the War of 1812, a vigorous nationalism increasingly came to characterize the political and popular culture of the United States. In all regions of the country, white men and women celebrated the achievements of the early leaders of the republic, the genius of the Constitution, and the success of the nation in withstanding serious challenges from both without and within. Party divisions faded.

But the broad nationalism of this supposed "era of good feelings" disguised some deep divisions. Indeed, philosophies of governance differed substantially from one region, and one group, to another. Battles continued between those who favored a strong central government committed to advancing the economic development of the nation and those who wanted a decentralization of power to open opportunity to more people. Battles continued as well over the role of slavery in American life—and in particular over the place of slavery in the new western territories. The Missouri Compromise of 1820 highlighted the danger of slavery to national unity and postponed the day of reckoning on that issue, but only for a time.

KEY TERMS/PEOPLE/PLACES/EVENTS

Adams-Onís Treaty 192
American System 199
Francis Cabot Lowell 187
Gibbons v. Ogden 197
Henry Clay 195

John Quincy Adams 191
McCulloch v. Maryland 196
Missouri Compromise 195
Monroe Doctrine 198
Panic of 1819 192

Seminole Wars 192
Stephen H. Long 190
Tallmadge Amendment 195
Worcester v. Georgia 197

RECALL AND REFLECT

1. How did the War of 1812 stimulate the national economy?
2. What were the reasons for the rise of sectional differences in this era? What attempts were made to resolve these differences? How successful were those attempts?
3. Why was the Monroe Doctrine proclaimed?
4. What was the significance of Andrew Jackson's victory in the election of 1828?

Design element: Stars and Stripes: McGraw Hill Education.

9 JACKSONIAN AMERICA

THE RISE OF MASS POLITICS
"OUR FEDERAL UNION"
THE REMOVAL OF NATIVE AMERICANS
JACKSON AND THE BANK WAR
THE CHANGING FACE OF AMERICAN POLITICS
POLITICS AFTER JACKSON

LOOKING AHEAD

1. How did the electorate expand during the Jacksonian era, and what were the limits of that expansion?
2. What events fed the growing tension between nationalism and states' rights, and what were the arguments on both sides of that issue?
3. What was the second party system, and how did its emergence change national politics?

MANY AMERICANS IN THE 1830s were growing apprehensive about the future of their expanding republic. Some feared that rapid economic and territorial growth would produce social chaos or overextension. They insisted that the country's first priority was to establish order and a clear system of authority. Others argued that the greatest danger facing the nation was the growth of inequality and privilege. They wanted to eliminate the favored status of powerful elites and make opportunity more widely available. Advocates of this latter vision seized control of the federal government in 1829 with the inauguration of Andrew Jackson.

The democratization of government over which Jackson presided came wrapped in the rhetoric of equality and aroused the hope and excitement of working people. Jackson was the child of Irish immigrants, raised in a farming family, and many in his inner circle had risen to prominence, in their eyes, by their own hard work, talent, and energy and not because of any accident of birth. They considered it their mission to ensure others would have the opportunity to do the same.

But Jacksonian democracy had its limits. While the seventh president believed in greater enfranchisement, both politically and economically, for white working men, both native-born and immigrant, he intentionally excluded women, indigenous peoples, and Black Americans, enslaved or free. His democracy was based explicitly on white supremacy and it came into being not just at the same time as widespread removal of Native Americans and the dramatic expansion of slavery *but because of them*.

THE RISE OF MASS POLITICS

On March 4, 1829, thousands of Americans from all regions of the country crowded before the U.S. Capitol to watch the inauguration of **Andrew Jackson**. After the ceremonies, they poured into a public reception at the White House, where, in their eagerness to shake the new president's hand, they overflowed the state rooms, trampled one another, soiled the carpets, and tore up the upholstery. Jackson's allies loved the sight of everyday citizens swarming the nation's house of power to see their new leader. "It was a proud day for the people," beamed Amos Kendall, one of Jackson's closest political associates. Others recoiled from it. Supreme Court Justice Joseph Story, a friend and colleague of John Marshall, complained that "The reign of King 'Mob' seems triumphant."

In fact, the "Age of Jackson" was much less a triumph of the common people than Kendall hoped and Story feared. But it did mark a change in American politics. Once restricted to a relatively small elite of white male property owners, politics now became open to virtually *all* the nation's white male citizens. In a limited political sense at least, the period had some claim to the title Jacksonians proudly gave it: the "Era of the Common Man."

Expanding Democracy

What some have called the "Age of Jackson" did not really bring economic equality. The distribution of wealth and property in the United States was little different at the end of the Jacksonian era than it had been at the start. But the extension of suffrage to new groups stimulated a transformation of American politics.

Until the 1820s, relatively few Americans had been permitted to vote. Most states restricted the franchise to white male property owners or taxpayers or both. But even before Jackson's election, the franchise was expanding. Change came first in Ohio and other new states of the West, which, on

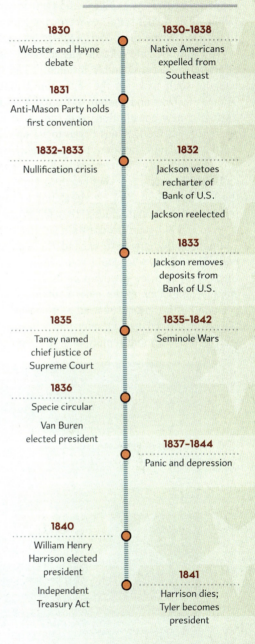

TIME LINE

1830
Webster and Hayne debate

1830–1838
Native Americans expelled from Southeast

1831
Anti-Mason Party holds first convention

1832–1833
Nullification crisis

1832
Jackson vetoes recharter of Bank of U.S.

Jackson reelected

1833
Jackson removes deposits from Bank of U.S.

1835
Taney named chief justice of Supreme Court

1835–1842
Seminole Wars

1836
Specie circular

Van Buren elected president

1837–1844
Panic and depression

1840
William Henry Harrison elected president

Independent Treasury Act

1841
Harrison dies; Tyler becomes president

joining the Union, adopted constitutions that guaranteed all adult white males—not just property owners or taxpayers—the right to vote and permitted all voters the right to hold public office. Older states, concerned about the loss of their population to the West, began to drop or reduce their own property ownership or taxpaying requirements.

The wave of state reforms was generally peaceful, but in Rhode Island democratization efforts created considerable instability. The Rhode Island constitution barred more than half the adult males in the state from voting in the 1830s. In 1840, lawyer and activist Thomas L. Dorr and a group of his followers formed a "People's party," held a convention, drafted a new constitution, and submitted it to a popular vote. It was overwhelmingly approved and the Dorrites began to set up a new government with Dorr as governor. The existing legislature, however, rejected the legitimacy of Dorr's constitution. And so, in 1842, two governments claimed power in Rhode Island. The old state government declared Dorr and his followers rebels and began to imprison them. Outraged and afraid for their safety, the Dorrites struck back by attempting to capture the state arsenal. Poorly organized and no match for the state militia, they failed and their cause withered. Yet the **Dorr Rebellion** spurred the old guard to draft a new state constitution with expanded suffrage.

Importantly, efforts to expand the vote still reflected prevailing assumptions among politicians about who could vote. Women and enslaved Blacks could vote nowhere; free Blacks only in a few northern cities. In the South, election laws heavily favored the planters and politicians of the older counties. For those who could cast a ballot, it was often not done in private but instead as a spoken vote, which meant that voters could be easily bribed or intimidated. Despite the persisting limitations, however, the number of voters increased much more rapidly than did the population as a whole.

Yale University Art Gallery

CANVASSING FOR A VOTE (1853). This lithograph of a painting by George Caleb Bingham depicts a politician (in top hat) speaking with potential voters outside a hotel in Arrow Rock, Missouri. The men represent different social classes and ages. Women and Blacks were barred from voting in the 1830s and 1840s, but political rights expanded substantially among white males.

One of the most striking political trends of the early nineteenth century was the change in the method of choosing presidential electors for the electoral college. In 1800, the legislatures had chosen the presidential electors (and thus determined those electors' votes for president) in ten states; the electors were chosen by the people in only six states. By 1828, electors were chosen by popular vote in every state but South Carolina, which meant that state electors in the college would cast their votes for president and vice president in accordance with the popular vote (as is common practice today). In short, these changes gave common voters a greater say in who won elections. In the presidential election of 1824, fewer than 27 percent of adult white males had voted. Only four years later, the figure was 58 percent; and in 1840, 80 percent.

TOCQUEVILLE AND *DEMOCRACY IN AMERICA*

The rapid growth of the electorate—and the emergence of political parties—was among the most striking events of the early nineteenth century. As the right to vote among white men spread widely in these years, it came to be touted as a public mark of freedom and democracy. One of the most important commentaries on this moment in American life was a book by the French aristocrat **Alexis de Tocqueville**. He spent two years in the United States in the 1830s watching the dramatic political changes during the age of Andrew Jackson. The French government had requested that he examine American prisons, which were thought to be more humane and effective institutions than those in Europe. But Tocqueville went far beyond the study of imprisonment and wrote a classic investigation of American life, *Democracy in America*. Tocqueville looked not just at the politics of the United States but also the daily lives of many groups of Americans and their cultures, their associations, and their visions of democracy. In France in the early decades of the nineteenth century, the fruits of democracy were largely restricted to landowners and aristocrats. Tocqueville recognized that traditional aristocracies were rapidly fading in the United States and that new elites could rise and fall no matter their backgrounds.

Tocqueville also realized that the rising democracy of America had many limits. Democracy was a powerful, visible force in the lives of most white men. Yet for many other Americans, it was a distant hope. Tocqueville wrote of these restrictions:

> He first who attracts the eye, the first in enlightenment, in power and in happiness, is the white man, the European, man par excellence; below him appear the Negro and the Indian. These two unfortunate races have neither birth, nor face, nor language, nor mores in common; only their misfortunes look alike. Both occupy an equally inferior position in the country that they inhabit; both experience the effects of tyranny; and if their miseries are different, they can accuse the same author for them.

Tocqueville's book helped expose the uneven character of American democracy for readers in France and other European nations. Only later did it become widely read and studied in the United States as a portrait of the emerging democracy in the United States.

THE LEGITIMIZATION OF PARTY

The high level of voter participation was only partly the result of an expanded electorate. It resulted as well from growing interest in politics, a strengthening of party organization, and increasing party loyalty. Although party competition had been part of American politics almost from the beginning, widespread acceptance of the idea of party had not. For more than thirty years, most Americans who had opinions about the nature of government considered parties evils to be avoided and thought the nation should seek a broad consensus without permanent factional lines. But in the 1820s and 1830s, those assumptions gave way to a new view that permanent, institutionalized parties were a desirable part of the political process and essential to democracy.

DEBATING THE PAST

Jacksonian Democracy

To many Americans in the 1820s and 1830s, Andrew Jackson was a champion of democracy, a symbol of the spirit of antielitism and egalitarianism that was sweeping American life. Historians, however, have disagreed sharply not only in their assessments of Jackson himself but in their portrayal of American society in his era.

The "progressive" historians of the early twentieth century tended to see Jacksonian politics as a forebear of their own battles against economic privilege and political corruption. Frederick Jackson Turner encouraged scholars to see Jacksonianism as a protest by the frontier against the conservative aristocracy of the East. Jackson represented those who wanted to make government responsive to the will of the people rather than to the power of special interests. The culmination of this progressive interpretation of Jacksonianism was Arthur M. Schlesinger's *The Age of Jackson* (1945). Less interested in the regional basis of Jacksonianism than the disciples of Turner had been, Schlesinger argued that Jacksonian democracy was an effort "to control the power of the capitalist groups, mainly Eastern, for the benefit of non-capitalist groups, farmers and laboring men, East, West, and South." He portrayed Jacksonianism as an early version of modern reform efforts to "restrain the power of the business community."

Richard Hofstadter, in an influential 1948 essay, sharply disagreed. Jackson, he argued, was the spokesman of rising entrepreneurs—aspiring businessmen who saw the road to opportunity blocked by the monopolistic power of eastern aristocrats. The Jacksonian leaders were less sympathetic to the aspirations of those below them than they were to the destruction of obstacles to their own success. Bray Hammond, writing in 1957, argued similarly that the Jacksonian cause was "one of enterprise against capitalist."

Other historians saw Jacksonianism less as a democratic reform movement than as a nostalgic effort to restore a lost past. Marvin Meyers's *The Jacksonian Persuasion* (1957) argued that Jackson and his followers looked with misgivings on the new industrial society emerging around them and yearned instead for a restoration of the agrarian, republican virtues of an earlier time.

In the 1960s, historians began taking less interest in Jackson and his supporters and more in the social and cultural bases of American politics in the time of Jackson. Lee Benson's *The Concept of Jacksonian Democracy* (1961) used quantitative techniques to demonstrate the role of religion and ethnicity in shaping party divisions. Edward Pessen's *Jacksonian America* (1969) portrayed the United States in the Jacksonian era as an increasingly stratified society. This inclination to look more closely at society than at formal "Jacksonianism" continued into the late twentieth and early twenty-first centuries. Sean Wilentz, in *Chants Democratic* (1984) and in *The Rise of American Democracy* (2005), examined the rise of powerful movements among ordinary white citizens who were attracted less to Jackson himself than to the notion of popular democracy.

Gradually, this attention to the nature of society led to major reassessments of Jackson himself and the nature of his regime. In *Fathers and Children* (1975), Michael Rogin portrays Jackson as a leader determined to secure the supremacy of white men in the United States. Alexander Saxton, in *The Rise and Fall of the White Republic* (1990), makes the related argument that "Jacksonian Democracy" was explicitly a white man's democracy that rested on the subjugation of enslaved people, women, and Native Americans. But the portrayal of Jackson as a champion of the common people still persisted. The most renowned biographer of Jackson, Robert V. Remini, argued that despite the flaws in his democratic vision, he was a genuine "man of the people." The journalist Jon Meacham reaches a similar conclusion in his Pulitzer Prize–winning *American Lion* (2009). Most recently, scholarship on Jacksonian democracy has centered on its exploitative nature. David and Jeanne Heidler, in *The Rise of Andrew Jackson* (2018), and Christina Snyder, in *Great Crossings* (2019), concentrated on the president's leading role in expanding slavery, undermining native cultures and forcing native peoples to relinquish their land, and establishing democracy that strictly limited citizenship to white men.

UNDERSTAND, ANALYZE, & EVALUATE

1. Who benefitted from Jacksonian democracy and who didn't? What do these differences tell you about the evolving nature of American society?
2. Why do so many Americans still view Jackson as a "man of the people" when we know he extended slavery and was responsible for the death and displacement of thousands of Native Americans?
3. How do we paint a portrait of Jackson that includes both his efforts to extend democracy for some Americans and deny it others?

The elevation of party occurred first at the state level, most prominently in New York. There, after the War of 1812, Martin Van Buren led a dissident political faction (known as the "Bucktails" for the bucktail they fixed to their caps) that challenged the established political elite led by the aristocratic governor, DeWitt Clinton. The Bucktails argued that Clinton's closed circle made genuine democracy impossible. Instead, they claimed, a more just society needed political parties, in which local networks of operatives, organizers, editors and voters from every corner of society would advocate for their candidates and issues. Any one party would give rise to at least one competitor and their battles for supremacy would force them to remain sensitive to the will of the people. Parties would also check and balance one another in much the same way as the different branches of government did.

By the late 1820s, this new idea of party had spread beyond New York. The election of Jackson in 1828, the result of a popular movement that stood apart from the usual political elites, seemed further to legitimize it. In the 1830s, finally, a fully formed two-party system began to operate at the national level. The anti-Jackson forces began to call themselves **Whigs**. Jackson's followers, once again, called themselves Democrats, which became a permanent name to what is now the nation's oldest political party.

PRESIDENT OF THE COMMON PEOPLE

Andrew Jackson had been born to recent Irish immigrants in 1767. From modest beginnings he went on to study law, then served as a representative, senator, and judge. By the early nineteenth century, he was prospering as a planter and merchant in Tennessee, serving in the militia, and earning notoriety for his armed campaigns against Native Americans in the Southeast and the British in New Orleans. Soon more than a hundred enslaved African Americans labored at his plantation and home, the Hermitage.

Unlike Thomas Jefferson, Jackson was no democratic philosopher. The Democratic Party, much less than the old Jeffersonian Republicans, embraced no clear or uniform ideological position. But Jackson himself did embrace a distinct and simple theory of democracy. Government, he said, should offer "equal protection and equal benefits" to all its white male citizens and favor no one region or class over another, with exceptions—women, enslaved Blacks, free Blacks, and Native Americans could expect no protection from the administration and indeed were left out of its democratic intents. Jackson would ultimately expel Native Americans from the Southeast and try (unsuccessfully) to ban antislavery literature from the mails. Jackson's brand of popular politics meant launching an assault on what he considered the citadels of the eastern aristocracy and extending opportunities to the rising classes of white men in the West and the South. (For historians' changing assessments of Jackson, see "Debating the Past: Jacksonian Democracy.")

Jackson's first target was the entrenched officeholders in the federal government, whom he bitterly denounced. They symbolized his suspicion of centralized power. Offices, he maintained, belonged to the people and not to a self-serving bureaucracy. But they also presented a threat to his own power and authority and removing them was a strategic move. Jackson could fill these halls of power with trusted advocates and reward supporters with offices. As one of Jackson's allies, William L. Marcy of New York, once quipped, "To the victors belong the spoils." This process of giving out jobs as political rewards became known as the **spoils system**. Although Jackson removed no more than one-fifth of existing federal officeholders, his embrace of the spoils system helped cement its place in party politics.

Jackson's supporters also worked to transform the process by which presidential candidates were selected. The Anti-Masonic Party and the Republican Party had held the first ever presidential nomination convention in 1831. In 1832, the president's followers staged their own national convention to renominate him. Through their convention, its founders believed, power in the party would arise directly from the people rather than from such elite political institutions as the congressional caucus.

"OUR FEDERAL UNION"

Jackson's commitment to extending power beyond entrenched elites was linked with efforts to reduce the functions of the federal government. A concentration of power in Washington would, he believed, restrict opportunity to people with political connections. But Jackson was also strongly committed to the preservation of the Union. Thus, at the same time as he was promoting an economic program to reduce the power of the national government, he was asserting the supremacy of the Union in the face of any potent challenge. For no sooner had he entered office than his own vice president–**John C. Calhoun**–began to champion a controversial constitutional theory: nullification.

CALHOUN AND NULLIFICATION

Once an outspoken protectionist, Calhoun had strongly supported the tariff of 1816. But by the late 1820s, he had come to believe that the tariff was responsible for the stagnation of South Carolina's economy, though the exhaustion of the state's farmland was the real reason for the decline. Some exasperated Carolinians were ready to consider a drastic remedy: secession.

With his future political hopes resting on how he met this challenge in his home state, Calhoun developed the theory of **nullification**. Drawing from the ideas of Madison and Jefferson and citing the Tenth Amendment to the Constitution, Calhoun argued that since the federal government was a creation of the states, the states—not the courts or Congress—were the final arbiters of the constitutionality of federal laws. If a state concluded that Congress had passed an unconstitutional law, then it could hold a special convention and declare the federal law null and void for itself. The nullification doctrine–and the idea of using it to nullify the 1828 tariff–quickly attracted broad support in South Carolina. But it did nothing to help Calhoun's standing within the new Jackson administration, in part because he had a powerful rival in **Martin Van Buren**.

The Rise of Van Buren

Van Buren had served briefly as governor of New York before becoming Jackson's secretary of state in 1829. He soon established himself as a member of both the president's official cabinet and his unofficial circle of political allies, known as the "Kitchen Cabinet." And Van Buren's power and influence with the president grew stronger still as a result of a quarrel over etiquette that drove a wedge between Jackson and Calhoun.

Peggy O'Neale was the daughter of a Washington tavern keeper with whom both Andrew Jackson and his friend John H. Eaton had taken lodgings while serving as senators from Tennessee. O'Neale was married, but rumors circulated in Washington in the mid-1820s that she and Senator John Eaton were having an affair. O'Neale's husband died in 1828, and she and Eaton were soon married. A few weeks later, Jackson named Eaton secretary of war and thus made the new Mrs. Eaton a cabinet wife. The rest of the administration wives, led by Mrs. Calhoun, would not receive her. Jackson, who blamed slanderous gossip for the death of his own wife, was furious and demanded that the members of the cabinet accept her into their social world. Calhoun, under pressure from his wife, flatly refused—which further antagonized Jackson who was still fuming from his vice-president's support for nullification. Seeing an opportunity to drive a wedge between the two and ingratiate himself with Jackson, Van Buren, a widower, openly befriended the Eatons. Calhoun and Jackson never patched up their working relationship and in fact it only worsened. Jackson eventually tapped Van Buren to run as his vice-presidential candidate in the 1832 election. Calhoun in turn resigned his office on December 28 of that year and was quickly appointed by the South Carolina legislature to fill a vacant seat in the Senate.

The Webster–Hayne Debate

In January 1830, in the midst of a routine debate over federal policy toward western lands, a senator from Connecticut suggested that all land sales and surveys be temporarily discontinued. Robert Y. Hayne, a young senator from South Carolina, charged that slowing down the growth of the West was simply a way for the East to retain its political and economic power. Hayne and other Southerners believed the South and West might form an alliance to check the strength of the East; that slavery would only thrive with new western lands opened up to the institution; and most crucially, that states, not the federal government, should control local matters like the sale of land and exercise the power to nullify any federal law attempting such control.

Daniel Webster, now a senator from Massachusetts, attacked Hayne (and through him Calhoun) for what he considered an assault on the integrity of the Union and the power of the federal government. Southern states, Webster charged, were putting crass economic and political concerns ahead of the survival of a strong nation unified by common respect for federal law. Hayne responded with a vigorous defense of nullification. Webster then spent two full afternoons delivering what became known as his "Second Reply to Hayne." He concluded with the ringing appeal: "Liberty and Union, now and for ever, one and inseparable!" The exchange over federal versus state power became known as the **Webster-Hayne debate**.

Both sides waited to hear what President Jackson thought of the argument. That became clear at the annual Democratic Party banquet in honor of Thomas Jefferson. After dinner, guests delivered a series of toasts. The president arrived with a written text in which he had underscored certain words: "Our Federal Union—It must be preserved." While he spoke, he looked directly at Calhoun. The diminutive Van Buren, who stood on his chair to see better, thought he saw Calhoun's hand shake and a trickle of wine run down his glass as he responded to the president's toast with his own: "The Union, next to our liberty most dear."

The Nullification Crisis

In 1832, the controversy over nullification mounted to a crisis when South Carolinians responded angrily to a congressional tariff bill that offered them no relief from the 1828 tariff of abominations. Almost immediately, the legislature summoned a state convention, which voted to nullify the tariffs of 1828 and 1832 and to forbid the collection of duties within the state. At the same time, South Carolina elected Hayne to serve as governor and Calhoun to replace Hayne as senator.

Jackson insisted that nullification was treason. He strengthened the federal forts in South Carolina and ordered a warship to Charleston. When Congress convened early in 1833, Jackson proposed a force bill authorizing the president to use the military to see that acts of Congress were obeyed. Violence seemed a real possibility.

Calhoun faced a predicament as he took his place in the Senate. Not a single state had come to South Carolina's defense. The timely intervention of Henry Clay, also newly elected to the Senate, averted a crisis. Clay devised a compromise by which the tariff would be lowered gradually so that by 1842 it would reach approximately the same level as in 1816. The compromise and the force bill were passed on the same day, March 1, 1833. Jackson signed them both. In South Carolina, the convention reassembled and repealed its nullification of the tariffs. But unwilling to allow Congress to have the last word, the convention nullified the force act—a purely symbolic gesture, since the tariff had already been amended. Calhoun and his followers claimed a victory for nullification, which had, they insisted, forced the revision of the tariff. But the episode taught Calhoun and his allies that no state could defy the federal government alone.

THE REMOVAL OF NATIVE AMERICANS

There had never been any doubt about Andrew Jackson's attitude toward the indigenous peoples whose lands were now encircled by the eastern states and territories of the United States. He wanted them off their land and moved westward. Since his early military expeditions in Florida, Jackson had harbored a deep hostility toward Native Americans. In this he was little different from most white Americans.

White Attitudes toward Native Peoples

In the eighteenth century, many whites shared Thomas Jefferson's view of indigenous peoples as "noble savages," with an inherent dignity that made civilization for them possible if they would only mimic white culture. But by the first decades of the nineteenth century, most were coming to view Native Americans simply as "savages" who had no real place in their world. At the same time they demanded the valuable land native nations still possessed to farm, hunt, and trap without interference or competition. In their effort to remove Native Americans from their land they looked to Jackson and the U.S. Army.

White settlers began to clamor louder for removal of native peoples in the face of growing unrest in the Northwest. In Illinois, an alliance of Sauk (or Sac) and Meskwaki under Black Hawk fought white settlers in 1831–1832 in an effort to overturn what Black Hawk considered an illegal cession of indigenous lands to the United States. The **Black Hawk War** was notable for its viciousness. The U.S. Army attacked the Native Americans even when they attempted to surrender, pursued them as they retreated, and slaughtered many of them.

Courtesy National Gallery of Art, Washington

BLACK HAWK AND FIVE OTHER SAUK PRISONERS After his defeat by white settlers in Illinois in 1832, Black Hawk and other Sauk warriors were captured and sent on a tour by Andrew Jackson, displayed to the public as trophies of war. This painting, by George Catlin, shows six Sauk prisoners the year of their capture at Jefferson Barracks near St. Louis. Black Hawk is third from left, holding the feather. The prisoners insisted on being portrayed with the cannon balls chained to their legs. • *Why did the prisoners demand that they be painted with chains?*

THE "FIVE CIVILIZED TRIBES"

Conflicts over native lands arose in the South as well. Western Georgia, Alabama, Mississippi, and Florida were home to what were known by whites as the **"Five Civilized Tribes"**—the Cherokee, Creek, Seminole, Chickasaw, and Choctaw. These groups had adopted practices and institutions based on Euro-American models of what defined a "civilized" people, including English literacy, organized government, agricultural economies, and even slavery. But it didn't matter much; white Southerners still longed for their land to fuel an expanding cotton economy. In 1830 Congress passed the **Indian Removal Act**, which authorized the financing of federal negotiations to relocate the southern nations to the West. If these "negotiations" failed, the threat of forced expulsion lay behind it. Some native peoples believed removal to be the least disagreeable option—better, perhaps, than the prospect of destitution, white encroachment, and violence. But others fought back.

The Cherokees turned to the courts to try to keep their land. They worked closely with William Worcester, a white Presbyterian minister, printer, and missionary to the Cherokee living in the state of Georgia who published the first newspaper in a native tongue, the *Cherokee Phoenix*, in 1828. A fervent believer in Cherokee sovereignty, Worcester refused a pardon and stayed in prison for violating what he deemed was an unjust law: an 1830 Georgia law mandating every white man be formally licensed by the state if they lived on Native American land. The case was now headed to the Supreme Court. Worcester's larger goal was a ruling on whether a state could dictate how native peoples should manage their land. In 1832 in *Worcester v. Georgia* the Supreme Court decided in favor of the Cherokee, asserting that no state had authority in native affairs. Jackson refused to obey the ruling.

CONSIDER THE SOURCE

LETTER FROM CHIEF JOHN ROSS TO THE SENATE AND HOUSE OF REPRESENTATIVES (1836)

Chief John Ross, leader of the Cherokee Nation, wrote the U.S. Congress to object to the Treaty of New Ochota. Signed by a minority faction within the tribe, the treaty's terms pledged the removal of Cherokee from all lands east of the Mississippi in return for money and land in present-day Oklahoma. Despite Ross's protests, the army forced the Cherokee westward on the "Trail of Tears." Thousands died on the route.

Red Clay Council Ground, Cherokee Nation, September 28, 1836

It is well known that for a number of years past we have been harassed by a series of vexations, which it is deemed unnecessary to recite in detail, but the evidence of which our delegation will be prepared to furnish. With a view to bringing our troubles to a close, a delegation was appointed on the 23rd of October, 1835, by the General Council of the nation, clothed with full powers to enter into arrangements with the Government of the United States, for the final adjustment of all our existing difficulties. The delegation failing to effect an arrangement with the United States commissioner, then in the nation, proceeded, agreeably to their instructions in that case, to Washington City, for the purpose of negotiating a treaty with the authorities of the United States.

After the departure of the Delegation, a contract was made by the Rev. John F. Schermerhorn, and certain individual Cherokees, purporting to be a "treaty, concluded at New Echota, in the State of Georgia, on the 29th day of December, 1835, by General William Carroll and John F. Schermerhorn, commissioners on the part of the United States, and the chiefs, headmen, and people of the Cherokee tribes of Indians." A spurious Delegation, in violation of a special injunction of the general council of the nation, proceeded to Washington City with this pretended treaty, and by false and fraudulent representations supplanted in the favor of the Government the legal and accredited Delegation of the Cherokee people, and obtained for this instrument, after making important alterations in its provisions, the recognition of the United States Government. And now it is presented to us as a treaty, ratified by the Senate, and approved by the President [Andrew Jackson], and our acquiescence in its requirements demanded, under the sanction of the displeasure of the United States, and the threat of summary compulsion, in case of refusal. It comes to us, not through our legitimate authorities, the known and usual medium of communication between the Government of the United States and our nation, but through the agency of a complication of powers, civil and military.

By the stipulations of this instrument, we are despoiled of our private possessions, the indefeasible property of individuals. We are stripped of every attribute of freedom and eligibility for legal self-defence. Our property may be plundered before our eyes; violence may be committed on our persons; even our lives may be taken away, and there is none to regard our complaints. We are denationalized; we are disfranchised. We are deprived of membership in the human family! We have neither land nor home, nor resting place that can be called our own. And this is effected by the provisions of a compact which assumes the venerated, the sacred appellation of treaty.

We are overwhelmed! Our hearts are sickened, our utterance is paralized, when we reflect on the condition in which we are placed, by the audacious practices of unprincipled men, who have managed their stratagems with so much dexterity as to impose on the Government of the United States, in the face of our earnest, solemn, and reiterated protestations.

The instrument in question is not the act of our Nation; we are not parties to its covenants; it has not received the sanction of our people. The makers of it sustain no office nor appointment in our Nation, under the designation of Chiefs, Head men, or any other title, by which they hold, or could acquire, authority to assume the reins of Government, and to make bargain and sale of our rights, our possessions, and our common country. And we are constrained solemnly to declare, that we cannot but contemplate the enforcement of the stipulations of this instrument on us, against our consent, as an act of injustice and oppression, which, we are well persuaded, can never knowingly be countenanced by the Government and people of the United States; nor can we believe it to be the design of these honorable and highminded individuals, who stand at the head of the Govt., to bind a whole Nation, by the acts of a few unauthorized individuals. And, therefore, we, the parties to be affected by the result, appeal with confidence to the justice, the magnanimity, the compassion, of your honorable bodies, against the enforcement, on us, of the provisions of a compact, in the formation of which we have had no agency.

UNDERSTAND, ANALYZE, & EVALUATE

1. On what grounds did Chief Ross object to the removal treaty?
2. What did the Cherokee stand to lose by the terms of removal, according to Chief Ross?
3. What options were available to native groups in their attempts to negotiate with the United States?

Source: Moulton, Gary E. (ed.), *The Papers of Chief John Ross, vol. 1, 1807–1839.* Norman, OK: University of Oklahoma Press, 1985.

In 1835, the U.S. government extracted a treaty from a minority faction of the Cherokee that ceded to Georgia the nation's land in that state in return for $5 million and a reservation west of the Mississippi. With removal inevitable, this Cherokee "Treaty Party" reasoned, a deal for cash and new land to the west was the best course of action. But the great majority of the 17,000 Cherokee, including their leader John Ross, did not recognize the treaty as legitimate. (See "Consider the Source: Letter from Chief John Ross.") Jackson sent an army of 7,000 under General Winfield Scott to round up the Cherokee and drive them westward. The president's directive showcased his disregard for native culture and native lives and led to a brutal campaign of extermination and forced removal.

Trail of Tears

In the face of military conflict, about 1,000 Cherokee fled to North Carolina, where the federal government eventually provided them with a small reservation in the Smoky Mountains that survives today. But the U.S. Army forced most of the rest to trek across hundreds of miles mostly on foot to "**Indian Territory**," in what later became Oklahoma, beginning in the winter of 1838. It was a march of extreme hardship: disease, starvation, and exposure claimed the lives of thousands. In the harsh new reservations, the survivors remembered the terrible journey as "The Trail Where They Cried," also dubbed the **Trail of Tears**.

The Trail of Tears now refers broadly to the federal government's brutal effort to relocate about 100,000 Native Americans from across the southeastern states into reservations in the West. As many as 15,000 died during the forced migration. Between 1830 and 1838, the U.S. Army compelled virtually all members of the Five Civilized Tribes to travel to

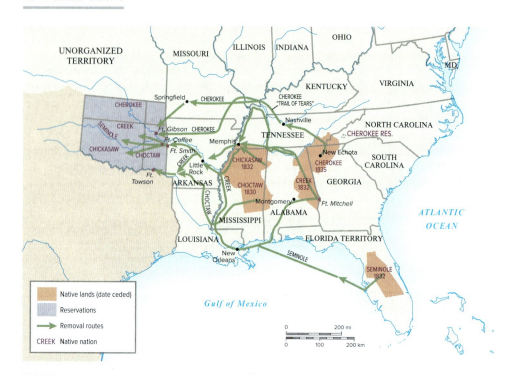

THE EXPULSION OF THE TRIBES, 1830–1842 Well before he became president, Andrew Jackson was famous for his military exploits against Native Americans. Once in the White House, he ensured that few native people would remain in the southern states of the nation, now that white settlement was increasing there. The result was a series of dramatic and bloody "removals" of indigenous societies out of their traditional lands and into new territories west of the Mississippi—mostly in Oklahoma. Note the very long distance many had to travel. • *Why was the route of the Cherokee, shown in the upper portion of the map, known as the Trail of Tears?*

Indian Territory. The Choctaw of Mississippi and western Alabama were the first to make the trek, beginning in 1830. The army moved out the Creek of eastern Alabama and western Georgia in 1836. A year later, the Chickasaw in northern Mississippi began their long forced march westward and the Cherokee, finally, a year after that.

Significantly, the Seminole in Florida were able to resist the U.S. Army, but even their success was limited. Like other native nations, the Seminole had agreed under pressure to a settlement by which they ceded their lands to the United States and agreed to move to Indian Territory within three years. Most did move west, but a substantial minority, under the leadership of the chieftain Osceola, balked and staged an uprising beginning in 1835 to defend their lands, the second of the Seminole Wars. Joining the Seminole in their struggle was a group of Blacks fleeing slavery, who had been living with them. Jackson sent troops to Florida to subdue or kill the resisters, but the Seminoles and their Black allies were masters of guerrilla warfare in the junglelike Everglades. Finally, in 1842, the government abandoned the war. By then, many of the Seminole had been either killed or forced westward.

THE MEANING OF REMOVAL

By the end of the 1830s, President Jackson and the U.S. Army had violently removed most of the indigenous societies east of the Mississippi. Native nations had collectively ceded over 100 million acres to the federal government and had received in return only about $68 million and 32 million acres in the far less hospitable lands west of the Mississippi, land that already had established native populations. There they lived, divided by tribe into a series of separate reservations, within a territory surrounded by a string of U.S. forts, whose climate and topography bore little relation to anything they had known before.

What, if any, were the alternatives to the forced removal of Native Americans? There was probably never any other likely possibilities. At this point in time whites viewed native people as inhuman, coveted their land, and saw little to be wrong with killing them to get what they wanted. Campaigns of extermination, or what we call genocide today, were not considered extreme. And Jackson himself was certainly not going to refute any negative views of native people or question his support of forced removal.

But there were historical alternatives to the brutal removal policy. The West already offered early examples of white settlers and native societies living side by side, if only for a short while. In the pueblos of New Mexico, in the fur-trading posts of the Pacific Northwest, and in parts of Texas and California, Native Americans and newcomers from Mexico, Canada, and the United States had created societies in which the various groups mingled intimately. Sometimes close contact between whites and indigenous peoples was beneficial to both sides, although it could also be cruel and exploitative. Still, some of the early multiracial societies of the West demonstrated that the cultures could interact in ways that were not immediately destructive and violent.

By the mid-nineteenth century, however, white Americans had adopted a different model. They believed that Native Americans could not be partners in the creation of any new societies in the West. Whites saw native peoples only as obstacles to be removed or killed.

JACKSON AND THE BANK WAR

Jackson was quite willing to use the full force of the federal government to manipulate Native Americans. But when it came regulating the economy and local affairs, he wanted less federal intervention. The president's interest in trimming congressional interventions in state affairs was shown by his 1830 veto of a bill providing a subsidy to the proposed Maysville Road in Kentucky. The bill was unconstitutional, Jackson argued, because the road in question lay entirely within Kentucky and was not, therefore, a part of "interstate commerce." Jackson also thought the bill unwise because it committed the government to what he considered extravagant expenditures. A similar resistance to congressional power lay behind Jackson's war against the Bank of the United States.

BIDDLE'S INSTITUTION

The Bank of the United States held a monopoly on federal deposits, provided credit to growing enterprises, issued banknotes that served as a dependable medium of exchange, and exercised a restraining effect on the less well-managed state banks. **Nicholas Biddle**, who ran the Bank from 1823 on, had done much to put the institution on a sound and prosperous basis. Nevertheless, many Americans—among them Andrew Jackson—were determined to destroy it.

Opposition to the Bank came from two very different groups: the "soft-money" and "hard-money" factions. Advocates of soft money consisted largely of state bankers and their allies. They objected to the Bank because it restrained state banks from issuing notes freely. The hard-money faction, which included Andrew Jackson, believed that gold or silver coin (specie) was the only safe currency, and they condemned *all* banks that issued banknotes, state or federal. Jackson inherently distrusted paper bills, which would become worthless if the specie that backed them dried up.

But his critiques of Biddle's bank went deeper than that. The Bank of the United States was the largest corporation in the country, partially public and partially private: it owed its existence to Congress but could float loans to state banks, businesses, and individuals as it saw fit. Jackson attacked it as both an overly centralized and domineering federal institution and also as a greedy private enterprise currying favor with the rich and powerful. Thus like many political issues then and now, the battle over the Bank became a focal point for broader antagonisms around privilege, governance, and power. Jackson made clear that he would not favor renewing the charter of the Bank of the United States, which was due to expire in 1836.

A Philadelphia aristocrat unaccustomed to politics, Biddle began granting banking favors to and seeking help from influential men. In particular, he relied on Daniel Webster, whom he named the Bank's legal counsel and director of the Boston branch. Webster helped Biddle enlist the support of Henry Clay as well. Clay, Webster, and other advisers persuaded Biddle to apply to Congress for a recharter bill in 1832, four years ahead of the expiration date. Congress passed the recharter bill, Jackson vetoed it, and the Bank's supporters in Congress failed to override the veto. The Bank question then emerged as the paramount issue of the 1832 election, just as Clay had hoped.

In 1832, Clay ran for president as the unanimous choice of the Whigs. But the "**Bank War**" failed to provide Clay with the winning issue he had hoped for. Jackson, with Van Buren as his running mate, won an overwhelming victory with 55 percent of the popular vote and 219 of the 286 electoral votes.

THE "MONSTER" DESTROYED

Jackson was now more determined than ever to destroy the "monster." He could not legally abolish the Bank before the expiration of its charter. But he could weaken it by removing the government's deposits from it. When his secretary of the treasury, believing that such an action would destabilize the financial system, refused to give the order, Jackson fired him and appointed a replacement. When the new secretary similarly procrastinated, Jackson fired him, too, and named a third: **Roger B. Taney**, the attorney general, a close friend and loyal ally of the president.

Taney began taking the government's deposits out of the Bank of the United States and putting them in a number of state banks. In response, Biddle called in loans and raised interest rates, explaining that without the government deposits the Bank's resources would be stretched thin which, he feared, would harm the economy. His actions shrunk or destroyed corporations, led to worker layoffs, and precipitated a short recession.

As financial conditions worsened in the winter of 1833–1834, supporters of the Bank sent petitions to Washington urging its rechartering. But the Jacksonians blamed the recession on Biddle. He had the money, they said, not the government. In the court of public opinion, claims of Biddle's overly broad power over the economy appeared to have been borne out. When the banker finally carried his contraction of credit too far and reversed himself to appease the business community, his hopes of winning a recharter of the Bank died in the process. Jackson had won a considerable political victory. But when the Bank of the United States expired in 1836, the country was left with a fragmented and chronically unstable banking system that would plague the economy for many years.

In the aftermath of the Bank War, Jackson moved against the most powerful remaining institution of economic nationalism: the Supreme Court. In 1835, when John Marshall died, the president appointed as the new chief justice his trusted ally Roger B. Taney. Taney did not bring a sharp break in constitutional interpretation, but he did help modify Marshall's vigorous nationalism. Perhaps the clearest indication of the new judicial climate was the celebrated case of *Charles River Bridge v. Warren Bridge* of 1837. The case involved a dispute between two Massachusetts companies over the right to build a bridge across the Charles River between Boston and Cambridge. Reversing the spirit of many Marshall decisions, Taney argued that a state could amend or nullify a contract if such action was necessary to advance the well-being of the community. The decision reflected one of the cornerstones of the Jacksonian idea: that the key to democracy was an expansion of economic opportunity, which would not occur if older corporations could maintain monopolies.

THE CHANGING FACE OF AMERICAN POLITICS

Jackson's forceful—some people claimed tyrannical—tactics in crushing the Bank of the United States helped galvanize a growing opposition coalition. The opposition began as a gathering of national political leaders opposed to Jackson's use of power. Denouncing the president as "King Andrew I," they began to refer to themselves as Whigs, after the party in England that traditionally worked to limit the power of the king. With the emergence of the Whigs, the nation once again had two competing political parties. What scholars now call the "second party system" had begun its relatively brief life.

Library of Congress Prints and Photographs Division [LC-USZ62-1562]

KING ANDREW THE FIRST This parody appeared sometime in 1833 in response to Andrew Jackson's withdrawal of federal funds from the Bank of the United States and his veto of its recharter the previous year. Jackson, trampling on a torn Constitution and the coat of arms of Pennsylvania, where the Bank was located, holds a veto in his left hand and a scepter in his right. • *How do the artist make fun of Jackson?*

Democrats and Whigs

The philosophy of the Democratic Party in the 1830s bore the stamp of Andrew Jackson. The federal government, the Democrats declared, should be limited in power. But what they really meant was that the federal government should only exercise its power for certain reasons. One permissible reason was to eliminate social and economic arrangements that entrenched privilege and stifled opportunity for all white men. It was also acceptable to use federal power to accomplish territorial expansion and removal of Native Americans, with the federal government leading efforts to take and sell land, protect white settlers living on it, and make new states out of it.

Jacksonian Democrats publicly celebrated "honest workers," "simple farmers," and "forthright businessmen" and contrasted them to the corrupt, monopolistic, aristocratic forces of established wealth. Radical members of the party—the so-called Loco Focos, mainly workingmen, small businessmen, and professionals in the Northeast—called for a vigorous assault on privileged elites and what they saw as the government's economic manipulations on their behalf. A Loco Foco protest against inflation turned into a riot of hungry New York workers in February 1837. The group's philosophies later found official expression in an 1840 act of Congress separating banking from the federal government.

In contrast, the political philosophy that became known as Whiggery favored the expansion of federal power and industrial and commercial development. Whigs believed the government needed to support and protect national institutions like the national bank. Whigs were slightly cautious about westward expansion, though they generally shared the Democrats' disdain of native culture and desire for their land. However, they worried that rapid territorial growth would produce instability in Congress.

The Whigs were strongest among the more substantial merchants and manufacturers of the Northeast, the wealthier planters of the South, and the ambitious farmers and rising commercial class of the West. The Democrats drew more support from smaller merchants and the workingmen of the Northeast, as well as southern and western planters who favored a predominantly agrarian economy. Whigs tended to be wealthier, to have more aristocratic backgrounds, and to be more commercially ambitious than the Democrats. But Whigs and Democrats alike were more interested in winning elections than in maintaining philosophical purity. And both parties made adjustments from region to region to attract the largest possible number of voters.

In New York, for example, the Whigs developed a popular following through a movement known as Anti-Masonry. The Anti-Mason Party had emerged in the 1820s in response to widespread resentment against the secret and exclusive, hence supposedly undemocratic, Society of Freemasons. Such resentment increased in 1826 when a former Mason, William Morgan, mysteriously disappeared from his home in Batavia, New York, shortly before he was scheduled to publish a book that would allegedly expose the secrets of Freemasonry. With help from a widespread assumption that Morgan had been abducted and murdered by vengeful Masons, Whigs seized on the Anti-Mason frenzy to launch spirited attacks on Jackson and Van Buren (both Freemasons), implying that the Democrats were connected with the antidemocratic conspiracy.

The Whig Party was more successful at defining its positions and attracting a constituency than it was at uniting behind a national leader. No one person was ever able to command the loyalties of the party in the way Jackson commanded those of the Democrats. Instead, Whigs tended to divide their allegiance among the "Great Triumvirate" of Henry Clay, Daniel Webster, and John C. Calhoun.

Clay won support from many who favored internal improvements and economic development through what he called the American System. But Clay's image as a devious political operator and his identification with the West were a liability. He ran for president three times and never won. Daniel Webster won broad support among the Whigs with his passionate speeches in defense of the Constitution and the Union; but his close connection with the Bank of the United States and the protective tariff, his reliance on rich men for financial support, and his excessive fondness for brandy prevented him from developing enough of a national constituency to win him his desired office. John C. Calhoun never considered himself a true Whig, and his identification with the nullification controversy in effect disqualified him from national leadership in any case. Yet he sided with Clay and Webster on the issue of the national bank, and he shared with them a strong animosity toward Andrew Jackson.

The Whigs competed relatively evenly with the Democrats in congressional, state, and local races, but they managed to win only two presidential elections in the more than twenty years of their history. Their problems became particularly clear in 1836. While the Democrats united behind Andrew Jackson's personal choice for president, Martin Van Buren, the Whigs could not agree on a single candidate. Instead, they ran several candidates in different regions, hoping they might separately draw enough votes from Van Buren to throw the election to the House of Representatives, where the Whigs might be better able to elect one of their candidates. In the end, however, Van Buren won easily, with 170 electoral votes to 124 for all his opponents combined.

POLITICS AFTER JACKSON

Martin Van Buren could not fill the shoes of his predecessor. He lacked Jackson's larger-than-life personality and magnetism as well as his political skill. Van Buren's administration suffered from economic difficulties that hurt both him and his party.

Van Buren and the Panic of 1837

Van Buren's success in the 1836 election was the result in part of a nationwide economic boom. Canal and railroad builders operated at a peak of activity. Prices were rising, credit was plentiful, and the land business, in particular, was booming. Between 1835 and 1837, the government sold nearly 40 million acres of public land, nearly three-fourths of it to speculators. Much of that land had been taken from Native Americans. These land sales, along with revenues the government received from the tariff of 1833, created a series of substantial federal budget surpluses and made possible a steady reduction of the national debt. From 1835 to 1837, the government for the first and only time in its history was out of debt, with a substantial surplus in the Treasury.

Congress and the administration now faced the question of what to do with the Treasury surplus. Support soon grew for returning the federal surplus to the states. An 1836 "distribution" act required the federal government to pay its surplus funds to the states each year in four quarterly installments as interest-free, unsecured loans. No one expected the "loans" to be repaid. The states spent the money quickly, mainly to promote the construction of highways, railroads, and canals. The distribution of the surplus thus gave further stimulus to the economic boom. At the same time, the withdrawal of federal funds strained the state banks in which they had been deposited by the government; the banks had to call in their own loans to make the transfer of funds to the state governments.

Library of Congress, Prints and Photographs Division
[LC-USZ62-89594]

HUMBUG GLORY BANK This image mocked opponents of the specie circular, and more broadly, ridiculed the issuance of currency not backed by gold or silver. It reflected Andrew Jackson's charge that banks, here led by "Honest Amos," would circulate worthless paper rather than gold or silver coin (specie) or currency backed by those precious metals.

Congress did nothing to check the speculative fever. Jackson, now in the last months of his presidency, feared that the government was selling land for state banknotes of questionable value. In 1836, he issued an executive order, the "**specie circular**." It ordered that the federal government, in payment for public lands, would accept only gold or silver coins or currency backed by gold or silver. The specie circular produced a financial crisis that greeted the first months of Van Buren's presidency, the **Panic of 1837**. Banks and businesses failed; unemployment grew; bread riots shook some of the larger cities; and prices fell, especially the price of land. Many railroad and canal projects failed; several of the debt-burdened state governments ceased to pay interest on their bonds, and a few repudiated their debts, at least temporarily. The worst depression in American history to that point, it lasted for five years and was a political catastrophe for Van Buren and the Democrats.

The Van Buren administration did little to fight the depression. In fact, some of the steps it took—borrowing money to pay government debts and accepting only specie for payment of taxes—may have made things worse. Other efforts failed in Congress: a "preemption" bill that would have given settlers the right to buy government land near them before it was opened for public sale and another bill that would have lowered the price of land. Van Buren did succeed in establishing a ten-hour workday on all federal projects via a presidential order, but he had few legislative achievements.

The most important and controversial measure in the president's program was a proposal for a new financial system. Under Van Buren's plan, known as the "independent treasury" or "subtreasury" system, government funds would be placed in an independent treasury in Washington and in subtreasuries in other cities. No private banks would have the government's money or name to use as a basis for speculation. Van Buren called a special session of Congress in 1837 to consider the proposal, but it failed in the House. In 1840, however, the administration finally succeeded in driving the measure through both houses of Congress.

THE LOG CABIN CAMPAIGN

As the campaign of 1840 approached, the Whigs realized that they would have to settle on a candidate for president. In December 1839, they held their first nominating convention. Passing over Henry Clay, they chose **William Henry Harrison**, a renowned soldier and a popular national figure. The Democrats again nominated Van Buren.

The 1840 campaign was the first in which the new and popular "penny press" carried news of the candidates to large audiences. (See "Patterns of Popular Culture: The Penny Press.") Such newspapers were deliberately livelier than the newspapers of the past, which had been almost entirely directed at the upper classes. The *Sun*, the first of the new breed, began publishing in 1833 and was from the beginning self-consciously egalitarian, targeting workers and farmers. It soon had the largest circulation in New York. Similar papers soon began appearing in other cities—reinforcing the increasingly democratic character of political culture and encouraging the inclination of both parties to try to appeal to ordinary voters as they planned their campaigns. That appeal often rested on sensationalized stories. No editor sought to practice what we would call today "objective journalism." A key goal was to sell papers, and sometimes that meant entertaining readers with fabricated accounts and lurid details that were not true.

The campaign of 1840 also illustrated how fully the spirit of party competition had established itself in the United States. The Whigs—who had emerged as a party largely because of their opposition to Andrew Jackson's common-people democracy—presented themselves in 1840 as the party of the common people. So, of course, did the Democrats. The Whig campaign was particularly effective in portraying William Henry Harrison, a wealthy member

Library of Congress Prints and Photographs Division [LC-USZ62-40740]

HARRISON AND REFORM This poster announced a meeting of supporters of William Henry Harrison. It conveys Harrison's presumably humble beginnings in a log cabin. In reality, Harrison was a wealthy, aristocratic man, but the unpopularity of the aristocratic airs of his opponent, President Martin Van Buren, persuaded the Whig Party that it would be good political strategy to portray Harrison as a humble "man of the people."

PATTERNS OF POPULAR CULTURE

THE PENNY PRESS

On September 3, 1833, a small newspaper appeared in New York City for the first time: the *Sun*, published by a young former apprentice from Massachusetts named Benjamin Day. Four pages long, it contained mostly trivial local news, with particular emphasis on sex, crime, and violence. It sold for a penny, launching a new age in the history of American journalism, the age of the "penny press."

Before the advent of the penny press, newspapers in the United States were far too expensive for most ordinary citizens to buy. But several important changes in both the business of journalism and the character of American society paved the way for Benjamin Day and others to challenge the established press. New technologies—the steam-powered cylinder printing press, new machines for making paper, railroads and canals for distributing issues to a larger market—made it possible to publish newspapers inexpensively and to sell them widely. A rising popular literacy rate, a result, in part, of the spread of public education, created a bigger reading public.

The penny press was also a response to the changing culture of the 1820s and 1830s. The spread of an urban market economy contributed to the growth of the penny press by drawing a large population of workers, artisans, and clerks into large cities, where they became an important market for the new papers. The spirit of democracy—symbolized by the popularity of Andrew Jackson and the rising numbers of white male voters across the country—helped create an appetite for journalism that spoke to and for "the people." The *Sun* and other papers like it were committed to feeding the appetites of the people of modest means, who constituted most of their readership. "Human interest stories" helped solidify their hold on the working public.

Within six months of its first issue, the *Sun* had the largest circulation in New York—8,000 readers, more than twice the number of its nearest competitors. James Gordon Bennett's *New York Herald*, which began publication in 1835, soon surpassed the *Sun* in popularity, with its lively combination of sensationalism and local gossip and its aggressive pursuit of national and international stories. The *Herald* pioneered a "letters to the editor" column, was the first paper to have regular reviews of books and the arts, and launched the first daily sports section. By 1860, its circulation of more than 77,000 was the largest of any daily newspaper in the world.

Not all the new penny papers were as sensationalistic as the *Sun* and the *Herald*. The *New-York Tribune*, founded in 1841 by Horace Greeley, prided itself on serious reporting and commentary. As serious as the *Tribune*, but more sober and self-consciously "objective" in its reportage, was *The New York Times*, which Henry Raymond founded in 1851. "We do not mean to write as if we were in a passion—unless that shall really be the case," the *Times* huffily proclaimed in its first issue, in an obvious reference to Greeley and his impassioned reportage, "and we shall make it a point to get into a passion as rarely as possible."

The newspapers of the penny press initiated the process of turning journalism into a profession. They were the first papers to pay their reporters, and they were also the first to rely heavily on advertisements, often devoting up to half their space to paid advertising. They tended to be sensationalistic and opinionated, but they were also usually aggressive in uncovering serious and important news—in police stations, courts, jails, streets, and private homes as well as in city halls, state capitals, Washington, and the world. •

Library of Congress, Chronicling America: Historic American Newspapers [sn83030311-18360101]

THE HERALD This 1836 front page of the *Herald*, which had begun publication the prior year, contains advertisements for medicines and insurance, markets the services of a doctor, a school, and a furniture maker, and announces the enlargement of the newspaper.

UNDERSTAND, ANALYZE, & EVALUATE

1. How were the penny press newspapers a product of the Jacksonian era?
2. Before the advent of the penny press, newspapers in the United States were aimed at a much narrower audience. Some published mainly business news, and others worked to advance the aims of a political party. What nationally circulated newspapers and other media today continue this tradition?

of the frontier elite with a considerable estate, as a simple man of the people who loved log cabins and hard cider. The Democrats had no effective defense against such tactics. But what really sunk their efforts was the depression. Voters blamed the Whigs for their economic troubles. Harrison won the election with 234 electoral votes to 60 for Van Buren and with a popular majority of 53 percent.

The Frustration of the Whigs

The Whigs found the four years after their resounding victory frustrating and divisive. The trouble began when their appealing new president died of pneumonia just one month after taking office. Vice President **John Tyler** of Virginia succeeded him.

Tyler was a former Democrat who had left the party in reaction to what he considered Jackson's excessively egalitarian program. But he still had some Democratic leanings that angered many Whigs. He refused to support Clay's attempt to recharter the Bank of the United States, and he vetoed several internal improvement bills sponsored by Clay and other congressional Whigs. A furious conference of congressional Whigs voted Tyler out of the party. Every cabinet member but Webster, who was serving as secretary of state, resigned. Five former Democrats took their places. When Webster, too, left the cabinet, Tyler appointed Calhoun, who had rejoined the Democratic Party, to replace him. The constant turmoil in the Van Buren administration limited it ability to implement any broad Whig agenda.

Whig Diplomacy

In the midst of these domestic controversies, anti-British factions in Canada launched an unsuccessful rebellion against the colonial government there in 1837. When the insurrection failed, some of the rebels took refuge near the United States border and chartered an American steamship, the *Caroline,* to ship them supplies across the Niagara River from New York. British authorities in Canada seized the *Caroline* and burned it, killing one American in the process. Resentment over the *Caroline* affair in the United States grew rapidly. At the same time, tensions flared over the boundary between Canada and Maine, which had been in dispute since the treaty of 1783. In 1838, rival groups of Americans and Canadians, mostly lumberjacks, began moving into the Aroostook River region in the disputed area, precipitating a violent brawl that became known as the "Aroostook War."

Several years later, in 1841, an American ship, the *Creole,* sailed from Virginia for New Orleans with more than 100 enslaved people aboard. En route the enslaved passengers mutinied, seized possession of the ship, and took it to the Bahamas. British officials there declared the mutineers free, and the London government refused to overrule them. Many Americans, especially Southerners, were furious.

At this critical juncture, a new government eager to reduce tensions with the United States came to power in Great Britain. It sent Lord Ashburton, an admirer of the United States, to negotiate an agreement on the Maine boundary and other matters. The result was the **Webster-Ashburton Treaty** of 1842, under which the United States received slightly more than half of the disputed area and agreed to a northern boundary as far west as the Rocky Mountains. Ashburton also eased the memory of the *Caroline* and *Creole* affairs by expressing regret and promising no future "officious interference" with American ships. Anglo-American relations improved significantly.

During the Tyler administration, the United States established its first diplomatic relations with China. In the 1844 Treaty of Wang Hya, American diplomats secured the same trading privileges as the British. In the next ten years, American trade with China steadily increased.

In their diplomatic efforts, at least, the Whigs were able to secure some important successes. But by the end of the Tyler administration, the party could look back on few other victories. And in the election of 1844, the Whigs lost the White House to James K. Polk, a Democrat with an explicit agenda of more aggressive westward expansion.

CONCLUSION

The election of Andrew Jackson in 1828 reflected the emergence of a new political world. Suddenly white men from every station in life could vote and join the political process. Even though the new laws governing political participation included only white men, the overall number of people who voted rose quickly. Along with this expansion of the electorate emerged a new spirit of party politics.

Jackson set out as president to entrench his party, the Democrats, in power. He attempted to closely control where and why the federal government intervened in society. He backed federal removal of Native Americans from their traditional lands and federal policies promoting westward expansion, which he thought was essential to promoting opportunities for white men and their families. He confronted the greatest challenge yet to American unity—the nullification crisis of 1832–33—with a strong assertion of the power and importance of the Union. Yet he demanded that Washington remain out of other sectors of American life. Most importantly, he worked to destroy the overarching power Bank of the United States, which he considered a corrupt vehicle of privilege, and refused to keep career federal bureaucrats on the payroll. These positions won him broad popularity and ensured his reelection in 1832 and the election of his designated successor, Martin Van Buren, in 1836.

But a new coalition of anti-Jacksonians, who called themselves the Whigs, launched a powerful new party that used much of the same anti-elitist rhetoric to win support for their own, much more nationalistic program. Their emergence culminated in the campaign of 1840 with the election of William Henry Harrison, the first Whig president. When his death led to the accidental presidency of John Tyler, however, further realignments were set in motion.

KEY TERMS/PEOPLE/PLACES/EVENTS

Alexis de Tocqueville 205
Andrew Jackson 203
Bank War 216
Black Hawk War 210
Daniel Webster 209
Dorr Rebellion 204
Five Civilized Tribes 211
Indian Removal Act 211
Indian Territory 213
John C. Calhoun 208
John Tyler 224
Martin Van Buren 208
Nicholas Biddle 215
nullification 208
Panic of 1837 220
Roger B. Taney 216
specie circular 220
spoils system 208
Trail of Tears 213
Webster-Ashburton Treaty 224
Webster-Hayne debate 209
Whigs 207
William Henry Harrison 220

RECALL AND REFLECT

1. What was Andrew Jackson's political philosophy, and how was it reflected in the policies and actions of his administration?
2. Who benefited under Jacksonian democracy? Who suffered?
3. How did Andrew Jackson change the office of the presidency?
4. Who supported and who opposed the Bank of the United States, and why? Who was right?
5. How and why did white attitudes toward Native Americans change, and how did these changes lead to the Indian Removal Act and the Trail of Tears?
6. How did Native Americans in the Southeast respond to white efforts to seize their land and remove them to the West?

10 | AMERICA'S ECONOMIC REVOLUTION

THE CHANGING AMERICAN POPULATION
TRANSPORTATION AND COMMUNICATIONS REVOLUTIONS
COMMERCE AND INDUSTRY
MEN AND WOMEN AT WORK
PATTERNS OF SOCIETY
THE AGRICULTURAL NORTH

LOOKING AHEAD

1. What were the factors sparking the U.S. economic revolution of the mid-nineteenth century?
2. How did the U.S. population change between 1820 and 1840, and how did it affect the nation's economy, society, and politics?
3. Why did the Industrial Revolution affect the northern economy and society differently than it did the southern economy and society?

WHEN THE UNITED STATES ENTERED the War of 1812, it was mostly an agrarian nation. There were, to be sure, some substantial cities in the United States and also modest but growing manufacturing centers, mainly in the Northeast. But the overwhelming majority of Americans were farmers and tradespeople.

By the time the Civil War began in 1861, however, the United States had transformed itself. Most farmers were now part of a national, and even international, market economy. Equally important, the United States was starting to challenge the industrial nations of Europe for supremacy in manufacturing. Although the majority of Americans were rural, the nation had experienced the beginning of its own Industrial Revolution.

TIME LINE

1817–1825
Erie Canal constructed

1830
Baltimore and Ohio Railroad begins operation

1830s
Immigration from Ireland and Germany begins

1834
Lowell Mills women strike

1837
Native American Association fights immigration

McCormick patents mechanical reaper

1844
Morse sends first telegraph message

1845
Native American Party formed

1846
Rotary press invented

1847
John Deere manufactures steel plows

1852
American Party (Know-Nothings) formed

THE CHANGING AMERICAN POPULATION

The American Industrial Revolution was the result of many factors: advances in transportation and communications, the growth of manufacturing technology, the development of new systems of business organization, and, perhaps above all, surging population growth.

Population Trends

Three trends characterized the American population during the **antebellum** period: rapid increase, movement westward, and the growth of towns and cities where demand for work was expanding.

The American population, 4 million in 1790, had reached 10 million by 1820 and 17 million by 1840. Improvements in public health played a role in this growth. Epidemics declined in both frequency and intensity and the death rate as a whole dipped. But the population increase was also a result of a high birthrate. In 1840, white women bore an average of 6.14 children each.

The Black population increased more slowly than the white population. After 1808, when the importation of enslaved labor became illegal, the proportion of Blacks to whites in the nation as a whole steadily declined. The slower increase of the Black population was also a result of its comparatively high death rate. Enslaved women typically had large families, but life was shorter for both enslaved and free Blacks than for whites—a result of the harsh working conditions in which Black Americans lived regardless of their status.

Immigration, choked off by wars in Europe and economic crises in the United States, contributed little to the American population in the first three decades of the nineteenth century. Of the total 1830 population of nearly 13 million, the foreign-born numbered fewer than 500,000. Soon, however, immigration began to grow once again. Famine and political unrest in European countries fueled people's desire to emigrate, while the transatlantic voyage became quicker and more affordable as steamships replaced older ships powered by wind alone.

Much of this new European immigration flowed into the rapidly growing cities of the Northeast. Urban growth was a result of substantial internal migration as well. As agriculture in New England and other areas grew less profitable, more and more people picked up stakes and moved—some to promising agricultural regions in the West but many to eastern cities.

Urban Growth and Immigration, 1840–1860

The growth of eastern cities accelerated dramatically between 1840 and 1860. The population of New York, for example, rose from 312,000 to 805,000, making it the nation's largest and most commercially important city. Philadelphia's population swelled over the same twenty-year period from 220,000 to 565,000; Boston's, from 93,000 to 177,000.

The rising agricultural economy of the West and South produced significant urban growth in these regions as well though not at the level as in the East. Cities developed in both slave and free states. Between 1820 and 1860 in states that permitted slavery, Baltimore emerged as the fourth largest city in the nation and New Orleans the sixth. Communities that had once been small villages or trading posts also grew dramatically: St. Louis was the eighth largest city, Louisville the twelfth. In states that banned slavery, Pittsburgh and Cincinnati boomed. All became centers of the growing carrying trade that connected the farmers of the Midwest with New Orleans and, through it, the cities of the Northeast. After 1830, however, an increasing proportion of this trade moved from the Mississippi River to the Great Lakes, creating important new port cities in free states, such as Buffalo, Detroit, Milwaukee, and Chicago.

Immigration from Europe swelled. Between 1840 and 1850, more than 1.5 million Europeans moved to the United States. In the 1850s, the number rose to 2.5 million. Almost half the residents of New York City in the 1850s were recent immigrants. In St. Louis, Chicago, and Milwaukee, the foreign-born outnumbered those of native birth. Comparatively few immigrants settled in the South.

The newcomers came from many different countries, but the overwhelming majority were from Ireland and Germany. By 1860, there were more than 1.5 million Irish-born and approximately 1 million German-born people in the United States. Many of the Irish were rural farmers escaping brutal poverty, British rule, and especially the Potato Famine that, from 1845 to 1852, rotted crops, caused widespread starvation and helped spread disease. The famine killed nearly one million Irish. Most Irish immigrants abandoned their agricultural roots and stayed in the eastern cities where they landed, becoming part of the unskilled labor force. The largest group of Irish immigrants comprised young, single women, who typically worked in factories or as domestics. Like the Irish, many German-speaking immigrants hungered for improved agricultural conditions, especially when wheat prices plummeted. But others came for explicitly political reasons. Many fled Europe in search of democracy after the failed revolutions of 1848. And those who were Jewish hoped to leave behind increasing anti-Semitism. Germans tended to arrive in family groups and on the whole with more money and skills than Irish immigrants, reflecting the fact that many were successful craftsmen displaced by Europe's industrial revolution. They generally moved on to the Northwest, where they opened businesses or established farms.

THE RISE OF NATIVISM

Many politicians, particularly Democrats, eagerly courted the support of the new arrivals. Other citizens, however, viewed the growing foreign population with alarm. Some people argued that the immigrants were racially inferior or that they corrupted politics by selling their votes. Others complained that they were stealing jobs from the native-born workforce. Protestants worried that the growing Irish population would increase the power of the Catholic Church. Older-stock Americans fretted that immigrants would become a radical force in politics. Out of these fears and prejudices emerged a number of secret societies to combat the "alien menace."

The first was the Native American Association, founded in 1837. It changed its name to the Native American Party in 1845. Five years later it joined ranks with other groups supporting **nativism** to form the Supreme Order of the Star-Spangled Banner, whose demands included banning Catholics and the foreign-born from holding public office, enacting more restrictive naturalization laws, and establishing literacy tests for voting. The order adopted a strict code of secrecy, which included a secret password: "I know nothing." Ultimately, members of the movement came to be known as the "**Know-Nothings**."

After the 1852 elections, the Know-Nothings created a new political organization that they called the American Party. It scored immediate success in the elections of 1854. The Know-Nothings did well in Pennsylvania and New York and actually won control of the state government in Massachusetts. Outside the Northeast, however, their progress was more modest. After 1854, the strength of the Know-Nothings declined and the party soon disappeared.

Library of Congress Prints & Photographs Division
[LC-USZC4-5004]

KNOW-NOTHING SOAP This advertising label alludes to the Know-Nothing nativist movement. It ironically adopts images of indigenous people to enforce the idea that only "native-born" citizens were welcome in the United States—even as Know-Nothings barred Native Americans themselves from their party.

TRANSPORTATION AND COMMUNICATIONS REVOLUTIONS

Just as the Industrial Revolution spurred an expanding population, it also gave rise to an efficient system of transportation and communications. The first half of the nineteenth century saw dramatic changes in both.

THE CANAL AGE

From 1790 until the 1820s, the so-called turnpike era, the United States had relied largely on roads for internal transportation. But roads alone were not adequate for the nation's expanding needs. In the 1820s and 1830s, Americans began to turn to other means of transportation as well.

Larger rivers like the Mississippi became increasingly important as steamboats replaced the slow barges that had previously dominated water traffic. The new riverboats carried the corn and wheat of northwestern farmers and the cotton and sugar of southwestern planters to New Orleans, where oceangoing ships took the cargoes on to eastern ports or abroad.

But this roundabout river–sea route satisfied neither western farmers nor eastern merchants, who wanted a new way to ship goods more directly and cheaply to the urban markets and ports of the Atlantic Coast. New highways cut across the mountains provided a partial solution to the problem. But the costs of hauling goods overland, although lower

CANALS IN THE NORTH, 1823–1860 This map illustrates a growing transportation network connecting the East and West. The great success of the Erie Canal, which opened in 1825, inspired decades of energetic canal building in many areas of the United States. But none of the new canals had anything like the impact of the original Erie Canal, and thus none of New York's competitors—among them Baltimore, Philadelphia, and Boston—were able to displace it as the nation's leading commercial center. • *How did the emergence of canals change the distribution of goods in the United States?*

than before, were still too high for anything except the most compact and valuable merchandise. And so interest grew in building canals—human-made waterways that connected bodies of water and were wide and deep enough for commercial vessels.

The job of financing canals fell largely to the states. New York was the first to act. It had the natural advantage of a good land route between the Hudson River and Lake Erie through the only break in the Appalachian chain. But the engineering tasks were still imposing. The more than 350-mile-long route was interrupted by high ridges and thick woods. After a long public debate over whether building a new route would be worth the financial risk, canal advocates prevailed and digging began on July 4, 1817.

The **Erie Canal** was the greatest construction project Americans had undertaken up until that time. The canal itself was basically a simple ditch forty feet wide and four feet deep, with towpaths along the banks for the horses or mules that were to draw the canal boats. Its construction involved hundreds of difficult cuts and fills to enable the canal to pass through hills and over valleys, stone aqueducts to carry it across streams, and eighty-eight locks of heavy masonry with great wooden gates to permit ascents and descents. The Erie Canal opened in October 1825 amid elaborate ceremonies and celebrations and traffic was soon so heavy that within about seven years tolls had repaid the entire cost of construction. By providing a route to the Great Lakes, the canal gave New York access to Chicago and the growing markets of the West. The Erie Canal also contributed to the decline of agriculture in New England. Now that it was so much cheaper for western farmers to ship their crops east, people farming marginal land in the Northeast found themselves struggling to compete.

The system of water transportation extended farther when Ohio and Indiana, inspired by the success of the Erie Canal, provided their own water connections between Lake Erie and the Ohio River. These canals made it possible to ship goods by inland waterways all the way from New York to New Orleans.

One of the immediate results of these new transportation routes was increased white settlement in the Northwest, because it was now easier for migrants to make the westward journey and to ship their goods back to southern and eastern markets. Much of the western produce continued to go downriver to New Orleans, but an increasing proportion went east to New York. And manufactured goods from throughout the East now moved in growing volume through New York and then to the West via the new water routes.

Rival cities along the Atlantic seaboard took alarm at New York's access to (and control over) so vast a market, largely at their expense. But they had limited success in catching up. Boston, its way to the Hudson River blocked by the Berkshire Mountains, did not even try to connect itself to the West by canal. Philadelphia, Baltimore, Richmond, and Charleston all aspired to build water routes to the Ohio Valley but never completed them. Some cities, however, saw opportunities in a different and newer means of transportation. Even before the canal age had reached its height, the era of the railroad was beginning.

THE EARLY RAILROADS

Railroads played a relatively small role in the nation's transportation system in the 1820s and 1830s because the industry itself was still developing. By 1804, both English and American inventors had experimented with steam engines for propelling land vehicles. In 1820, John Stevens ran a locomotive and cars around a circular track on his New Jersey estate. And in 1825, the Stockton and Darlington Railroad in England became the first line to carry general traffic.

Universal History Archive/Universal Images Group/Getty Images

RACING ON THE RAILROAD Peter Cooper designed and built the first steam-powered locomotives in the United States in 1830 for the Baltimore and Ohio Railroad. On August 28 of that year, he raced his locomotive (the "Tom Thumb") against a horse-drawn railroad car. This sketch depicts the moment when Cooper's engine overtook the horse-drawn railroad car.

American entrepreneurs quickly grew interested in the English experiment. The first company to begin actual operations was the **Baltimore and Ohio Railroad**, which opened a thirteen-mile stretch of track in 1830. In New York, the Mohawk and Hudson began running trains along the sixteen miles between Schenectady and Albany in 1831. By 1836, more than a thousand miles of track had been laid in eleven states.

THE TRIUMPH OF THE RAILS

Railroads gradually supplanted canals and all other forms of transport. In 1840, the total railroad trackage of the country was under 3,000 miles. By 1861, over 31,500 miles of tracks spread across the nation. New England and the mid-Atlantic states contained about 10,000 miles. The South, excluding Maryland and Delaware, featured about 9,500 miles. And the Midwest, from Ohio over to Minnesota and Iowa, had slightly over 11,000 miles. Railroads spanned the Mississippi at several points across great iron bridges. Chicago eventually became the rail center of the West, securing its place as the dominant city of that region.

The emergence of the great train lines diverted traffic from the main water routes—the Erie Canal and the Mississippi River. By lessening the dependence of the West on the Mississippi, the railroads also helped weaken further the connection between the Northwest and the South.

Railroad construction required massive amounts of capital. Some came from private sources, but much of it came from government funding. State and local governments invested in railroads, but even greater assistance came from the federal government in the form of public land grants. By 1860, Congress had allotted over 30 million acres to eleven states to assist railroad construction. Much of it was originally indigenous land that the federal government had seized.

It would be difficult to exaggerate the impact of the rails on the American economy, on American society, even on American culture. Where railroads went, towns, ranches, and farms grew up rapidly along their routes. Areas once cut off from markets during winter found that the railroad could transport goods year-round. Most of all, the railroads cut the time and cost of shipment and travel. In the 1830s, traveling from New York to Chicago by lake and canal took roughly three weeks. By railroad in the 1850s, the same trip took less than two days.

The railroads were much more than a fast and economically attractive form of transportation. They were an engine for technological advances and economic growth and, for many citizens, a visible sign of American progress.

234 · CHAPTER 10

RAILROAD GROWTH, 1850–1860 These two maps illustrate the dramatic growth of American railroads in the 1850s. Note the particularly extensive increase in mileage in the Northeast and the upper Midwest (known at the time as the Old Northwest). Note, too, the relatively smaller increase in railroad mileage in the South. Railroads forged a close economic relationship between the upper Midwest and the Northeast and weakened the Midwest's relationship with the South. • *How did the growth of railroads contribute to the South's growing sense of insecurity within the Union?*

THE TELEGRAPH

Like the railroad, the telegraph was a dramatic advance over traditional methods and a symbol of national progress. Before the telegraph, communication over great distances could be achieved only by direct, physical contact. That meant that virtually all long-distance communication relied on the mail, which traveled first on horseback and coach and later by railroad. There were obvious disadvantages to this system, not the least of which was the difficulty in coordinating the railroad schedules. By the 1830s, experiments with many methods of improving long-distance communication had been conducted, among them a procedure for using the sun and reflective devices to send light signals as far as 187 miles.

In 1832, Samuel F. B. Morse—a professor of art with an interest in science—began experimenting with a different system. Fascinated with the possibilities of electricity, Morse set out to find a way to send signals along an electric cable. Technology did not yet permit the use of electric wiring to send reproductions of the human voice or any complex information. But Morse realized that electricity itself could serve as a communication device—that pulses of electricity could themselves become a kind of language. He experimented at first with a numerical code, in which each number would represent a word on a list available to recipients. Gradually, however, he became convinced of the need to find a more universal telegraphic "language," and he developed what became the **Morse code**, in which alternating long and short bursts of electric current represented individual letters.

In 1843, Congress appropriated $30,000 for the construction of an experimental telegraph line between Baltimore and Washington; in May 1844 it was complete, and Morse succeeded in transmitting the news of James K. Polk's nomination for the presidency over the wires. By 1860, more than 50,000 miles of wire connected most parts of the country; a year later, the Pacific Telegraph, with 3,595 miles of wire, opened between New York and San Francisco. By then, nearly all the independent lines had joined in one organization, the **Western Union Telegraph Company**. The telegraph spread rapidly across Europe as well, and in 1866, the first transatlantic cable was laid, allowing telegraphic communication between the United States and Europe.

One of the first beneficiaries of the telegraph was the growing system of rails. Wires often ran alongside railroad tracks and telegraph offices were often located in railroad stations. The telegraph allowed railroad operators to communicate directly with stations in cities, small towns, and even rural hamlets—to alert them to schedule changes and warn

Villorejo/Alamy Stock Photo

THE TELEGRAPH The telegraph provided rapid communication across the country—and eventually oceans—for the first time. Samuel F. B. Morse was one of a number of inventors who helped create the telegraph, but he was the most commercially successful of the rivals.

them about delays and breakdowns. Among other things, this new form of communication helped prevent accidents by alerting stations to problems that engineers in the past had to discover for themselves.

New Technology and Journalism

Another beneficiary of the telegraph was American journalism. The wires delivered news in a matter of hours—not days, weeks, or months, as in the past—across the country and the world. Where once the exchange of national and international news relied on the cumbersome exchange of newspapers by mail, now it was possible for papers to share their reporting. In 1846, newspaper publishers from around the nation formed the Associated Press to promote cooperative news gathering by wire.

Other technological advances spurred the development of the American press. In 1846, Richard Hoe invented the steam-powered cylinder rotary press, making it possible to print newspapers much more rapidly and cheaply than had been possible in the past. Among other things, the rotary press spurred the dramatic growth of mass-circulation newspapers. The *New York Sun,* the most widely circulated paper in the nation, had 8,000 readers in 1834. By 1860, its successful rival the *New York Herald*—benefiting from the speed and economies of production the rotary press made possible—had a circulation of 77,000.

COMMERCE AND INDUSTRY

By the middle of the nineteenth century, the United States had developed the beginnings of a modern capitalist economy with an advanced industrial capacity. This economy had developed along highly unequal lines—benefiting some classes and some regions far more than others.

The Expansion of Business, 1820–1840

American business grew rapidly in the 1820s and 1830s in part because of important innovations in management and changes in the law. Individuals or limited partnerships continued to operate most businesses and the dominant figures were still the great merchant capitalists, who generally had sole ownership of their enterprises. In some larger businesses, however, the individual merchant capitalist was giving way to the corporation. Corporations, which had the advantage of combining the resources of a large number of shareholders, began to develop particularly rapidly in the 1830s, when some legal obstacles to their formation were removed. Previously, a corporation could obtain a charter only by a special act of a state legislature and only then if it directly benefitted the "public good," as in the case of a school or church. But by the 1830s, states began passing general incorporation laws under which almost any group could secure a charter merely by paying a fee. The laws also permitted a system of limited liability, in which individual stockholders risked losing only the value of their own investment—and not the full value of the corporation's larger losses as in the past—if the enterprise failed. These changes made possible much larger manufacturing and business enterprises.

The Emergence of the Factory

The most profound economic development in the mid-nineteenth-century United States was the rise of the factory. Before the War of 1812, most manufacturing took place within households or in small workshops. Later in the nineteenth century, New England textile manufacturers began using new water-powered machines that allowed them to bring their operations together under a single roof. This **factory system**, as it came to be known, soon penetrated the shoe industry and other industries as well.

Between 1840 and 1860, American industry experienced particularly dramatic growth. For the first time, the value of manufactured goods was roughly equal to that of agricultural products. More than half of the approximately 140,000 manufacturing establishments in the country in 1860, including most of the larger enterprises, were located in the Northeast. Not surprisingly, the Northeast produced more than two-thirds of the manufactured goods and employed nearly three-quarters of the men and women working in manufacturing.

Advances in Technology

Even the most highly developed industries were still relatively immature. American cotton manufacturers, for example, produced goods of coarse grade; fine items continued to come from England. By the 1840s, significant advances were occurring.

Among the most important was in the manufacturing of machine tools—the tools used to make machinery parts. The government supported much of the research and development of machine tools, often in connection with supplying the military. For example, a government armory in Springfield, Massachusetts, developed two important tools—the turret lathe (used for cutting screws and other metal parts) and the universal milling machine (which replaced the hand chiseling of complicated parts and dies)—early in the nineteenth century. The precision grinder (which became critical to, among other things, the construction of sewing machines) was designed in the 1850s to help the army produce standardized rifle parts. By the 1840s, the machine tools used in the factories of the Northeast were already better than those in most European factories.

One important result of better machine tools was that the principle of interchangeable parts spread into many industries. Eventually, interchangeability would revolutionize watch and clock making, the manufacturing of locomotives, the creation of steam engines, and the making of many farm tools and guns. It would also help make possible bicycles, sewing machines, typewriters, cash registers, and eventually the automobile.

Industrialization was also profiting from new sources of energy. The production of coal, most of it mined around Pittsburgh in western Pennsylvania, leaped from 50,000 tons in 1820 to 14 million tons in 1860. The new power source, which replaced wood and water power, made it possible to locate mills away from running streams and thus permitted the wider expansion of the industry.

These great industrial advances owed much to American inventors. In 1830, the number of inventions patented was 544; in 1860, it stood at 4,778. Several industries provide particularly vivid examples of how a technological innovation could produce major economic change. In 1839, Charles Goodyear, a New England hardware merchant, discovered a method of vulcanizing rubber (treating it to give it greater strength and elasticity); by 1860, his process had found over 500 uses and had helped create a major American rubber industry. It became a critical ingredient in a wide variety of goods, such as shoes, waterproof clothing, life preservers, umbrellas, hats, and balls. In 1846, Elias Howe of Massachusetts constructed a sewing machine; Isaac Singer made improvements on it, and the Howe-Singer machine was soon being used in the manufacture of ready-to-wear clothing.

Library of Congress, Prints and Photographs Division [LC-USZ62-69112]

THE ENVIRONMENTAL COSTS OF INDUSTRIALIZATION Nineteenth-century factories like this print works in Manchester contributed to unprecedented levels of air pollution.

Industrialization was not without environmental costs. It created unprecedented levels of water and air pollution that eventually contributed to growing public awareness about the need to protect the environment and citizens and triggered early efforts at reform. To stop toxic runoff from cattle processing plants, for example, Wisconsin passed the Slaughterhouse Offal Act of 1862 that prohibited dumping slaughter wastes in surface water. By 1861 Chicago and Cincinnati had both implemented smoke laws aimed at decreasing the soot, ash, and heavy smog produced by coal and iron factories, railroads, and ships.

Rise of the Industrial Ruling Class

The merchant capitalists remained figures of importance in the 1840s. In such cities as New York, Philadelphia, and Boston, influential mercantile groups operated shipping lines to southern ports and dispatched fleets of trading vessels to Europe and Asia. But merchant capitalism was declining by the middle of the century. This was partly because British competitors were stealing much of the United State's export trade, but mostly because there were greater opportunities for profit in manufacturing than in trade. That was one reason why industries developed first in the Northeast: an affluent merchant class with the money and the will to finance them already existed there. They supported the emerging industrial capitalists and soon became the new aristocrats of the Northeast, with far-reaching economic and political influence.

MEN AND WOMEN AT WORK

In the 1820s and 1830s, factory labor came primarily from the native-born population. After 1840, the growing immigrant population became the most important new source of workers.

Recruiting a Native Workforce

Recruiting a labor force was not an easy task in the early years of the factory system. Ninety percent of the American people in the 1820s still lived and worked on farms. Many urban residents were skilled artisans who owned and managed their own shops, and the available unskilled workers were not numerous enough to meet industry's needs. But dramatic improvements in agricultural production, particularly in the Midwest, meant that each region no longer had to feed itself; it could import the food it needed. As a result, rural people from relatively unprofitable farming areas of the East began leaving the land to work in the factories.

Two systems of recruitment emerged to bring this new labor supply to the expanding textile mills. One, common in the mid-Atlantic states, brought whole families from the farm to work together in the mill. The second system, common in Massachusetts and New England in general, enlisted young white women, mostly farmers' daughters in their late teens and early twenties. It was known as the **Lowell or Waltham system**, after the towns in which it first emerged. Many of these women worked for several years, saved their wages, and then returned home to marry and raise children. Others married men they met in the factories or in town. Most eventually stopped working in the mills and took up domestic roles instead.

Labor conditions in these early years of the factory system, hard as they often were, were significantly better than they would later become. The Lowell workers, for example, were generally well fed, carefully supervised, and housed in clean boardinghouses and dormitories, which the factory owners maintained. (See "Consider the Source: *Handbook to Lowell*.") Wages for the Lowell workers were relatively generous by the standards of the time. The women even published a monthly magazine, the *Lowell Offering*.

Yet even these relatively well-treated workers found the transition from farm life to factory work difficult. Forced to live among strangers in a regimented environment, many women had trouble adjusting to the nature of factory work. However uncomfortable women may have found factory work, they had few other options. Work in the mills was in many cases virtually the only alternative to returning to farms that could no longer support them.

The factory system of Lowell did not, in any case, survive for long. In the competitive textile market of the 1830s and 1840s, manufacturers found it difficult to maintain the high living standards and reasonably attractive working conditions of before. Wages declined; the hours of work lengthened; the conditions of the boardinghouses deteriorated. In 1834, mill workers in Lowell organized a union—the **Factory Girls Association**—which staged a strike to protest a 25 percent wage cut. Two years later, the association struck again—against a rent increase in the boardinghouses. Both strikes failed. Eight years later, the Lowell women, led by the militant **Sarah Bagley**, created the Female Labor Reform Association, which grew to around 500 members in five months. It was one of the first American labor organizations created by women. Members published the *Voice of Industry* to air their grievances and political goals, which included a ten-hour day and improvements in conditions in the mills. The new association also asked state governments for legislative investigation of conditions in the mills. Although mill owners reduced the workday by 30 minutes, larger labor reforms would have to wait. The association dissolved in 1848 because the character of the factory workforce was changing again, lessening the urgency of their demands. Many mill girls were now gradually moving into other occupations: teaching, domestic service, or homemaking. And textile manufacturers were turning to a less demanding labor supply: immigrants.

THE IMMIGRANT WORKFORCE

The increasing supply of immigrant workers after 1840 was a boon to manufacturers and other entrepreneurs. These new workers, because of their growing numbers and their unfamiliarity with their new country, had less leverage than the women they displaced and as a result they often experienced far worse working conditions. Poorly paid construction

CONSIDER THE SOURCE

HANDBOOK TO LOWELL (1848)

Strict rules governed the working life of the young white women who worked in the textile mills in Lowell, Massachusetts, in the first half of the nineteenth century. Equally strict rules regulated their time away from work (what little leisure time they enjoyed) in the company-supervised boardinghouses in which they lived. The excerpts from the *Handbook to Lowell* from 1848 that follow suggest the tight supervision under which the Lowell mill girls worked and lived.

FACTORY RULES

REGULATIONS TO BE OBSERVED by all persons employed in the factories of the Hamilton Manufacturing Company. The overseers are to be always in their rooms at the starting of the mill, and not absent unnecessarily during working hours. They are to see that all those employed in their rooms are in their places in due season, and keep a correct account of their time and work. They may grant leave of absence to those employed under them, when they have spare hands to supply their places and not otherwise, except in cases of absolute necessity.

All persons in the employ of the Hamilton Manufacturing Company are to observe the regulations of the room where they are employed. They are not to be absent from their work without the consent of the overseer, except in cases of sickness, and then they are to send him word of the cause of their absence. They are to board in one of the houses of the company and give information at the counting room, where they board, when they begin, or, whenever they change their boarding place; and are to observe the regulations of their boarding-house.

Those intending to leave the employment of the company are to give at least two weeks' notice thereof to their overseer.

All persons entering into the employment of the company are considered as engaged for twelve months, and those who leave sooner, or do not comply with all these regulations, will not be entitled to a regular discharge.

The company will not employ anyone who is habitually absent from public worship on the Sabbath, or known to be guilty of immorality.

A physician will attend once in every month at the counting-room, to vaccinate all who may need it, free of expense.

Anyone who shall take from the mills or the yard, any yarn, cloth or other article belonging to the company will be considered guilty of stealing and be liable to prosecution.

Payment will be made monthly, including board and wages. The accounts will be made up to the last Saturday but one in every month, and paid in the course of the following week.

These regulations are considered part of the contract, with which all persons entering into the employment of the Hamilton Manufacturing Company, engage to comply.

BOARDING-HOUSE RULES

REGULATIONS FOR THE BOARDING-HOUSES of the Hamilton Manufacturing Company. The tenants of the boarding-houses are not to board, or permit any part of their houses to be occupied by any person, except those in the employ of the company, without special permission.

They will be considered answerable for any improper conduct in their houses, and are not to permit their boarders to have company at unseasonable hours.

The doors must be closed at ten o'clock in the evening, and no person admitted after that time, without some reasonable excuse.

The keepers of the boarding-houses must give an account of the number, names and employment of their boarders, when required, and report the names of such as are guilty of any improper conduct, or are not in the regular habit of attending public worship.

The buildings, and yards about them, must be kept clean and in good order; and if they are injured, otherwise than from ordinary use, all necessary repairs will be made, and charged to the occupant.

The sidewalks, also, in front of the houses, must be kept clean, and free from snow, which must be removed from them immediately after it has ceased falling; if neglected, it will be removed by the company at the expense of the tenant.

It is desirable that the families of those who live in the houses, as well as the boarders, who have not had the kine pox, should be vaccinated, which will be done at the expense of the company, for such as wish it.

Some suitable chamber in the house must be reserved, and appropriated for the use of the sick, so that others may not be under the necessity of sleeping in the same room.

JOHN AVERY, Agent.

UNDERSTAND, ANALYZE, & EVALUATE

1. What do these rules suggest about the everyday lives of the mill workers?
2. What do the rules suggest about the company's attitude toward the workers? Do the rules offer any protections to the employees, or are they all geared toward benefiting the employer?
3. Why would the company enforce such strict rules? Why would the mill workers accept them?

Source: *Handbook to Lowell*, 1848.

gangs, made up increasingly of Irish immigrants, performed the heavy, unskilled work on turnpikes, canals, and railroads. Many of them lived in flimsy shanties, in grim conditions that endangered the health of their families (and reinforced native prejudices toward the "shanty Irish"). Irish workers began to predominate in the New England textile mills as well in the 1840s. Employers began paying piece rates rather than a daily wage and used other devices to speed up production and exploit the labor force more efficiently. The factories themselves were becoming large, noisy, unsanitary, and often dangerous places to work; the average workday was extending to twelve, often fourteen hours; and wages were declining. Women and children, whatever their skills, earned less than most men.

THE FACTORY SYSTEM AND THE ARTISAN TRADITION

Factories were also displacing the trades of skilled artisans. Artisans were as much a part of the older, republican vision of the United States as sturdy yeoman farmers. Independent craftspeople clung to a vision of economic life that was very different from that promoted by the new capitalist class. The **artisans** embraced not just the idea of individual, acquisitive success but also a sense of a "moral community." Skilled artisans valued their independence, their stability, and their relative equality within their economic world.

Some artisans made successful transitions into small-scale industry. But others found themselves unable to compete with the new factory-made goods. In the face of this competition, skilled workers in cities such as Philadelphia, Baltimore, Boston, and New York formed societies for mutual aid. During the 1820s and 1830s, these craft societies began to combine on a citywide basis and set up central organizations known as trade unions. In 1834, delegates from six cities founded the National Trades' Union, and in 1836, printers and cordwainers (makers of high-quality shoes and boots) set up their own national craft unions.

Virtually all the early craft unions excluded women. As a result, women began establishing their own, new protective unions in the 1850s. They built on the efforts made by women to organize in Lowell and, unfortunately, experienced similar results. The new female unions, like their male counterparts, had little power in dealing with employers and enacted no long-term reforms. They did, however, serve an important role as mutual aid societies for women workers.

Hostile laws and hostile courts handicapped the unions, as did the Panic of 1837 and the depression that followed. But some artisans managed to retain control over their productive lives.

FIGHTING FOR CONTROL

Industrial workers made continuous efforts to improve their lives. They tried, with little success, to persuade state legislatures to pass laws setting a maximum workday and regulating child labor. Their greatest legal victory came in Massachusetts in 1842, when the state supreme court, in *Commonwealth v. Hunt*, declared that unions were lawful organizations and that the strike was a permissible weapon. Other state courts gradually accepted the principles of the Massachusetts decision, but employers tended to ignore them.

Many factors combined to inhibit the growth of better working standards. Among the most important obstacles was the flood into the country of immigrant laborers, who were usually willing to work for lower wages than native workers. Because they were so numerous, manufacturers had little difficulty replacing disgruntled or striking workers with eager immigrants. Ethnic divisions often led workers to channel their resentments into internal bickering among one another rather than into their shared grievances. Another obstacle was the sheer strength of the industrial capitalists, who possessed not only economic but also political and social power.

PATTERNS OF SOCIETY

The Industrial Revolution made the United States both dramatically wealthier and increasingly unequal. It transformed social relationships at almost every level.

THE RICH AND THE POOR

The commercial and industrial growth of the United States greatly elevated the average income of Americans. But this increasing wealth was distributed highly unequally. In fact, substantial groups of the population—enslaved people, free Blacks, Native Americans, landless farmers, and many of the unskilled workers on the fringes of the manufacturing system—shared hardly at all in the economic growth. Even among the rest of the population, disparities of income were growing. Merchants and industrialists accumulated enormous fortunes, creating a new culture of wealth, especially in the urban areas in which they lived.

In large cities, people of great wealth gathered together in neighborhoods of astonishing opulence. They looked increasingly for ways to display their wealth—in great mansions, showy carriages, lavish household goods, and the elegant cultural establishments they patronized. They founded exclusive clubs and developed elaborate social rituals. New York

Everett Historical/Shutterstock

CENTRAL PARK Daily carriage rides allowed the wealthy to take in fresh air while showing off their finery to their neighbors.

developed a particularly elaborate high society. The construction of Central Park, which began in the 1850s, was in part a result of pressure from the members of high society, who wanted an elegant setting for their daily carriage rides.

At the same time a significant population of poor people also developed in the growing urban centers. These people were almost entirely without resources, often homeless and dependent on charity or crime, or both, for survival. Substantial numbers of people actually starved to death or died of exposure. Some of these "paupers," as contemporaries called them, were widows and orphans stripped of the family structures that allowed most working-class Americans to survive. Some were people suffering from alcoholism or mental illness, unable to work. Others were recent immigrants who were victims of native prejudice—barred from all but the most menial employment because of ethnicity.

Those who had the poorest prospects in the North were free Blacks. Most major urban areas had significant Black populations. Some of these African Americans were descendants of families who had lived in the North for generations. Others were formerly enslaved and had escaped or been released by those who had held them in slavery. In material terms, at least, life was not always much better for them in the North than it had been in slavery. Most had access to very menial jobs at best. In most parts of the North, Blacks could not vote, attend public schools, or use any of the public services available to white residents. They lived with the daily threat of violence, which sometimes came in the form of riots. Even so, nearly all Black Americans preferred life in the North, however arduous, to life in slavery.

SOCIAL AND GEOGRAPHICAL MOBILITY

Despite the contrasts between wealth and poverty in the antebellum United States, the waves of large-scale working class revolt that periodically swept through European nations did not take place. Yet class conflict was still a major theme. Through their unions, mutual aid societies, strikes, and petitions for better working conditions, poor Americans voiced their frustration and anger over the state of their lives. Sometimes that rage manifested itself in pitched battles between workers competing for scarce jobs, as was often the case between the Irish and free Blacks.

Part of the explanation for why there was no class warfare, as in Europe, is that life was generally better for most foreign-born industrial workers in the United States than it had been in their country of origin. They also found that it was possible to move up the economic ladder, especially when compared to opportunities in much of Europe. A few workers—a very small number, but enough to support the dreams of others—managed to move from poverty to riches by dint of work, ingenuity, and luck. And a much larger number of workers managed to step at least one notch up the ladder—for example, becoming in the course of a lifetime a skilled, rather than an unskilled, laborer.

Closely related to social mobility was geographical mobility. Some workers saved money, moved West, bought land, and farmed it. But few urban workers or the poor in general could afford such a relocation. Much more common was the movement of laborers from one industrial town to another. These migrants, often the victims of layoffs, were constantly looking for better opportunities elsewhere even if that search seldom produced something better. This rootlessness among a large and distressed segment of the workforce made effective organization and protest difficult.

Middle-Class Life

In between the very rich and the very poor in antebellum society was the middle class, a small but growing segment of society that encompassed people with dependable jobs and stable income who made enough money to buy adequate food, stylish clothing, and comfortable homes. Economic development created many opportunities for the middle class to own or work in shops or businesses, engage in trade, enter professions, and administer organizations. In earlier times, when landownership had been the only real basis of wealth, society had been divided between those with little or no land (people Europeans generally called peasants) and a landed gentry (which in Europe usually became an inherited aristocracy). Once commerce and industry became a source of wealth, these rigid distinctions broke down; many people could become prosperous without owning land but by providing valuable goods or services.

Middle-class life in the antebellum years rapidly established itself as an influential cultural form in the urban United States. Solid, substantial middle-class houses lined city streets, larger in size and more elaborate in design than the cramped, functional row houses in working-class neighborhoods—but also far less lavish than the great houses of the very rich. Middle-class people tended to own their homes, often for the first time. Workers and artisans remained mostly renters.

Middle-class women usually remained in the household, although increasingly they were also able to hire servants—usually young, unmarried immigrant women. In an age when doing the family's laundry could take an entire day, one of the aspirations of middle-class women was to escape from some of the drudgery of housework.

New household inventions greatly improved the character of life in middle-class homes. Perhaps the most important was the invention of the cast-iron stove, which began to replace fireplaces as the principal vehicle for cooking in the 1840s. These wood- or coal-burning devices were hot, clumsy, and dirty by later standards, but compared to the inconvenience and danger of cooking on an open hearth, they seemed a great luxury. Stoves gave cooks greater control over food preparation and allowed them to prepare several things at once.

Middle-class diets were changing rapidly, and not just because of the wider range of cooking that the stove made possible. The expansion and diversification of American agriculture and the ability of distant farmers to ship goods to urban markets by rail greatly increased the variety of food available in cities. Fruits and vegetables were difficult to ship over long distances in an age with little refrigeration, but families had access to a greater variety of meats, grains, and dairy products than in the past. A few wealthy households acquired iceboxes, which allowed them to keep meat and dairy products fresh for several days. Most families, however, did not yet have any refrigeration. For them, preserving food meant curing meat with salt and preserving fruits in sugar. Diets were generally much heavier and starchier than they are today, and middle-class people tended to be considerably stouter than would be considered healthy or fashionable now.

The style of middle-class homes came to be very different from those of workers and artisans. The spare, simple styles of eighteenth-century homes gave way to the much more elaborate, even baroque household styles of the Victorian era—increasingly characterized by crowded, even cluttered rooms, dark colors, lush fabrics, and heavy furniture and draperies. Middle-class homes also became larger. It became less common for children to share beds and for all members of a family to sleep in the same room. Parlors and dining rooms separate from the kitchen—once a luxury—became the norm among the middle class. Some urban middle-class homes had indoor plumbing and indoor toilets by the 1850s—a significant advance over outdoor wells and privies.

The Changing Family

The new industrializing society profoundly changed the nature of the family. Among the changes was the movement of families from farms to urban areas. Sons and daughters in urban households were much more likely to leave the family in search of work than they

Fotosearch/Archive Photos/Getty Images

FAMILY TIME, 1842 This illustration for *Godey's Lady's Book*, a magazine whose audience was better-off white women, offers an idealized image of family life. The father reads to his family from a devotional text; two servants off to the side listen attentively as well. What does this image communicate about the roles of the household members?

had been in the rural world. This was largely because of the shift of income-earning work out of the home. In the early decades of the nineteenth century, the family itself had been the principal unit of economic activity. Now most income earners left home each day to work in a shop, mill, or factory. A sharp distinction began to emerge between the public world of the workplace and the private world of the family, which was now dominated by housekeeping, child rearing, and other domestic concerns.

There was a significant decline in the birthrate, particularly in urban areas and in middle-class families. In 1800, the average American woman could be expected to give birth to approximately seven children. By 1860, the average woman bore five children.

The "Cult of Domesticity"

The growing separation between the workplace and the home sharpened distinctions between the social roles of men and women. Those distinctions affected not only the growing middle class but also factory workers and farmers.

With fewer legal and political rights than men, most women remained under the virtually absolute authority of their husbands. They were seldom encouraged to pursue education above the primary level. Women students were not accepted in any college or university until 1836, when Georgia Women's College (now Wesleyan College) in Macon, Georgia, received a charter. The next year Mary Lyon opened Mount Holyoke Female Seminary (now Mount Holyoke College) in South Hadley, Massachusetts, as an academy for women.

However unequal the positions of men and women in the preindustrial era, those positions had generally been defined within the context of a household in which all members played important economic roles. In the middle-class family of the new industrial society, by contrast, the husband was assumed to be the principal, usually the only, income producer. The image of women changed from one of contributors to the family economy to one of guardians of the "domestic virtues." Middle-class women learned to place a higher value on keeping a clean, comfortable, and well-appointed home; on entertaining; and on dressing elegantly and stylishly.

Within their own separate sphere, middle-class women began to develop a distinctive female culture. A "lady's" literature began to emerge. Romantic novels written for female readers focused on the private sphere that middle-class women now inhabited, as did women's magazines that focused on fashions, shopping, homemaking, and other purely domestic concerns.

This **cult of domesticity**, as some scholars have called it, provided many middle-class women greater material comfort than they had enjoyed in the past and placed a higher value on their "female virtues." At the same time, it left women increasingly detached from the public world, with fewer outlets for their interests and energies. Except for teaching and nursing, work by women outside the household gradually came to be seen as a lower-class preserve.

Working-class women continued to work in factories and mills, but under conditions far worse than those that the original, more "respectable" women workers of Lowell and Waltham had experienced. Domestic service became another frequent source of employment for them. They didn't have the time or money to engage in the same pursuits as middle-class women.

Leisure Activities

Leisure time was scarce for all but the wealthiest Americans. Most people worked long hours every day without any vacation. For the lucky, Sunday was a day off, set aside for rest and maybe church-going. Not surprisingly, holidays took on a special importance, as suggested by the strikingly elaborate celebrations of the Fourth of July in the nineteenth century. The celebrations were not just expressions of patriotism, but a way of enjoying one of the few nonreligious holidays from work available to most Americans.

For urban people, leisure was something to be seized in what few free moments they had. Men gravitated to taverns for drinking, talking, and game-playing after work. Women gathered in one another's homes for conversation and card games. For educated people, reading became one of the principal leisure activities. Newspapers and magazines proliferated rapidly and books became staples of affluent homes. In contrast, rural Americans, because of the seasonal nature of farm work, enjoyed more free time in the late fall and winter. They pursued similar pastimes as urbanites, but within the home.

A public culture of leisure emerged too, especially in larger cities. Theaters became popular and attracted audiences that crossed class lines. Minstrel shows—in which white actors wearing blackface mimicked and ridiculed Black culture—became popular staples among white audiences. They roared their approval of stock characters acting out racist

Bettmann/Getty Images

P. T. BARNUM AND TOM THUMB P. T. Barnum stands next to his star Charles Stratton, whose stage name was General Tom Thumb after the fairy-tale character. Stratton joined Barnum's touring company as a child, singing, dancing, and playing roles such as Cupid and Napoleon Bonaparte. The adult Stratton and Barnum became business partners.

stereotypes of unintelligent Blacks who shunned work and cared only for song and dance. Also popular were Shakespeare's plays, reworked to appeal to American audiences. Tragedies were given happy endings; comedies were interlaced with regional humor; lines were rewritten with American dialect; and scenes were abbreviated or cut so that the play could be one of several in an evening's program. So familiar were many Shakespearean plots that audiences took delight in seeing them parodied in productions such as *Julius Sneezer* and *Hamlet and Egglet*.

Public sporting events—boxing, horse racing, cockfighting (already becoming controversial), and others—often attracted considerable audiences. Baseball, not yet organized into professional leagues, was beginning to attract large crowds when played in city parks or fields. A particularly exciting event in many communities was the arrival of the circus.

Popular tastes in public spectacle tended toward the bizarre and the fantastic. Relatively few people traveled; and in the absence of film, radio, television, or even much photography, Americans hungered for visions of unusual phenomena. People going to the theater or the circus or the museum wanted to see things that amazed and even frightened them. The most celebrated provider of such experiences was the famous and unscrupulous showman P. T. Barnum, who opened the American Museum in New York in 1842—not a showcase for art or nature, but an exhibit of "human curiosities" that included people with dwarfism, Siamese twins, magicians, and ventriloquists. Barnum was a genius in publicizing his ventures with garish posters and elaborate newspaper announcements. Later, in the 1870s, he launched the famous circus for which he is still best remembered.

Lectures were one of the most popular forms of entertainment in the nineteenth-century United States. Men and women flocked in enormous numbers to lyceums, churches, schools, and auditoriums to hear lecturers explain the latest advances in science, describe their visits to exotic places, provide vivid historical narratives, or rail against the evils of alcohol or slavery. Messages of social uplift and reform attracted rapt audiences, particularly among women.

THE AGRICULTURAL NORTH

Even in the rapidly urbanizing and industrializing Northeast, and more so in what nineteenth-century Americans called the Northwest, most people remained tied to the agricultural world. But agriculture, like industry and commerce, was becoming increasingly a part of the new capitalist economy.

Northeastern Agriculture

The story of agriculture in the Northeast after 1840 is one of decline and transformation. Farmers of this section of the country could no longer compete with the new and richer soil of the Northwest. In 1840, the leading wheat-growing states were New York, Pennsylvania, Ohio, and Virginia. In 1860, they were Illinois, Indiana, Wisconsin, Ohio, and Michigan. Illinois, Ohio, and Missouri also supplanted New York, Pennsylvania, and Virginia as growers of corn. In 1840, the most important cattle-raising areas in the country were New York, Pennsylvania, and New England. By the 1850s, the leading cattle states were Illinois, Indiana, Ohio, and Iowa in the Northwest and Texas in the Southwest.

Some eastern farmers responded to these changes by moving west themselves and establishing new farms. Still others moved to mill towns and became laborers. Some farmers, however, remained on the land and turned to what was known as "truck farming"—supplying food to the growing cities. They raised vegetables or fruit and sold their produce in nearby towns. Supplying milk, butter, and cheese to local urban markets also attracted many farmers in central New York, southeastern Pennsylvania, and various parts of New England.

The Old Northwest

Life was different in the states of the Old Northwest (now known as the Midwest). In the two decades before the Civil War, this section of the country experienced steady industrial growth, particularly in and around Cleveland (on Lake Erie) and Cincinnati, the center of meatpacking in the Ohio Valley. Farther west, Chicago was emerging as the national center of the agricultural machinery and meatpacking industries. Most of the major industrial activities of the Old Northwest either served agriculture (for example, producing farm machinery) or relied on agricultural products (flour milling, meatpacking, whiskey distilling, the making of leather goods, and others).

Some areas of the Old Northwest were not yet dominated by whites. Native Americans remained the most numerous inhabitants of large portions of the upper third of the Great Lakes states until after the Civil War. In those areas, hunting and fishing, along with some sedentary agriculture, remained the principal economic activities.

For the settlers who populated the lands farther south, the Old Northwest was primarily an agricultural region. Its rich lands made farming highly lucrative. Thus the typical citizen of the Old Northwest was not the industrial worker or poor, marginal farmer, but the owner of a reasonably prosperous family farm.

Industrialization, in both the United States and Europe, provided the greatest boost to agriculture. With the growth of factories and cities in the Northeast, the domestic market for farm goods increased dramatically. The growing national and worldwide demand for farm products resulted in steadily rising farm prices. For most farmers, the 1840s and early 1850s were years of increasing prosperity.

The expansion of agricultural markets also had profound effects on sectional alignments in the United States. The Old Northwest sold most of its products to the Northeast and became an important market for the products of eastern industry. A strong economic relationship was emerging between the two sections that was profitable to both—and that was increasing the isolation of the South within the Union.

By 1850, the growing western white population was moving into the prairie regions on both sides of the Mississippi. These farmers cleared forest lands or made use of fields that indigenous inhabitants had cleared many years earlier. And they developed a timber industry to make use of the remaining forests. Although wheat was the staple crop of the region, other crops—corn, potatoes, and oats—and livestock were also important.

The Old Northwest also increased production by adopting new agricultural techniques. Farmers began to cultivate new varieties of seed, notably Mediterranean wheat, which was hardier than the native type; and they imported better breeds of animals, such as hogs and sheep from England and Spain. Most important were improved tools and farm machines. The cast-iron plow, invented in 1814, had the advantage of being more durable than older wooden plows, more capable of breaking up hard and stony fields, and eventually capable of being fixed with replaceable parts. But it was still ineffective at churning up the thick sod and clay soils found throughout the Midwest. It was replaced in 1847 by the steel plow, manufactured by the John Deere company in Moline, Illinois, which quickly became a farming staple.

Two new machines heralded a coming revolution in grain production. The automatic reaper, the invention of Cyrus H. McCormick of Virginia, took the place of sickle and

Oxford Science Archive/Print Collector/Hulton Archive/Getty Images

CYRUS MCCORMICK'S AUTOMATIC REAPER

cradle, which required hand labor. Pulled by a team of horses, it had a row of horizontal knives on one side for cutting wheat; the wheels drove a paddle that bent the stalks over the knives, with the cut stalks falling onto a moving belt and into the back of the vehicle. The reaper enabled a crew of six or seven men to harvest in a day as much wheat as fifteen men could harvest using the older methods. McCormick, who had patented his device in 1834, established a factory at Chicago in 1847. By 1860, more than 100,000 reapers were in use. Almost as important to the grain grower was the thresher—a machine that separated the grain from the wheat stalks—which appeared in large numbers after 1840. (Before that, farmers generally flailed grain by hand or used farm animals to tread the stalks, releasing the grains.) The Jerome I. Case factory in Racine, Wisconsin, manufactured most of the threshers. (Modern "harvesters" later combined the functions of the reaper and the thresher.)

The Old Northwest was the most self-consciously democratic section of the country. But its democracy was of a relatively conservative type—capitalistic, property-conscious, middle-class, and for white men. Abraham Lincoln, an Illinois Whig, voiced the optimistic economic opinions of many of the people of his section. "I take it that it is best for all to leave each man free to acquire property as fast as he can," said Lincoln. "When one starts poor, as most do in the race of life, free society is such that he knows he can better his condition; he knows that there is no fixed condition of labor for his whole life."

Rural Life

Life for farming people varied greatly from one region to another. In the more densely populated areas east of the Appalachians and in the easternmost areas of the Old Northwest, farmers made extensive use of typical community institutions—churches, schools, stores, and taverns. As white settlement moved farther west, farmers became more isolated and had to struggle to find any occasions for contact with people outside their own families.

Religion drew farm communities together more than any other force in remote communities. Town or village churches were popular meeting places, both for services and for social events—most of them dominated by women. Even in areas with no organized churches, farm families—and women in particular—gathered in one another's homes for prayer meetings, Bible readings, and other religious activities. Weddings, baptisms, and funerals also united communities.

But religion was only one of many reasons for interaction. Farm people joined together frequently to share tasks such as barn raising. Large numbers of families gathered at harvest time to help bring in crops, husk corn, or thresh wheat. Women came together to share domestic tasks, holding "bees" in which groups of women made quilts, baked goods, preserves, and other products.

Despite the many social gatherings farm families managed to create, they had much less contact with popular culture and public life than people who lived in towns and cities. Most rural people treasured their links to the outside world—letters from relatives and friends in distant places, newspapers and magazines from cities they had never seen, catalogs advertising merchandise that their local stores never had. Yet many also valued the relative autonomy that a farm life gave them. One reason many rural Americans looked back nostalgically on country life once they moved to the city was that they sensed that in the urban world they had lost some control over the patterns of their daily lives.

CONCLUSION

Between the 1820s and the 1850s, the American economy experienced the beginnings of an industrial revolution—a change that transformed almost every area of life in fundamental ways.

The American industrial revolution was a result of many things: population growth, advances in transportation and communication, new technologies that spurred the development of factories and mass production, the recruiting of a large industrial labor force, and the creation of corporate bodies capable of managing large enterprises. The new economy expanded the ranks of the wealthy, helped create a growing middle class, and introduced high levels of inequality.

Culture in the industrializing areas of the North changed, too, as did the structure and behavior of the family, the role of women, and the way people used their leisure time and encountered popular culture. The changes helped widen the gap in experience and understanding between the generation of the Revolution and the generation of the mid-nineteenth century. They also helped widen the gap between North and South.

KEY TERMS/PEOPLE/PLACES/EVENTS

antebellum 228
artisan 241
Baltimore and Ohio
 Railroad 233
Commonwealth v. Hunt 242
cult of domesticity 246
Erie Canal 232

Factory Girls
 Association 239
factory system 237
industrialization 237
Know-Nothings 230
Lowell or Waltham
 system 239

Morse code 235
nativism 230
Sarah Bagley 239
Western Union
 Telegraph Company 235

RECALL AND REFLECT

1. What were the political responses to immigration in the United States in the mid-nineteenth century? Do you see any parallels to responses to immigration today?
2. Why did the rail system supplant the canal system as the nation's major transportation network?
3. How did the industrial workforce change between the 1820s and the 1840s? What were the effects on American society of changes in the workforce?
4. How did the Industrial Revolution and the factory system change family life and women's social and economic roles in the United States?
5. How did agriculture in the North change as a result of growing industrialization and urbanization?

11 | COTTON, SLAVERY, AND THE OLD SOUTH

THE COTTON ECONOMY
SOUTHERN WHITE SOCIETY
SLAVERY: THE "PECULIAR INSTITUTION"
BLACK CULTURE UNDER SLAVERY

LOOKING AHEAD

1. How did slavery shape the southern economy and society?
2. Why was there less industrialization in the South than the North?
3. How did enslaved people challenge their status? How successful were their efforts?

THE SOUTH, LIKE THE NORTH, experienced significant growth in the middle years of the nineteenth century. The southern agricultural economy boomed. Trade in such staples as sugar, rice, tobacco, and above all cotton made the South a major force in international commerce. It also made the South integral to the emerging capitalist world of the United States and its European trading partners.

Yet despite all these changes, the South experienced a much less fundamental transformation in these years than did the North. It had begun the nineteenth century as a primarily agricultural region; it remained overwhelmingly so in 1860. It had begun the century with few important cities and little industry; and so it remained sixty years later. In 1800, a plantation system dependent on the labor of enslaved people had dominated the southern economy; by 1860, that system had only strengthened its grip on the economy and culture. By the outbreak of the Civil War, few southern white leaders could imagine the health and prosperity of their region without slavery.

TIME LINE

1800 — Gabriel Prosser's revolt

1808 — Foreign trade of enslaved people outlawed

1820s — Depression in tobacco prices begins; High cotton production in Southwest

1822 — Denmark Vesey's Revolt

1831 — Nat Turner rebellion

1833 — John Randolph frees 400 enslaved people

1837 — Cotton prices plummet

1846 — De Bow's Commercial Review founded

1849 — Cotton production boom

THE COTTON ECONOMY

The most important economic development in the mid-nineteenth-century South was the shift of economic power from the "upper South" (the original southern states along the Atlantic Coast) to the "lower South" (the expanding agricultural regions in the new states of the Southwest). That shift reflected the growing dominance of cotton and slavery in the region's economy.

THE RISE OF KING COTTON

Much of the upper South continued to rely on the cultivation of tobacco. But the market for that crop was historically unstable and tobacco rapidly exhausted the land on which it grew. As early as the 1780s, tobacco farmers in the Chesapeake Bay region were shifting to other crops, especially grain. Fifty years later the center of tobacco cultivation was slowly moving westward, into the Piedmont area. Yet even as late as 1860, Maryland and eastern Virginia were still producing a lot of tobacco.

The southern regions of the coastal South—South Carolina, Georgia, and parts of Florida—continued to rely on the cultivation of rice, a more stable and lucrative crop than tobacco. But rice demanded substantial irrigation and an exceptionally long growing season (nine months), so its cultivation remained restricted to a relatively small area. Sugar growers along the Gulf Coast similarly enjoyed a reasonably profitable market for their crop. But sugar cultivation brought its own challenges, chief among them the requirement for a large labor force able to take on the back-breaking work required to produce the commodity. Additionally, there were sizeable equipment costs, the need for a certain number of frost-free days, and major competition from the great sugar plantations of the Caribbean. Not surprisingly, then, only relatively

COTTON, SLAVERY, AND THE OLD SOUTH · 255

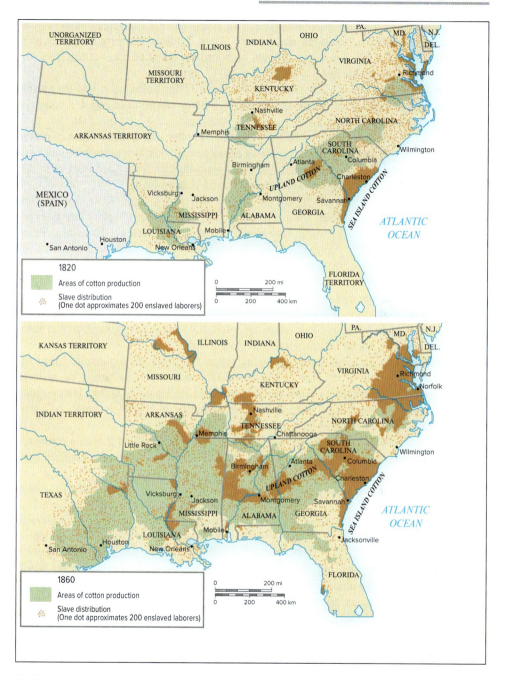

SLAVERY AND COTTON IN THE SOUTH, 1820 AND 1860 These two maps show the remarkable spread of cotton cultivation in the South in the decades before the Civil War. Both maps show the areas of cotton cultivation (the green-colored areas) as well as areas with large enslaved populations (the brown-dotted areas). Note how in the top map, which represents 1820, cotton production is concentrated largely in the East, with a few areas scattered among Alabama, Mississippi, Louisiana, and Tennessee. Slavery is concentrated along the Georgia and South Carolina coast, areas in which long-staple cotton was grown, with only a few other areas of highly dense slave populations. By 1860, however, the South had changed dramatically. Cotton production had spread throughout the lower South, from Texas to northern Florida, and slavery had moved with it. Slavery was also much denser in the tobacco-growing regions of Virginia and North Carolina, which had also grown. • *How did this economic shift affect the white South's commitment to slavery?*

wealthy planters could afford to grow sugar and its cultivation did not spread much beyond a small area in southern Louisiana and eastern Texas. Another lucrative crop was long-staple cotton, but like rice and sugar, it could grow only in a limited area—in this case, the coastal regions of the Southeast.

The decline of the tobacco economy and the limits of the sugar, rice, and long-staple cotton economies might have forced the region to shift its attention to other, nonagricultural pursuits had it not been for the growing importance of a new product that soon overshadowed all else: short-staple cotton. This was a hardier and coarser strain of cotton that could be grown successfully in a variety of climates and soils. It was harder to process, however, than long-staple cotton because its seeds were difficult to remove from the fiber. The invention of the cotton gin in 1793 largely solved that problem.

Demand for cotton increased rapidly in Britain in the 1820s and 1830s and in New England in the 1840s and 1850s. From the western areas of South Carolina and Georgia, production moved into Alabama and Mississippi and then into northern Louisiana, Texas, and Arkansas. By the 1850s, cotton had become the linchpin of the southern economy. By the time of the Civil War, cotton constituted nearly two-thirds of the total export trade of the United States.

Cotton production boomed in the lower South. Some began to call it the "**cotton kingdom.**" The prospect of tremendous profits drew settlers to the lower South by the thousands. Some were wealthy planters from the older states, but most were small slaveholders or non slaveholding farmers who hoped to move into the planter class.

A similar shift, if an involuntary one, occurred in the enslaved population. Since the 1820s and continuing until the start of the Civil War, hundreds of thousands of enslaved men and women were forced to move across state lines and into the cotton states—either accompanying slaveholders who were themselves migrating to the lower South or (more often) through sale to planters already there.

This "**second middle passage**," as the historian Ira Berlin has called it, was a traumatic experience for perhaps a million dislocated Black Americans. Enslaved families were routinely broken up and scattered across the expanding cotton kingdom. Marched over hundreds of rugged miles, tied or chained together, they arrived in unfamiliar and usually forbidding territory where they were made to construct new plantations and work in cotton fields. The sale of enslaved laborers to the lower South became an important economic activity for whites in the upper South, where agricultural production was slowly declining.

Even as slavery slowly became the lifeblood of the southern economy and society, it was never strictly a southern phenomenon. In the North during the Colonial era, the enslaved population numbered about 40,000, mostly concentrated in coastal cities and inland farms. In New York City in 1740, for example, about 20 percent of the entire population was enslaved. Northern sea merchants had long profited handsomely by trafficking in enslaved labor. After the American Revolution, however, slavery in the North slowly died out. The Northwest Ordinance of 1787 outlawed slavery in the Northwest Territory and the states that developed in this area—Ohio, Indiana, Michigan, Illinois, Wisconsin, and Minnesota—were always "free" states. By 1804, all other northern states had outlawed slavery, although some had done so through gradual emancipation. In 1807 Congress enacted a law banning the importation of enslaved men and women from foreign lands to the United States. Despite these prohibitions, the North continued to indirectly support slavery. Its commerce and industry relied on crops and cotton produced by enslaved people in the South, and thus it largely tolerated southern slavery, at least until the late antebellum era.

Southern Trade and Industry

While slave-based agriculture boomed, other forms of economic activity were slow to develop in the South. Flour milling and textile and iron manufacturing grew, particularly in the upper South, but industry remained relatively insignificant in comparison with the agricultural economy. The total value of southern textile manufactures in 1860 was $4.5 million—a threefold increase over the value of those goods twenty years before. But the value of cotton exports alone was approximately $200 million.

The limited nonfarm commercial sector that did develop in the South was largely intended to serve the needs of the plantation economy. Banks clustered in cities and had few if any branches in rural areas. They hired brokers or "factors" to travel into the countryside and provide planters with credit and advice on marketing and selling their crops. Other obstacles to economic development included the South's inadequate transportation system: canals were almost nonexistent; most roads were crude and unsuitable for heavy transport; and railroads, although they expanded substantially in the 1840s and 1850s, failed to tie the region together effectively. In fact, a key aim of many southern railroads was to link larger plantations to urban markets. Not surprisingly, then, the principle means of large-scale transportation was by steamboat and other larger watercraft along major rivers or the coast to Charleston, New Orleans, Savannah, Houston, Mobile, and other port cities. While this system benefited planters, who typically located near major waterways, it made it difficult for industry to grow in landlocked areas far from cities.

The South, therefore, was becoming more and more dependent on the industrial manufacturers, merchants, and professionals of the North. Concerned by this trend, some Southerners began to advocate economic independence for the region, among them James D. B. De Bow of New Orleans, whose magazine, *De Bow's Commercial Review,* called for southern commercial expansion and economic independence from the North. Yet even *De Bow's Commercial Review* was filled with advertisements from northern manufacturing firms; and its circulation was far smaller in the South than such northern magazines as *Harper's Weekly.*

SOUTHERN WHITE SOCIETY

An important question about antebellum southern history is why the region did so little to develop a larger industrial and commercial economy of its own. The explanation rests on the dominant place of slavery in southern society.

Part of the reason that the South lagged behind the North's pace and spread of industrialization was the great profitability of the region's agricultural system. In the Northeast, many people had turned to manufacturing as the agricultural economy of the region declined. But in the South, the agricultural economy was booming and ambitious entrepreneurs and capitalists had little incentive to look elsewhere. Another reason was that wealthy Southerners had so much capital invested in land and enslaved people that they had little money leftover for other investments.

The South's resistance to transforming its economy was above all tied to a culture that blindly celebrated the wealth and profit wrung from the labor of enslaved people. Southern planters and intellectuals fed themselves lies depicting slavery as a benevolent institution in which the enslaved were well cared for, even loved. They and nearly every white Southerner viewed Blacks as too dumb, passionate, and dangerous to live in freedom and were

CONSIDER THE SOURCE

SENATOR JAMES HENRY HAMMOND DECLARES, "COTTON IS KING" (1858)

James Henry Hammond, a U.S. senator from South Carolina, was a leading advocate for the view that cotton was of overwhelming significance to the economy of the South and the nation. He famously made his point in 1858 in his "Cotton Is King" speech.

If we never acquire another foot of territory for the South, look at her. Eight hundred and fifty thousand square miles. As large as Great Britain, France, Austria, Prussia, and Spain. Is not that territory enough to make an empire that shall rule the world? . . . With the finest soil, the most delightful climate, whose staple productions none of those great countries can grow, we have three thousand miles of continental shore line, so indented with bays and crowded with islands, that, when their shore lines are added, we have twelve thousand miles. Through the heart of our country runs the great Mississippi, the father of waters, into whose bosom are poured thirty-six thousand miles of tributary streams; and beyond we have the desert prairie wastes, to protect us in our rear. Can you hem in such a territory as that? . . .

[. . .] Upon our muster-rolls we have a million of militia. In a defensive war, upon an emergency, every one of them would be available. At any time, the South can raise, equip, and maintain in the field, a larger army than any Power of the earth can send against her, and an army of soldiers—men brought up on horseback, with guns in their hands.

[. . .] It appears, by going to the reports of the Secretary of the Treasury, which are authentic, that last year the United States exported in round numbers $279,000,000 worth of domestic produce, excluding gold and foreign merchandise re-exported. Of this amount $158,000,000 worth is the clear produce of the South; . . .

[. . .] [W]e have nothing to do but to take off restrictions on foreign merchandise and open our ports, and the whole world will come to us to trade. They will be too glad to bring and carry for us, and we never shall dream of a war. Why the South has never yet had a just cause of war. Every time she has drawn her sword it has been on the point of honor, and that point of honor has been mainly loyalty to her sister colonies and sister States, who have ever since plundered and calumniated her.

But if there were no other reason why we should never have war, would any sane nation make war on cotton? Without firing a gun, without drawing a sword, should they make war on us we can bring the whole world to our feet. The South is perfectly competent to go on, one, two, or three years without planting a seed of cotton. I believe that if she was to plant but half her cotton for three years to come, it would be an immense advantage to her. I am not so sure but that after three total years' abstinence she would come out stronger than ever she was before and better prepared to enter afresh upon her great career of enterprise. . . . England would topple headlong and carry the whole civilized world with her, save the South. No, you dare not make war on cotton. No power on earth dares to make war upon it. Cotton *is* king.

In all social systems there must be a class to do the menial duties, to perform the drudgery of life. That is, a class requiring but a low order of intellect and but little skill. Its requisites are vigor, docility, fidelity. . . . Fortunately for the South, she found a race adapted to that purpose to her hand—a race inferior to her own, but eminently qualified in temper, in vigor, in docility, in capacity to stand the climate, to answer all her purposes. We use them for our purpose, and call them slaves. We found them slaves by the "common consent of mankind," which, according to Cicero, *"lex naturae est"*; the highest proof of what is Nature's law. We are old-fashioned at the South yet; it is a word discarded now by "ears polite." I will not characterize that class at the North by that term; but you have it; it is there; it is everywhere; it is eternal.

UNDERSTAND, ANALYZE, & EVALUATE

1. How did Hammond view the South in a global context? What do you think of this assessment?
2. What justifications did Hammond offer for slavery? Describe the comparison Hammond drew between northern wage labor and southern slavery.

Source: Hammond, James Henry, Speech on the Kansas-Lecompton Constitution, U.S. Senate, March 4, 1858, in *Congressional Globe*, 35th Cong., 1st Sess., Appendix, 70–71, in Regina Lee Blaszczyk and Philip B. Scranton (eds.), *Major Problems in American Business History*, Boston: Houghton Mifflin Company, 2006, 149–154.

---

terrified of the prospect of dwelling near any emancipated Black person. Most believed that white liberty and freedom depended on Black enslavement.

The central place of slavery in southern society cannot be underestimated. In 1860, the total slaveholding families in the South constituted about one-quarter of the total white population. But in the lower South, the figure was one-third or more while in states like Mississippi and South Carolina it was more like one-half. More importantly, many non-slaveholders regularly hired enslaved laborers and relied heavily on them in all sorts of ways–planting, clearing woods and forests, building fences and barns and homes, to name just a few. Indeed, the percentage of whites dependent on the use of enslaved labor was much higher than the percentage of whites who actually enslaved people.

Slavery was a key part of southern politics. Pro-slavery planters dominated the ranks of southern governors, judges, and legislators, making it nearly impossible for anyone who disagreed with them to be elected or even have their voices heard in public. As the cotton economy surged in the late 1850s, Senator **James Henry Hammond** of South Carolina spoke for many whites when he championed slavery's value to southern society and politics. Slavery, and the cotton crop it made profitable, needed to be protected at all costs. (See "Consider the Source: Senator James Henry Hammond Declares, 'Cotton Is King.'")

THE PLANTER CLASS

The **planter class** was not nearly as leisured and genteel as the aristocratic myth its members perpetrated suggested. Most planters did not own a vast plantation with hundreds of enslaved workers and a mansion with large pillars and a deep front porch. Growing staple crops was tough work. Planters focused on the stubborn basics of running their business: buying and selling enslaved labor, anticipating fluctuations in markets, arranging for the transportation of the harvest, controlling costs, and winning a profit at year's end. Sometimes they managed poorly and lost everything. Indeed, planters were just as much competitive capitalists as were the industrialists of the North. Many lived rather modestly, their wealth so heavily invested in land and enslaved workers that there was often little left for personal comfort. And they tended to move frequently as new and presumably more productive areas opened up to cultivation.

Still, wealthy southern white men strove to create and sustain their image as aristocrats. They tended to avoid such "coarse" occupations as trade and commerce—even though they bought and sold enslaved people. Those who did not become planters often gravitated toward the military.

The "Southern Lady"

The popular portrayal of wealthy southern women as highly educated belles bedecked in flowing hoop skirts who spent their days playing the harp and anticipating their next social was almost entirely fictional. Southern white women had less access to education than their northern counterparts. The few female "academies" in the South trained women primarily to be suitable wives. The southern white birthrate remained nearly 20 percent higher than that of the nation as a whole, but infant mortality in the region was also higher. Most southern white women lived on farms with little access to the public world and thus had little opportunity to look beyond their roles as wives, mothers, and workers. Those with farms of modest size or smaller often experienced a fuller engagement in the economic life of the family. These white women engaged in spinning, weaving, and cooking but also participated in hunting, planting, tending to crops, and harvesting.

Slavery shaped the lives of southern white women in important ways. As many as 40 percent of slaveholders were women in the decades before the Civil War. Many were married but some were single or widows. The richest were often gifted enslaved help at birth or as young girls. Those who couldn't afford enslaved workers still aspired to own them and rented them when possible.

Southern white women joined southern white men in ordering and supervising the work and life of enslaved people. The slave labor system spared some of them from certain kinds of arduous work, but it also tied them directly to a world of physical and sexual exploitation. Male slaveholders often raped enslaved women. The children of these unions forced white women to confront the extreme violence and suffering that governed the lives of the enslaved as well as themselves.

Beneath the Planter Class

A small middle-class of white Southerners encompassed merchants, store owners, and lawyers who usually lived in towns and cities. These professionals owned modest homes, formed social clubs, attended houses of worship (typically evangelical Christian but sometimes Jewish and, in Baltimore, Mobile, and New Orleans, Catholic), and sent their children to a local school if there was one or perhaps an academy. They often owned some enslaved men and women, but fewer than the planters.

The other category of the southern white middle class consisted of farmers who owned medium-sized or small plots of land as well as some enslaved workers, typically fewer than ten. They grew crops for the market and were just as ruthless and exploitative as planters in trying to amass fortunes in the slave economy. They generally lived outside of urban areas, built small homes, and enjoyed fewer social and educational opportunities than the professionals.

Beneath the white middle class was a broad and diverse stratum of poor white Southerners who typically did not own enslaved workers. Many lived in the midst of the plantation system itself. Small farmers depended on the local plantation aristocracy for access to cotton gins, markets for their modest crops and their livestock, and credit in time of need. In many areas, the poorest resident of a county might easily be a cousin of the richest aristocrat.

Poor whites also found jobs in the South's small industrial economy, particularly in textile mills and iron foundries and as deckhands and stevedores in the shipping sector. Some lived in isolated areas like the Appalachian ranges east of the Mississippi, in the Ozarks to the west of the river, and in other "hill country" or "backcountry" areas. They practiced a simple form of subsistence agriculture and were not directly embedded in the cotton economy. Others occupied the infertile lands of the pine barrens, the red hills, and the swamps and lived in genuine squalor. Many owned no land and supported themselves by foraging, hunting, or working as common laborers for their neighbors. Their degradation resulted partly from dietary deficiencies and disease. Forced to resort at times to eating clay (hence the tendency of more-affluent whites to refer to them disparagingly as "clay eaters"), they chronically suffered from pellagra, hookworm, and malaria. Planters and small farmers alike held them in contempt.

Poor southern whites were uneven in their support of slavery. Some viewed the institution as benefitting only the rich and questioned laws and public monies dedicated to preserving and extending it. What bound most Southerners together was not a universal embrace of slavery or plantation agriculture but a perception of racial differences. However poor and miserable white Southerners might be, they could still look down on the Black population of the region and share a sense of racial supremacy with fellow whites.

SLAVERY: THE "PECULIAR INSTITUTION"

White Southerners often referred to slavery as the "**peculiar institution**," a phrase calling attention to American slavery's distinctiveness. The South in the mid-nineteenth century was one of the few areas in the Western world—along with Brazil, Cuba, and Puerto Rico—where slavery still existed. Slavery, more than any other single factor, isolated the South from the rest of American society and much of the world.

Within the South itself, slavery was the institution that most shaped everyday life. It chained whites and Blacks together in a brutal, inhumane relationship that bred hatred, suspicion, and resentment between them. Every aspect of southern culture—family, religion, education, politics, economics—bore the stamp of slavery. The right to protect, extend, and continue slavery without interference from abolitionists or northern free states or the federal government ultimately became the main reason that the South would fight the Civil War. (See "Debating the Past: Analyzing Slavery.")

SLAVERY AND PUNISHMENT

Southern **slave codes** aimed to closely control enslaved people. They forbade the enslaved to hold property, leave a slaveholder's premises without permission, be out after dark, congregate with other enslaved people anywhere other than at church, carry firearms, testify in court against white people, or strike a white person even in self-defense. They prohibited whites from teaching the enslaved to read or write and contained no provisions to legalize marriages or divorces of enslaved people. If a slaveholder killed an enslaved person while punishing them, the act was generally not considered a crime. Enslaved men and women, however, faced the death penalty for killing or even resisting a white person and for inciting revolt.

Enforcement of the codes was often swift and severe. White slaveholders handled most transgressions by those they held in slavery themselves and rarely chose the least injurious punishment. Whippings, beatings, and brandings were not uncommon.

LIFE UNDER SLAVERY

Although the majority of slaveholders were small farmers, the majority of enslaved people lived on plantations of medium or large size, with substantial enslaved workforces. Regardless

DEBATING THE PAST

ANALYZING SLAVERY

No issue in American history has produced a more spirited debate than the nature of plantation slavery. The debate began well before the Civil War, when **abolitionists** strove to expose slavery to the world as a dehumanizing institution while southern defenders tried to depict it as a benevolent and paternalistic system. But by the late nineteenth century, with white Americans eager for sectional conciliation, most white northern and southern chroniclers of slavery began to accept a romanticized and unthreatening picture of the Old South and its peculiar institution.

The first major scholarly examination of slavery was Ulrich B. Phillips's *American Negro Slavery* (1918), which portrayed slavery as an essentially benign institution in which kindly slaveholders looked after submissive and generally contented African Americans. Phillips's apologia for slavery remained the authoritative work on the subject for nearly thirty years.

In the 1940s, challenges to Phillips began to emerge. Melville J. Herskovits disputed Phillips's contention that Black Americans retained little of their African cultural inheritance. Herbert Aptheker published a chronicle of revolts by enslaved people as a way of refuting Phillips's claim that Blacks were submissive and content.

A somewhat different challenge to Phillips emerged in the 1950s from historians who emphasized the brutality of the institution. Kenneth Stampp's *The Peculiar Institution* (1956) and Stanley Elkins's *Slavery* (1959) described a labor system that did serious physical and psychological damage to its victims. They portrayed slavery as something like a prison, in which men and women had virtually no space to develop their own social and cultural lives. Elkins compared the system to Nazi concentration camps and likened the childlike "Sambo" personality of slavery to tragic distortions of character produced by the Holocaust.

In the early 1970s, an explosion of new scholarship on slavery shifted the emphasis away from the damage the system inflicted on Black Americans and toward the striking success of the enslaved themselves in building a culture of their own. John Blassingame in 1973 argued that "the most remarkable aspect of the whole process of enslavement is the extent to which the American-born slaves were able to retain their ancestors' culture." Herbert Gutman, in *The Black Family in Slavery and Freedom* (1976), challenged the prevailing belief that slavery had weakened or even destroyed the Black family. On the contrary, Gutman argued, the Black family survived slavery with impressive strength, although with some significant differences from the prevailing form of the white family. Eugene Genovese's *Roll, Jordan, Roll* (1974) revealed how Black Americans manipulated white paternalist assumptions to build a large cultural space of their own where they could develop their own family life, social traditions, and religious patterns.

Significantly, the late twentieth century also saw a focus on enslaved women. Deborah Gray White's *Ar'n't I a Woman?: Female Slaves in the Plantation South* (1999) and Stephanie Camp's *Closer to Freedom: Enslaved Women and Everyday Resistance in the Plantation South* (2004) offer valuable portraits of Black women struggling to protect themselves and their families from slavery, challenging the limits by which slavery attempted to define them, and undercutting the violence and degradation of bondage.

At the turn of the twenty-first century, studies by Walter Johnson and Ira Berlin marked an at least partial return to the

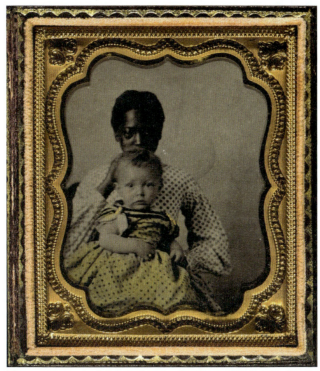

Library of Congress, Prints & Photographs Division [LC-USZC4-5251]

NURSING THE MASTER'S CHILD This 1855 photograph of an enslaved woman and a slaveholder's child is documentary evidence of the complex relationships that historians of slavery have studied.

"damage" approach to slavery of the 1970s. Walter Johnson's *Soul by Soul* (2000) examines the South's largest slave market, New Orleans. For whites, he argues, purchasing enslaved men and women was a way of fulfilling the middle-class male fantasy of mastery. For the enslaved themselves, the trade was dehumanizing and destructive to Black families and communities. Ira Berlin's *Many Thousands Gone* (2000) and *Generation of Captivity* (2003)—among the most important studies of slavery in a generation—similarly emphasize the dehumanizing character of the slave market and show that, whatever white slaveowners might say, slavery was less a social system than a commodification of human beings.

More recently scholars have embedded slavery at the center of the growth of American capitalism. Sven Beckert, *Empires of Cotton* (2014), Ed Baptist, *The Half Has Never Been Told* (2016), Diane Berry, *The Price for Their Pound of Flesh* (2017), and Gerald Horne, *The Dawning of the Apocalypse* (2020) all ask readers to see the different ways that slavery lay at the heart of the nation's economic growth. •

UNDERSTAND, ANALYZE, & EVALUATE

1. Why would historians seek not to emphasize the overwhelming harm and damage done to Blacks living under slavery?
2. What might be some reasons for the return to the focus on the "damage" thesis of slavery in the works by Walter Johnson and Ira Berlin?

Library of Congress Prints and Photographs Division Washington, D.C. 20540 USA
[LC-USZC4-2524] [LC-USZ62-41839] [LC-USZCN4-253]

THE LASH This illustration, by H. L. Stephens, is a harsh reminder of the ever present risk of physical violence and suffering that permeated southern slavery. Published in 1863 in Philadelphia, it was part of a set of cards distributed by abolitionists to dramatize and protest the inhumanity of slavery.

of where they lived, enslaved people subsisted on a diet calculated to keep them alive at minimum cost to the slaveholder. It consisted mainly of cornmeal, salt pork, molasses, and, on rare occasions, fresh meat or poultry. To add to their plate enslaved people cultivated small gardens when possible. They wore flimsy clothes and shoes and lived in small rough cabins provided by the slaveholder. A doctor retained by the owner might provide some basic medical care, but enslaved women—as "healers," midwives, or simply as mothers—were the more important providers of medical care.

Enslaved people faced lives of barbaric and unceasing labor. No job was seen as too arduous or demeaning for them. They toiled from sun-up to sun-down and even longer during harvest time. Enslaved women worked particularly hard. They generally labored in the fields with the men and also handled cooking, cleaning, and child rearing for their own families.

On a small plantation, the same enslaved people might do both field work and housework. But on the largest estates, there would generally be a separate domestic staff: Black nursemaids, housemaids, cooks, butlers, and coachmen. These household servants had a marginally easier life—physically at least—than did field hands. But they often resented their isolation from the other enslaved people on the plantation and the lack of privacy and increased discipline that came with working in such close proximity to the slaveholder's family. Black women especially dreaded domestic work because they were more vulnerable to being raped by slaveholders and white overseers, which happened regularly.

Life and Adventures of Jonathan Jefferson Whitlaw, by Frances Trollope, 1836

CLEAR STARCHING IN LOUISIANA This 1837 etching by French artist Auguste Hervieu depicts a plantation mistress verbally abusing an enslaved woman and child. Hervieu traveled to the United States with British writer and abolitionist Frances Trollope. This illustration is from Trollope's 1836 novel *The Life and Adventures of Jonathan Jefferson Whitlaw: Or, Scenes on the Mississippi*, a work that exposed the degrading effects of slavery on both enslaved people and slaveholders.

SLAVERY IN THE CITIES

The conditions of urban slavery differed significantly from those in the countryside. On the relatively isolated plantations, enslaved people had little contact with free Blacks and lower-class whites. Those who held them in slavery or employed them could maintain a fairly direct watch over them. In the city, however, the line between slavery and freedom was less distinct. An urban slaveholder often could not supervise his enslaved people closely and at the same time use them profitably. Even if they slept at night in carefully watched backyard barracks, they moved about independently during the day, often alone, performing errands or jobs of various kinds and coming into contact with a range of Southerners.

There was a considerable market in the South for common laborers, particularly since, unlike in the North, there were few European immigrants to perform menial chores. As a result, slaveholders often hired out enslaved people for such tasks. Particularly skilled workers such as blacksmiths or carpenters were in demand. Some enslaved people on contract worked in mining and lumbering (often far from cities), but others worked on the docks and on construction sites, drove wagons, and performed numerous unskilled jobs in cities and towns. Enslaved women and children worked in the region's few textile mills.

FREE BLACKS

Over 250,000 free Blacks lived in the slaveholding states by the start of the Civil War, more than half of them in Virginia and Maryland. In some cases, they had earned money to buy their own and their families' freedom. It was most often urban Blacks, with their greater freedom of movement and activity, who could take that route. One example was **Elizabeth Keckley**, an enslaved woman who bought freedom for herself and her son with proceeds from sewing. She later became a seamstress, personal servant, and companion to Mary Todd Lincoln in the White House. But few slaveholders had any incentive or inclination to give up their enslaved workers, so this route was open to relatively few people.

Some enslaved people were set free by slaveholders who had moral qualms about slavery, or by a slaveholder's will after his death—for example, the more than 400 enslaved people held by John Randolph of Roanoke and freed following his death in 1833. From the 1830s on, however, state laws governing slavery became stricter, in part in response to the fears Nat Turner's revolt created among white Southerners. The new laws made **manumission**, or the ability of a slaveholder to free those he enslaved, much more difficult and in some cases practically impossible.

A few free Blacks attained wealth and prominence. Some even held people in slavery themselves, usually relatives whom they had bought to ensure their ultimate emancipation. In a few cities—such as New Orleans, Natchez, and Charleston—free Black communities managed to grow with relatively little interference from whites and with some economic stability. Most southern free Blacks, however, lived in abject poverty. Yet, great as the hardships of freedom were, Blacks preferred it to slavery.

THE SLAVE TRADE

The domestic slave trade—the selling and buying of enslaved people by white Americans—was one of the most terrible dimensions to the peculiar institution. It operated mainly through the efforts of professional slave traders, who sold Black Americans in central markets found in Natchez, New Orleans, Mobile, and Charleston. A sound young field hand could fetch a price that might vary in the 1840s and 1850s from $500 to $1,700, depending on fluctuations in the market and the health and age of the individual.

The domestic slave trade utterly dehumanized enslaved people. Those for sale were treated like livestock: potential buyers poked and prodded their bodies in search of signs of infirmity

Library of Congress Prints & Photographs Division [LC-DIG-ppmsca-11746]

THE BUSINESS OF SLAVERY The offices of slave dealers were familiar sights on the streets of pre–Civil War southern cities and towns. They provide testimony to the way in which slavery was not just a social system but a business, deeply woven into the fabric of southern economic life.

or infection and made them bare their teeth and stick out their tongues to determine how well they were fed. Slave traders and buyers cared nothing about who went up for sale: they regularly separated children from parents, parents from each other, and husband from wife.

While the domestic slave trade operated legally within the South, the importation of enslaved people from Africa and other foreign countries had been banned since 1808. However, some enslaved people continued to be smuggled into the United States as late as 1860. Reopening the international slave trade was a steady goal of southern planters. "If it is right to buy slaves in Virginia and carry them to New Orleans," William L. Yancey asked his fellow delegates to the annual commercial convention of slaveholders held in Montgomery, Alabama, in 1858, "why is it not right to buy them in Cuba, Brazil, or Africa." The convention that year voted to repeal all the laws against slave imports, but the federal government never acted on their proposal.

The continued smuggling of the enslaved was not without resistance. In 1839, a group of 53 Africans who had been abducted by Portuguese slave hunters and shipped to Cuba took charge of a ship, the *Amistad*, that was transporting them to a Caribbean plantation. Their goal was to return to their homelands in Africa. The Africans had no experience with sailing and they tried to compel the crew to steer them across the Atlantic. Instead, *Amistad* sailed up the Atlantic Coast until it was captured by a ship of the U.S. Revenue Service. Many Americans, including President Van Buren, thought the enslaved Africans should be returned to Cuba. But at the request of a group of abolitionists, former president John Quincy Adams went before the Supreme Court to argue that they should be freed. Adams claimed that the foreign slave trade was illegal and thus the *Amistad* rebels could not be returned to slavery. The Court accepted his argument in 1841, and those who survived were returned to Africa, with funding from American abolitionists.

Two years later, another group of enslaved people revolted onboard a ship and took control of it—this time an American vessel bound from Norfolk, Virginia, to New Orleans, Louisiana—and steered it (and its 135 enslaved passengers) to the British Bahamas, where slavery was illegal and the mutineers were given sanctuary. Such shipboard revolts were rare, but they were symbols of continued efforts to overcome slavery.

BLACK CULTURE UNDER SLAVERY

Fleeing the United States by commandeering a ship was a very rare if very dramatic example of how enslaved people responded to their condition. Much more common was an elaborate process of cultural adaptation and resistance. Blacks developed their own culture within slavery that helped them to sustain a sense of shared pride and unity.

RELIGION

By the early nineteenth century, many enslaved Black Americans were Christians. Some had converted voluntarily, others were coerced by their owners and Protestant and Catholic missionaries. Slaveholders often expected those they enslaved to join their denominations and worship under the supervision of white ministers. A distinct religion among enslaved people was not supposed to exist.

But it did. Enslaved Blacks throughout the South developed their own versions of Christianity, at times incorporating into it such practices as voodoo or other polytheistic religious traditions of Africa. Black American religion reflected the influence of African customs and practices. Enslaved people's prayer meetings routinely involved fervent chanting, spontaneous exclamations, and ecstatic conversion experiences. Black religion was generally more joyful and affirming than that of many white denominations. And above all, it emphasized dreams of freedom and deliverance. In their prayers and songs and sermons, Black Christians imagined the day when the Lord would "call us home," "deliver us to freedom," or "take us to the Promised Land." While whites generally chose to interpret such language merely as the expression of hopes for life after death, Blacks used the images of Christian salvation to express their own hopes for liberty in the present world.

Importantly, free Blacks usually worshipped in congregations of their own making. Working with missionaries from the Black denominations founded in the early nineteenth century in the North, especially the African Methodist Episcopal Church and the African Methodist Episcopal Zion Church, they built offshoots in their towns and cities. Some also constructed independent Black Baptist churches. Regardless of their affiliation, these houses of worship became critical sources of aid and support for their members.

Language and Song

Having arrived in the United States speaking many different African languages, the first generations of enslaved people sometimes struggled to communicate with one another. To overcome these barriers, they created sophisticated shared vocabularies and utilized complex sentences that mixed African words from different native languages with those drawn from English. Features of this early adaption survived in Black speech for many years.

Black field workers often used songs to pass the time; since they sang them in the presence of the whites, the songs usually contained relatively innocuous words or used coded phrases to communicate anger or sadness. But Black Americans also created more politically challenging music in the relative privacy of their own religious services. It was there that the tradition of the "spiritual" emerged, which expressed faith in a loving and forgiving God, sorrow over the plight of Black Americans, and hope for freedom.

Family

Like religion, language and song, the enslaved family was a crucial institution of Black culture in the South. What we now call the "nuclear family" consistently emerged as a kinship model among African Americans, but not the only one.

Black women generally began bearing children at younger ages than whites, often as early as fourteen or fifteen (sometimes as a result of rape by slaveholders). Their communities did not condemn premarital pregnancy in the way white society did, and Black couples would sometimes begin living together before marrying. It was customary, however, for couples to eventually marry in a ceremony involving formal vows. Husbands and wives on neighboring plantations sometimes visited each other with the permission of those that held them in slavery, but often such visits had to be in secret, at night. Family ties among the enslaved were generally no less strong than those of whites.

When marriages did not survive, it was often because of circumstances over which Blacks had no control. Up to a third of all Black families were split apart by the slave trade. Extended kinship networks were strong and important and helped compensate for the breakup of nuclear families. An enslaved person forced suddenly to move to a new area, far from his or her family, might create fictive kinship ties and become "adopted" by a family in the new community. Even so, the impulse to maintain contact with a spouse or children remained strong long after the breakup of a family. Indeed, one of the most frequent causes of flight from the plantation was an enslaved person's attempt to reunite with a husband, wife, or child. After the Civil War, white and Black newspapers were filled with notices from those formerly enslaved seeking to reconnect with family members separated during bondage.

Resistance

The dominant response of Blacks to slavery was a combination of adaptation and resistance. At the extremes, slavery could produce two very different reactions, each of which served as the basis for powerfully debasing stereotype in white society. One extreme was what became known as the "Sambo"—the shuffling, grinning, head-scratching, deferential enslaved person who acted out what was recognized as the role the white world expected of them. But any Sambo pattern of behavior was a charade put on by Blacks, a façade assumed in the presence of whites to manipulate them. The other extreme was the rebel—the Black American who completely shunned both acceptance or accommodation of slavery and instead forever resisted white authority.

Actual slave revolts were rare, but the knowledge that they were possible struck terror into the hearts of white Southerners. In 1800, **Gabriel Prosser** gathered 1,000 rebellious enslaved people outside Richmond; but two enslaved men gave away the plot and the Virginia militia stymied the uprising before it could begin. Prosser and thirty-five others were executed. In 1822, the Charleston free Black **Denmark Vesey** and his followers—rumored to total 9,000—made preparations for revolt; but again word leaked out and suppression and retribution followed. On a summer night in 1831, **Nat Turner**, an enslaved preacher, led a band of enslaved and free Black people armed with guns and axes from house to house in Southampton County, Virginia. They killed sixty white men, women, and children before being overpowered by state and federal troops. More than a hundred Blacks were executed in the aftermath.

For the most part, resistance by enslaved people to slavery took less violent forms. Many tried to run away. A sizable number—estimates run as high as 100,000—actually managed to escape to the North or to Canada, especially after sympathetic whites and free Blacks began organizing secret escape routes, known as the "underground railroad," to assist them

MPI/Archive Photos/Getty Images

HARRIET TUBMAN WITH SOME WHO ESCAPED SLAVERY Harriet Tubman (ca. 1820–1913) was born into slavery in Maryland. In 1849, when the man who held her in slavery died, she escaped to Philadelphia to avoid being sold out of state. Over the next ten years, she assisted first members of her own family and then up to 300 other enslaved men and women escape from Maryland to freedom. She is shown here, on the far left, with some of those she had helped to freedom.

in flight. But the odds were strongly against a successful escape. The hazards of distance and unfamiliar terrain were serious obstacles, as were the white "slave patrols," which stopped wandering Blacks on sight and demanded to see travel permits. Despite all the obstacles to success, however, Blacks continued to flee slaveholders in large numbers.

CONCLUSION

"While the North was creating a complex and rapidly developing commercial-industrial economy, the South was expanding and investing heavily in its agrarian economy dependent on the labor of enslaved people." Southern industry was small, concentrated in textiles, and centered in urban areas. Southern whites flooded new agricultural areas in the Deep South in search of land and opportunity, typically bringing enslaved laborers with them or renting and buying them once settled. They created a booming cotton economy that greatly enriched the planter class and made it the dominant force within southern society—both as holders of vast numbers of enslaved people and as patrons, creditors, politicians, landlords, and marketers for the large number of poor whites who lived on the edge of the planter world.

The differences between the North and the South were a result of differences in natural resources, social structure, climate, and culture. Above all, they reflected the South's deep investment in the brutal and inhumane system of slavery, which blocked the kind of social fluidity that an industrializing society usually requires. Notwithstanding the dehumanizing features of the peculiar institution, enslaved men and women managed to create a culture of their own that testified to their ability to survive and blunt the cruel edges of white subjugation.

KEY TERMS/PEOPLE/PLACES/EVENTS

abolitionist 262
Amistad 267
cotton kingdom 256
Denmark Vesey 269
Elizabeth Keckley 265

Gabriel Prosser 269
James Henry Hammond 259
manumission 266
Nat Turner 269
peculiar institution 261

planter class 259
second middle passage 256
slave codes 261

RECALL AND REFLECT

1. Why did cotton become the leading crop of the South?
2. Why did industry fail to develop in the South to the extent that it did in the North?
3. How did slavery function economically and socially?
4. What was the effect of slavery on white slaveholders? On those enslaved? On whites who held no enslaved people? On free Blacks?
5. Through what means did the enslaved create and maintain a distinct African American culture?

12 | ANTEBELLUM CULTURE AND REFORM

THE ROMANTIC IMPULSE
REMAKING SOCIETY
THE CRUSADE AGAINST SLAVERY

LOOKING AHEAD

1. How did an American national culture of art, literature, philosophy, and communal living develop in the nineteenth century?
2. What did social and moral reformers want to change about the United States? What was motivating them?
3. Why did the crusade against slavery become the most important issue of the reform movement?

THE UNITED STATES IN THE EARLY AND MID-NINETEENTH CENTURY was growing rapidly in size, population, and economic complexity. Most Americans were excited by the new possibilities these changes produced. But many people were also painfully aware of the problems that accompanied them.

A key result of these conflicting attitudes was the emergence of movements to "reform" the nation—to refine and improve it. Some reforms rested on an optimistic faith in human nature, a belief that within every individual resided a spirit that was basically good, which society should attempt to unleash. A second impulse was deeply religious. The Second Great Awakening in particular and evangelical Protestantism more generally fired an intense spirit of perfectionism, which preached that believers could, by the grace of God and by self-discipline, purge themselves of sin. One of the largest reform movements was temperance. Tens of thousands of men and women swore off alcohol and campaigned to outlaw liquor as part of an effort to establish greater levels of social order and control over a world they feared was spinning out of control.

One issue would eventually eclipse all others. Public efforts to end slavery began as a small movement, easily overshadowed by temperance. Opposition to slavery actually worked against the popular impulse to stabilize society, as it called for ripping out one of the most powerful engines of the economy and the centerpiece of southern society. By the end of the 1840s, the fight to uproot slavery had become the most influential reform movement of all.

TIME LINE

1821 New York constructs first penitentiary

1826 Cooper's *The Last of the Mohicans*

1830 Joseph Smith publishes the Book of Mormon

1831 The *Liberator* begins publication

1833 American Antislavery Society founded

1837 Horace Mann appointed secretary of Massachusetts Board of Education

1840 Liberty Party formed

1841 Brook Farm founded

1845 Frederick Douglass's autobiography

1848 Women's rights convention at Seneca Falls, N.Y.
Oneida Community founded

1850 Hawthorne's *The Scarlet Letter*

1851 Melville's *Moby Dick*

1852 Stowe's *Uncle Tom's Cabin*

1854 Thoreau's *Walden*

1855 Whitman's *Leaves of Grass*

THE ROMANTIC IMPULSE

"In the four quarters of the globe," wrote the English wit Sydney Smith in 1820, "who reads an American book? or goes to an American play? or looks at an American picture or statue?" The answer, he laughed, was no one.

American intellectuals were painfully aware of the low regard in which Europeans held their culture and they tried to create an artistic life that would express their own nation's special virtues. At the same time, many of the United States' cultural leaders were striving for another kind of liberation, which was—ironically—largely an import from Europe: the spirit of romanticism. In literature, philosophy, art, even politics and economics, American intellectuals were committing themselves to the liberation of the human spirit.

NATIONALISM AND ROMANTICISM IN AMERICAN PAINTING

Despite Sydney Smith's off-putting question, a great many people in the United States were, in fact, looking at American paintings—and they were doing so because they believed Americans were creating important new artistic traditions of their own.

American artists sought to capture the power of nature by portraying some of the nation's grandest landscapes. The first great school of American painters—known as the **Hudson River school**—first emerged in New York in the early 1820s. Frederic Church, Thomas Cole, Thomas Doughty, Asher Durand, and others painted spectacular vistas of the Hudson Valley. They focused on capturing the "unspoiled" state of the natural world, which they believed was an untapped source of wisdom and personal fulfillment. Perhaps in response to harsh descriptions of the lack of American art from critics across the Atlantic, they suggested that the United States, unlike Europe, still possessed "wild nature" and was a nation of greater promise than the overdeveloped lands of the Old World. To be sure, the Hudson River painters spoke mainly for themselves—white men of influence—and less so for the poor, women, and Native and Black Americans. Theirs was a particular self-serving idea of nature, which was not yet settled by civilized men and existed for the taking.

In later years, some of the Hudson River painters traveled farther west. Their enormous canvases of great natural wonders—the Yosemite Valley, Yellowstone, the Rocky Mountains—touched a passionate chord among the public. Some of the most famous of their paintings—particularly the works of Albert Bierstadt and Thomas Moran—traveled around the country attracting enormous crowds.

An American Literature

The effort to create a distinctively American literature made considerable progress in the 1820s through the work of the first great American novelist: James Fenimore Cooper. What most distinguished his work was its evocation of the American West. Cooper had a lifelong fascination with the human relationship to nature and the challenges (and dangers) of the United States' expansion westward. His most important novels—among them *The Last of the Mohicans* (1826) and *The Deerslayer* (1841)—examined the experiences of rugged white frontiersmen in relation to Native Americans, pioneers, violence, and the law. Cooper's heroes evoked the ideal of the independent individual characterized by a natural inner goodness—an ideal that many Americans feared was in jeopardy.

Another, later group of American writers displayed more clearly the influence of romanticism. **Walt Whitman's** book of poems *Leaves of Grass* (1855) celebrated democracy, the liberation of the individual spirit, and the pleasures of the flesh. In crafting verse free from traditional, restrictive conventions, he also expressed a yearning for emotional and physical release and personal fulfillment—a yearning perhaps rooted in part in his own experience as (what we call today) a gay man living in a society profoundly intolerant of unconventional sexuality.

Less exuberant was **Herman Melville.** He would come to be recognized as the greatest American writer of his era but, sadly, never received much critical acclaim or made much money from his writings during his lifetime. *Moby Dick,* published in 1851, is Melville's most important novel. It tells the story of Ahab, the powerful, driven captain of a whaling vessel, and his obsessive search for Moby Dick, the great white whale that had once maimed him. It is a story of courage and the strength of human will. But it is also a tragedy of pride and revenge. In some ways it is an uncomfortable metaphor for the harsh, individualistic, achievement-driven culture of the nineteenth-century United States.

Literature in the Antebellum South

The South experienced a literary flowering of its own in the mid-nineteenth century and it produced writers and artists who were, like their northern counterparts, concerned with defining the nature of the United States. But white Southerners tended to produce very different images of what society was and should be.

The southern writer Edgar Allan Poe penned stories and poems that were primarily sad and macabre. His first book, *Tamerlane and Other Poems* (1827), received little recognition. But later works, including his most famous poem, "The Raven" (1845), established him as a major, if controversial, literary figure. Poe evoked images of individuals rising above the narrow confines of intellect and exploring the deeper—and often painful and horrifying—world of the spirit and emotions.

Other southern novelists of the 1830s (among them Nathaniel Beverley Tucker, William Alexander Caruthers, and John Pendleton Kennedy) produced historical romances and eulogies for the plantation system of the upper South. The most distinguished of the region's authors was William Gilmore Simms. For a time, his work expressed a broad nationalism that transcended his regional background; but by the 1840s, he too became a strong defender of southern institutions—especially slavery—against the encroachments of the North. There was, he believed, a unique and honorable quality to southern life, including slavery, that fell to intellectuals like himself to defend.

One group of southern writers, however, produced works that were more broadly American. These writers often lived in the cotton frontier in the southwest. Instead of romanticizing their subjects, they were deliberately and sometimes painfully realistic, seasoning their sketches with a robust, vulgar humor that was new to American literature. These southern realists established a tradition of American regional humor that was ultimately to find its most powerful voice in Mark Twain.

Born in Virginia in 1815 but moving to Alabama as a young man in search of a place to practice law, Joseph G. Baldwin hung his shingle in Livingston, Alabama. He served a brief term in the state legislature before turning his skills to writing. In *The Flush Times of Alabama and Mississippi: A Series of Sketches*, published in 1853, he penned a series of sketches poking fun at legal politics and ordinary people trying to make it in the booming cotton economy. Another lawyer, Augustus B. Longstreet, satirized everyday life on the frontier in his home state of Georgia. Longstreet, who was also a minister, uncle to Confederate General James Longstreet, and president of four different southern educational institutions (Emory College, Centenary College, the University of Mississippi, and the University of South Carolina), published *George Sketches* in 1835. It offered humorous sketches of poor whites ("crackers" and "dirt eaters") and unscrupulous horse traders as well as fist fights, gambling, and drunken shooting matches.

THE TRANSCENDENTALISTS

One of the outstanding expressions of the romantic impulse in the United States came from a group of New England writers and philosophers known as the **transcendentalists**. Borrowing heavily from German and English writers and philosophers, the transcendentalists promoted a theory of the individual that rested on a distinction between what they called "reason" and "understanding." Reason, as they defined it, had little to do with rationality. It was, rather, the individual's innate capacity to grasp beauty and truth by giving full expression to the instincts and emotions. Understanding, by contrast, was the use of intellect in the narrow, artificial ways imposed by society; it involved the repression of instinct and the victory of externally imposed learning. Every person's goal, therefore, should be the cultivation of "reason"—and, thus, liberation from "understanding." Each individual should strive to "transcend" the limits of the intellect and allow the emotions, the "soul," to create an "original relation to the Universe."

Transcendentalist philosophy emerged first among a small group of intellectuals centered in Concord, Massachusetts, and led by **Ralph Waldo Emerson**. A Unitarian minister in his youth, Emerson left the clergy in 1832 to devote himself to writing, teaching, and lecturing. In "Nature" (1836), Emerson wrote that in the quest for self-fulfillment, individuals should work for a communion with the natural world: "in the woods, we return to reason and

Library of Congress Prints and Photographs Division Washington, D.C. 20540 USA [LC-USZ62-47039]

MARGARET FULLER Margaret Fuller was in the forefront of efforts to probe how contemporary gender roles limited the free expression of women's souls and abilities.

faith.... Standing on the bare ground,—my head bathed by the blithe air, and uplifted into infinite space,—all mean egotism vanishes.... I am part and particle of God." In other essays, he was even more explicit in advocating a commitment to individuality and the full exploration of inner capacities.

Emerson's stress on individuality and the search for inner meaning apart from society inspired Margaret Fuller. In 1840 she became the first editor of the transcendentalist journal *The Dial,* where she argued for the important relationship between the discovery of the "self" and the questioning of the prevailing gender roles of her era. In *Women in the Nineteenth Century*, published in 1845 and considered an important work in the early history of feminism, Fuller wrote, "Many women are considering within themselves what they need and what they have not." She urged readers, especially women, to set aside conventional thinking about the place of women in society.

Henry David Thoreau went even further than Emerson and Fuller in repudiating the repressive forces of society, which produced, he said, "lives of quiet desperation." Each individual should work for self-realization by resisting pressures to conform to society's expectations and responding instead to his or her own instincts. Thoreau's own effort to free himself—immortalized in *Walden* (1854)—led him to build a small cabin in the Concord woods on the edge of Walden Pond. Though never fully removed from modern society, he lived in partial seclusion for two years as part of an effort to question what he considered society's excessive interest in material comforts. In his 1849 essay "Resistance to Civil Government," he extended his critique of artificial constraints in society to government, arguing that when government required an individual to violate his or her own morality, it had no legitimate authority. The proper response was "civil disobedience" or "passive resistance"—a public refusal to obey unjust laws. It was a belief that would undergird the thinking of leaders of the modern civil rights movement in the 1950s and 1960s, especially Dr. Martin Luther King, Jr.

The Defense of Nature

As Emerson's and Thoreau's tributes to nature suggest, a small but influential group of Americans in the nineteenth century feared the impact of capitalism on the integrity of the natural world. "The mountains and cataracts, which were to have made poets and painters," complained the essayist and future Supreme Court Justice Oliver Wendell Holmes in the early 1800s, "have been mined for anthracite and dammed for water power."

To the transcendentalists and others, nature was not just a setting for economic activity, as many farmers, miners, and others believed. It was the source of deep, personal human inspiration—the vehicle through which individuals could best realize the truth within their own souls. Genuine spirituality, they argued, did not come from formal religion but through communion with the natural world.

In making such claims, the transcendentalists were among the first Americans to anticipate the conservation movement of the late nineteenth century and the environmental movement of the twentieth century. They had no scientific basis for their defense of the wilderness and little sense of the twentieth-century notion of the interconnectedness of species. But they did believe in, and articulate, an essential unity between humanity and nature—a spiritual unity, they believed, without which civilization would be impoverished. They looked at nature, they said, "with new eyes," and with those eyes they saw that "behind nature, throughout nature, spirit is present."

Visions of Utopia

Although transcendentalism was at its heart an individualistic philosophy, it helped spawn one of the most famous nineteenth-century experiments in communal living: Brook Farm. The dream of the Boston transcendentalist George Ripley, Brook Farm was established in 1841 as an experimental community in West Roxbury, Massachusetts. There, according to Ripley, individuals would gather to create a new society that would permit every member to have full opportunity for self-realization. All residents would share equally in the labor of the community so that all could share as well in the leisure, which was essential for cultivation of the self. The tension between the ideal of individual freedom and the demands of a communal society, however, eventually took its toll on Brook Farm. Many residents became disenchanted and left. When a fire destroyed the central building of the community in 1847, the experiment dissolved.

Among the original residents of Brook Farm was the writer **Nathaniel Hawthorne**, who expressed his disillusionment with the experiment and, to some extent, with transcendentalism in a series of novels. In *The Blithedale Romance* (1852), he wrote scathingly of Brook Farm itself. In other novels—most notably *The Scarlet Letter* (1850) and *The House of the Seven Gables* (1851)—he wrote equally passionately about the price individuals pay for cutting themselves off from society. Egotism, he claimed (in an indirect challenge to the transcendentalist faith in the self), was the "serpent" that lay at the heart of human misery.

Brook Farm was only one of many experimental communities in the years before the Civil War. The Scottish industrialist and philanthropist Robert Owen founded an experimental community in Indiana in 1825, which he named New Harmony. It was to be a "Village of Cooperation," in which every resident worked and lived in total equality. The community was an economic failure, but the vision that had inspired it continued to enchant some Americans. Dozens of other "Owenite" experiments were established in other locations in the ensuing years.

Redefining Gender Roles

Inspired by the transcendentalist emphasis on liberating the individual from the constraints of social convention, many of the new utopian communities revised the traditional relationship between men and women. Some even experimented with radical redefinitions of gender roles.

One of the more radical of the utopian colonies of the nineteenth century was the Oneida Community, established in 1848 in upstate New York by John Humphrey Noyes. The **Oneida "Perfectionists,"** as residents of the community called themselves, rejected traditional notions of family and marriage. All residents, Noyes declared, were "married" to all other residents; there were to be no permanent conjugal ties. But Oneida was not, as horrified critics often claimed, an experiment in unrestrained "free love." It was a place where the community carefully monitored sexual behavior, women were protected from unwanted childbearing, and children were raised communally, often seeing little of their own parents. The Oneidans took pride in what they considered the liberation of women from the demands of male "lust" and from the traditional restraints of family. Their numbers were never large—only once did they count more than 300 members—but still the community itself lasted 30 years.

The **Shakers,** too, redefined traditional gender roles. Founded by "Mother" Ann Lee in the 1770s, the society of the Shakers attracted a particularly large following in the mid-nineteenth century and established more than twenty communities throughout the Northeast and Northwest in the 1840s. They derived their name from a unique religious ritual—in which members of a congregation would "shake" themselves free of sin while performing a loud chant and an ecstatic dance. The most distinctive feature of Shakerism, however, was its commitment to absolute material simplicity and the conviction that any kind of frills and trappings were distractions from God and each other. Part of this simplicity included a rejection of any kind of carnal relations, meaning that all members had to voluntarily adopt celibacy, which obviously meant that no one could be born into the faith.

Shakerism attracted about 6,000 members in the 1840s, more women than men. They lived in communities where contact between men and women was strictly limited. Members endorsed the idea of sexual equality, although women exercised the greater power. The Shakers were not, however, motivated only by a desire to escape the burdens of traditional gender roles. They were also trying to create a society set apart from the chaos and disorder they believed had come to characterize American life. In that, they were much like other dissenting religious sects and utopian communities of their time.

The Mormons

Among the most important religious efforts to create a new and more ordered society was that of the Church of Jesus Christ of Latter-day Saints—the Mormons. The original Mormons were white men and women, many of whom felt that social changes had left them economically marginalized and in doubt about the fundamental tenets of traditional

Christianity. They discovered in Mormonism a new history of the world, a new establishment of the Christian church, a new book of scripture and set of prophets to join those of the Bible, a new call to prepare for the Second Coming, and a new promise that—just as God had once been a man—all believers could become like God himself.

Mormonism began in upstate New York through the efforts of **Joseph Smith**. In 1830, when he was just twenty-four, he organized the church and published a remarkable document, the Book of Mormon, named for the ancient prophet who he claimed had been its chief editor. It was, he said, a translation of a set of golden tablets he had found in the hills of New York, revealed to him by Moroni, an angel of God and the book's last prophet. The Book of Mormon told the story of two ancient civilizations in America, whose people had anticipated the coming of Christ and were rewarded when Jesus actually came to America after his resurrection. Ultimately, both civilizations collapsed because of their rejection of Christian principles. But Smith believed their history as righteous societies could serve as a model for building a new holy community in the United States.

In 1831, gathering a small group of believers around him, Smith began searching for a sanctuary for his new community of "saints," an effort that would continue unhappily for more than fifteen years. Time and again, the Latter-day Saints, as they called themselves, attempted to establish peaceful communities. Time and again, they met with persecution from their neighbors, who were suspicious of their radical religious doctrines—their claims of new prophets, new scripture, and divine authority. Opponents were also concerned by their rapid growth and their increasing political strength. Near the end of his life, Joseph Smith introduced the practice of polygamy (giving a man the right to take several wives), which became public knowledge after Smith's death. From then on, polygamy became a central target of anti-Mormon opposition.

Everett Historical/Shutterstock

MORMONS HEADING WEST This lithograph by William Henry Jackson imagines the physical challenges that Mormon pioneers faced in their journey to Utah in 1850. Many of the men are shown pulling their families and possessions on handcarts.

Driven from their original settlements in Independence, Missouri, and Kirtland, Ohio, the Mormons founded a new town in Illinois that they named Nauvoo. In the early 1840s, it became an imposing and economically successful community that frightened non-Mormons. Tension emerged from within Mormonism as well, including a dissenter who published a newspaper story that exposed Smith's teaching on polygamy. Smith, after ordering his followers to destroy the offending press, was arrested and imprisoned in nearby Carthage. There, an angry mob attacked the jail and fatally shot him. The Mormons soon abandoned Nauvoo and, under the leadership of Smith's successor, Brigham Young, traveled—12,000 strong, in one of the largest voluntary single-group migrations in American history—across the Great Plains and the Rocky Mountains. They established several communities in Utah, including the present Salt Lake City, where, finally, the Mormons were able to create a lasting settlement.

REMAKING SOCIETY

Central to romanticism and transcendentalism was a reform impulse—the drive to improve people's lives and health. Seeking to uncover the divinity of the individual inspired larger quests to perfect the world itself. These quests became reform movements to remake mainstream society—movements which often were led by women or strongly influenced by them.

Revivalism, Morality, and Order

Along with romanticism and transcendentalism, Protestant revivalism was another powerful source of the popular effort to rewrite society. It was the movement that had begun with the Second Great Awakening early in the century and had, by the 1820s, evolved into a powerful force for social reform.

The new evangelicals embraced the optimistic belief that every individual was capable of salvation through his or her own efforts; and that the saved had an obligation to proselytize and repair not just the lives of others but society as a whole. Put another way, revivalism became not only a means of personal salvation but also an effort to reform culture at large.

Evangelical Protestantism greatly strengthened the crusade against drunkenness. No social vice, **temperance** advocates argued, was more responsible for crime, disorder, and poverty than the excessive use of alcohol. Women complained that men spent money their families needed on alcohol and that drunken husbands often beat and abused their wives. Temperance also appealed to those who were alarmed by immigration; drunkenness, many nativists believed, was responsible for violence and disorder in immigrant communities. By 1840, temperance had become the largest and most powerful social movement with more than a million followers who had signed a formal pledge to forgo hard liquor.

Health, Science, and Phrenology

For some Americans, the search for individual and social perfection led to an interest in new theories of health and knowledge. Urban living and industrializing society produced lives of poverty and filth. Disease broke out regularly, proliferated rapidly, and infected thousands. The ill often never got better, dying painfully. Part of the problem was the state of medicine at the time, which was not yet aware of the nature of bacterial infections.

Not surprisingly, many Americans turned to popular theories for improving health. Affluent men and especially women flocked to health spas for the celebrated "water cure,"

Library of Congress Prints and Photographs Division [LC-DIG-ppmsca-32719]

THE DRUNKARD'S PROGRESS This 1846 lithograph by Nathaniel Currier depicts what temperance advocates argued was the inevitable consequence of alcohol consumption. Beginning with an apparently innocent "glass with a friend," the young man rises step by step to the summit of drunken revelry, then declines to desperation and suicide while his abandoned wife and child grieve.

which purported to improve health through immersing people in hot or cold baths or wrapping them in wet sheets. Other people adopted new dietary theories. Sylvester Graham, a Connecticut-born Presbyterian minister and committed reformer, won many followers with his prescriptions for eating fruits, vegetables, and bread made from coarsely ground flour—a prescription not unlike some dietary theories today—and for avoiding meat. (The graham cracker is made from a kind of flour named for him.)

Perhaps strangest of all to modern sensibilities was the widespread belief in the new "science" of phrenology, which appeared first in Germany and became popular in the United States beginning in the 1830s through the efforts of Orson and Lorenzo Fowler, publishers of the *Phrenology Almanac*. Phrenologists argued that the shape of an individual's skull was an important indicator of character and intelligence. They made elaborate measurements of bumps and indentations to calculate the size (and, they claimed, the strength) of different areas of the brain. Phrenology seemed to provide a way of measuring an individual's fitness for various positions in life and to promise an end to the arbitrary process by which people matched their talents to occupations and responsibilities. In the hands of racists, phrenology tragically became a means of enforcing white supremacy by claiming that Anglo-Saxons had the most perfect attributes and thus deserved to be in positions of power over nonwhites, whose supposedly different skull sizes and shapes revealed their limits and deficits. Phrenology and especially its racist applications are now universally rejected.

MEDICAL SCIENCE

In an age of rapid technological and scientific advances, medicine sometimes seemed to lag behind. In part, that was because of the character of the medical profession, which—in the absence of any significant regulation or prescribed pathway of schooling—attracted many poorly educated people and not a few quacks. Efforts to regulate the profession were beaten back until 1847, when 250 men from medical colleges and societies from across the country founded the American Medical Society "to promote the science and art of medicine and the betterment of public health."

The biggest problem facing American medicine, however, was the absence of basic knowledge about how disease worked. The great medical achievement of the eighteenth century—the development of a vaccination against smallpox—came from no broad theory of infection but instead the brilliant insights of Onesimus, an enslaved person owned by Cotton Mather, who introduced it based on knowledge and techniques that he learned in his native Africa. The development of anesthetics in the nineteenth century similarly came not from medical doctors but from New England dentist, William Morton. Beginning in 1844, Morton began experimenting with sulfuric ether as a way to help patients endure the painful extraction of teeth. Using ether quickly became adopted by surgeons seeking to sedate their patients.

In the absence of any broad acceptance of scientific methods and experimental practice in medicine, it was very difficult for even the most talented doctors to make progress in treating disease. Even so, halting progress toward the discovery of the germ theory did occur in the antebellum United States. In 1843, the Boston essayist, poet, and physician Oliver Wendell Holmes published a study of large numbers of cases of "puerperal fever" (septicemia in children) and concluded that the disease could be transmitted from one person to another. This discovery of contagion met with a storm of criticism but was later vindicated by the clinical success of the Hungarian physician Ignaz Semmelweis, who noticed that infection seemed to be spread by medical students who had been working with diseased corpses. Once he began requiring students to wash their hands and disinfect their instruments, the infections virtually disappeared.

Education

One of the most important reform movements of the mid-nineteenth century was the effort to produce a system of universal public education. As of 1830, no state had such a system. Soon after that, however, interest in public education began growing rapidly.

The greatest of the educational reformers was **Horace Mann**, the first secretary of the Massachusetts Board of Education, which was established in 1837. To Mann and his followers, education was the only way to preserve democracy, for an educated electorate was essential to the workings of a free political system. Mann reorganized the Massachusetts school system, lengthened the academic year (to six months), doubled teachers' salaries, broadened the curriculum, and introduced new methods of professional training for teachers. Other states followed by building new schools, creating teachers' colleges, and offering many children access to education for the first time. By the 1850s, the principle (although not yet the reality) of tax-supported elementary schools was established in every state.

The quality of public education continued to vary widely. In some places—Massachusetts, for example—educators were generally capable men and women, often highly trained. In other areas, however, barely literate teachers and severely limited funding hindered education. Among the highly dispersed population of the West, many children had no access to schools at all. In the South, where there were very few schools, all African Americans were barred from public education, and only about a third of white children of school age were enrolled in 1860. In the North, 72 percent were enrolled, but even there, many students attended classes only briefly and casually.

Among the goals of educational reformers was teaching children the social values of thrift, order, discipline, punctuality, and respect for authority. These values partnered nicely with the widespread hopes that schools could, on the one hand, produce men and women who would become successful capitalists, and also that they could assimilate immigrants, making them into reliably Protestant Americans.

The interest in education contributed to the growing movement to educate Native Americans. Some reformers believed that indigenous peoples could be "civilized" if they could be taught the ways of the white world. Efforts by missionaries and others to educate Native Americans and encourage them to assimilate were particularly prominent in such areas of the Far West as Oregon, where conflicts between white settlers and native peoples had not yet become acute. Nevertheless, the great majority of Native Americans remained outside the reach of white educational reform.

Despite limitations and inequities, the achievements of the school reformers were impressive. By the beginning of the Civil War, the United States had one of the highest literacy rates of any nation in the world: 94 percent of the population of the North, and 83 percent of the white population of the South.

Rehabilitation

The belief in the potential of the individual also sparked the creation of new institutions to help individuals with disabilities. Among them was the Perkins School for the Blind in Boston. Nothing better exemplified the romantic reform spirit of the era than the conviction of those who founded Perkins. They believed that even society's supposedly most disadvantaged members could be helped to discover their own inner strength and wisdom.

Similar ideas produced changes in prisons. Inmates were typically housed in cramped cells, fed only scraps and morsels, denied medical care, and forced to endure beatings and ridicule from guards. They were seen as morally defective and socially deviant. Prison reformers advocated for a new system of criminal justice based on the idea of the penitentiary, or place where people could be "penitent" for their crimes. In a vision that slowly was accepted by legislatures and the public, reformers taught that criminals were not necessarily defective or deviant but instead were capable of improvement. Key to helping criminals to improve was placing them in clean individual cells, offering a rigid regime of daily tasks and duties, and isolating them from society and its temptations until they were capable of living responsibly.

Some of the same impulses that produced changes in prison culture underlay the emergence of a new approach toward Native Americans. For several decades, the dominant thrust of the United States' policy toward the nation's native inhabitants had been relocation—getting indigenous people out of the way of white civilization. But among some whites, there had also been another intent: to move indigenous groups to places where they would be allowed to develop to a point at which assimilation into white society might be possible.

It was a small step from the idea of relocation to the idea of the reservation. Just as prisons, asylums, and orphanages would provide society with an opportunity to train and uplift outcasts and unfortunates within white society, so the reservations might provide a way to undertake what one official called "the great work of regenerating the Indian race." These optimistic goals failed to meet the expectations of the reformers.

The Rise of Feminism

Many women who became involved in reform movements in the 1820s and 1830s came to resent the social and legal restrictions that limited their participation. Out of their concerns emerged the first American feminist movement. Sarah and Angelina Grimké, sisters who became active and outspoken abolitionists, ignored claims by men that their activism was inappropriate to their gender. "Men and women were created equal," they argued. "They are both moral and accountable beings, and whatever is right for man to do, is right for women to do." Other reformers—**Harriet Beecher Stowe**, **Lucretia Mott**, **Elizabeth Cady Stanton**, and **Susan B. Anthony**—similarly pressed the boundaries of "acceptable" female behavior.

In 1840, American female delegates arrived at a world antislavery convention in London, only to be turned away by the men who controlled the proceedings. Angered at the rejection, several of the delegates became convinced that their first duty as reformers should now be to elevate the status of women. Over the next several years, Mott, Stanton, and others began drawing pointed parallels between the plight of women and the plight of the enslaved; and in 1848, in **Seneca Falls**, New York, they organized a convention to discuss the question of women's rights. Out of the meeting came the Declaration of Sentiments and Resolutions, which stated that "all men and women are created equal," and that women no less than men are endowed with certain inalienable rights. (See "Consider the Source: Declaration of Sentiments and Resolutions, Seneca Falls, New York.") In advocating not only for fuller equality but specifically for their right to vote, the authors positioned themselves on the outward edge of feminism in the antebellum era. It was an extreme claim for the time, but one that helped begin a movement for woman suffrage that would survive until the battle was finally won in 1920.

Many of the women involved in these feminist efforts were Quakers. Quakerism had long embraced the ideal of sexual equality and had tolerated, indeed encouraged, the emergence of women as preachers and community leaders. Of the women who drafted the Declaration of Sentiments, all but Elizabeth Cady Stanton were Quakers.

Feminists benefited greatly from their association with other reform movements, most notably abolitionism, but they also suffered as a result. The demands of women were usually assigned a secondary position to what many considered the far greater issue of the rights of the enslaved.

Struggles and Successes of Black Women

Among the leading voices for women's rights was **Sojourner Truth**, a Black woman born into slavery in Ulster County in New York in 1799 who escaped to freedom at age 29. Born Isabella Baumfree, she changed her name after becoming a Methodist in 1843 and believed that God had called her to testify to the truths of freedom from sin and the evils of slavery. A staunch abolitionist, she attended the Ohio Women's Rights Convention in Akron, Ohio, in 1851, and delivered an impassioned and well-publicized call for equal rights for women and Blacks. Afterward Truth became a leading spokesperson for both causes.

While Black women like Sojourner Truth campaigned publicly for women's and Blacks' civil rights, others attempted to reform society from within their religious traditions. Like white clerics, Black preachers in African American churches widely banned female congregants from becoming ordained or obtaining a license to preach and often required them to seek special permission to serve as class and prayer leaders. Indeed, no Black denomination formally recognized a woman as a cleric until the African Methodist Episcopal Church ordained Julia Foote in 1895. Still, Black women sought to preach throughout the colonial and antebellum eras. Among the first was **Jarena Lee**, born free in 1783 in Cape May, New Jersey. At twenty-one years old, then living in Philadelphia, she preached in public with such verve and passion that she earned an invitation from Rev. Richard Allen to speak at his church. Yet few other ministers welcomed her, a slight which Lee struggled to understand

CONSIDER THE SOURCE

DECLARATION OF SENTIMENTS AND RESOLUTIONS, SENECA FALLS, NEW YORK (1848)

On July 19 and 20, 1848, leaders of the women's rights movement gathered in Seneca Falls, New York, to host a national conversation about "the social, civil, and religious conditions and rights of women." They outlined their grievances and goals in the Declaration of Sentiments and Resolutions, which helped shape a national reform movement:

When, in the course of human events, it becomes necessary for one portion of the family of man to assume among the people of the earth a position different from that which they have hitherto occupied, but one to which the laws of nature and of nature's God entitle them, a decent respect to the opinions of mankind requires that they should declare the causes that impel them to such a course.

We hold these truths to be self-evident: that all men and women are created equal; that they are endowed by their Creator with certain inalienable rights; that among these are life, liberty, and the pursuit of happiness; that to secure these rights governments are instituted, deriving their just powers from the consent of the governed. Whenever any form of government becomes destructive of these ends, it is the right of those who suffer from it to refuse allegiance to it, and to insist upon the institution of a new government, laying its foundation on such principles, and organizing its powers in such form, as to them shall seem most likely to effect their safety and happiness. Prudence, indeed, will dictate that governments long established should not be changed for light and transient causes; and accordingly all experience hath shown that mankind are more disposed to suffer, while evils are sufferable, than to right themselves by abolishing the forms to which they are accustomed. But when a long train of abuses and usurpations, pursuing invariably the same object, evinces a design to reduce them under absolute despotism, it is their duty to throw off such government, and to provide new guards for their future security. Such has been the patient sufferance of the women under this government, and such is now the necessity which constrains them to demand the equal station to which they are entitled. The history of mankind is a history of repeated injuries and usurpations on the part of man toward woman, having in direct object the establishment of an absolute tyranny over her. To prove this, let facts be submitted to a candid world.

He has never permitted her to exercise her inalienable right to the elective franchise. He has compelled her to submit to laws, in the formation of which she had no voice. He has withheld from her rights which are given to the most ignorant and degraded men—both natives and foreigners.

Having deprived her of this first right of a citizen, the elective franchise, thereby leaving her without representation in the halls of legislation, he has oppressed her on all sides. He has made her, if married, in the eye of the law, civilly dead.

He has taken from her all right in property, even to the wages she earns.

He has made her, morally, an irresponsible being, as she can commit many crimes with impunity, provided they be done in the presence of her husband. In the covenant of marriage, she is compelled to promise obedience to her husband, he becoming, to all intents and purposes, her master—the law giving him power to deprive her of her liberty, and to administer chastisement.

He has so framed the laws of divorce, as to what shall be the proper causes, and in case of separation, to whom the guardianship of the children shall be given, as to be wholly regardless of the happiness of women—the law, in all cases, going upon a false supposition of the supremacy of man, and giving all power into his hands.

After depriving her of all rights as a married woman, if single, and the owner of property, he has taxed her to support a government which recognizes her only when her property can be made profitable to it.

He has monopolized nearly all the profitable employments, and from those she is permitted to follow, she receives but a scanty remuneration. He closes against her all the avenues to wealth and distinction which he considers most honorable to himself. As a teacher of theology, medicine, or law, she is not known.

He has denied her the facilities for obtaining a thorough education, all colleges being closed against her.

He allows her in church, as well as state, but a subordinate position, claiming apostolic authority for her exclusion from the ministry, and, with some exceptions, from any public participation in the affairs of the church.

He has created a false public sentiment by giving to the world a different code of morals for men and women, by which moral delinquencies which exclude women from society, are not only tolerated, but deemed of little account in man.

He has usurped the prerogative of Jehovah himself, claiming it as his right to assign for her a sphere of action, when that belongs to her conscience and to her God.

He has endeavored, in every way that he could, to destroy her confidence in her own powers, to lessen her self-respect, and to make her willing to lead a dependent and abject life.

Now, in view of this entire disfranchisement of one-half the people of this country, their social and religious degradation—in view of the unjust laws above mentioned, and because women do feel themselves aggrieved, oppressed, and fraudulently deprived of their most sacred rights, we insist that they have immediate admission to all the rights and privileges which belong to them as citizens of the United States.

UNDERSTAND, ANALYZE, & EVALUATE

1. What central claim about the relationship of men and women lies at the heart of this declaration? What evidence did the authors produce to support their claim?
2. With what demand did the authors conclude their resolution? How would you have reacted to this text?

Source: Stanton, Elizabeth Cady, *A History of Woman Suffrage*, vol. 1. Rochester, NY: Fowler and Wells, 1889, 70–71.

---

theologically. As she argued in 1833, "If the man may preach, because the Savior died for him, why not the women, seeing he died for her also? Is he not a whole Savior, instead of a half one, as those who hold it wrong for a woman to preach, would seem to make it appear? Did not Mary *first* preach the risen Savior? Then did not Mary, a woman, preach the gospel?"

A more radical contemporary of Lee's was **Rebecca Cox Jackson**. Growing up a free woman in Philadelphia during the early 1800s, she lived much of her life with her brother, Joseph Cox, an African Methodist Episcopal minister. Following instructions given to her by a heavenly spirit in 1830, Jackson began to host prayer meetings that quickly surged in popularity. She stirred controversy by tossing aside convention and inviting men and women to worship side by side. She earned a temporary reprieve, however, after a visit by Rev. Morris Brown, who succeeded Rev. Richard Allen as bishop of the African Methodist Episcopal Church. Brown came to one of Jackson's meetings with the idea of silencing her, but left thoroughly impressed by her preaching and ordered that she be left alone. In 1833 Jackson embarked on a preaching tour outside Philadelphia but met with new and greater resistance. Her insistence on her right to preach, open refusal to join a church, and radical views on sexuality that included celibacy within marriage angered area clerics and, Jackson claimed, motivated some to assault her. Eventually she broke ranks with the free Black church movement and joined a Shaker group in Watervliet, New York. In 1851 she returned to Philadelphia and founded a Shaker community composed mainly of Black women.

Lee and Jackson rejected the limitations placed on their preaching because of their gender and race. Like other Black women, they found confirmation for their efforts not in any church rule or clerical pronouncement but rather through their personal interpretation of the Bible and, more important, an unflagging conviction of their dignity as human beings and belief that God had called them to preach. Though denied official recognition as preachers, they touched the lives of many and represented a vital dimension in the religious lives of northern Blacks.

THE CRUSADE AGAINST SLAVERY

The antislavery movement was not new to the mid-nineteenth century. Nor was it primarily a domestic crusade. Indeed, the struggle to end slavery took root in countries around the world. (See "America in the World: The Abolition of Slavery.") But in the United States, it was not until the 1830s that the antislavery movement begin to gather the force that would ultimately enable it to overshadow virtually all other efforts at social reform.

EARLY OPPOSITION TO SLAVERY

In the early years of the nineteenth century, white people who opposed slavery were, for the most part, a calm and genteel lot, expressing moral disapproval but doing little else. To the extent that there was an organized antislavery movement, it centered on the resettlement of American Blacks in Africa or the Caribbean. In 1816, a group of prominent white Virginians organized the American Colonization Society (ACS), which proposed a gradual freeing of enslaved workers, with slaveholders receiving compensation. The liberated men and women would be transported out of the country and helped to establish a new society of their own. Predictably, the ACS was no favorite of slaveholders, who mostly opposed any effort to end slavery, through colonization or otherwise. Still, the ACS found small pockets of support in the South and even received money from Congress and the legislatures of Virginia and Maryland. The height of success for the ACS was resettling free Blacks in a colony in West African that in 1847 became the nation of **Liberia**. About 13,000 American Blacks had emigrated by 1867.

But the ACS was in the end a negligible force in the fight to end slavery. There were far too many Blacks in the United States in the nineteenth century to be transported to Africa by any conceivable program. The ACS met resistance, in any case, from Blacks themselves, many of whom were now three or more generations removed from Africa and, despite their loathing of slavery, had no wish to emigrate. They viewed themselves as entitled to fair treatment as Americans.

BLACK ABOLITIONISTS

Abolitionism obviously had a particular appeal to the free Black population of the North. These free Blacks typically lived in conditions of poverty and oppression and faced a barrage of local customs and state laws that frequently reminded them of their lowly social positions. For all their problems, however, northern Blacks were fiercely proud of their freedom and sensitive to the plight of those who remained in bondage. Many in the 1830s came to support white leaders such as William Lloyd Garrison. But they also rallied to leaders of their own.

Among the earliest and fiercest Black abolitionists was **David Walker**, who preceded even Garrison in publicly calling for an uncompromising opposition to slavery on moral grounds. In 1829, Walker, a free Black man who had moved from North Carolina to Boston, published a harsh pamphlet—*An Appeal to the Coloured Citizens of the World*—that described slavery as a sin that would draw divine punishment if not abolished. "America is more our country than it is the whites'—we have enriched it with our blood and tears." He urged the enslaved to "kill [those who held them in slavery] or be killed."

Most Black critics of slavery were somewhat less violent in their rhetoric but equally uncompromising in their commitment to abolition. The greatest Black abolitionist of all—and one of the most electrifying orators of his time, Black or white—was **Frederick Douglass**. Born into slavery in Maryland, Douglass escaped to Massachusetts in 1838 and became an outspoken leader of the antislavery movement. In order to avoid capture following his escape, he fled to England in 1845, where he lectured extensively on the evils of slavery. Returning two years later, Douglass formally purchased his freedom from his Maryland owner and founded an antislavery newspaper, *The North Star*, in Rochester, New York. He achieved wide renown as well for his autobiography, *Narrative of the Life of Frederick Douglass* (1845), in which he presented a damning picture of slavery and demanded full social and economic equality for Blacks.

Douglass delivered one of his most important speeches on July 5, 1852, in Rochester, New York, to the Rochester Ladies Anti-Slavery Society. In what is now titled "What to the Slave is the 4th of July," the address pushed his audience and the nation to see the moral hypocrisy of celebrating a day of freedom in a nation permitting slavery.

Digital image courtesy of the Getty's Open Content Program

FUGITIVE SLAVE LAW CONVENTION Abolitionists gathered in Cazenovia, New York, in August 1850 to consider how to respond to the law recently passed by Congress requiring northern states to return those who had fled slavery to their owners. Frederick Douglass is seated just to the left of the table in this photograph of some of the participants. The gathering was unusual among abolitionist gatherings in including substantial numbers of African Americans.

AMERICA IN THE WORLD

THE ABOLITION OF SLAVERY

The United States formally abolished slavery through the Thirteenth Amendment of the Constitution in 1865, in the aftermath of the Civil War. But the effort to abolish slavery did not begin or end in North America. Emancipation in the United States was part of a worldwide antislavery movement that began in the late eighteenth century and continued through the end of the nineteenth.

The end of slavery, like the end of monarchies and established aristocracies, was one of the ideals of the Enlightenment, which inspired new concepts of individual freedom. As Enlightenment ideas spread throughout the Western world in the seventeenth and eighteenth centuries, people on both sides of the Atlantic began to examine slavery anew. Many Enlightenment thinkers, including most of the founders of the American republic, believed that freedom was appropriate for white people but not for people of color. But others came to believe that all human beings had an equal claim to liberty, and their views became the basis for an escalating series of antislavery movements.

Opponents of slavery first targeted the slave trade—the vast commerce in human beings that had grown up in the seventeenth and eighteenth centuries and had come to involve large parts of Europe, Africa, the Caribbean, and North and South America. In the aftermath of the revolutions in America, France, and Haiti, the attack on the slave trade quickly gained momentum. Its central figure was the English reformer William Wilberforce, who spent years attacking Britain's connection with the slave trade on moral and religious grounds. After the Haitian Revolution, Wilberforce and other antislavery activists denounced slavery on the grounds that its continuation would create more slave revolts. In 1807, he persuaded Parliament to pass a law ending the slave trade within the entire British Empire. The British example foreshadowed many other nations to make the slave trade illegal as well: the United States in 1808, France in 1814, Holland in 1817, Spain in 1845. Trading in enslaved people persisted within countries and colonies where slavery remained legal (including the United States), and some illegal slave trading continued throughout the Atlantic World. But the international sale of the enslaved steadily declined after 1807. The last known shipment of enslaved people across the Atlantic—from Africa to Cuba—occurred in 1867.

Ending the slave trade was a great deal easier than ending slavery itself, in which many people had major investments and on which much agriculture, commerce, and industry depended. But pressure to abolish slavery grew steadily throughout the nineteenth century, with Wilberforce once more helping to lead the international outcry against the institution. In Haiti, the slave revolts that began in 1791 eventually abolished not only slavery but also French rule. In some parts of South America, slavery came to an end with the overthrow of Spanish rule in the 1820s. Simón Bolívar, the great leader of Latin American independence, considered abolishing slavery an important part of his mission, freeing those who joined his armies and insisting on constitutional prohibitions of slavery in several of the constitutions he helped frame. In 1833, the British parliament passed a law abolishing slavery throughout the British Empire and compensated slaveholders for freeing those they held in slavery. France abolished slavery in its empire in 1848, after years of agitation from abolitionists. In the Caribbean, Spain followed

Library of Congress Prints and Photographs Division
[LC-USZC4-5321]

ANTISLAVERY MESSAGE The image of an enslaved man praying to God was popular in both British and American antislavery circles. It first appeared as the seal of the Committee for the Abolition of the Slave Trade, a British abolitionist group formed in 1787, accompanied by the quote, "Am I not a man and a brother?" This example from 1837 was used to illustrate John Greenleaf Whittier's antislavery poem "Our Countrymen in Chains."

Britain in slowly eliminating slavery from its colonies. Puerto Rico abolished slavery in 1873; and in the face of increasing slave resistance and the declining profitability of slave-based plantations, in 1886 Cuba became the last colony in the Caribbean to end slavery. Brazil was the last nation in the Americas to abolish slavery, ending the system in 1888. The Brazilian military began to turn against slavery after the valiant participation of enslaved soldiers in Brazil's war with Paraguay in the late 1860s; eventually, educated Brazilians began to oppose the system too, arguing that it obstructed economic and social progress.

In the United States, the power of world opinion—and the example of Wilberforce's movement in England—became an important influence on the abolitionist movement as it gained strength in the 1830s. American abolitionism, in turn, helped reinforce the movements abroad. Frederick Douglass, the formerly enslaved American turned abolitionist, became a major figure in the international antislavery movement and was a much-admired and much-sought-after speaker in England and Europe in the 1840s and 1850s. The United States was one of the last countries in the Western world to ban slavery and had to fight a four-year war with itself to do so. Its efforts, importantly, were part of a worldwide movement toward emancipation. •

UNDERSTAND, ANALYZE, & EVALUATE

1. Why did opponents of slavery focus first on ending the slave trade, rather than abolishing slavery itself? Why was ending the slave trade easier than ending slavery?
2. How do William Wilberforce's arguments against slavery compare with those of the abolitionists in the United States?

"What have I, or those I represent, to do with your national independence? Are the great principles of political freedom and of natural justice, embodied in that Declaration of Independence, extended to us? . . . What, to the American slave, is your 4th of July? I answer: a day that reveals to him, more than all other days in the year, the gross injustice and cruelty to which he is the constant victim. . . . There is not a nation on earth guilty of practices, more shocking and bloody, than are the people of these United States at this very hour."

GARRISON AND ABOLITIONISM

In 1830, with pro-slavery ideology spreading rapidly in the South and the antislavery movement struggling, a new figure emerged: **William Lloyd Garrison**. Born in Massachusetts in 1805, Garrison was an assistant to the New Jersey Quaker Benjamin Lundy in the 1820s, helping to publish the leading antislavery newspaper of the time. Garrison grew impatient with his employer's moderate tone and in 1831 returned to Boston to found his own newspaper, *The Liberator*.

Like many white abolitionists, Garrison initially focused on persuading white Southerners to embrace his cause by stressing the negative impact slavery had on the nation's soul and economy. But he became more radical, offering a philosophy that was so simple that it was genuinely revolutionary. Opponents of slavery, he said, should not talk about the evil influence of slavery on white society but rather the damage the system did to the enslaved. And they should, therefore, reject "gradualism" and demand the immediate abolition of slavery and the extension of all the rights of American citizenship to both enslaved and free African Americans. Garrison wrote in a relentless, uncompromising tone. "I am aware," he wrote in the very first issue of *The Liberator*, "that many object to the severity of my language; but is there not cause for severity? I will be as harsh as truth, and as uncompromising as justice. . . . I am in earnest—I will not equivocate—I will not excuse—I will not retreat a single inch—and I will be heard."

Garrison soon attracted a large group of followers throughout the North, enough to enable him to found the New England Antislavery Society in 1832 and, a year later, after a convention in Philadelphia, the American Antislavery Society.

ANTI-ABOLITIONISM

The rise of abolitionism laid bare currents of racism long swirling in the North and the South. Whites in both regions widely held Blacks in contempt and viewed them as intellectual and moral inferiors. The rise of a more radical abolitionism in the 1830s sparked fears of hordes of free Blacks competing for jobs and land and claiming full equality with whites. Even while many whites in the free states viewed slavery as morally wrong and harmful to Black Americans and American society generally, very few viewed Blacks as their equals.

Waves of mob violence swept across the North in the 1830s. Protestors targeted abolitionists but also free Black communities, torching homes and businesses and lynching men and women. A mob in Philadelphia attacked the abolitionist headquarters there in 1834, burning it to the ground and beginning a bloody race riot. Another mob seized Garrison on the streets of Boston in 1835 and threatened to hang him. He was saved from death when police rescued him and locked him in jail for his own safety. Elijah Lovejoy, the editor of an abolitionist newspaper in Alton, Illinois, was victimized repeatedly and finally killed when he tried to defend his printing press from attack.

Violence greeted abolitionism in the South, too. Ardent pro-slavery supporters openly targeted abolitionists, publishing their whereabouts in local newspapers. They put bounties on antislavery activists' heads, beat people in the street for carrying abolitionist literature, and chased down abolitionists with promises of death.

That so many men and women continued to embrace abolitionism in the face of such vicious opposition suggests that abolitionists were not people who took their political commitments lightly. More importantly, the racial violence that abolitionism laid bare revealed that slavery would not be ended easily or peacefully. It also demonstrated that any campaign for Black civil rights would face enormous and bloody challenges.

ABOLITIONISM DIVIDED

By the mid-1830s, the unity of the abolitionist crusade began to crack. One cause was the violence of anti-abolitionists, which persuaded some members of the abolition movement that a more moderate approach was necessary. Another cause was the growing radicalism of William Lloyd Garrison, who shocked even many of his own allies (including Frederick Douglass) by attacking not only slavery but the government itself. The Constitution, he said, was "a covenant with death and an agreement with hell." In 1840, Garrison precipitated a formal division within the American Antislavery Society by insisting that women be permitted to participate in the movement on terms of full equality and be promoted to leadership positions. Some feared that this policy was too radical for the times and would hurt the movement's chances for progress. Garrison continued to arouse controversy after 1840 with more radical stands: an extreme pacifism that rejected even defensive wars; opposition to all forms of coercion—not just slavery, but also prisons and asylums; and finally, in 1843, a call for northern disunion from the South.

From 1840 on, therefore, abolitionism spoke with a number of different voices. The radical and uncompromising Garrisonians remained influential. But so were others, committed to a more moderate approach, who argued that abolition could be accomplished only as the result of a long, patient, peaceful struggle. The moderates appealed to the conscience of slaveholders; and when that produced no results, turned to political action, seeking to induce the northern states and the federal government to pass laws to aid the cause. Abolitionists helped fund the legal battle over the Spanish slave vessel, *Amistad*. After the Supreme Court (in *Prigg v. Pennsylvania,* 1842) ruled that states need not aid in enforcing the 1793 law requiring the return of those fleeing slavery to their owners, abolitionists won passage in several northern states of "personal liberty laws," which forbade state officials to assist in the capture and return of those who had fled slavery. The antislavery societies also petitioned Congress to abolish slavery in places where the federal government had jurisdiction—in the territories and in the District of Columbia—and to prohibit the interstate slave trade.

Antislavery sentiment underlay the formation in 1840 of the Liberty Party, which ran Kentucky antislavery leader James G. Birney for president. It supported abolitionism and the idea of "Free Soil," which meant keeping slavery out of any new territories. Some Free-Soilers were concerned about the welfare of Blacks; others cared nothing about slavery but simply wanted the West to be reserved as a place of opportunity for whites. The Free-Soil movement would ultimately do what abolitionism never could: attract the support of large numbers of the white population of the North.

The most powerful of all abolitionist propaganda was Harriet Beecher Stowe's novel *Uncle Tom's Cabin,* published as a book in 1852. It sold more than 300,000 copies within a year of publication and was reissued again and again. It succeeded in bringing the message of abolitionism to an enormous new audience—not only those who read the book but also those who watched countless theater companies reenact it across the nation. In both the North and the South, her novel helped inflame sectional tensions to a new level of passion.

Stowe's novel emerged not just out of abolitionist politics but also from a popular tradition of sentimental novels written by, and largely for, women. (See "Patterns of Popular Culture: Sentimental Novels.") Stowe artfully integrated the emotional conventions of the

PATTERNS OF POPULAR CULTURE

SENTIMENTAL NOVELS

"America is now wholly given over to a damned mob of scribbling women," Nathaniel Hawthorne complained in 1855, "and I should have no chance of success while the public taste is occupied with their trash." Hawthorne, one of the leading novelists of his time, was complaining about the most popular form of fiction in the mid-nineteenth-century United States—not his own dark and brooding works, but the "sentimental novel," a genre of literature written and read mostly by middle-class women.

In an age when affluent women occupied primarily domestic roles, and in which finding a favorable marriage was the most important thing many women could do to secure or improve their lots in life, the sentimental novel gave voice to both female hopes and female anxieties. The plots of sentimental novels were usually filled with character-improving problems and domestic trials, but most of them ended with the heroine securely and happily married. They were phenomenally successful, many of them selling more than 100,000 copies each—far more than almost any other books of the time.

Sentimental heroines were almost always beautiful and endowed with specifically female qualities—"all the virtues," one novelist wrote, "that are founded in the sensibility of the heart: Pity, the attribute of angels, and friendship, the balm of life, delight to dwell in the female breast." Women were highly sensitive creatures, the sentimental writers believed, incapable of disguising their feelings, and subject to fainting, mysterious illnesses, trances, and, of course, tears—things rarely expected of men. But they were also capable of a kind of nurturing love and natural sincerity that was hard to find in the predominantly male public world. In Susan Warner's *The Wide,* *Wide World* (1850), for example, the heroine, a young girl named Ellen Montgomery, finds herself suddenly thrust into the "wide, wide world" of male competition after her father loses his fortune. She is unable to adapt to this world, but she is saved in the end when she is taken in by wealthy relatives, who will undoubtedly prepare her for a successful marriage. They restore to her the security and comfort to which she had been born and without which she seemed unable to thrive.

Sentimental novels accepted uncritically the popular assumptions about women's special needs and desires, and they offered stirring tales of how women satisfied them. But sentimental novels were not limited to romanticized images of female fulfillment through protection and marriage. They hinted as well at the increasing role of women in reform movements. Many such books portrayed women dealing with social and moral problems—and using their highly developed female sensibilities to help other women escape from their troubles. Women were particularly suitable for such reform work, the writers implied, because they were specially gifted at helping and nurturing others.

The most famous sentimental novelist of the nineteenth century was Harriet Beecher Stowe. Most of her books—*The Minister's Wooing, My Wife and I, We and Our Neighbors,* and others—portrayed the travails and ultimate triumphs of women as they became wives, mothers, and hostesses. But Stowe was and remains best known for her 1852 antislavery novel, *Uncle Tom's Cabin,* one of the most influential books ever published in the United States. The story centers on the character of Tom, an aging Black enslaved man who ostensibly accepts his status. Tom holds a powerful Christian faith and an

UNCLE TOM'S CABIN Uncle Tom's Cabin did much to inflame public opinion in both the North and the South in the last years before the Civil War. At the time, however, Stowe was equally well known as one of the most successful American writers of sentimental novels.

many of the same dilemmas that the female heroines of other sentimental novels encounter in their struggles to find security and tranquility in their lives.

Another way in which women were emerging from their domestic sphere was by becoming consumers for the expanding products of the United States' industrializing economy. The female characters in sentimental novels searched not just for love, security, and social justice; they also searched for luxury and for the pleasure of buying some favored item. In *The Wide, Wide, World*, Susan Warner illustrated this aspect of the culture of the sentimental novel—and the desires of the women who read them—with her description of the young Ellen Montgomery buying a Bible in an elegant bookstore: "Such beautiful Bibles she had never seen; she pored in ecstasy over their varieties of type and binding, and was very evidently in love with them all." •

unwavering conviction in the goodness of God and his promise of deliverance from evil. In this regard, Stowe positions Christianity as being opposed to slavery and raises questions about Southern claims that God favored the peculiar institution.

Uncle Tom's Cabin is a sentimental novel, too. Stowe's critique of slavery is based partly on her belief in the importance of domestic values and family security. Slavery's violation of those values, and its denial of that security, is what made it so abhorrent to her. The simple, decent Uncle Tom faces

UNDERSTAND, ANALYZE, & EVALUATE

1. How did the lives of the heroines of the sentimental novels compare with the lives of real women of the nineteenth century? What made them so popular?
2. How did the sentimental novels encourage women's participation in public life? Did the novels reinforce prevailing attitudes toward women or broaden the perception of women's "proper role"?
3. *Uncle Tom's Cabin* is probably one of the best-known works of American fiction. Why was this novel so powerful?

sentimental novel with the political ideas of the abolitionist movement, and to sensational effect. By embedding the antislavery message within a familiar literary form in which women were the key protagonists, her novel brought that message to an enormous new audience.

Even divided, abolitionism remained a powerful influence on the life of the nation. Only a relatively small number of people before the Civil War ever accepted the abolitionist position that slavery must be entirely eliminated in a single stroke. But the crusade that Douglass and Garrison had launched, and that thousands of committed men and women kept alive for three decades, was a constant, visible reminder of how deeply the institution of slavery was dividing the United States.

CONCLUSION

The rapidly changing society of the antebellum United States encouraged interest in a wide range of reforms. Writers, artists, intellectuals, and others drew heavily from new European notions of personal liberation and fulfillment—a set of ideas often known as romanticism. But they also strove to create a truly American culture. The literary and artistic life of the nation expressed the rising interest in personal liberation—in giving individuals the freedom to explore their own souls and to find in nature a full expression of their divinity. It also called attention to some of the nation's glaring social problems.

Reformers, too, made use of the romantic belief in the divinity of the individual. They flocked to religious revivals, worked on behalf of "moral" reforms such as temperance, supported education, and articulated some of the first statements of modern feminism. And in the North, they rallied against slavery. Free Blacks and their allies demanded an end to the peculiar institution. As abolitionism grew in popularity and power and inspired violent counter-efforts, it became apparent that the end of slavery would not come easily. Or without bloodshed.

KEY TERMS/PEOPLE/PLACES/EVENTS

David Walker 287
Elizabeth Cady Stanton 283
Frederick Douglass 287
Harriet Beecher Stowe 283
Henry David Thoreau 275
Herman Melville 273
Horace Mann 281
Hudson River school 272
Jarena Lee 283

Joseph Smith 278
Liberia 286
Lucretia Mott 283
Nathaniel Hawthorne 276
Oneida "Perfectionists" 277
Ralph Waldo Emerson 274
Rebecca Cox Jackson 285
Seneca Falls
 Convention 283

Shakers 277
Sojourner Truth 283
Susan B. Anthony 283
temperance 279
transcendentalism 274
Walt Whitman 273
William Lloyd
 Garrison 290

RECALL AND REFLECT

1. What is "romanticism" and how was it expressed in American literature and art?
2. How did religion affect reform movements, and what was the effect of these movements on religion?
3. What were the aims of the women's movement of the nineteenth century? How successful were women in achieving these goals?
4. What arguments and strategies did the abolitionists use in their struggle to end slavery? Who opposed them and why?

Design element: Stars and Stripes: McGraw Hill Education.

13 | THE IMPENDING CRISIS

LOOKING WESTWARD
EXPANSION AND WAR
THE SECTIONAL DEBATE
THE CRISES OF THE 1850s

LOOKING AHEAD

1. How did the annexation of western territories lead to deeper divisions between the North and the South?
2. What were the major arguments for and against slavery?
3. What compromises attempted to resolve the conflicts over the expansion of slavery into new territories? Why did they fail?

UNTIL THE 1840s, NATIONAL POLITICAL TENSIONS that could lead to disunion remained relatively contained. Few, other than Black writers and clerics and radical abolitionists, predicted a civil war.

But midcentury brought a rash of explosive issues that politicians struggled—and ultimately failed—to resolve peacefully. In the North the abolitionist movement picked up steam and inspired legions of supporters, the most aggressive of whom sought to fight slavery with the sword as well as the pen. The South birthed a generation of militant pro-slavery spokesmen who brooked no compromise over a state's right to build a society based on slavery. From the West emerged burning controversies over the political fate of the territories and whether they would enter the Union as either slave or free states. Partisans recruited sympathizers from across the nation and even took up arms to secure the victory of their position.

TIME LINE

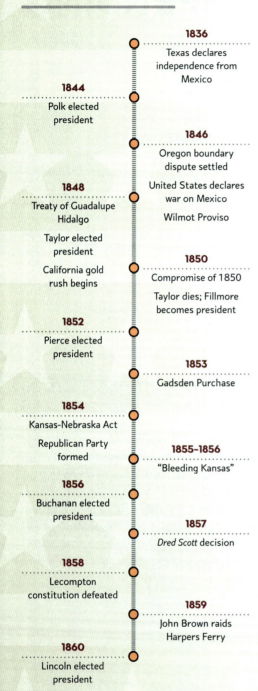

1836 — Texas declares independence from Mexico

1844 — Polk elected president

1846 — Oregon boundary dispute settled; United States declares war on Mexico; Wilmot Proviso

1848 — Treaty of Guadalupe Hidalgo; Taylor elected president; California gold rush begins

1850 — Compromise of 1850; Taylor dies; Fillmore becomes president

1852 — Pierce elected president

1853 — Gadsden Purchase

1854 — Kansas-Nebraska Act; Republican Party formed

1855–1856 — "Bleeding Kansas"

1856 — Buchanan elected president

1857 — *Dred Scott* decision

1858 — Lecompton constitution defeated

1859 — John Brown raids Harpers Ferry

1860 — Lincoln elected president

LOOKING WESTWARD

More than a million square miles of new territory came under the control of the United States during the 1840s. By the end of the decade, the nation possessed all the territory of the present-day United States except Alaska, Hawaii, and a few relatively small areas acquired later through border adjustments. Many factors accounted for this great new wave of expansion, but one of the most important was an ideology known as **Manifest Destiny**.

MANIFEST DESTINY

Manifest Destiny reflected both the growing pride that characterized American nationalism in the mid-nineteenth century and the idealistic vision of social perfection that fueled so much of the reform energy of the time. It rested on the idea that the United States was destined—by God and by history—to expand its boundaries over a vast area. It also reflected long-standing racism. For all of its national and religious rhetoric, Manifest Destiny operated as a license for white men to continue their forcible acquisition of indigenous land, kill or remove native people who stood in their way, and decide if slavery would be permitted in their new territories and states.

By the 1840s, publicized by the inexpensive newspapers dubbed the "penny press," the idea of Manifest Destiny had spread throughout the nation. The most ardent advocates of Manifest Destiny envisioned a vast new "empire of liberty" that would include Canada, Mexico, the Caribbean and Pacific islands, and ultimately much of the rest of the world. Countering this bombast were politicians such as Henry Clay, and others, who warned that territorial expansion would reopen the painful controversy over slavery. Their voices, however, could not compete with the enthusiasm over expansion in the 1840s, which began with issues surrounding the potential statehood of Texas and Oregon.

Americans in Texas

Twice in the 1820s, the United States had offered to purchase Texas from the Republic of Mexico, and twice Mexico refused. Mexican officials wanted to both keep the frontier region and to populate it because they believed it to be threatened by nomadic indigenous groups like the Comanche as well as possibly by Spain. As a result, they enacted what would be viewed in hindsight as a curious policy—namely, a colonization law that offered cheap land and a four-year exemption from taxes to any American willing to move into Texas. Thousands of Americans flocked into the region, the great majority of them white Southerners and those they enslaved, intent on establishing cotton plantations. By 1830, there were about 7,000 Americans living in Texas, more than twice the number of Mexicans there.

Most of the settlers came to Texas through the efforts of American intermediaries, who received sizable land grants from Mexico in return for bringing new residents into the region. The most successful was **Stephen F. Austin**, a young immigrant from Missouri who established the first legal American settlement in Texas in 1822. Austin and others created centers of power in the region that competed with the Mexican government. Not surprisingly, in 1830 the Mexican government barred any further American immigration into the region. But Americans kept flowing into Texas anyway.

Friction between the American settlers and the Mexican government was already growing in the mid-1830s when instability in Mexico resulted in General **Antonio López de Santa Anna** seizing power as a dictator. He increased the powers of the Mexican government at the expense of the state governments, a measure that Texans from the United States assumed was aimed specifically at them. Sporadic fighting between Americans and Mexicans in Texas erupted into a major conflict in 1835. The next year, the American settlers defiantly proclaimed their independence from Mexico.

Santa Anna led a large army into Texas to quell the uprising. Mexican forces annihilated an American garrison at the **Alamo** mission in San Antonio after a famous, if futile, defense by a group of Texas "patriots" that included, among others, the renowned frontiersman and former Tennessee congressman Davy Crockett. Another garrison at Goliad suffered the same fate. By the end of 1836, the rebellion appeared to have collapsed. But General **Sam Houston** managed to keep a small force of Texans together. And on April 21, 1836, at the

Kathie Rees/EyeEm/Getty Images

THE LONE STAR FLAG Texas was an independent republic for nine years. The tattered banner pictured here was one of the republic's original flags.

Battle of San Jacinto, he defeated the Mexican army and took Santa Anna prisoner. Santa Anna, under pressure from his captors, signed a treaty giving Texas independence.

A number of Mexican residents of Texas (*Tejanos*) had fought with the Americans in the revolution. But soon after Texas won its independence, their positions grew difficult. The Americans did not trust them, feared that they were agents of the Mexican government, and drove many of them out of the new republic. Most of those who stayed had to settle for a politically and economically subordinate status.

One of the first acts of the new president of Texas, Sam Houston, was to send a delegation to Washington with an offer to join the Union. But President Jackson, fearing that adding a large new slave state to the Union would increase sectional tensions, blocked annexation and even delayed recognizing the new republic until 1837.

Spurned by the United States, Texas cast out on its own. England and France, concerned about the surging power of the United States, saw Texas as a possible check on its growth and began forging ties with the new republic. At that point, President Tyler persuaded Texas to apply for statehood again in 1844. But northern senators, fearing the admission of a new slave state, defeated the application. Statehood would have to wait until after the election of President Polk.

OREGON

Control of what was known as "Oregon country" in the Pacific Northwest was also a major political issue in the 1840s. Both Britain and the United States claimed sovereignty over the region. Unable to resolve their conflicting claims diplomatically, they agreed in an 1818 treaty to allow citizens of each country equal access to the territory. This "joint occupation" continued for twenty years.

At the time of the treaty neither Britain nor the United States had established much of a presence in Oregon country. White settlement in the region consisted largely of scattered American and Canadian fur trading posts. But American interest in Oregon grew substantially in the 1820s and 1830s.

By the mid-1840s, white Americans substantially outnumbered the British in Oregon. They had also devastated much of the native population, in part through a measles epidemic that spread through the Cayuse people, in part through bloody raids. American settlements were sprouting up along the Pacific Coast and the new settlers were urging the U.S. government to take possession of the disputed Oregon country.

THE WESTWARD MIGRATION

The migrations into Texas and Oregon were part of a larger movement that took hundreds of thousands of white and Black Americans into the far western regions of the continent between 1840 and 1860. The largest number of migrants were from the Old Northwest. Most were relatively young white people who had traveled in family groups. Few were wealthy, but many were relatively prosperous. Poor people who could not afford the trip on their own usually had to join other families or groups as laborers—men as farm or ranch hands; women as domestic servants, teachers, or in some cases, prostitutes. Groups heading for areas where mining or lumbering was the principal economic activity consisted mostly of men. Those heading for farming regions traveled mainly as families.

To start their journey, migrants generally gathered in one of several major depots in Iowa and Missouri (Independence, St. Joseph, or Council Bluffs), joined a wagon train led

WESTERN TRAILS IN 1860 As settlers began the long process of exploring and establishing farms and businesses in the West, major trails began to develop to facilitate travel and trade between the region and the more thickly settled areas to the east. Note how many of the trails led to California and how few of them led into any of the far northern regions of U.S. territory. Note, too, the important towns and cities that grew up along these trails. • *What forms of transportation later performed the functions that these trails performed prior to the Civil War?*

by hired guides, and set off with their belongings piled in covered wagons and livestock trailing behind. The major route west was the 2,170 mile **Oregon Trail**, which stretched from Independence across the Great Plains and through the South Pass of the Rocky Mountains. From there, migrants moved north into Oregon or south (along the California Trail) to the northern California coast. Other migrations moved along the Santa Fe Trail, southwest from Independence into New Mexico.

However they traveled, overland migrants faced an arduous journey. Most journeys lasted five or six months (from May to November), and travelers felt the pressure to get through the Rockies before the snows began, often not an easy task given the very slow pace of most wagon trains. To save their horses for pulling the wagons, most people walked most of the way. Diseases, including cholera, decimated and slowed many groups traveling west. The women, who did the cooking and washing at the end of the day, generally worked harder than the men, who usually rested when the caravan halted.

Despite the traditional image of westward migrants as rugged individualists, most travelers experienced the journey as a communal experience. Many expeditions consisted of groups of friends, neighbors, or relatives who had decided to pull up stakes and move west together. The intensity of the journey strengthened existing bonds and forged new ones. It was a rare expedition in which there were not some internal conflicts before the trip was over; but those who made the journey successfully generally learned the value of cooperation.

Only a few expeditions experienced attacks by Native Americans. In the twenty years before the Civil War, fewer than 400 migrants (slightly more than one-tenth of 1 percent) died in conflicts with native people. In fact, Native Americans were usually more helpful than dangerous to the white migrants. They often served as guides and traded horses, clothing, and fresh food with the travelers.

EXPANSION AND WAR

The growing number of white Americans in the lands west of the Mississippi put great pressure on the government in Washington to annex Texas, Oregon, and other territory. And in the 1840s, these expansionist pressures helped push the slavery question to the forefront of political debate and move the United States closer to war.

THE DEMOCRATS AND EXPANSION

In preparing for the election of 1844, the two leading candidates—Henry Clay of the Whig Party and Martin Van Buren of the Democratic Party—both avoided taking a strong public stand on the controversial annexation of Texas. Sentiment for expansion was actually mild within the Whig Party, and Clay had no difficulty securing the nomination despite his noncommittal position. But many southern Democrats strongly supported annexation, and the party passed over Van Buren to nominate **James K. Polk**, who shared their enthusiasm.

Polk had represented Tennessee in the House of Representatives for fourteen years, four of them as Speaker, and had subsequently served as governor. But by 1844, he had been out of public office for three years. What made his victory possible was his open support for the position, expressed in the Democratic platform, "that the re-occupation of Oregon and the re-annexation of Texas at the earliest practicable period are great American measures." By combining the Oregon and Texas questions, the Democrats hoped to appeal to both northern and southern expansionists—and they did. Polk carried the election, 170 electoral votes to 105.

Polk entered office with a clear set of goals and with plans for attaining them. John Tyler accomplished the first of Polk's ambitions for him in the last days of his own presidency. Interpreting the election returns as a mandate for the annexation of Texas, the outgoing president won congressional approval for it. That December, Texas became a state.

Polk himself resolved the Oregon question. The British minister in Washington had brusquely rejected a compromise that would establish the U.S.-Canadian border at the 49th parallel. Incensed, Polk again asserted the American claim to all of Oregon. There was loose talk of war on both sides of the Atlantic—in the United States, war mongers chanted "Fifty-four forty or fight!" (a reference to where the Americans ultimately hoped to draw the northern boundary of their part of Oregon). But neither country really wanted war. Finally, the British government accepted Polk's original proposal to divide the territory at the 49th parallel. On June 15, 1846, the Senate approved a treaty that fixed the boundary there.

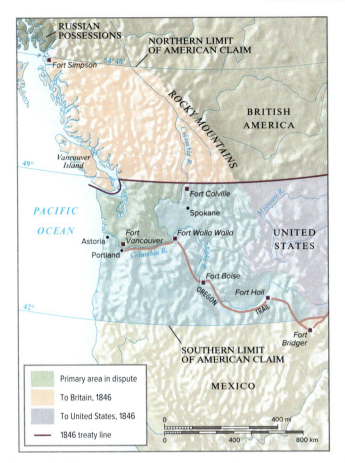

THE OREGON BOUNDARY, 1846 One of the last major boundary disputes between the United States and Great Britain involved the territory known as Oregon—the large region on the Pacific Coast north of California (which in 1846 was still part of Mexico). For years, the United States and Britain had overlapping claims on the territory. The British claimed land as far south as the present state of Oregon, while the Americans claimed land extending well into what is now Canada. Tensions over the Oregon border at times rose to the point that many Americans were demanding war, some using the slogan "Fifty-four forty or fight!" referring to the latitude of the northernmost point of the American claim. • *How did President James K. Polk defuse the crisis?*

THE SOUTHWEST AND CALIFORNIA

One of the reasons the Senate and the president had agreed so readily to the British offer to settle the Oregon question was that their attention was turning to new tensions emerging in the Southwest. As soon as the United States admitted Texas to statehood in 1845, the Mexican government broke off diplomatic relations with Washington. Mexican–American relations grew still worse when a dispute developed over the boundary between Texas and Mexico. Texans claimed the Rio Grande as their western and southern border. Mexico, although still not conceding the loss of Texas, argued nevertheless that the border had always been the Nueces River, to the north of the Rio Grande. Polk accepted the Texas claim, and in the summer of 1845 he sent a small army under General Zachary Taylor to Texas to protect the new state against a possible Mexican invasion.

Part of the area in dispute was New Mexico, whose Spanish and indigenous residents lived in a multiracial society that by the 1840s had endured for nearly a century and a half.

In the 1820s, the Mexican government had invited American traders into the region, hoping to speed development of the province. But New Mexico, like Texas, soon became more American than Mexican, particularly after a flourishing commerce developed between Santa Fe and Independence, Missouri.

White Americans were also increasing their interest in California. In this vast region lived members of several western Native American tribes and perhaps 7,000 Mexicans. Gradually, however, white Americans began to arrive: first maritime traders and captains of Pacific whaling ships, who stopped to barter goods or buy supplies; then merchants, who established stores, imported goods, and developed a profitable trade with the Mexicans and native people; and finally pioneering farmers, who entered California from the east and settled in the Sacramento Valley. Some of these new settlers began to dream of bringing California into the United States.

President Polk soon came to share their dream and committed himself to acquiring both New Mexico and California for the United States. At the same time that he dispatched the troops under Taylor to Texas, he sent secret instructions to the commander of the Pacific naval squadron to seize the California ports if Mexico declared war. Representatives of the president quietly informed Americans in California that the United States would respond sympathetically to a revolt against Mexican authority there.

The Mexican War

Having appeared to prepare for war, Polk turned to diplomacy by dispatching a special minister to try to buy off the Mexicans. But Mexican leaders rejected the American offer to purchase the disputed territories. On January 13, 1846, as soon as he heard this news, Polk ordered Taylor's army in Texas to move across the Nueces River, where it had been stationed, to the Rio Grande. For months, the Mexicans refused to fight. But finally, according to disputed American accounts, some Mexican troops crossed the Rio Grande and attacked a unit of American soldiers. On May 13, 1846, Congress declared war by votes of 40 to 2 in the Senate and 174 to 14 in the House.

Whig critics charged that Polk had deliberately maneuvered the country into the conflict and had staged the border incident that had precipitated the declaration. Opposition intensified as the war continued and as the public became aware of the rising casualties and expense.

Victory did not come as quickly as Polk had hoped. The president ordered Taylor to cross the Rio Grande, seize parts of northeastern Mexico, beginning with the city of Monterrey, and then march on to Mexico City itself. Taylor captured Monterrey in September 1846, but he let the Mexican garrison evacuate without pursuit. Polk now began to fear that Taylor lacked the tactical skill for the planned advance against Mexico City. He also feared that, if successful, Taylor would become a powerful political rival (as, in fact, he did).

In the meantime, Polk ordered other offensives against New Mexico and California. In the summer of 1846, a small army under Colonel Stephen W. Kearny captured Santa Fe with no opposition. He then proceeded to California, where he joined a conflict already in progress that was being staged jointly by American settlers, a well-armed exploring party led by John C. Frémont, and the American navy: the so-called Bear Flag Revolt. Kearny brought the disparate American forces together under his command, and by the autumn of 1846 he had completed the conquest of California.

But Mexico still refused to concede defeat. At this point, Polk and General Winfield Scott, the commanding general of the army and its finest soldier, launched a bold new campaign. Scott assembled an army at Tampico, which the navy transported down the

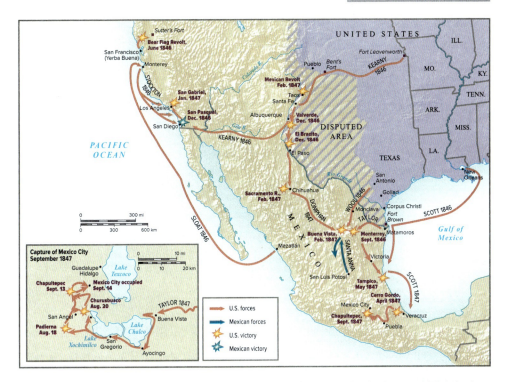

THE MEXICAN WAR, 1846-1848 Shortly after the settlement of the Oregon border dispute with Britain, the United States entered a war with Mexico over another contested border. This map shows the movement of Mexican and American troops during the fighting, which extended from the area around Santa Fe south to Mexico City and west to the coast of California. Note the American use of its naval forces to facilitate a successful assault on Mexico City, and others on the coast of California. Note, too, how unsuccessful the Mexican forces were in their battles with the United States. Mexico won only one battle—a relatively minor one at San Pasqual near San Diego—in the war. • *How did President Polk deal with the popular clamor for the United States to annex much of present-day Mexico?*

Mexican coast to Veracruz. With an army that never numbered more than 14,000, Scott advanced 260 miles along the Mexican National Highway toward Mexico City, kept American casualties low, and never lost a battle before finally seizing the Mexican capital. A new Mexican government took power and announced its willingness to negotiate a peace treaty.

President Polk continued to encourage those who demanded that the United States annex much of Mexico itself. At the same time, he was growing anxious to get the war finished quickly. Polk sent a special presidential envoy, Nicholas Trist, to negotiate a settlement. On February 2, 1848, he reached agreement with the new Mexican government on the **Treaty of Guadalupe Hidalgo**, by which Mexico agreed to cede California and New Mexico to the United States and acknowledge the Rio Grande as the boundary of Texas. In return, the United States promised to assume any financial claims its new citizens had against Mexico and to pay the Mexicans $15 million. Trist had obtained most of Polk's original demands, but he had not satisfied the new, more expansive dreams of acquiring additional territory in Mexico itself. Polk angrily claimed that Trist had violated his instructions, but he soon realized that he had no choice but to accept the treaty to silence a bitter battle growing between ardent expansionists demanding the annexation of "All Mexico!" and antislavery leaders charging that the expansionists were conspiring to extend slavery to new realms. The president submitted the Trist treaty to the Senate, which approved it by a vote of 38 to 14.

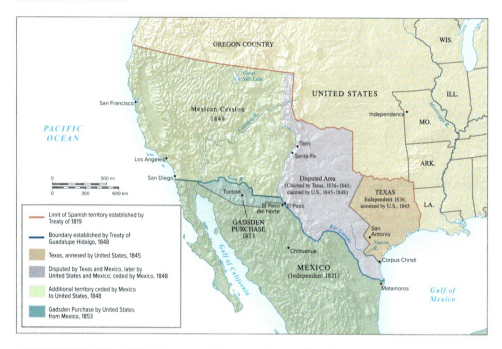

SOUTHWESTERN EXPANSION, 1845–1853 The annexation of much of what is now Texas in 1845, the much larger territorial gains won in the Mexican War in 1848, and the purchase of additional land from Mexico in 1853 completed the present continental border of the United States. • *What great event shortly after the Mexican War contributed to a rapid settlement of California by migrants from the eastern United States?*

THE SECTIONAL DEBATE

James Polk tried to be a president whose policies transcended sectional divisions. But conciliating the sections was becoming an ever more difficult task, and Polk gradually earned the enmity of Northerners and Westerners alike, who believed his policies favored the South at their expense.

SLAVERY AND THE TERRITORIES

In August 1846, while the Mexican War was still been in progress, Polk asked Congress to appropriate $2 million for purchasing peace with Mexico. The prospect of peace raised the question of whether slavery would be allowed in any newly acquired territory. Representative David Wilmot of Pennsylvania, an antislavery Democrat, introduced an amendment to the appropriation bill prohibiting slavery in any territory acquired from Mexico. The so-called **Wilmot Proviso** passed the House but failed in the Senate. (See "Consider the Source: Wilmot Proviso.") Southern militants contended that all Americans had equal rights in the new territories, including the right to move those they held in slavery (which they considered property) into them.

As the sectional debate intensified, President Polk supported a proposal to extend the Missouri Compromise line through the new territories to the Pacific Coast, banning slavery north of the line and permitting it south of the line. Others supported a plan, originally known as "squatter sovereignty" and later by the more dignified phrase "**popular sovereignty**," that would allow the people of each territory to decide the status of slavery there. The debate over these various proposals dragged on for many months.

The presidential campaign of 1848 dampened the controversy for a time as both Democrats and Whigs tried to avoid the slavery question. When Polk, in poor health, declined to run again, the Democrats nominated Lewis Cass of Michigan, a dull, aging party regular. The Whigs nominated General **Zachary Taylor** of Louisiana, hero of the Mexican War but a man with no political experience. Opponents of slavery found the choice of candidates unsatisfying, and out of their discontent emerged the new **Free-Soil Party**, whose candidate was former president Martin Van Buren.

Taylor won a narrow victory. But while Van Buren failed to carry a single state, he polled an impressive 291,000 votes (10 percent of the total), and the Free-Soilers elected ten members to Congress. The emergence of the Free-Soil Party as an important political force signaled the inability of the existing parties to contain the political passions slavery was creating. It was also an early sign of the coming collapse of the second party system in the 1850s.

THE CALIFORNIA GOLD RUSH

By the time Taylor took office, the pressure to resolve the question of slavery in the far western territories had become more urgent as a result of dramatic events in California. In January 1848, a foreman working in a sawmill owned by John Sutter (one of California's leading ranchers) found traces of gold in the foothills of the Sierra Nevada. Within months, news of the discovery had spread throughout the nation and much of the world. Almost immediately, thousands of white people began flocking to California in a frantic search for gold.

The atmosphere in California at the peak of the gold rush was one of almost crazed excitement and greed. Most migrants to the Far West prepared carefully before making the journey. But the California migrants (known as "**Forty-niners**") threw caution to the winds, abandoning farms, jobs, homes, and families, piling onto ships and overrunning the overland trails. The overwhelming majority of the Forty-niners (perhaps 95 percent) were white men, and the society they created in California was unusually fluid and volatile.

The gold rush also attracted some of the first Chinese migrants to the western United States. News of the discoveries created great excitement in China, particularly in impoverished areas. It was, of course, extremely difficult for a poor Chinese peasant to get to the United States; but many young, adventurous people (mostly men) decided to go anyway—in the belief that they could quickly become rich and then return to China. Emigration brokers loaned many migrants money for passage to California, which the migrants were to pay off out of their earnings there.

The gold rush produced a serious labor shortage in California, as many male workers left their jobs and flocked to the gold fields. That created opportunities for many people who needed work (including Chinese immigrants). It also led to a frenzied exploitation of Native Americans that resembled slavery in all but name. A new state law permitted the arrest of "loitering" or orphaned indigenous people and subsequent assignment to a term of "indentured" labor.

The gold rush was of critical importance to the growth of California, but not for the reasons most of the migrants hoped. There was substantial gold in the hills of the Sierra Nevada, and many people got rich from it. But only a tiny fraction of the Forty-niners ever found gold. Some disappointed migrants returned home after a while; however, many stayed in California and swelled both the agricultural and urban populations of the territory. By 1856, for example, San Francisco—whose population had been 1,000 before the gold rush—was the home of over 50,000 people. By the early 1850s, California itself, which had always

CONSIDER THE SOURCE

WILMOT PROVISO (1846)

To counter rising tensions over the question of whether territory acquired from Mexico would be slave or free, Representative David Wilmot of Pennsylvania spearheaded an effort to ban slavery from that territory forever. His amendment passed the House twice but failed in the Senate because of heated opposition from northern pro-slavery politicians.

Provided, that, as an express and fundamental condition to the acquisition of any territory from the Republic of Mexico by the United States, by virtue of any treaty which may be negotiated between them, and to the use by the Executive of the moneys herein appropriated, neither slavery nor involuntary servitude shall ever exist in any part of said territory, except for crime, whereof the party shall first be duly convicted.

UNDERSTAND, ANALYZE, & EVALUATE

1. What condition did the proviso impose on future territory?
2. Why did this simple provision prove so controversial? What were its consequences?

Library of Congress Prints & Photographs Division [LC-DIG-ppmsca-32195]

LOOKING FOR GOLD In this 1850s lithograph, the unnamed artist presents the West as a world of abundance and great wealth available to all with just a little bit of pluck and luck. Gold is there for the taking, and happiness befalls all who pan for riches. Absent is any sense of failure, hard work, and suffering.

had a diverse population, had become even more heterogeneous. The gold rush had attracted not just white Americans but also Chinese, South Americans, Mexicans, free Blacks, and enslaved workers who accompanied white southern migrants. Conflicts over gold intersected with racial and ethnic tensions to make the territory an unusually turbulent place.

Rising Sectional Tensions

Zachary Taylor believed statehood could become the solution to the issue of slavery in the territories. As long as the new lands remained territories, the federal government was responsible for deciding the fate of slavery within them. But once they became states, he thought, their own governments would be able to settle the slavery question. At Taylor's urging, California quickly adopted a constitution that prohibited slavery, and in December 1849 Taylor asked Congress to admit California as a free state.

Congress balked, in part because several other controversies concerning slavery were complicating the debate. One was the effort of antislavery forces to abolish slavery in the District of Columbia. Another was the emergence of personal liberty laws in northern states, which barred courts and police officers from returning those who had run away from slavery to their former owners, in defiance of the Constitution's Fugitive Slave Clause. But the biggest obstacle to the president's program was the white South's fear that new free states would create an antislavery majority in Congress. The number of free and slave states was equal in 1849—fifteen each. The admission of California would upset the balance; and New Mexico, Oregon, and Utah—all candidates for statehood—might upset it further.

Library of Congress, Prints and Photographs Division Washington, D.C. 20540 USA [LC-USZ62-11138]

STIRRING THE POT OF SECTIONAL TENSIONS James Baillie's 1850 lithograph, *The Hurly-Burly Pot*, warns of sharpening antagonisms between the North and the South and the rising threat of disunion. He targets the most vocal partisans: from the North, abolitionist William Lloyd Garrison, Free-Soil promoter David Wilmot, and journalist Horace Greeley; and from the South, states' rights promoter Senator John C. Calhoun. Like the witches in Shakespeare's *Macbeth*, the first three dance around a bubbling cauldron, adding to it sacks branded "Free Soil," "Abolition," and "Fourierism" (tossed in by Greeley, a supporter of utopian socialist Charles Fourier). Behind them looms John Calhoun, who crows about this act of treason: "For success to the whole mixture, we invoke our great patron Saint Benedict Arnold."

Even many otherwise moderate southern leaders now began to talk about secession from the Union. In the North, every state legislature but one adopted a resolution demanding the prohibition of slavery in the territories.

THE COMPROMISE OF 1850

Faced with this mounting crisis, moderates and unionists spent the winter of 1849–1850 trying to frame a great compromise. The aging Henry Clay, who was spearheading the effort, believed that no compromise could last unless it settled all the issues in dispute. As a result, he took several measures that had been proposed separately, combined them into a single piece of legislation, and presented it to the Senate on January 29, 1850. Among the bill's provisions were the admission of California as a free state; the formation of territorial governments in the rest of the lands acquired from Mexico, without restrictions on slavery; the abolition of the slave trade, but not slavery itself, in the District of Columbia; and a new and more effective fugitive slave law. These resolutions launched a debate that raged for seven months.

Finally at midyear, the climate for compromise improved. President Taylor suddenly died and Vice President Millard Fillmore of New York took his place. Fillmore, who understood the importance of flexibility, supported Clay's compromise and persuaded northern Whigs to do so as well. When the bill failed, **Stephen A. Douglas**, a Democratic senator from Illinois and the acknowledged leader of northwestern Democrats, stepped in. He divided the omnibus bill into individual bills, which allowed congressmen to vote or abstain on each. Douglas also gained support with complicated backroom deals linking the compromise to the sale of government bonds and the construction of railroads in the states of politicians

SLAVE AND FREE TERRITORIES UNDER THE COMPROMISE OF 1850 The acquisition of vast new western lands raised the question of the status of slavery in new territories organized for statehood by the United States. Tension between the North and the South on this question led in 1850 to a great compromise, forged in Congress, to settle this dispute. The compromise allowed California to join the Union as a free state and introduced the concept of "popular sovereignty" for other new territories. • *How well did the Compromise of 1850 work?*

who opposed any of the bills. As a result of his efforts, by mid-September Congress had enacted all the components of the compromise.

The **Compromise of 1850** was a victory of individual self-interest. Still, members of Congress hailed the measure as a triumph of statesmanship; and Millard Fillmore boasted it was a just settlement of the sectional problem, "in its character final and irrevocable." Rarely has a politician been so wrong.

THE CRISES OF THE 1850s

For a few years after its passage, the Compromise of 1850 actually seemed to work. Sectional conflict appeared to fade amid booming prosperity and growth. But the relative harmony was a façade. While the level of tension subsided, the tensions themselves remained. And chief among these tensions, as always, was the place of slavery in American society.

The Uneasy Truce

In the run-up to the presidential election of 1852, both major parties endorsed the Compromise of 1850 and nominated candidates not strongly identified with sectional passions. The Democrats chose the obscure New Hampshire politician Franklin Pierce, and the Whigs chose the military hero General Winfield Scott. But the sectional question quickly became a divisive influence in the election and the Whigs were the principal victims. They suffered massive defections by antislavery members who were angered by the party's evasiveness on their core issue. Many of them flocked to the Free-Soil Party, whose antislavery presidential candidate, John P. Hale, repudiated the Compromise of 1850. The divisions among the Whigs helped produce a victory for the Democrats in 1852.

Franklin Pierce attempted to maintain harmony by avoiding divisive issues, particularly slavery. But it was an impossible task. Northern opposition to the Fugitive Slave Act intensified quickly after 1850. Mobs formed in some northern cities to prevent enforcement of the fugitive slave law and several northern states also passed their own laws barring the deportation of those who had fled slavery. White Southerners watched with growing anger and alarm as the one element of the Compromise of 1850 that they had considered a victory seemed to become meaningless through northern defiance.

"Young America"

One of the ways Franklin Pierce hoped to dampen sectional controversy was through his support of a movement in the Democratic Party known as **"Young America."** Its adherents saw the expansion of American democracy throughout the world as a way to divert attention from the controversies over slavery. The great liberal and nationalist revolutions of 1848 in Europe stirred them to dream of a republican Europe with governments based on the model of the United States. They dreamed as well of acquiring new territories in the Western Hemisphere.

But efforts to extend the nation's domain could not avoid becoming entangled with the sectional crisis. Pierce had been pursuing diplomatic attempts to buy Cuba from Spain (efforts begun in 1848 by Polk). In 1854, however, a group of Pierce's envoys sent him a private document from Ostend, Belgium, making a case for seizing Cuba by force. When the Ostend Manifesto, as it became known, was leaked to the public, antislavery Northerners charged the administration with conspiring to bring a new slave state into the Union.

The South, for its part, opposed all efforts to acquire new territory that would not support a slave system. The kingdom of Hawaii agreed to join the United States in 1854, but the treaty died in the Senate because it contained a clause prohibiting slavery in the islands. A powerful movement to annex Canada to the United States similarly foundered, at least in part because of slavery.

Slavery, Railroads, and the West

What fully revived the sectional crisis, however, was one of the same issues that had produced it in the first place: slavery in the territories. By the 1850s, the line of substantial white settlement had moved beyond the boundaries of Missouri, Iowa, and what is now Minnesota into a great expanse of plains, which many white Americans had once believed was unfit for cultivation. Now it was becoming apparent that large sections of this region were, in fact, suitable for farming. In the states of the Old Northwest, prospective white settlers urged the government to open the area to them, provide territorial governments, and dislodge local native societies. There was relatively little opposition from any segment of white society to this proposed violation of indigenous rights. But the interest in further settlement raised two issues that did prove highly controversial and that gradually became entwined with each other: railroads and slavery.

As the nation expanded westward, broad support began to emerge for building a transcontinental railroad. The problem was where to place it—and in particular, where to locate the railroad's eastern terminus, where the line could connect with the existing rail network east of the Mississippi. Northerners favored Chicago, while Southerners supported St. Louis, Memphis, or New Orleans. The transcontinental railroad had also become part of the struggle between the North and the South.

Pierce's secretary of war, Jefferson Davis of Mississippi, removed one obstacle to a southern route. Surveys indicated that a railroad with a southern terminus would have to pass through an area in Mexican territory. But in 1853, Davis sent James Gadsden, a southern railroad builder, to Mexico, where he persuaded the Mexican government to accept $10 million in exchange for a strip of land that today comprises parts of Arizona and New Mexico. The so-called **Gadsden Purchase** only accentuated the sectional rivalry as it added more slave territory.

The Kansas–Nebraska Controversy

The momentum for an intercontinental railroad continued to build, amid the great debate over where to put it. Senator Douglas from Illinois wanted the transcontinental railroad to run north through his state, but he also recognized, as many did, that a northern route through the territories would run mostly through Native American populations and he wanted to avoid possible conflict with them. As a result, he introduced a bill in January 1854 to organize (and thus open to white settlement and railroads) a huge new territory, known as Nebraska, west of Iowa and Missouri from the still unorganized territory of the Louisiana Purchase.

Douglas knew the South would oppose his bill because organized territories over time become states, and the proposed territory was north of the Missouri Compromise line (36°3′) and hence closed to slavery since 1820. Initially, Douglas attempted to appease Southerners by including a provision that territorial legislatures would decide the status of slavery. This meant that, in theory, the region could choose to open itself to slavery, effectively repealing the Missouri Compromise. When southern Democrats demanded more,

Douglas also agreed to divide the area into two territories—Nebraska and Kansas—instead of one. The new, second territory (Kansas) was thought more likely to become a slave state. In its final form, the measure was known as the **Kansas-Nebraska Act**. President Pierce supported the bill, and after a strenuous debate, it became law in May 1854 with unanimous support from the South and partial support from northern Democrats.

No piece of legislation in American history produced so many immediate and ominous political consequences. It split and destroyed the Whig Party. It divided the northern Democrats (many of whom were appalled at the effective repeal of the Missouri Compromise) and drove many of them from the party. Most important, it spurred the creation of a new party that was frankly sectional in composition and creed. People in both major parties who opposed Douglas's bill began to call themselves Anti-Nebraska Democrats and Anti-Nebraska Whigs. In 1854, they formed a new organization and named it the Republican Party and it instantly became a major force in American politics. In the elections of that year, the Republicans won enough seats in Congress to permit them, in combination with allies among the Know-Nothings, to organize the House of Representatives.

"Bleeding Kansas"

White settlers began moving into Kansas almost immediately after the passage of the Kansas-Nebraska Act. In the spring of 1855, elections were held for a territorial legislature. There were only about 1,500 legal voters in Kansas at that point, but thousands of Missourians, some traveling in armed bands into Kansas, swelled the vote to over 6,000. As a result, pro-slavery forces elected a majority to the legislature, which immediately legalized slavery. Outraged free-staters elected their own delegates to an independent constitutional convention, which met at Topeka and adopted a constitution excluding slavery. They then chose their own governor and legislature and petitioned Congress for statehood. President Pierce denounced them as traitors and threw the full support of the federal government behind the pro-slavery territorial legislature. A few months later, a pro-slavery federal marshal assembled a large posse, consisting mostly of Missourians, to arrest the free-state leaders, who had set up their headquarters in Lawrence. The posse sacked the town, burned the "governor's" house, and destroyed several printing presses. Retribution came quickly.

Among the most fervent abolitionists in Kansas was **John Brown**, a fiercely committed ideologue who had moved to Kansas to fight to make it a free state. After the events in Lawrence, he gathered six followers (including four of his sons) and in one night murdered five pro-slavery men near Dutch Henry's crossing on Pottawatomie Creek in Franklin County on May 24, 1856. Brown lost his oldest son, Frederick, in the raid. The "Pottawatomie Massacre," as newspapers coined the killings, fired passions on both sides even more. Brown left that fall to raise money for abolitionists only to return to lead more raids on pro-slavery forces. He left Kansas for good in early 1859.

"Bleeding Kansas" emerged as a powerful symbol of the sectional controversy. Northerners and Southerners alike feared that the political stakes over the fate of slavery had risen to the point where white men would kill each other over its outcome. Compromise seemed increasingly unlikely, but not yet impossible.

Another symbol of escalating tensions between the North and South soon appeared, this time in the U.S. Senate. In May 1856, **Charles Sumner** of Massachusetts, a strong antislavery leader, rose to give a speech titled "The Crime Against Kansas." In it he singled out Senator Andrew P. Butler of South Carolina, an outspoken defender of slavery. The South Carolinian was, Sumner hissed, the "Don Quixote" of slavery, having "chosen a

Library of Congress Prints & Photographs Division [LC-DIG-ppmsca-23763]

JOHN BROWN Even in this formal photographic portrait (taken in 1859, the last year of his life), John Brown conveys the fierce sense of righteousness that fueled his extraordinary activities in the fight against slavery.

mistress . . . who, though ugly to others, is always lovely to him, though polluted in the sight of the world, is chaste in his sight . . . the harlot slavery."

The pointedly sexual references and the general viciousness of the speech enraged Butler's nephew, Preston Brooks, a member of the House of Representatives from South Carolina. Several days after the speech, Brooks approached Sumner at his desk in the Senate chamber during a recess and beat him repeatedly on the head and shoulders with a heavy cane. Sumner, trapped in his chair, tried to escape and even tore the desk from the bolts holding it to the floor. Then he collapsed, bleeding and unconscious. So severe were his injuries that he was unable to return to the Senate for four years. Throughout the North, he became a martyr to the cause of antislavery and a symbol of the barbarism of the South. In the South, Preston Brooks became a hero. Censured by the House, he resigned his seat and returned to South Carolina—only to stand successfully for reelection.

THE FREE-SOIL IDEOLOGY

What had happened to produce such deep hostility between the two sections? In part, the tensions were reflections of each section's different economic and territorial interests. But they were also reflections of a hardening of ideas in both the North and the South.

In the North, assumptions about the proper structure of society came to center on the belief in "free soil" and "free labor." Most white Northerners came to understand that the existence of slavery was dangerous because of what it threatened to do to whites. At the heart of American democracy, they argued, was the right of all citizens to own land, to control their own labor, and to have access to opportunities for advancement. Slavery, as a system of coerced labor, made a mockery of this belief. At the same time, many—but far from all—northern whites shared a conviction that slavery was morally wrong. Northern Blacks also embraced free labor ideology, but fused to it a staunch antislavery conviction.

According to this northern vision, the South was the antithesis of democracy—a closed, static society with an entrenched aristocracy. While the North was growing and prospering, the South was stagnating and openly rejecting the values of individualism and progress. More worrisome, it conspired to extend slavery throughout the nation and destroy the openness of northern capitalism. The only solution to this "slave power conspiracy" was to fight the spread of slavery and extend the nation's democratic (i.e., free labor) ideals to all sections of the country.

This ideology, which lay at the heart of the new Republican Party, also strengthened the commitment of Republicans to the Union itself. Since the idea of continued growth and progress was central to the free-labor vision, the prospect of dismemberment of the nation was to the Republicans unthinkable.

The Pro-Slavery Argument

In the meantime, a very different ideology was emerging in the South. It was a result of many things: the Nat Turner uprising in 1831, which terrified southern whites; the expansion of the cotton economy into the Deep South, which made slavery unprecedentedly lucrative; and the growth of the Garrisonian abolitionist movement, with its strident attacks on southern society. The popularity of Harriet Beecher Stowe's *Uncle Tom's Cabin* was perhaps the most glaring evidence of the power of those attacks, but other abolitionist writings had been antagonizing white Southerners for years.

In response to these pressures, a number of white Southerners produced a new intellectual defense of slavery. Professor Thomas R. Dew of the College of William and Mary helped begin that effort in 1832. Twenty years later, apologists for slavery summarized their views in an anthology that gave their ideology its name: *The Pro-Slavery Argument*. John C. Calhoun stated its essence in 1837: Slavery was "a good—a positive good." It was good for those enslaved, because they enjoyed better conditions than industrial workers in the North; good for southern society, because it was the only way whites and Blacks could live together in peace; and good for the entire country because the southern economy, based on slavery, was the key to universal prosperity.

Above all, southern apologists argued, slavery was good because it served as the basis for the southern way of life—and this was of life was superior to any other in the United States, perhaps in the world. White Southerners looking at the North saw a spirit of greed, debauchery, and destructiveness. "The masses of the North are venal, corrupt, covetous, mean and selfish," wrote one Southerner. Others wrote with horror about the factory system and crowded, pestilential cities teeming with unruly immigrants. In contrast, the South was a stable, orderly society free from the feuds between capital and labor plaguing the North. It protected the welfare of its workers. And it allowed the aristocracy to enjoy a refined and accomplished cultural life. It was, in short, an ideal social order in which all elements of the population were secure and content. The defense of slavery rested, too, on increasingly elaborate arguments about the biological inferiority of Black Americans, who were, white Southerners claimed, inherently unfit to take care of themselves, let alone exercise the rights of citizenship.

Buchanan and Depression

In this unpromising political climate, the presidential campaign of 1856 began. Democratic Party leaders wanted a candidate who, unlike President Pierce, was not closely associated with the explosive question of "Bleeding Kansas." They chose James Buchanan of Pennsylvania, who as Minister to the United Kingdom had been safely out of the country during the recent controversies. The Republicans, participating in their first presidential contest, endorsed a Whiggish program of internal improvements, thus combining the idealism of antislavery with the economic aspirations of the North. The Republicans nominated John C. Frémont, who had made a national reputation as an explorer of the Far West and who had no political record. The Know-Nothing Party was beginning to break apart, but it nominated former president Millard Fillmore, who also received the endorsement of a small remnant of the Whig Party.

After a heated campaign, Buchanan won a narrow victory over Frémont and Fillmore. Whether because of age and physical infirmities or because of a more fundamental weakness of character, he became a painfully timid and indecisive president at a critical moment in history. In the year Buchanan took office, a financial panic struck the country, followed by a depression that lasted several years. In the North, the depression strengthened the Republican Party, because distressed manufacturers, workers, and farmers came to believe that the hard times were the result of the unsound policies of southern-controlled, proslavery Democratic administrations. They expressed their frustrations by moving into an alliance with antislavery elements and into the Republican Party.

THE *DRED SCOTT* DECISION

On March 6, 1857, the U.S. Supreme Court inserted itself into the sectional tensions with one of the most controversial and notorious decisions in its history—*Dred Scott v. Sandford*. Dred Scott was an enslaved Missourian, once owned by an army surgeon who had taken Scott with him into Illinois and Wisconsin, where slavery was forbidden. In 1846, after the surgeon died, Scott sued his slaveholder's widow for freedom on the grounds that his residence in free territory had liberated him from slavery. The claim was well grounded in Missouri law, and in 1850 the circuit court in which Scott filed the suit declared him free. By now, John Sanford, the brother of the surgeon's widow, was claiming ownership of Scott, and he appealed the circuit court ruling to the state supreme court, which reversed the earlier decision. When Scott appealed to the federal courts, Sanford's attorneys claimed that Scott had no standing to sue because he was not a citizen.

The Supreme Court (which misspelled Sanford's name in its decision) was so divided that it was unable to issue a single ruling on the case. The thrust of the various rulings, however, was a stunning defeat for the antislavery movement. Chief Justice Roger Taney, who wrote one of the majority opinions, declared that Scott could not bring a suit in the federal courts because he was not a citizen. Blacks had no claim to citizenship, Taney argued. Those held in slavery were property, and the Fifth Amendment prohibited Congress from taking property without "due process of law." Consequently, Taney concluded, Congress possessed no authority to pass a law depriving persons of their enslaved property in the territories. He therefore declared the Missouri Compromise to be unconstitutional.

The **Dred Scott decision** did not challenge the right of an individual state to prohibit slavery within its borders, but the statement that the federal government was powerless to act on the issue was a drastic and startling one. Southern whites were elated: the highest tribunal in the land had sanctioned parts of the most extreme southern argument. In the North, the ruling produced widespread dismay. The decision, the *New York Tribune* wrote, "is entitled to just so much moral weight as would be the judgment of a majority of those congregated in any Washington bar-room." Republicans threatened that when they won control of the national government, they would reverse the decision—by "packing" the Court with new members.

Deadlock over Kansas

President Buchanan timidly endorsed the *Dred Scott* decision. At the same time, he tried to resolve the controversy over Kansas by supporting its admission to the Union as a slave state. In response, the pro-slavery territorial legislature called an election for delegates to a constitutional convention. The free-state residents refused to participate, claiming that the legislature had discriminated against them in drawing district lines. As a result, the pro-slavery forces won control of the convention, which met in 1857 at Lecompton, framed a constitution legalizing slavery, and refused to give voters a chance to reject it. When an election for a new territorial legislature was called, the antislavery groups turned out in force and won a majority. The new antislavery legislature promptly submitted the Lecompton constitution to the voters, who rejected it by more than 10,000 votes.

Both sides had resorted to fraud and violence, but it was clear nevertheless that a majority of the people of Kansas opposed slavery. Buchanan, however, pressured Congress to admit Kansas under the pro-slavery Lecompton constitution. Stephen A. Douglas and other northern and western Democrats refused to support the president's proposal, which died in the House of Representatives. Finally, in April 1858, Congress approved a compromise: the Lecompton constitution would be submitted to the voters of Kansas again. If it was approved, Kansas would be admitted to the Union; if it was rejected, statehood would be postponed. Again, Kansas voters decisively rejected the Lecompton constitution. Not until the closing months of Buchanan's administration in 1861 did Kansas enter the Union—as a free state.

The Emergence of Lincoln

Given the gravity of the sectional crisis, the congressional elections of 1858 took on a special importance. Of particular note was the U.S. Senate contest in Illinois, which pitted Stephen A. Douglas, now the most prominent northern Democrat, against **Abraham Lincoln**, who was largely unknown outside Illinois.

Lincoln was a successful lawyer who had long been involved in state politics. He had served several terms in the Illinois legislature and one undistinguished term in Congress. But he was not a national figure like Douglas, and so he tried to increase his visibility by engaging Douglas in a series of debates. The Lincoln-Douglas debates attracted enormous crowds and received wide attention.

At the heart of the debates was a basic difference on the issue of slavery. Douglas appeared to have no moral position on the issue, Lincoln claimed. He stated that Douglas did not care whether slavery was "voted up, or voted down." Lincoln's opposition to slavery was more fundamental. If the nation could accept that Blacks were not entitled to basic human rights, he argued, then it could accept that other groups—immigrant laborers, for example—could be deprived of rights, too. And if slavery were to extend into the western territories, he argued, opportunities for poor white laborers to better their lots there would be lost. The nation's future, Lincoln argued (reflecting the central idea of the Republican Party), rested on the spread of free labor.

Lincoln believed slavery was morally wrong, but he was not an abolitionist. That was in part because he could not envision an easy alternative to slavery in the areas where it already existed. He also shared the prevailing view among northern whites that Blacks were not prepared to live on equal terms with whites. But even while Lincoln accepted the inferiority of Black people, he continued to believe that they were entitled to some basic rights. "I have no purpose to introduce political and social equality between the white and the black races. . . . But I hold that . . . there is no reason in the world why the negro is not entitled to all the natural rights enumerated in the Declaration of Independence, the right to life, liberty, and the pursuit of happiness. I hold that he is as much entitled to these as the white man." Lincoln and his party would "arrest the further spread" of slavery. They would not directly challenge it where it already existed but would trust that the institution would gradually die out there of its own accord.

Douglas's popular sovereignty position satisfied his followers sufficiently to produce a Democratic majority in the state legislature, which returned him to the Senate, but with little enthusiasm. Lincoln, by contrast, lost the election but emerged with a growing following both in and beyond the state. And outside Illinois, the elections went heavily against the Democrats. The party retained control of the Senate but lost its majority in the House, with the result that the congressional sessions of 1858 and 1859 were bitterly deadlocked.

John Brown's Raid

The battles in Congress, however, were almost entirely overshadowed by an event that enraged and horrified the South. In the fall of 1859, John Brown, the antislavery radical whose bloody actions in Kansas had inflamed the crisis there, staged an even more dramatic episode, this time in the South itself. With private encouragement and financial aid from some prominent abolitionists, he made elaborate plans to seize a mountain fortress in Virginia from which, he believed, he could foment a slave insurrection in the South. On October 16, he and a group of eighteen followers attacked and seized control of a U.S. arsenal in **Harpers Ferry**, Virginia. But the slave uprising Brown hoped to inspire did not occur, and he quickly found himself besieged in the arsenal by citizens, local militia companies, and, before long, U.S. troops under the command of Robert E. Lee. After ten of his men were killed, Brown surrendered. He was promptly tried in a Virginia court for treason and sentenced to death. He and six of his followers were hanged.

No other single event did more than the Harpers Ferry raid to convince white Southerners that they could not live safely in the Union. Many Southerners believed (incorrectly) that John Brown's raid had the support of the Republican Party, and it suggested to them that the North was now committed to producing a slave insurrection.

The Election of Lincoln

As the presidential election of 1860 approached, the Democratic Party was torn apart by a battle between Southerners, who demanded a strong endorsement of slavery, and westerners, who supported the idea of popular sovereignty. When the party convention met in April in Charleston, South Carolina, and endorsed popular sovereignty, delegates from eight states in the lower South walked out. The remaining delegates could not agree on a presidential candidate and finally adjourned after agreeing to meet again in Baltimore. The decimated convention at Baltimore nominated Stephen Douglas for president. In the meantime, disenchanted southern Democrats met in Richmond and nominated John C. Breckinridge of Kentucky.

The Republican leaders, in the meantime, were trying to broaden their appeal in the North. The platform endorsed such traditional Whig measures as a high tariff, internal improvements, a homestead bill, and a Pacific railroad to be built with federal financial assistance. It supported the right of each state to decide the status of slavery within its borders. But it also insisted that neither Congress nor territorial legislatures could legalize slavery in the territories. The Republican convention chose Abraham Lincoln as the party's presidential nominee. Lincoln was appealing because of his growing reputation for eloquence, because of his firm but moderate position on slavery, and because his relative

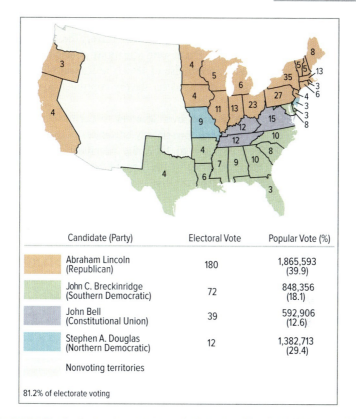

THE ELECTION OF 1860 The stark sectional divisions that helped produce the Civil War were clearly visible in the results of the 1860 presidential election. Abraham Lincoln, the antislavery Republican candidate, won virtually all the free states. Stephen Douglas, a northern Democrat with no strong position on the issue of slavery, won two of the border states, and John Bell, a supporter of both slavery and union, won others. John Breckinridge, a strong pro-slavery southern Democrat, carried the entire Deep South. Lincoln won under 40 percent of the popular vote but, because of the four-way division in the race, managed to win a clear majority of the electoral vote. • *What impact did the election of Lincoln have on the sectional crisis?*

obscurity ensured that he would have none of the drawbacks of other, more prominent (and therefore more controversial) Republicans.

In the November election, Lincoln won the presidency with a majority of the electoral votes but only about two-fifths of the fragmented popular vote. In other words, his victory was far from decisive. And his party, moreover, failed to win a majority in Congress. Even so, many white Southerners interpreted the election of Lincoln as the death knell to their power and influence in the Union. And within a few weeks of Lincoln's victory, the process of disunion began—a process that would quickly lead to a prolonged and bloody war.

CONCLUSION

In the decades following the War of 1812, a vigorous nationalism pervaded much of American life and helped smooth over the growing differences among the very distinct societies emerging in the United States. During the 1850s, however, the forces that had worked to hold the nation together in the past fell victim to new and more divisive pressures.

Driving the sectional tensions of the 1850s was a battle over national policy toward the place of slavery within the western territories. Should slavery be permitted in the new states? Who should make the decision? There were strenuous efforts to craft compromises and solutions to this dilemma: the Compromise of 1850, the Kansas-Nebraska Act of 1854, and others. But despite these efforts, positions on slavery continued to harden in both the North and the South.

Bitter battles in the territory of Kansas over whether to permit slavery there; growing agitation by abolitionists in the North and pro-slavery advocates in the South; the Supreme Court's controversial *Dred Scott* decision in 1857; the popularity of *Uncle Tom's Cabin* throughout the decade; and the emergence of the Republican Party—openly and centrally opposed to slavery: all worked to destroy the hopes for compromise and push the South toward secession.

In 1860, all pretense of common sentiment collapsed when no political party presented a presidential candidate capable of attracting national support. The Republicans nominated Abraham Lincoln of Illinois, a little-known politician recognized for his eloquent condemnations of slavery in a Senate race two years earlier. The Democratic Party split apart, with northern and southern wings each nominating different candidates. Lincoln won the election easily, but with less than 40 percent of the popular vote. And almost immediately after his victory, the states of the South began preparing to secede from the Union.

KEY TERMS/PEOPLE/PLACES/EVENTS

Abraham Lincoln 315
Alamo 297
Antonio López de Santa Anna 297
Charles Sumner 311
Compromise of 1850 309
Dred Scott decision 314
Forty-niners 305
Free-Soil Party 305

Gadsden Purchase 310
Harpers Ferry 316
James K. Polk 300
John Brown 311
Kansas-Nebraska Act 311
Manifest Destiny 296
Oregon Trail 299
popular sovereignty 304
Sam Houston 297

Stephen A. Douglas 308
Stephen F. Austin 297
Tejanos 298
Treaty of Guadalupe Hidalgo 303
Wilmot Proviso 304
Young America 309
Zachary Taylor 305

RECALL AND REFLECT

1. How were the boundary disputes over Oregon and Texas resolved? Why were the resolutions in the two cases so different?
2. How did Polk's decisions and actions as president intensify the sectional conflict?
3. What was the issue at stake in "Bleeding Kansas," and how did events in Kansas reflect the growing sectional division between the North and the South?
4. What was the *Dred Scott* decision? What was the decision's impact on the sectional crisis?
5. How did the growing sectional crisis affect the nation's major political parties?

Design element: Stars and Stripes: McGraw Hill Education.

14 | THE CIVIL WAR

THE SECESSION CRISIS
THE MOBILIZATION OF THE NORTH
THE MOBILIZATION OF THE SOUTH
STRATEGY AND DIPLOMACY
CAMPAIGNS AND BATTLES

LOOKING AHEAD

1. What was the role of slavery in causing the Civil War?
2. How did the war change American society?
3. Why did the North win the Civil War?
4. When the Civil War ended, what questions about freedom and democracy remained unanswered?

THE UNITED STATES FACED A TIME OF RECKONING by the end of 1860. As had been long predicted by Black critics of American society like David Walker and Frederick Douglass, the country was being ripped apart by the institution of slavery. Politicians now leapt at one another's throats, fighting over its future. The second party system had collapsed, no longer able to contain sectional differences and find compromises over the meaning and expansion of the peculiar institution. The federal government was no longer a remote agency. Forced to resolve the status of territories and settle the question of whether they would be open to slavery, Washington became a more direct influence in the lives of Americans. The presidential election of 1860 centered on the question of slavery and President Lincoln seemed to be in favor of at least containing it where it existed. These developments brought sectional tensions to a boil and forced Americans to confront the possibility that the nation would soon divide into two separate countries. They precipitated the bloodiest war in the nation's history whose outcome—always in doubt until the end—remade the meaning of American democracy.

TIME LINE

1861
Confederate States of America formed
Davis president of Confederacy
Conflict at Fort Sumter
First Battle of Bull Run

1862
Battles of Shiloh, Antietam, Second Bull Run
Confederacy enacts military draft

1863
Emancipation Proclamation
Battle of Gettysburg
Vicksburg surrenders
Union enacts military draft
New York City antidraft riots

1864
Battle of the Wilderness
Lincoln reelected

1865
Lee surrenders to Grant
13th Amendment

THE SECESSION CRISIS

Almost as soon as news of Abraham Lincoln's election victory reached the South, militant leaders began to demand an end to the Union.

THE WITHDRAWAL OF THE SOUTH

South Carolina, long the hotbed of Southern separatism, seceded first, on December 20, 1860. By the time Lincoln took office, six other Southern states—Mississippi (January 9, 1861), Florida (January 10), Alabama (January 11), Georgia (January 19), Louisiana (January 26), and Texas (February 1)—had withdrawn from the Union. In February 1861, representatives of the seven seceded states met at Montgomery, Alabama, and formed a new nation—the **Confederate States of America**. Two months earlier, President James Buchanan had told Congress that no state had the right to secede from the Union but that the federal government also had no authority to stop a state from doing so.

Why did the South secede? The degree to which slavery and white supremacy were the primary causes of the Civil War has historically been a topic of passionate debate among Americans. Sadly, for far too long many have avoided the painful truth that the desire to protect slavery was at the heart of the South's secession. Worse still, they accepted histories that erased this fact. That is simply wrong. In 1861, white Southerners chose to die rather than accept a world where slavery could be outlawed. Their actions reflected an unwavering belief in racial hierarchy: they claimed that, as white Americans, they were intellectually, biologically, and morally superior to Black Americans. To be sure, many white Northerners shared all or part of this racism but they refused to allow the country to become permanently divided because of the South's insistence on protecting slavery. (See "Consider the Source: Ordinances of Secession.")

The Failure of Compromise

The seceding states immediately confiscated poorly protected federal property within their boundaries. But they did not at first have sufficient military power to seize two fortified offshore military installations: **Fort Sumter**, in the harbor of Charleston, South Carolina, garrisoned by a small force under Major Robert Anderson; and Fort Pickens, in Pensacola, Florida. President Buchanan refused to yield Fort Sumter when South Carolina demanded it. Instead, in January 1861, he ordered an unarmed merchant ship to proceed to Fort Sumter with additional troops and supplies. Confederate guns turned it back. Still, few were ready to concede that a war had begun. Instead, politicians in Washington scrambled to knit together yet another compromise to quiet tensions.

Gradually, the compromise efforts came to focus on a proposal advanced by John J. Crittenden of Kentucky. Known as the Crittenden Compromise, it proposed reestablishing the Missouri Compromise line and extending it westward to the Pacific. Slavery would be prohibited north of the line and permitted south of it. Southerners in the Senate seemed willing to accept the plan. But the compromise required the Republicans to abandon their most fundamental position—that slavery would not be allowed to expand—and they would have none of it. Whether the failure to compromise and find common ground between Northern and Southern politicians triggered the Civil War has been a topic of debate among historians for generations.

When Abraham Lincoln arrived in Washington, talk of secession and possible war filled the air. In his inaugural address, Lincoln insisted that acts of force or violence to support secession were insurrectionary and that the government would "hold, occupy, and possess" all federal property in the seceded states—a not-so-vague reference to Fort Sumter. Fort Sumter was running short of supplies, forcing Lincoln to act. He decided to send in a relief expedition, but told South Carolina authorities that he would not authorize troops or munitions to accompany it—unless the supply ships were to be met with resistance. The new Confederate government did not waste any time in responding. It ordered General P. G. T. Beauregard, commander of Confederate forces at Charleston, to take the fort. When Anderson refused to give up, Beauregard bombarded it for two days. On April 14, 1861, Anderson surrendered. The fighting had begun.

Almost immediately, four more slave states seceded from the Union and joined the Confederacy: Virginia (April 17), Arkansas (May 6), Tennessee (May 7), and North Carolina (May 20). The four remaining slave states, Maryland, Delaware, Kentucky, and Missouri—under heavy political pressure from Washington—remained in the Union.

The Opposing Sides

The North had many important material advantages, most notably an advanced industrial system able by 1862 to manufacture almost all of the North's war materials. The South had its own network of industries—textiles and iron, for example—but this network wasn't as widespread. In addition, the North had a superior transportation system, with more and better railroads than did the South. During the war, in fact, the Confederate railroad system steadily deteriorated and by early 1864 had almost collapsed.

But the South also had advantages. Its armies were, for the most part, fighting a defensive war on familiar land with local support. The Northern armies, on the other hand, were fighting mostly within the South amid hostile local populations and had to maintain long lines of communication and supply. The commitment of the white population of the South to the war was, with limited exceptions, clear and firm throughout much of the early years

CONSIDER THE SOURCE

ORDINANCES OF SECESSION (1860/1861)

As southern states declared their independence from the Union, they defended their actions in "Ordinances of Succession." The need to protect slavery dominated these public statements.

South Carolina (seceded December 20, 1860; published ordinance December 24, 1860)

The ends for which the Constitution was framed are declared by itself to be "to form a more perfect union, establish justice, insure domestic tranquility, provide for the common defence, promote the general welfare, and secure the blessings of liberty to ourselves and our posterity."

These ends it endeavored to accomplish by a Federal Government, in which each State was recognized as an equal, and had separate control over its own institutions. The right of property in slaves was recognized by giving to free persons distinct political rights, by giving them the right to represent, and burdening them with direct taxes for three-fifths of their slaves; by authorizing the importation of slaves for twenty years; and by stipulating for the rendition of fugitives from labor.

Mississippi (seceded January 9, 1861; published ordinance January 26, 1861)

Our position is thoroughly identified with the institution of slavery—the greatest material interest of the world. Its labor supplies the product which constitutes by far the largest and most important portions of commerce of the earth. These products are peculiar to the climate verging on the tropical regions, and by an imperious law of nature, none but the Black race can bear exposure to the tropical sun. These products have become necessities of the world, and a blow at slavery is a blow at commerce and civilization. That blow has been long aimed at the institution, and was at the point of reaching its consummation. There was no choice left us but submission to the mandates of abolition, or a dissolution of the Union, whose principles had been subverted to work out our ruin.

Texas (seceded February 1, 1861; published ordinance February 2, 1861)

Texas abandoned her separate national existence and consented to become one of the Confederated Union to promote her welfare, insure domestic tranquility and secure more substantially the blessings of peace and liberty to her people. She was received into the confederacy with her own constitution, under the guarantee of the federal constitution and the compact of annexation, that she should enjoy these blessings. She was received as a commonwealth holding, maintaining and protecting the institution known as negro slavery—the servitude of the African to the white race within her limits—a relation that had existed from the first settlement of her wilderness by the white race, and which her people intended should exist in all future time.

That in this free government **all white men are and of right ought to be entitled to equal civil and political rights** [emphasis in original]; that the servitude of the African race, as existing in these States, is mutually beneficial to both bond and free, and is abundantly authorized and justified by the experience of mankind, and the revealed will of the Almighty Creator, as recognized by all Christian nations; while the destruction of the existing relations between the two races, as advocated by our sectional enemies, would bring inevitable calamities upon both and desolation upon the fifteen slave-holding states.

UNDERSTAND, ANALYZE, & EVALUATE

1. How do southern states defend slavery in these ordinances?
2. In what ways is slavery described as more than simply an economic institution? How does it lay at the foundation of southern society?
3. Why have many white Americans ignored the central role of slavery in the the Civil War? How has that avoidance affected the lives of Black Americans?

Source: *The Avalon Project: Documents in Law, History, and Diplomacy,* Yale Law School, Yale University, https://avalon.law.yale.edu/subject_menus/major.asp

---

of fighting. In the North, opinion was more divided and support for the war remained shaky until the end. A major Southern victory at any one of several crucial moments might have proved decisive in breaking the North's will to continue the struggle. Finally, the dependence of the English and French textile industries on Southern cotton inclined many leaders in those countries to initially favor the Confederacy; and Southerners hoped, with some reason, that one or both might intervene on their behalf.

GOING TO WAR

Northerners generally viewed the war as a fight to restore the Union and preserve the American democratic experiment. They questioned how the nation might endure if parts of it could secede at every undesirable election result. Indeed, if the Constitution and

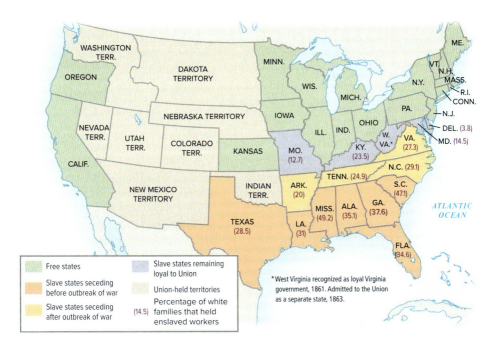

THE ORDER OF SECESSION Slave states did not secede all at once, nor did every slave state leave the Union. States where the highest percentage of white families held enslaved laborers led the charge out of the Union. As seen here, the South was not uniformly enthusiastic about leaving the Union, and not all states that permitted slavery supported the Confederacy. • *Why did slave states secede at different times? What does that tell us about the path to disunion?*

federal government could not be upheld, then what level of legal society could ever stand? They also believed the nation to be the beacon of liberty to the old autocracies of Europe; if it collapsed so did the "last, best hope" for the world. Most white Northerners understood that divisions over slavery had caused the Civil War, but they did not fight, at first, to abolish it. Later, when it became clear that abolishing slavery would help end the rebellion, emancipation joined reunion as a public war aim. In contrast, Black Northerners immediately viewed the war as a struggle for emancipation.

Southerners convinced themselves that they were fighting to preserve what they called sacred American values. They claimed the right of states to secede because they found the federal government oppressive, just as the colonies had claimed the right to declare independence from an oppressive Britain in the Revolutionary War. This was despite the fact that such action was clearly treasonous according to the U.S. Constitution. They also presented their defense of slavery as a broader defense of individual property rights, shamelessly defining Black human beings as "property." At heart Confederates were fighting to keep slavery viable. Confederate vice president Alexander Stephens confessed as much when he publicly identified slavery as the "cornerstone" of his new nation.

Two portions of the American population that never wavered in their understanding of the war were enslaved people and free Blacks. They immediately recognized that the conflict was rooted in slavery and that its outcome would directly impact them and the future of democracy for all people. As Frederick Douglass, himself formerly enslaved, put it, the "civil war was not a mere strife for territory and dominion, but a contest of civilization against barbarism."

THE MOBILIZATION OF THE NORTH

In the North, the war produced considerable discord, frustration, and suffering. But it also produced prosperity and economic growth. With the South now gone from Congress, the Republican Party enjoyed almost unchallenged supremacy for the first few years of conflict. During the war, it aggressively promoted economic development.

Economic Nationalism

Two 1862 acts assisted the rapid development of the West. The **Homestead Act** permitted any citizen or prospective citizen to purchase 160 acres of public land for a small fee after living on it for five years. The Morrill Land Grant College Act transferred substantial public acreage to the state governments, which could now sell the land and use the proceeds to finance public education. This act led to the creation of many new state colleges and universities, the so-called land-grant institutions. Congress also passed a series of tariff bills that by the end of the war had raised duties to the highest level in the nation's history—a great boon to domestic industries eager for protection from foreign competition, but a hardship for many farmers and other consumers. Without the seceding states to block their legislation, Congress bent to the political will of the northern and western factions.

Congress also moved to spur completion of a transcontinental railroad. It created two new federally chartered corporations: the Union Pacific Railroad Company, which was to build westward from Omaha, and the Central Pacific, which was to build eastward from California. The two projects met in the middle and completed the link in 1869 at Promontory

Point, Utah. Equally important, Congress mandated a single standard gauge for all railroad track, thus creating a unified network of rail transport.

The National Bank Acts of 1863–1864 created a new national banking system. Existing or newly formed banks could join the system if they had enough capital and were willing to invest one-third of it in government securities. In return, they could issue U.S. Treasury notes as currency. This eliminated much (although not all) of the chaos and uncertainty surrounding the nation's currency.

More difficult was financing the war itself. The government took three approaches: levying taxes, issuing paper currency, and borrowing. Congress levied new taxes on almost all goods and services and in 1861 levied an income tax for the first time. But taxation raised only a small proportion of the money needed to fight the war, and strong popular resistance prevented the government from raising the rates.

At least equally controversial was the printing of paper currency, or "**greenbacks**." The new currency was backed not by gold or silver but (as today) simply by the good faith and credit of the government. The value of the greenbacks fluctuated according to the fortunes of the Northern armies. Early in 1864, with the war effort bogged down, a greenback dollar was worth only 39 percent of a gold dollar. But at the close of the war, with confidence high, it was worth 67 percent of a gold dollar.

By far the largest source of financing for the war was loans. The Treasury persuaded ordinary citizens to buy over $400 million worth of bonds. Public bond purchases constituted only a small part of the government's borrowing, which in the end totaled $2.6 billion, most of it from banks and large financial interests.

Raising the Union Armies

At the beginning of 1861, the regular army of the United States consisted of only 16,000 troops, many of them stationed in the West. The Union, like the Confederacy, had to raise its army mostly from scratch. After the surrender of Fort Sumter, on April 15, 1861, Lincoln directed individual state militias to help him create a new federal force of 75,000 soldiers by providing men for an initial term of three months. This proved to be far from enough. When Congress convened in July 1861, it authorized the enlistment of 500,000 volunteers for three-year terms. Lincoln issued another call for volunteers from the states on July 1, 1863, for an additional 300,000 men.

This voluntary system of recruitment produced adequate forces only briefly, during the first flush of enthusiasm for the war. By March 1863, Congress was forced to pass a national draft law. Virtually all young adult males were eligible to be drafted, although a man could escape service by hiring someone to go in his place or by paying the government a fee of $300.

To many who were accustomed to a remote and inactive national government, **conscription** was strange and threatening. Objection to the draft law was widespread, particularly among laborers, immigrants, and Democrats opposed to the war (known as "Peace Democrats" or "Copperheads" by their opponents). Occasionally resistance to the draft erupted into violence. One notable disturbance took place in New York City in July 1863, where demonstrators rioted for four days after the first names were selected for conscription. Irish workers, already angry because Black strikebreakers had been used against them in a recent longshoremen's strike, were at the center of the violence. They blamed Black Americans for the war, which they thought was being fought largely for their benefit. The rioters lynched several Black Americans and burned down Black homes, businesses, and even an orphanage, leaving more than 100 dead. Only the arrival of federal troops directly from the Battle of Gettysburg halted the violence.

CONSIDER THE SOURCE

LETTER FROM A REFUGEE (1862)

John Boston, enslaved in Maryland, escaped to a New York regiment of Union soldiers camped in Union Hill, Virginia. He wrote this letter to his wife, Elizabeth, whom he left behind. It is unclear if she ever received the letter.

My Dear Wife it is with grate joy I take this time to let you know Whare I am
   I am now in Safety in the 14th Regiment of Brooklyn . . . this Day i can Adress you thank god as a free man I had a little truble in giting away But as the lord led the Children of Isrel to the land of Canon So he led me to a land Whare fredom Will rain in spite Of earth and hell Dear you must make your Self content i am free from al the Slavers Lash and as you have chose the wise plan of Serving the lord I hope you will pray Much and i Will try by the help of god To Serv him With all my hart I am With a very nice man and have All that hart Can Wish But My Dear I Cant express my grate desire that i Have to See you i trust the time Will Come When We Shal meet again And if We dont met on earth We Will Meet in heven Whare Jesas ranes . . .

UNDERSTAND, ANALYZE, & EVALUATE

1. What does this letter tell us about how enslaved people experienced and understood slavery?
2. Now that John is no longer enslaved, what are his biggest hopes?
3. How might this letter serve the antislavery cause?
4. What does this letter tell us about the religious faith of enslaved people?

Source: John Boston to his wife Elizabeth, January 12, 1862, enclosed in a letter from Major General George B. McClellan to the Honorable Edwin Stanton; Letters Received, 1805–1889; Records of the Adjutant General's Office, 1762–1984, Record Group 94; National Archives Identifier 783102, https://www.archives.gov/exhibits/eyewitness/html.php?section=9

Wartime Politics

When Abraham Lincoln arrived in Washington from Illinois, with little experience in national government, many powerful Republicans assumed they might be able to control him with little effort. But Lincoln was not easily influenced. The new president understood his own (and his party's) weaknesses and he assembled a cabinet representing every faction of the Republican Party and every segment of Northern opinion. He moved boldly to use the war powers of the presidency, with the understanding that the Founding Fathers intended the government to be able to defend itself from treason and internal enemies. And so he initially sent troops into battle without asking Congress for a declaration of war, arguing that the conflict was a domestic insurrection, and unilaterally proclaimed a naval blockade of the South.
   Among Lincoln's greatest political challenge was to encourage and sustain popular support for the war in the North. His administration pioneered new tools of persuasion. They mass-produced pro-war pamphlets, posters, speeches, and songs. They enlisted sympathetic religious leaders to assure followers that God meant for the Union to be preserved. And they mobilized a significant corps of photographers—organized by the renowned Mathew Brady, one of the first important photographers in American history—to take pictures of the war. The photographs that resulted from this effort—new to warfare—were among the

grimmest ever made to that point, many of them displaying the vast numbers of dead on the Civil War battlefields. For some Northerners, the images of death contributed to a revulsion for the war. But for most they were evidence of the sacrifice that had been made for the preservation of the Union and thus spurred the nation on to victory. (Southerners used similar propaganda in the Confederacy, although less effectively.)

By the time of the 1864 election, however, the North was in political turmoil. The Republicans had suffered heavy losses in the midterm elections of 1862 and in response party leaders tried to create a broad coalition of all the groups that supported the war. They called the new organization the Union Party, but it was, in reality, little more than the Republican Party plus a small faction of pro-war Democrats. They nominated Lincoln for a second term and Andrew Johnson of Tennessee, a "War Democrat" who had opposed his state's decision to secede, for the vice presidency.

The Democrats nominated **George B. McClellan**, the former Union general. He presided over a party whose platform included a peace plank that called the war a failure and demanded its immediate end. It was co-authored by Clement Vallandigham, a lawyer from Ohio who served three terms (1856-1862) in the House of Representatives, where he was a leader of the Copperheads or "Peace Demoncrats." In 1863 he was convicted by an Army court martial for his strong opposition to the war and fled to Canada before eventually returning to Ohio. McClellan repudiated the peace demand, but the Democrats were clearly the peace party and tried to profit from growing war weariness in the North. For a while, the tactic worked and Lincoln's prospects for reelection looked doubtful.

At this crucial moment, however, several Northern military victories, particularly the capture of Atlanta, Georgia, early in September 1864, rejuvenated Northern morale and boosted Republican prospects. With the overwhelming support of Union troops, Lincoln won reelection comfortably, with 212 electoral votes to McClellan's 21.

THE POLITICS OF EMANCIPATION

Despite their shared support of the war and their general agreement on most economic matters, the Republicans disagreed sharply with one another on the issue of slavery. "Radical Republicans"—led in Congress by such men as Representative Thaddeus Stevens of Pennsylvania and Senators Charles Sumner of Massachusetts and Benjamin Wade of Ohio— wanted to use the war to abolish slavery immediately and completely. "Conservative Republicans" favored a more cautious approach, in part to placate the slave states that remained, precariously, within the Union.

Nevertheless, momentum began to gather behind emancipation early in the war. In 1861, Congress passed the **Confiscation Acts**, which declared that all enslaved people used for "insurrectionary" purposes (that is, in support of the Confederate military effort) would be considered freed. Subsequent laws in the spring of 1862 abolished slavery in the District of Columbia and the western territories and provided for the compensation of owners. In July 1862, Radicals pushed through Congress the second Confiscation Act, which declared free those held in slavery by people supporting the insurrection and authorized the president to employ Black Americans as soldiers.

While northern politicians debated emancipation, enslaved people didn't wait for any formal announcement about the political fate of slavery. Starting slowly at the conflict's outset but growing as it went on, they fled slavery and sought refuge across Union lines. Thousands eventually made their way into army encampments, often entire families, finding freedom for the first time. (See "Consider the Source: Letter from a Refugee.")

CONSIDER THE SOURCE

THE EMANCIPATION PROCLAMATION (1863)

President Lincoln signed the Emancipation Proclamation on January 1, 1863, as a wartime effort to free all enslaved people living in the Confederate States of America. Here is an excerpt.

"That on the first day of January, in the year of our Lord one thousand eight hundred and sixty-three, all persons held as slaves within any State or designated part of a State, the people whereof shall then be in rebellion against the United States, shall be then, thenceforward, and forever free; and the Executive Government of the United States, including the military and naval authority thereof, will recognize and maintain the freedom of such persons, and will do no act or acts to repress such persons, or any of them, in any efforts they may make for their actual freedom.

Now, therefore I, Abraham Lincoln, President of the United States, by virtue of the power in me vested as Commander-in-Chief, of the Army and Navy of the United States in time of actual armed rebellion against the authority and government of the United States, and as a fit and necessary war measure for suppressing said rebellion, do, on this first day of January, in the year of our Lord one thousand eight hundred and sixty-three, and in accordance with my purpose so to do publicly proclaimed for the full period of one hundred days, from the day first above mentioned, order and designate as the States and parts of States wherein the people thereof respectively, are this day in rebellion against the United States, the following, to wit:

Arkansas, Texas, Louisiana, (except the Parishes of St. Bernard, Plaquemines, Jefferson, St. John, St. Charles, St. James Ascension, Assumption, Terrebonne, Lafourche, St. Mary, St. Martin, and Orleans, including the City of New Orleans) Mississippi, Alabama, Florida, Georgia, South Carolina, North Carolina, and Virginia, (except the forty-eight counties designated as West Virginia, and also the counties of Berkley, Accomac, Northampton, Elizabeth City, York, Princess Ann, and Norfolk, including the cities of Norfolk and Portsmouth[)], and which excepted parts, are for the present, left precisely as if this proclamation were not issued.

And by virtue of the power, and for the purpose aforesaid, I do order and declare that all persons held as slaves within said designated States, and parts of States, are, and henceforward shall be free; and that the Executive government of the United States, including the military and naval authorities thereof, will recognize and maintain the freedom of said persons.

And I hereby enjoin upon the people so declared to be free to abstain from all violence, unless in necessary self-defence; and I recommend to them that, in all cases when allowed, they labor faithfully for reasonable wages.

And I further declare and make known, that such persons of suitable condition, will be received into the armed service of the United States to garrison forts, positions, stations, and other places, and to man vessels of all sorts in said service.

And upon this act, sincerely believed to be an act of justice, warranted by the Constitution, upon military necessity, I invoke the considerate judgment of mankind, and the gracious favor of Almighty God.

UNDERSTAND, ANALYZE, & EVALUATE

1. Why did Lincoln sign the Emancipation in 1863 and not at the start of the war?
2. Why aren't *all* enslaved people freed by the Proclamation?
3. Do you find Lincoln to be motivated by moral grounds or military ambition in signing the Emancipation Proclamation?
4. What does the Emancipation Proclamation tell us about Lincoln's political skill and leadership?

Source: *The National Archive, The Emancipation Proclamation, Transcript of the Emancipation Proclamation,* https://www.archives.gov/exhibits/featured-documents/emancipation-proclamation/transcript.html.

As the war progressed, many in the North slowly accepted emancipation as a central war aim; nothing less, they believed, would justify the enormous sacrifices the struggle had required. As a result, the Radicals gained increasing influence within the Republican Party—a development that did not go unnoticed by the president, who decided to seize the leadership of the rising antislavery sentiment himself.

On September 22, 1862, after the Union victory at the Battle of Antietam, the president announced his intention to use his war powers to issue an executive order freeing all enslaved people in the Confederacy. On January 1, 1863, he formally signed the **Emancipation Proclamation**, which declared forever free enslaved people living inside the Confederacy. The proclamation did not apply to the Union slave states; nor did it affect those parts of the Confederacy already under Union control (Tennessee, western Virginia, and southern Louisiana). It applied, in short, only to enslaved people over whom the Union had no control. Still, the document was of great importance. It clearly established that the war was being fought to preserve the Union *and* eliminate slavery. Eventually, as federal armies occupied much of the South, the proclamation became a practical reality and led directly to the freeing of thousands of enslaved people. (See "Consider the Source: The Emancipation Proclamation.")

Almost everywhere federal soldiers went in the Confederacy, enslaved people flocked to them, seeking freedom. Some of them joined the Union army, others simply stayed with the troops until they could find their way to free states. When the Union captured New Orleans and much of southern Louisiana, enslaved people refused to work for their former slaveholders, even though the Union occupiers had not made any provisions for liberating them.

By the end of the war, two Union slave states (Maryland and Missouri) and three Confederate states occupied by Union forces (Tennessee, Arkansas, and Louisiana) had abolished slavery. In 1865, Congress finally approved and the states ratified the **Thirteenth Amendment**, which abolished slavery in all parts of the United States. After more than two centuries, legalized slavery finally ceased to exist in the United States.

Black Americans and the Union Cause

By the war's end, about 186,000 emancipated Blacks served as soldiers, sailors, and laborers for the Union forces. Yet in the first months of the war, Black Americans were largely excluded from the military. A few Black regiments did eventually take shape in some of the Union-occupied areas of the Confederacy. But once Lincoln issued the Emancipation Proclamation, Black enlistment boomed and the Union military began actively to recruit Black soldiers and sailors in both the North and, where possible, the South.

Some of these men were organized into fighting units, of which the best known was probably the Fifty-fourth Massachusetts Infantry, which (like most Black regiments) had a white commander: Robert Gould Shaw, a member of an aristocratic Boston family. It was an elite unit whose ranks included two of Frederick Douglass' sons, Charles and Lewis. In the Second Battle of Fort Wagner, located on Morris Island in the harbor of Charleston, the 54th suffered terrible losses but displayed bravery that dispelled any racist notions about Black soldiers' inability or unwillingness to fight and die for the Union. Over 250 of the 600 members of the 54th were either wounded, captured, or killed. Lewis Douglass was injured and recovered. The dead included Robert Shaw.

Most Black soldiers were assigned menial tasks behind the lines, such as digging trenches and transporting water. Even though many fewer Blacks than whites died in combat, the mortality rate for African Americans was actually higher than for white soldiers because so many Black soldiers died of diseases contracted while working long, arduous hours in unsanitary conditions. Conditions for Blacks and whites were unequal in other ways as well. Until 1869, Black soldiers were paid a third less than were white soldiers. Black soldiers captured by the Confederates were sent back to their slaveholders (if they were formally enslaved) or executed. In 1864, Confederate soldiers led by General Nathan Bedford Forrest overran Union troops in the Battle of Fort Pillow. Forrest ordered the killing of over 260 Black Union soldiers and their white officers after capturing them. The mass execution subsequently became known as "The Fort Pillow Massacre."

However dangerous the conditions they faced or menial the tasks they were given, most Black soldiers felt enormous pride in their service—pride they retained throughout their lives, and that was often held by their descendants as well. Many moved from the army into politics and other forms of leadership (in both the North and, after the war, the Reconstruction South).

WOMEN, NURSING, AND THE WAR

Thrust into new and often unfamiliar roles by the war, women took over positions vacated by men, such as teachers, salesclerks, office workers, mill and factory hands, and above all,

Library of Congress Prints & Photographs Division [LC-DIG-ppmsca-34584]

AFRICAN AMERICAN TROOPS   Although most of the Black soldiers who enlisted in the Union army performed noncombat jobs behind the lines, some Black combat regiments—members of one of which are pictured here—fought with great success and valor in critical battles. • *What might this photo suggest about the links between citizenship and military service?*

nurses. Nursing had only begun to be recognized as vital to health care during the Crimean War in 1854, through the work and advocacy of Florence Nightingale in Britain. Dorothea Dix, who had led American prison and mental health reform efforts in the antebellum era, was appointed Superintendent of Female Nurses of the Union Army two months after the fighting began. She was given authority to create a nursing corps (though neither she nor her appointees were given any formal rank). Dix specified that nurses be between 35 and 50 years old, have "matronly experience," be of good character and health, and be willing to accept payment of 40 cents per day and commit to a minimum initial contract of three months. Between 5,000 and 10,000 women served as nurses, the most famous being Clara Burton and Louisa May Alcott. By contrast, there was no Confederate nursing corps, though four states had their own relief organization in which women served. More commonly, Confederate women served on a personal basis, attending to the wounded near them.

Female nurses encountered considerable resistance from male doctors and military officers, many of whom thought it inappropriate for women to take care of male strangers. The **United States Sanitary Commission**, an organization of civilian volunteers, mobilized large numbers of female nurses to serve in field hospitals. It countered such arguments by presenting nursing in domestic terms: as a profession that made use of the same maternal, nurturing roles women played as wives and mothers.

Most women, of course, did not serve as nurses or care for the injured on the battlefield. At home, many supported the war effort by writing letters to soldiers, keeping their spirits up by telling them that everything was good at home and they should stay in the army and finish the fight. They raised money, collected supplies, and supported families who lost husbands or sons.

Some women came to see the war as an opportunity to gain support for their own goals. Elizabeth Cady Stanton and Susan B. Anthony together founded the Women's National Loyal League in 1863 and worked simultaneously for the abolition of slavery and the awarding of suffrage to women. It was an important precursor to more formal efforts by women to win the vote in the early twentieth century.

THE MOBILIZATION OF THE SOUTH

Early in February 1861, representatives of the seven seceding states met at Montgomery, Alabama, to create a new Southern nation. When Virginia seceded several months later, the leaders of the Confederacy moved to Richmond. There were, of course, important differences between the new Confederate nation and the nation it had left. But there were also significant similarities.

THE CONFEDERATE GOVERNMENT

The Confederate constitution was almost identical to the Constitution of the United States, with several significant exceptions. It explicitly acknowledged the sovereignty of the individual states (although not the right of secession). And it specifically sanctioned slavery and made its abolition (even by one of the states) practically impossible.

The constitutional convention at Montgomery named a provisional president and vice president: **Jefferson Davis** of Mississippi and **Alexander H. Stephens** of Georgia, who were later elected by the general public, without opposition, for six-year terms. The Confederate government, like the Union government, was dominated throughout the war by men of the political center. Davis was a reasonably able administrator and encountered little interference from the generally tame members of his unstable cabinet. But he rarely provided genuinely national leadership, spending too much time on routine items and micro-managing the war effort.

Although there were no formal political parties in the Confederacy, its politics were fiercely divided nevertheless. There were pockets of white Southern opposition to secession and war altogether, but they were few. Most openly supported the war, but as in the North, were critical of the government and the military, particularly as the tide of battle turned against the South.

States' rights had become so important to many white Southerners that they questioned efforts to exert national authority, even those necessary to win the war. States' rights enthusiasts obstructed conscription and restricted Davis's ability to impose martial law and suspend habeas corpus. Governors Joseph Brown of Georgia and Zebulon M. Vance of North Carolina even tried at times to keep their own troops apart from the Confederate forces.

But the national government was not impotent. States' rights sentiment was a significant handicap, but the South nevertheless took important steps in the direction of centralization. National Confederate leadership experimented, successfully for a time, with a "food draft," which permitted soldiers to feed themselves by seizing crops and livestock from farms in the area in which they marched or fought. The Confederacy also seized control of railroads and shipping and impressed enslaved workers.

Money and Manpower

Financing the Confederate war effort was a monumental task. The Confederate congress tried at first to requisition funds from the individual states, but the states were as reluctant to tax their citizens as the congress was. In 1863, the congress enacted an income tax. But taxation produced only about 1 percent of the government's total income. Borrowing was not much more successful. The Confederate government issued bonds in such vast amounts that the public lost faith in them, and efforts to borrow money from Europe, using cotton as collateral, fared no better.

As a result, the Confederacy had to pay for the war through the least stable, most destructive form of financing: paper currency, which it began issuing in 1861. By 1864, the Confederacy had issued the staggering total of $1.5 billion in paper money. The result was a disastrous inflation—a 9,000 percent increase in prices in the course of the war (in contrast to 80 percent in the North).

Like the United States, the Confederacy first raised armies by calling for volunteers. And, as in the North, by the end of 1861 voluntary enlistments were declining. In April 1862, the Confederate Congress enacted the Conscription Act, which subjected all white males between the ages of eighteen and thirty-five to military service for three years.

The Conscription Act was not without its critics. It actually increased dissatisfaction and dissent in the Appalachian regions of the Confederacy (eastern Tennessee, western North Carolina, and northern Alabama) not strongly aligned with slaveholders. One provision of the Act particularly angered poor white men and their families. Southern draftees, like men in the North, could avoid service if they furnished a substitute, an expensive proposition that typically could only be afforded by the very wealthy. The provision aroused such opposition from poorer whites that it was repealed in 1863.

Even so, conscription worked for a time, in part because enthusiasm for the war was intense and widespread among white men in most of the South. At the end of 1862, about

500,000 soldiers were in the Confederate army, not including the many enslaved men and women forced by the military to cook, launder, and build housing and fortifications.

Still, by 1864 the Confederate government faced a critical manpower shortage. The South was suffering from intense war weariness, and many Southerners had concluded that defeat was inevitable. Nothing could attract or retain an adequate army any longer. In a frantic final attempt to raise men, the congress authorized the conscription of 300,000 enslaved men, but the war ended before the government could attempt this experiment.

Economic and Social Effects of the War

The war cut off Southern planters and producers from Northern markets, and a Union blockade of Confederate ports eventually reduced the sale of cotton outside of the region and overseas by nearly 95 percent. As a result, the South redoubled efforts to diversify and expand its crops, bolster its textile industry, and make armaments and wartime provisions. Nothing came close to replacing cotton, however. Indeed, in the South, production of all goods during the war declined by more than a third (while in the North, production grew). Above all, the fighting itself wreaked havoc on the South. As the war continued, the shortages, the inflation, and the carnage slowly destabilized Southern society. Resistance to conscription, food impressment, and taxation rose throughout the Confederacy, as did hoarding and black-market commerce. Even before emancipation, the war had far-reaching effects on the lives of enslaved people. Confederate leaders enforced slave codes with particular severity, but many enslaved people—especially those near the front—escaped those who enslaved them and crossed Union lines.

The war turned the meaning of work upside down for white Southern women. With fathers, brothers, and husbands all gone to war—some to never return—many assumed greater responsibility for raising crops, caring for animals, hunting, and fishing. Others turned to mills and manufacturing jobs to earn money for food. They also found new responsibilities supporting the Confederate cause by helping relief associations, writing letters, sewing uniforms and blankets, and volunteering in hospitals.

STRATEGY AND DIPLOMACY

The military initiative in the Civil War lay mainly with the North, since it needed to destroy the Confederacy as quickly as possible. Southerners needed only to survive; however, survival required diplomatic recognition from other nations, especially Great Britain, and the financial and military aid that would accompany it. Lincoln and his diplomatic corps worked tirelessly to keep foreign powers and their resources out of the Confederacy while the Union navy blockaded Southern ports.

The Commanders

The most important Union military leader was Abraham Lincoln. He ultimately succeeded as commander in chief because he recognized the North's material advantages, and he realized that the proper objective of his armies was to destroy the Confederate armies' ability to fight. Despite his missteps and inexperience, the North was fortunate to have Lincoln, but the president struggled to find a general as well suited for his task as Lincoln was for his.

From 1861 to 1864, Lincoln tried repeatedly to find a chief of staff capable of orchestrating the Union war effort. He turned first to General Winfield Scott, the ailing seventy-four-year-old hero of the Mexican War who had already contributed strategic advice to the president, but Scott was no longer physically capable of leading an army. Lincoln then appointed the young George B. McClellan, the commander of the Union forces in the East, the Army of the Potomac. But McClellan was overly cautious, extremely worried about being outnumbered in battle and unable to take advantage where he could attack and win. Lincoln returned McClellan to his previous command in March 1862. For most of the rest of the year, Lincoln had no chief of staff at all. When he eventually appointed General Henry W. Halleck to the post, he found him ineffectual as well. Not until March 1864 did Lincoln finally find a general he trusted to command the war effort, **Ulysses S. Grant**. Grant shared Lincoln's belief in unremitting combat and in targeting not only enemy armies but their resources as well.

Lincoln's handling of the war effort drew constant scrutiny from the Committee on the Conduct of the War, a joint investigative committee of the two houses of Congress. Established in December 1861 and chaired by Senator Benjamin E. Wade of Ohio, the committee complained constantly of timid Northern generals, which Radicals on the committee falsely attributed to a secret sympathy among the officers for slavery. The committee's efforts often seriously interfered with the conduct of the war.

Southern military leadership centered on President Davis, a trained soldier who nonetheless failed to create an effective central command system. Early in 1862, Davis named General **Robert E. Lee** as his principal military adviser. But in fact, Davis had no intention of sharing control of strategy with anyone. After a few months, Lee left Richmond to command forces in the field, and for the next two years, Davis planned strategy alone. In February 1864, he named General Braxton Bragg as a military adviser, but Bragg never provided much more than technical advice.

Men of markedly similar backgrounds controlled the war in both the North and the South. The U.S. Military Academy at West Point produced scores of influential leaders on both sides. Union Generals George McClellan, Ulysses S. Grant, Willian Tecumseh Sherman, George Thomas, George Meade, Winfield Hancock, Philip Sheridan, and George Custer all had West Point pedigrees. The Confederacy's high command was a who's who of West Pointers: Generals Robert E. Lee, Thomas Jonathan Jackson, James Longstreet, Albert Sidney Johnson, Joseph Johnson, Braxton Bragg, John Bell Hood, AP Hill, George Picket, and Lewis Armistead. Many had trained as engineers, had experience fighting Native Americans, and were skeptical of volunteer officers, who rose about them as the war went on. Amateur officers played an important role in both armies as commanders of volunteer regiments. In both the North and the South, such men were usually economic or social leaders in their communities. Sometimes this system produced officers of real ability, but often it did not.

THE ROLE OF SEA POWER

The Union had an overwhelming advantage in naval power. The navy played two important roles in the war: enforcing a blockade of the Southern coast and assisting the Union armies in field operations.

The blockade, which began in the first weeks of the war, kept most oceangoing ships out of Confederate ports, but for a time small blockade runners continued to slip through. Gradually, however, federal forces tightened the blockade by seizing the Confederate ports

Library of Congress Prints and Photographs Division [LC-USZ61-903]; Library of Congress Prints and Photographs Division [LC-DIG-cwpb-04406]

GRANT (LEFT) AND LEE These portraits are significant not only because they capture the generals as wartime leaders but also because they became stock images in Lost Cause mythology that developed in the South after the war. Grant was heavily criticized for his violent, merciless, and unrelenting assault on the South's people and landscape; Lee was sainted as a man of honor, restraint, and bearing. Both representations were vital in depicting the North as an aggressive invader; the South as having fought a valiant, worthy battle; and slavery as incidental to the conflict. • *Why did white Southerners seek to imagine the war in ways that ignored slavery as the primary cause?*

themselves. The last important port in Confederate hands—Wilmington, North Carolina—fell to the Union early in 1865.

The Confederates made a bold attempt to break the blockade with an ironclad warship, constructed by plating with iron a former U.S. frigate, the *Merrimac*. On March 8, 1862, the refitted *Merrimac*, renamed the *Virginia*, left Norfolk to attack a blockading squadron of wooden ships at nearby Hampton Roads. It destroyed two of the ships and scattered the rest. But the Union government had already built ironclads of its own, including the *Monitor*, which arrived only a few hours after the *Virginia*'s dramatic foray. The next day, it met the *Virginia* in battle. Neither vessel was able to sink the other, but the *Monitor* put an end to the *Virginia*'s raids and preserved the blockade.

The Union navy was particularly important in the western theater of the war, where the major rivers were navigable by large vessels. The navy transported supplies and troops and joined in attacking Confederate strong points. The South had no significant navy of its own and could defend against the Union gunboats only with ineffective fixed land fortifications.

EUROPE AND THE DISUNITED STATES

Judah P. Benjamin, the Confederate secretary of state for most of the war, was an intelligent but undynamic man who attended mostly to routine administrative tasks. William Seward, his counterpart in Washington, gradually became one of the outstanding American secretaries of state. He had invaluable assistance from Charles Francis Adams, the American minister to London. This gap between the diplomatic skills of the Union and the Confederacy was a decisive factor in the war.

At the beginning of the conflict, the sympathies of the ruling classes of England and France lay largely with the Confederacy, partly because the two nations imported much Southern cotton; but it was also because they were eager to weaken the United States, an increasingly powerful rival to them in world commerce. France was unwilling to take sides in the conflict unless England did so first. And in England, the government was reluctant to act because there was powerful popular support for the Union—particularly from the large and influential English antislavery movement. After Lincoln issued the Emancipation Proclamation, antislavery groups worked feverishly for the Union. Southern leaders hoped to counter the strength of the British antislavery forces by arguing that access to Southern cotton was vital to the English and French textile industries. But English manufacturers had a surplus of both raw cotton and finished goods on hand in 1861 and could withstand a temporary loss of access to American cotton. Later, as the supply began to diminish, both England and France managed to keep at least some of their mills open by importing cotton from Egypt, India, and elsewhere. Equally important, even the 500,000 English textile workers thrown out of jobs as a result of mill closings continued to support the North. And as the conflict wore on, the British actually became dependent on supplies of grain from Northern farms.

No European nation ever offered diplomatic recognition, financial support, or military aid to the Confederacy. No nation wanted to antagonize the United States unless the Confederacy seemed likely to win, and the South never came close enough to victory to convince its potential allies to support it. In the end, the Confederacy was on its own.

Even so, tension, and on occasion near hostilities, continued between the United States and Britain. The Union government was angry when Great Britain, France, and other nations declared themselves neutral early in the war, thus implying that the two sides to the conflict had equal stature. Leaders in Washington insisted that the conflict was simply a domestic insurrection, not a war between two legitimate governments.

A more serious crisis, the so-called *Trent* affair, began in late 1861. Two Confederate diplomats, James M. Mason and John Slidell, had slipped through the then-ineffective Union blockade to Havana, Cuba, where they boarded an English steamer, the *Trent,* for England. Waiting in Cuban waters was the American frigate *San Jacinto,* commanded by the impetuous Charles Wilkes. Acting without authorization, Wilkes stopped the British vessel, arrested the diplomats, and carried them in triumph to Boston. The British government demanded the release of the prisoners, reparations, and an apology. Lincoln and Seward were aware that Wilkes had violated maritime law. Unwilling to risk war with England, they eventually released the diplomats with an indirect apology.

CAMPAIGNS AND BATTLES

In the absence of direct intervention by the European powers, the two contestants were left to resolve the conflict between themselves. They did so in four long years of bloody combat. More than 750,000 Americans died in the course of the Civil War, far more than the 112,000 who perished in World War I or the 405,000 who died in World War II. There were nearly 2,000 deaths for every 100,000 people during the Civil War. In World War I, the comparable figure was only 109, and in World War II it was 241. In many respects, the Civil War was the bloodiest war in modern times.

The vast majority of Civil War soldiers were volunteers. Recruiters for the Union often pulled in groups of men from the same town or ethnic group. There were entire companies and even regiments of Irish Americans and German Americans and, later, Black Americans. Initially, these soldiers and their officers were haphazardly trained and unprepared for battle. Many officers jokingly referred to their troops as armed mobs.

Because individual states outfitted their soldiers at the outset of the war, Union uniforms were at first far from uniform, ranging from the dark blue jackets and light blue pants of the regular army to dark blue and red "Zouave" uniforms based on the French colonial regiments in Algeria. "Federals" did not consistently wear blue uniforms until 1862, when the federal government provided them. Like his Union counterpart, the average Confederate soldier rarely had proper military training and tended to serve with comrades from the same area. Regiments wore different uniforms, and even after the Confederacy adopted gray, the government was never able to clothe every soldier.

THE TECHNOLOGY OF WAR

Much of what happened on the battlefield in the Civil War was a result of new technologies. The most obvious change was the nature of the armaments. Most important was the rifled musket, which had a much greater range than anything that came before it and thus increased death rates and the size of the killing field. It was used by about 90 percent of the infantry regiments on both sides. These rifles were the cause of most of the wounds, amputations, and deaths suffered in combat. Also significant were greatly improved cannons and artillery, a result of earlier advances in iron and steel technology.

The deadliness of the new weapons encouraged armies on both sides to build elaborate fortifications and trenches to protect themselves from enemy fire. The sieges of Vicksburg and Petersburg, the defense of Richmond, and many other military events all involved the construction of vast fortifications around both cities and attacking armies. (They were the predecessors to the vast network of trenches of World War I.) While both armies still practiced conventional linear tactics to organize soldiers on battlefield, they also began to experiment with these larger defense efforts.

Important to the conduct of the war were two other relatively new technologies: the railroad and the telegraph. The railroad was particularly important in mobilizing millions of soldiers and transferring them to the front. Transporting such enormous numbers of soldiers, and the supplies necessary to sustain them, by foot or by horse and wagon would have been almost impossible. But, ironically, railroads also limited the mobility of the armies. Commanders were forced to organize their campaigns at least in part around the location of the railroads rather than on the basis of the best topography or most direct land route to a destination. The dependence on the rails—and the resulting necessity of concentrating huge numbers of men in a few places—also encouraged commanders to prefer great battles with large armies to smaller engagements with fewer troops.

The impact of the telegraph on the war was limited both by the scarcity of qualified telegraph operators and by the difficulty of bringing telegraph wires into the fields where battles were being fought. Things improved somewhat after the new U.S. Military Telegraph Corps, begun in 1861 and guided by Thomas Scott and Andrew Carnegie, trained and employed hundreds of operators. Gradually, too, both the Union and Confederate armies began to string telegraph wires along the routes of their troops, so that field commanders were able to stay in close touch with one another during battles.

Naval technologies changed too. The use of ironclad ships—and even torpedoes and submarine technology, which made fleeting appearances in the 1860s—marked the beginning of the end for wooden warships. It heralded the start of a revolution in naval warfare that would lead to the development of modern fleets far deadlier and capable than their predecessors.

THE OPENING CLASHES, 1861

The Union and the Confederacy fought their first major battle of the war in northern Virginia. A Union army of over 30,000 men under General Irvin McDowell was stationed just outside Washington. About thirty miles away, at Manassas, sat a slightly smaller Confederate army commanded by General P. G. T. Beauregard. If the Northern army could destroy the Southern one, Union leaders believed, the war might end at once. In mid-July, McDowell marched his inexperienced troops toward Manassas. Beauregard moved his troops behind Bull Run, a small stream north of Manassas, and called for reinforcements, which reached him the day before the battle.

On July 21, in the **First Battle of Bull Run**, or First Battle of Manassas, McDowell almost succeeded in dispersing the Confederate forces. But the Southerners managed to fight off a last strong Union assault and then counterattacked savagely. Surprised, the Union troops panicked, broke ranks, and ran. Caught off guard himself, and unable to readily identify his soldiers because of mismatched and irregular uniforms, McDowell failed to reorganize the troops. Instead he ordered a retreat to Washington—which quickly became a wild, chaotic withdrawal complicated by the presence along the route of many civilians, who had ridden down from the capital, picnic baskets in hand, to watch the battle from nearby hills. The Confederates, disorganized by the turn of events, did not pursue. But they hailed the outcome of the battle as a smashing success and a sure sign of future victories. By contrast, the battle was a severe blow to Union morale and to the president's confidence in his officers.

Elsewhere, Union forces achieved some small but significant victories in 1861. A Union force under General George B. McClellan moved east from Ohio into western Virginia. By the end of 1861, it had "liberated" the antisecession mountain people of the region, who created their own state government loyal to the Union; the state was admitted to the Union as West Virginia in 1863.

THE WESTERN THEATER

After the battle at Bull Run, military operations in the East settled into a long and frustrating stalemate. The first decisive operations occurred in the western theater in 1862. Here Union forces were trying to seize control of the southern part of the Mississippi River from both the north and south, moving down the river from Kentucky and up from the Gulf of Mexico toward New Orleans.

In April, a Union squadron commanded by Flag Officer David G. Farragut smashed past weak Confederate forts near the mouth of the Mississippi and from there sailed up to New Orleans. The city was virtually defenseless because the Confederate high command had expected the attack to come from the north. The surrender of New Orleans on April 25, 1862, was an important turning point in the war. From then on, the mouth of the Mississippi was closed to Confederate trade, and the South's largest city and most important banking center was in Union hands.

Farther north in the western theater, Confederate troops under General Albert Sidney Johnston were stretched out in a long defensive line around two forts in Tennessee, Fort Henry and Fort Donelson. Early in 1862, Ulysses S. Grant attacked Fort Henry, whose defenders, awed by the ironclad riverboats accompanying the Union army, surrendered with almost no resistance. Grant then moved both his naval and ground forces to Fort Donelson, where the Confederates put up a stronger fight but finally, on February 16, surrendered. Grant thus gained control of river communications and forced Confederate troops out of Kentucky and half of Tennessee.

With about 40,000 men, Grant now advanced south along the Tennessee River. At Shiloh, Tennessee, he met a force almost equal to his own, commanded by Albert Sidney Johnston and P. G. T. Beauregard. The result was the Battle of **Shiloh**, April 6-7, 1862. In the first day's fighting (during which Johnston was killed), the Southerners drove Grant back to the river. But the next day, reinforced by 25,000 fresh troops, Grant recovered the lost ground and forced Beauregard to withdraw. After the narrow Union victory at Shiloh, Northern forces occupied Corinth, Mississippi, and took control of the Mississippi River as far south as Memphis.

General Braxton Bragg, now in command of the Confederate army in the West, gathered his forces at Chattanooga, in eastern Tennessee, where he faced a Union army trying to capture the city. The two armies maneuvered for advantage inconclusively in northern Tennessee and southern Kentucky for several months until December 31-January 2, when they finally clashed in the Battle of Murfreesboro, or Stone's River. Bragg was forced to withdraw to the South in defeat.

By the end of 1862, Union forces had made considerable progress in the West. But the heart of the war remained in the East.

THE VIRGINIA FRONT, 1862

During the winter of 1861-1862, George B. McClellan, commander of the Army of the Potomac, concentrated on training his army of 150,000 men near Washington. Finally, he designed a spring campaign to capture the Confederate capital at Richmond. But instead of heading overland directly, McClellan chose a complicated route that he thought would circumvent the Confederate defenses. The navy would carry his troops down the Potomac to a peninsula east of Richmond, between the York and James Rivers; the army would approach the city from there in what became known as the Peninsular campaign.

McClellan set off with 100,000 men, reluctantly leaving 30,000 members of his army behind, under General Irvin McDowell, to protect Washington. McClellan eventually persuaded Lincoln to send him the additional men. But before the president could do so, a Confederate army under General **Thomas J. ("Stonewall") Jackson** staged a rapid march north through the Shenandoah Valley, as if preparing to cross the Potomac and attack Washington. Lincoln postponed sending reinforcements to McClellan, retaining McDowell's corps to head off Jackson. In the Valley campaign of May 4-June 9, 1862, Jackson defeated two separate Union forces and slipped away before McDowell could catch him.

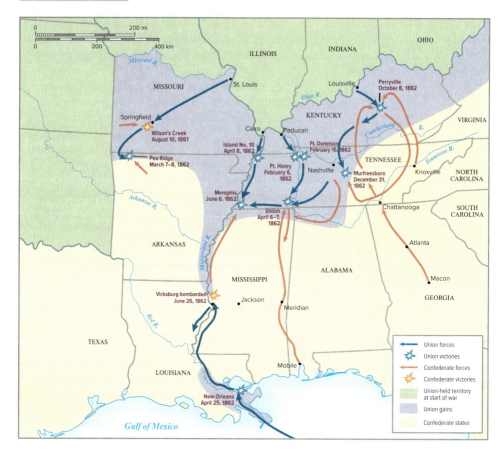

THE WAR IN THE WEST, 1861–1863 While the Union armies in Virginia were meeting with repeated frustrations, the Union armies in the West were scoring notable successes in the first two years of the war. This map shows a series of Union drives in the western Confederacy. Flag Officer David Farragut's ironclads led to the capture of New Orleans—a critical Confederate port—in April 1862, while forces farther north under the command of Ulysses S. Grant drove the Confederate army out of Kentucky and western Tennessee. These battles culminated in the Union victory at Shiloh, which led to Union control of the upper Mississippi River. • *Why was control of the Mississippi so important to both sides?*

Meanwhile, McClellan was fighting Confederate troops under General Joseph E. Johnston outside Richmond in the two-day Battle of Fair Oaks, or Seven Pines (May 31–June 1), and holding his ground. Johnston, badly wounded, was replaced by Robert E. Lee, who then recalled Stonewall Jackson from the Shenandoah Valley. With a combined force of 85,000 to face McClellan's 100,000, Lee launched a new offensive, known as the Battle of the Seven Days (June 25–July 1), in an effort to cut McClellan off from his base on the York River. But McClellan fought his way across the peninsula and set up a new base on the James.

McClellan was now only twenty-five miles from Richmond and in a good position to renew the campaign. But despite continuing pressure from Lincoln, he did not advance. Hoping to force a new offensive against Richmond along the direct overland route he had always preferred, Lincoln finally ordered the army to move back to northern Virginia and join up with a smaller force under General John Pope. As the Army of the Potomac left the peninsula by water, Lee moved north with the Army of northern Virginia to strike Pope before McClellan could join him. Pope was as rash as McClellan was cautious, and he attacked the approaching Confederates without waiting for the arrival of all of McClellan's troops. In the Second Battle of Bull Run, or Manassas (August 29–30), Lee threw back the assault and routed Pope's army, which fled to Washington. With hopes for an overland campaign against Richmond now in disarray, Lincoln removed Pope from command and put McClellan back in charge of all the federal forces in the region.

Lee soon went on the offensive again, heading north through western Maryland, and McClellan moved out to meet him. McClellan had the good luck to get a copy of Lee's orders, which revealed that a part of the Confederate army, under Stonewall Jackson, had separated from the rest to attack Harpers Ferry. But instead of attacking quickly before the Confederates could recombine, McClellan stalled, again giving Lee time to pull most of his forces back together behind **Antietam** Creek, near the town of Sharpsburg. There, on September 17, McClellan's 87,000-man army repeatedly attacked Lee's force of 50,000, with staggering casualties on both sides. Late in the day, just as the Confederate line seemed ready to break, the last of Jackson's troops arrived from Harpers Ferry to reinforce it. McClellan might have broken through with one more assault. Instead, he allowed Lee to retreat into Virginia. Technically, Antietam was a Union victory; but in reality, McClellan had squandered an opportunity to destroy much of the Confederate army. In November, Lincoln finally removed McClellan from command for good.

McClellan's replacement, General Ambrose E. Burnside, was a short-lived mediocrity. He tried to move toward Richmond by crossing the Rappahannock River at Fredericksburg. There, on December 13, he launched a series of attacks against Lee, all of them bloody, all of them hopeless. After losing a large part of his army, he withdrew to the north bank of the Rappahannock. He was relieved at his own request. The year 1862 ended, therefore, with a series of frustrations in the East for the Union.

THE PROGRESS OF THE WAR

Why did the Union—with its much greater population and its much better transportation system and technology than the Confederacy—make so little decisive progress in the first two years of the war? Had there been a crushing victory by the North early in the war—for example, a routing at the First Battle of Bull Run—the conflict might have ended quickly. But none occurred.

Many Northerners blamed the military stalemate on timid or incompetent Union generals, and there was some truth to that view. But the more important reason for the drawn-out conflict was that it was not a war of traditional tactics and military strategy. It was, even if the leaders of both sides were not yet fully aware of it, a war of attrition, and the Confederacy could survive only if the Union quit fighting. Winning or losing battles here and there would not determine the outcome of the war. What would bring the war to a conclusion was the steady and intentional destruction of the enemy's resources. More than two bloody years of fighting were still to come. Those last years witnessed the slow, steady deterioration of the Confederacy's ability to maintain the war, and the consistent growth of the resources that made the Union armies steadily stronger.

1863: YEAR OF DECISION

At the beginning of 1863, General Joseph Hooker was commanding the still-formidable Army of the Potomac, which remained north of the Rappahannock, opposite Fredericksburg. Taking part of his army, Hooker crossed the river above Fredericksburg and moved toward the town and Lee's army. But at the last minute, he drew back to a defensive position

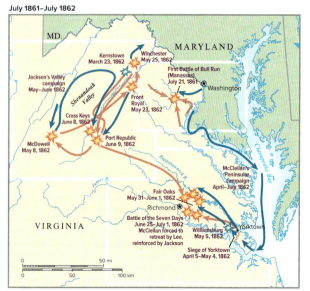

THE VIRGINIA THEATER, 1861–1863 Much of the fighting during the first two years of the Civil War took place in what became known as the Virginia theater—although the campaigns in this region eventually extended north into Maryland and Pennsylvania. The Union hoped for a quick victory over the newly created Confederate army. But as these maps show, the Southern forces consistently thwarted such hopes. The map at top left shows the battles of 1861 and the first half of 1862, almost all of them won by the Confederates. The map at lower left shows the last months of 1862, during which the Southerners again defeated the Union in most of their engagements—although Northern forces drove the Confederates back from Maryland in September. The map on the right shows the troop movements that led to the climactic battle of Gettysburg in 1863. • *Why were the Union forces unable to profit more from material advantages during these first years of the war?*

in a desolate area of brush and scrub trees known as the Wilderness. Lee divided his forces for a dual assault on the Union army. In the Battle of Chancellorsville, May 1–5, Stonewall Jackson attacked the Union right and Lee himself charged the front. Hooker barely managed to escape with his army. Lee had frustrated Union objectives, but he had not destroyed the Union army. And his ablest officer, Jackson, was fatally wounded in the course of the battle.

While the Union forces were suffering repeated frustrations in the East, they were winning some important victories in the West. In the spring of 1863, Ulysses S. Grant was driving at Vicksburg on the Mississippi River. Vicksburg was well protected on land and had good artillery coverage of the river itself. But in May, Grant boldly moved men and supplies—over land and by water—to an area south of the city, where the terrain was reasonably good. He then attacked Vicksburg from the rear. Six weeks later, on July 4, Vicksburg—whose residents were by then starving as a result of the prolonged siege—surrendered. At almost the same time, the other Confederate strongpoint on the river, Port Hudson,

THE SIEGE OF VICKSBURG, MAY–JULY 1863 In the spring of 1863, Grant began a campaign to win control of the final piece of the Mississippi River still controlled by the Confederacy. To do that required capturing the Southern stronghold at Vicksburg—a well-defended city sitting above the river. Vicksburg's main defenses were to the north, so Grant boldly moved men and supplies around the city and attacked it from the south. Eventually, he cut off the city's access to the outside world, and after a six-week siege, its residents finally surrendered.
• What impact did the combined victories at Vicksburg and Gettysburg have on Northern commitment to the war?

Louisiana, also surrendered to a Union force that had moved north from New Orleans. With the achievement of one of the Union's basic military aims—control of the whole length of the Mississippi—the Confederacy was split in two, with Louisiana, Arkansas, and Texas cut off from the other seceded states. The victories on the Mississippi were one of the great turning points of the war.

During the siege of Vicksburg, Lee proposed an invasion of Pennsylvania, which would, he argued, divert Union troops north. Further, he argued, if he could win a major victory on Northern soil, England and France might come to the Confederacy's aid. The war-weary North might even quit the war before Vicksburg fell.

In June 1863, Lee moved up the Shenandoah Valley into Maryland and then entered Pennsylvania. The Union Army of the Potomac, commanded first by Hooker and then (after June 28) by General George C. Meade, moved north, too. The two armies finally encountered each other at the small town of **Gettysburg**, Pennsylvania. There, on July 1–3, 1863, they fought the most celebrated battle of the war.

Meade's army established a strong, well-protected position on the hills south of the town. Lee attacked, but his first assault on the Union forces on Cemetery Ridge failed. A day later, he ordered a second, larger effort. In what is remembered as Pickett's Charge, a force of 15,000 Confederate soldiers advanced for almost a mile across open country while being swept by Union fire. Only about 5,000 made it up the ridge, and this remnant finally had

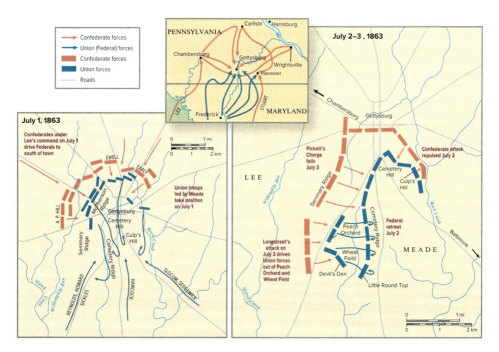

GETTYSBURG, JULY 1–3, 1863 Gettysburg was the most important single battle of the Civil War. Had Confederate forces prevailed at Gettysburg, the future course of the war might well have been very different. The map on the left shows the distribution of Union and Confederate forces at the beginning of the battle, July 1, after Lee had driven the Northern forces south of town. The map on the right reveals the pattern of the attacks on July 2 and 3. Note, in particular, Pickett's bold and costly charge, whose failure on July 3 was the turning point in the battle and, some chroniclers have argued, the war. • *Why did Robert E. Lee believe that an invasion of Pennsylvania would advance the Confederate cause?*

to surrender or retreat. By now, Lee had lost nearly a third of his army. On July 4, the same day as the surrender of Vicksburg, Lee withdrew from Gettysburg. The retreat was another major turning point in the war. Never again were the weakened Confederate forces able seriously to threaten Northern territory. Months later Lincoln visited the battlefield site and delivered one of the most famous speeches in American history.

Before the end of the year, there was one more important turning point, this time in Tennessee. After occupying Chattanooga on September 9, Union forces under General William Rosecrans began an unwise pursuit of Bragg's retreating Confederate forces. The two armies engaged in western Georgia, in the Battle of Chickamauga (September 19-20). Union forces could not break the Confederate lines and retreated back to Chattanooga.

Bragg now began a siege of Chattanooga itself, seizing the heights nearby and cutting off fresh supplies to the Union forces. Grant came to the rescue. In the Battle of Chattanooga (November 23-25), the reinforced Union army drove the Confederates back into Georgia. Union forces had now achieved a second important objective: control of the Tennessee River. Grant's leadership and success, demonstrated once again, convinced Lincoln to consider him for larger purposes.

The Last Stage, 1864-1865

By the beginning of 1864, President Lincoln had appointed General Ulysses S. Grant as chief of all the Union armies. Grant, like Lincoln, believed in using the North's great advantage in troops and material resources to overwhelm the South. He planned two great offensives for 1864. In Virginia, the Army of the Potomac would advance toward Richmond and force Lee into a decisive battle. In Georgia, the western army, under General **William T. Sherman**, would advance east toward Atlanta and destroy the remaining Confederate force, now under the command of Joseph E. Johnston.

The Northern campaign began when the Army of the Potomac, 115,000 strong, plunged into the rough, wooded Wilderness area of northwestern Virginia in pursuit of Lee's 75,000-man army. After avoiding an engagement for several weeks, Lee turned Grant back in the Battle of the Wilderness (May 5-7). Without stopping to rest or reorganize, Grant resumed his march toward Richmond and met Lee again in the bloody Battle of Spotsylvania Court House (May 8-21) in which 12,000 Union troops and a large, but unknown, number of Confederates fell. Grant kept moving, but victory continued to elude him. Lee kept his army between Grant and the Confederate capital and on June 1-3 repulsed the Union forces again, just northeast of Richmond, at Cold Harbor.

Grant now moved his army east of Richmond and headed south toward the railroad center at Petersburg. If he could seize Petersburg, he could cut off the capital's communications with the rest of the Confederacy. But Petersburg had strong defenses; and once Lee came to the city's relief, the assault became a prolonged siege.

In Georgia, meanwhile, Sherman was facing less ferocious resistance. With 90,000 men, he confronted Confederate forces of 60,000 under Johnston. As Sherman advanced, Johnston tried to delay him by maneuvering. The two armies fought only one real battle—Kennesaw Mountain, northwest of Atlanta, on June 27—where Johnston scored an impressive victory. Even so, he was unable to stop the Union advance toward Atlanta, prompting Davis to replace him with General John B. Hood. Sherman took the city on September 2 and burned it.

Hood now tried unsuccessfully to draw Sherman out of Atlanta by moving back up through Tennessee and threatening an invasion of the North. Sherman sent Union troops

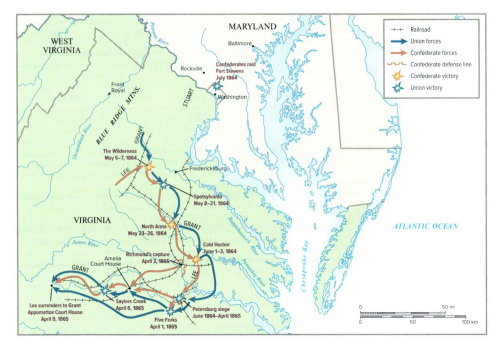

VIRGINIA CAMPAIGNS, 1864–1865 From the Confederate defeat at (and retreat from) Gettysburg until the end of the war, most of the eastern fighting took place in Virginia. By now, Ulysses S. Grant was commander of all Union forces and had taken over the Army of the Potomac. Although Confederate forces won a number of important battles during the Virginia campaign, the Union army grew steadily stronger and the southern forces steadily weaker. Grant believed that the Union strategy should reflect the North's greatest advantage: its superiority in men and equipment. • *What effect did this decision have on the level of casualties?*

to reinforce Nashville. In the Battle of Nashville on December 15–16, 1864, Northern forces practically destroyed what was left of Hood's army.

Meanwhile, Sherman had left Atlanta to begin his soon-to-be-famous "March to the Sea." Living off the land, destroying government buildings, military installations, and any manufacturing that could support the Confederacy, his army cut across southern Georgia. Sherman sought to deprive the Confederate army of war materials and railroad communications as a way to impoverish it and break the will of the soldiers. By December 20, he had reached Savannah, which surrendered two days later. Early in 1865, Sherman continued his destructive march, moving northward through South Carolina. He was virtually unopposed until he was well inside North Carolina, where a small force under Johnston could do no more than cause a brief delay.

In April 1865, Grant's Army of the Potomac—still engaged in the prolonged siege at Petersburg—finally captured a vital railroad junction southwest of the town. Without rail access to the South, and plagued by heavy casualties and massive desertions, Lee informed the Confederate government that he could no longer defend Richmond. Within hours, Jefferson Davis, his cabinet, and as much of the white population as could find transportation fled. That night, mobs roamed the city, setting devastating fires. And the next morning, Northern forces entered the Confederate capital. With them was Abraham Lincoln, who walked through the streets of the burned-out city surrounded by Black men and women

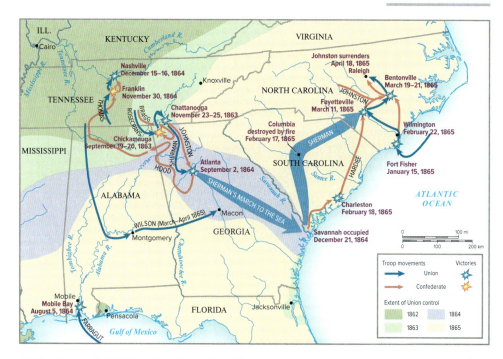

SHERMAN'S MARCH TO THE SEA, 1864–1865 While Grant was wearing down Lee in Virginia, General William Tecumseh Sherman was moving south across Georgia. After a series of battles in Tennessee and northwest Georgia, Sherman captured Atlanta and then marched unimpeded to Savannah, on the Georgia coast—deliberately devastating the towns and plantations through which his troops marched. Note that after capturing Savannah by Christmas 1864, Sherman began moving north through the Carolinas. A few days after Lee surrendered to Grant at Appomattox, Confederate forces farther south surrendered to Sherman. • *What did Sherman believe his devastating March to the Sea would accomplish?*

cheering him as the "Messiah" and "Father Abraham." In one particularly stirring moment, the president turned to someone who had been enslaved kneeling on the street before him and said: "Don't kneel to me. . . . You must kneel to God only, and thank Him for the liberty you will enjoy hereafter."

With the remnant of his army, now about 25,000 men, Lee began moving west in the forlorn hope of finding a way around the Union forces so that he could move south and link up with Johnston in North Carolina. But the Union army pursued him and blocked his escape route. Lee finally recognized that further bloodshed was futile. He arranged to meet Grant at a private home in the small town of **Appomattox Court House**, Virginia, where on April 9 he surrendered what was left of his forces. Nine days later, near Durham, North Carolina, Johnston surrendered to Sherman. The long war was now effectively over. Jefferson Davis was captured in Georgia. A few Southern diehards continued to fight, but even their resistance collapsed before long. General George Meade's shout, "It's all over," described in simple terms one of the most momentous events in the nation's history.

The war ensured the permanence of the Union but left many other issues far from settled. What would happen to the formerly enslaved? Would the rights of full citizenship be extended to Black Americans and would whites accept them as equals? Could the South and the North ever truly reconcile? The end of the war was the beginning of generations of struggle to determine the legacy of the Civil War.

The Metropolitan Museum of Art, New York, Gift of Mrs. Frank B. Porter, 1922

THE QUESTION OF VICTORY Depicting the capture of a Confederate officer and two enlisted men by Union Brigadier General Francis Channing Barlow on June 21, 1864, at Petersburg, Virginia, Winslow Homer asks what comes after Union victory. In the cocky posture and steely glare of the Confederate officer facing Barlow, Homer raises questions about whether white Southerners will abide by laws outlawing slavery and establishing Black Americans as citizens.

CONCLUSION

The American Civil War began with high hopes and high ideals on both sides. In the North and the South alike, thousands of men enthusiastically enlisted in local regiments and went off to war. Four years later, over 750,000 were dead and many more maimed and traumatized for life. A fight for "principles" and "ideals"—a fight few people had thought would last more than a couple of months—had become one of the longest wars, and by far the bloodiest war, in American history, before or since.

During the first two years of fighting, the Confederate forces seemed to have all the advantages. They were fighting on their own soil. Their troops seemed more committed to the cause than those of the North. Their commanders were exceptionally talented, while Union forces were, for a time, erratically led. Gradually, however, the Union's advantages began to assert themselves. The North had a stabler political system led by one of the greatest leaders in the nation's history. It had a much larger population, a far more developed industrial economy, superior financial institutions, and a better railroad system. By the middle of 1863, the tide of the war had shifted; over the next two years, Union forces gradually wore down the Confederate armies before finally triumphing in 1865.

The war strengthened the North's economy, giving a spur to industry and railroad development. It greatly weakened the South's by destroying millions of dollars of property and depleting the region's young male population. Southerners had gone to war in part because of their fears of growing Northern dominance. Ironically, the war itself confirmed and strengthened that dominance.

But most of all, the Civil War was a victory for millions of enslaved Blacks. Thousands of Blacks freed themselves: deserting those who had enslaved them, seeking refuge behind Union lines, and at times serving and fighting in the Union army. The war produced Abraham Lincoln's Emancipation Proclamation and, later, the Thirteenth Amendment to the Constitution, which abolished slavery altogether. The future of freed Blacks was unsure, but slavery was finally ended.

KEY TERMS/PEOPLE/PLACES/EVENTS

Alexander H. Stephens 331
Antietam 341
Appomattox Court House 347
Confederate States of America 320
Confiscation Acts 327
conscription 325
Emancipation Proclamation 329
First Battle of Bull Run 338
Fort Sumter 321
George B. McClellan 327
Gettysburg 344
greenbacks 325
Homestead Act 324
Jefferson Davis 331
Robert E. Lee 334
Shiloh 339
Thirteenth Amendment 329
Thomas J. ("Stonewall") Jackson 339
Ulysses S. Grant 334
United States Sanitary Commission 331
William T. Sherman 345

RECALL AND REFLECT

1. What was the role of slavery in the start of the war?
2. Why did the South think it could win a war against the North?
3. How did Black Americans affect the war?
4. How did foreign countries impact the outcome of the war?
5. Why did the North win the Civil War? What questions did it settle, what did it leave open?

15 | RECONSTRUCTION AND THE NEW SOUTH

THE PROBLEMS OF PEACEMAKING
RADICAL RECONSTRUCTION
THE SOUTH IN RECONSTRUCTION
THE GRANT ADMINISTRATION
THE ABANDONMENT OF RECONSTRUCTION
THE NEW SOUTH

LOOKING AHEAD

1. Who were the winners and losers in Reconstruction?
2. How did Reconstruction change Black life?
3. Why were the civil rights of Black citizens under attack during Reconstruction?

FEW PERIODS IN THE HISTORY of the United States have inspired as much controversy as the era of Reconstruction—the years following the Civil War during which Americans attempted to reunite their shattered nation. To many white Southerners, Reconstruction was a vicious and destructive experience, a period when vindictive Northerners inflicted humiliation on the defeated South and enshrined formerly enslaved people into positions of power over them. Northern defenders of Reconstruction, in contrast, argued that their policies were the only way to prevent unrepentant Confederates from restoring southern society to what it had been before the war.

To most Black Americans at the time, and to many people of all backgrounds since, Reconstruction was notable for other reasons. Neither a vicious tyranny, as white Southerners charged, nor a thoroughgoing reform, as many Northerners hoped, it was instead an important first step in the effort to secure Black civil rights and economic power. Importantly, Reconstruction did not provide Black Americans with the enduring protections or the material resources to ensure anything like real equality. Most Black men and women lacked the formal power to overturn their oppression for many decades.

And yet for all its shortcomings, Reconstruction did make it possible for Black Americans to create new institutions like schools, colleges, and churches. It also furnished important legal precedents that lay the groundwork for the modern civil rights movement.

THE PROBLEMS OF PEACEMAKING

The end of the Civil War brought a new problem: making peace. President Lincoln had begun to think about how best to reunite the nation well before April 1865, but his assassination prevented him from implementing his vision. President Johnson was left with the task of negotiating a reunited United States.

THE AFTERMATH OF WAR AND EMANCIPATION

More than 750,000 thousand Americans had died in the Civil War and many others returned home wounded or sick. The South was reeling: towns had been gutted, plantations burned, fields neglected, bridges and railroads destroyed. Many white Southerners—stripped of those they had held in slavery and of capital invested in worthless Confederate bonds and currency—had almost no personal property. Starvation and homelessness were not uncommon.

While the physical conditions were bad for Southern whites, they were far worse for the 3.5 million men and women emerging from bondage. As soon as the war ended, hundreds of thousands left plantations in search of family members, work, and land. Few owned much more than the clothes on their backs.

COMPETING NOTIONS OF FREEDOM

For most white Southerners, freedom meant the ability to control their own destinies without interference from the North or the federal government. In the immediate aftermath of the war, this notion of freedom motivated white Southerners to try to restore their society to its antebellum form, which included preserving local and regional autonomy and imposing control over the newly freed Black population.

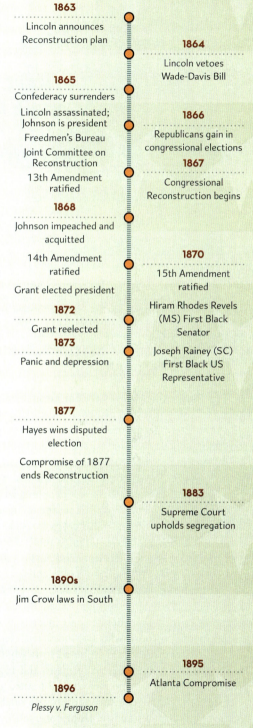

TIME LINE

1863
Lincoln announces Reconstruction plan

1864
Lincoln vetoes Wade-Davis Bill

1865
Confederacy surrenders
Lincoln assassinated; Johnson is president
Freedmen's Bureau
Joint Committee on Reconstruction
13th Amendment ratified

1866
Republicans gain in congressional elections

1867
Congressional Reconstruction begins

1868
Johnson impeached and acquitted
14th Amendment ratified
Grant elected president

1870
15th Amendment ratified
Hiram Rhodes Revels (MS) First Black Senator
Joseph Rainey (SC) First Black US Representative

1872
Grant reelected

1873
Panic and depression

1877
Hayes wins disputed election
Compromise of 1877 ends Reconstruction

1883
Supreme Court upholds segregation

1890s
Jim Crow laws in South

1895
Atlanta Compromise

1896
Plessy v. Ferguson

Library of Congress Prints and Photographs Division [LC-USZC4-4593]

RICHMOND, 1865 By the time Union forces captured Richmond in early 1865, the Confederate capital had been under siege for months and much of the city lay in ruins, as this photograph reveals. On April 4, President Lincoln, accompanied by his son Tad, visited Richmond. As he walked through the streets of the shattered city, hundreds of those emancipated from slavery emerged from the rubble to watch him pass. "No triumphal march of a conqueror could have equalled in moral sublimity the humble manner in which he entered Richmond," a Black soldier serving with the Union army wrote. "It was a great deliverer among the delivered. No wonder tears came to his eyes."

For Black Americans, freedom meant independence from white control. This freedom required some measure of economic independence, which for many meant owning land. For a short while during the war, Union generals and federal officials aided this effort, awarding confiscated land to those who had formally been enslaved to work it.

When Union forces occupied the Sea Islands of South Carolina and their main harbor, Port Royal in November 1862, the islands' white property owners fled to the mainland. Ten thousand formerly enslaved people seized control of the vacated land, marking the beginning of the "Port Royal Experiment," in which formerly enslaved Blacks were permitted to farm the land and raise crops for sale. Some saved enough money to purchase the land from the federal government, a total of about 33,000 acres. Union officials and Northern missionaries recruited teachers, nurses, and doctors to build schools and hospitals for the newly freed people of the Sea Islands.

A broader redistribution of land occurred later in the war. At the urging of Black leaders, General William Sherman issued Special Field Order No. 15 on January 16, 1865, granting former Confederate land in coastal Georgia and South Carolina (including the Sea Islands) to those who had been enslaved in the region. Within five months, nearly 400,000 acres had been distributed to newly freed people, most of it in 40-acre plots.

In the war's immediate aftermath, the federal government established the Bureau of Refugees, Freedmen, and Abandoned Lands. The **Freedmen's Bureau**, as it became known, helped feed, clothe, educate, and provide medical care for those freed from slavery. It also settled land disputes and set labor contracts between freedmen and white property owners.

Headed by General Oliver O. Howard, the Freedman's Bureau, although operating on a shoestring budget with fewer than 1,000 agents, emerged as a key federal institution shaping life for both Blacks and whites in the South after the war. Howard emerged as a leading spokesperson for Black education and was the namesake of Howard University, founded in 1867 in Washington, D.C., to educate Blacks. He served as Howard's president from 1869 to 1874.

The Freedmen's Bureau, for a while at least, also supported the redistribution of land, overseeing the allocation of 850,000 acres of confiscated land to those who had been enslaved. In his famous "Circular 13," General Howard instructed his agents to lease the land in 40-acre plots with the intention of eventually selling it to its occupants. A small number of Blacks purchased land outright under the Southern Homestead Act of 1866, which made available for sale 46 million acres, in 160-acre plots, of public land in Alabama, Arkansas, Florida, Louisiana, and Mississippi. (The act was repealed before many were able to take advantage of it.)

Southern Blacks demanded that the federal government protect their new rights as citizens. Aware that their freedoms would be challenged, often violently, by white Southerners, they turned to Washington, and the U.S. army, for support. (See "Consider the Source: Southern Blacks Demand Federal Aid.")

PLANS FOR RECONSTRUCTION

Political control of **Reconstruction** rested in the hands of the Republicans, who were deeply divided into three major groups. Conservatives insisted that the South accept abolition, but proposed few other conditions for the readmission of the seceded states. The **Radicals**, led by Representative **Thaddeus Stevens** of Pennsylvania and Senator **Charles Sumner** of Massachusetts, urged a much harsher course, including disenfranchising large numbers of Southern whites, protecting Black civil rights, confiscating the property of wealthy whites who had aided the Confederacy, and distributing the land among the freedmen. Finally the Moderates rejected the most stringent demands of the Radicals but supported extracting at least some concessions from the South on Black rights.

Ultimately two plans for Reconstruction emerged while Lincoln was still alive. Lincoln himself favored a lenient policy, believing that Southern Unionists (mostly former Whigs) could become the nucleus of new, loyal state governments. In December 1863 he introduced what came to be known as the "Ten Percent Plan," which allowed Southern states to be readmitted to the Union once 10 percent of eligible voters—defined as those present on the voter roles of the 1860 election—pledged an oath of loyalty to the government and accepted the abolition of slavery. These loyal voters could then elect representatives who would fashion new state constitutions and governments. Lincoln's plan offered full amnesty to all Southerners except high-ranking military officers and government leaders of the Confederacy. He also proposed extending suffrage to African Americans who were educated, owned property, or had served in the Union army. Three Southern states—Louisiana, Arkansas, and Tennessee, all under Union occupation—reestablished loyal governments under the Lincoln formula in 1864.

Outraged at the mildness of Lincoln's program, the Radical Republicans refused to admit representatives from the three "reconstructed" states to Congress. In July 1864, they pushed their own plan through Congress: the **Wade-Davis Bill**. Named for Senator Benjamin Wade of Ohio and Representative Henry Davis of Maryland, it called for the president to appoint a provisional governor for each conquered state. In contrast to Lincoln's Ten

CONSIDER THE SOURCE

SOUTHERN BLACKS DEMAND FEDERAL AID (1865)

Shortly after the war ended, groups of formerly enslaved people gathered in conventions to petition the federal government for support of their civil rights, including the use of the U.S. Army to subdue hostile ex-Confederates. In this example, African Americans from Virginia air their hopes and fears only three months after Appomattox.

We, the undersigned members of a Convention of colored citizens of the State of Virginia, would respectfully represent that, although we have been held as slaves, and denied all recognition as a constituent of your nationality for almost the entire period of the duration of your Government, and that by your *permission* we have been denied either home or country, and deprived of the dearest rights of human nature; yet when you and our immediate oppressors met in deadly conflict upon the field of battle—the one to destroy and the other to save your Government and nationality, we, with scarce an exception, in our inmost souls espoused your cause, and watched, and prayed, and waited, and labored for your success. . . .

When the contest waxed long, and the result hung doubtfully, you appealed to us for help, and how well we answered is written in the rosters of the two hundred thousand colored troops now enrolled in your Service; and as to our undying devotion to your cause, let the uniform acclamation of escaped prisoners, "whenever we saw a black face we felt sure of a friend," answer.

Well, the war is over, the rebellion is "put down," and we are *declared* free! Four fifths of our enemies are paroled or amnestied, and the other fifth are being pardoned, and the President has, in his efforts at the reconstruction of the civil government of the States, late in rebellion, left us entirely at the mercy of these subjugated but unconverted rebels, in *everything* save the privilege of bringing us, our wives and little ones, to the auction block. *We know* these men—know them well—and we assure you that, with the majority of them, loyalty is only "lip deep," and that their professions of loyalty are used as a cover to the cherished design of getting restored to their former relations with the Federal Government, and then, by all sorts of "unfriendly legislation," to render the freedom you have given us more Intolerable than the slavery they intended for us.

We warn you in time that our only safety is in keeping them under Governors of the *military persuasion* until you have so amended the Federal Constitution that it will prohibit the States from making any distinction between citizens on account of race or color. In one word, the only salvation for us besides the power of the Government, is in the *possession of the ballot*. Give us this, and we will protect ourselves. . . . But, 'tis said we are ignorant. Admit it. Yet who denies we *know* a traitor from a loyal man, a gentleman from a rowdy, a friend from an enemy? . . . All we ask is an *equal chance* with the white *traitors* varnished and japanned with the oath of amnesty. Can you deny us this and still keep faith with us? . . .

We are "sheep in the midst of wolves," and nothing but the military arm of the Government prevents us and all the *truly* loyal white men from being driven from the land of our birth. Do not then, we beseech you, give to one of these "wayward sisters" the rights they abandoned and forfeited when they rebelled until you have secured our rights by the aforementioned amendment to the Constitution.

UNDERSTAND, ANALYZE, & EVALUATE

1. What did the authors of this petition emphasize in the first two paragraphs, and why did they feel this was important?
2. Why did the authors emphasize that they had been "declared" free? What dangers to their prospect of freedom did they observe?
3. What federal legal responses did they propose? Can you recognize these suggestions in the constitutional changes that came with Reconstruction?

Source: "Proceedings of the Convention of the Colored People of Virginia, Held in the City of Alexandria, August 2, 3, 4, 5, 1865," Alexandria, Va., 1865, in W. L. Fleming (ed.), *Documentary History of Reconstruction*. Cleveland, Ohio, 1906, vol. 1, 195–196; located in "Southern Blacks Ask for Help (1865)," in Thomas A. Bailey and David M. Kennedy (eds.), *The American Spirit*, vol. 1, 7th ed., Lexington, Mass., 1991, 466–467.

Percent Plan, the Wade-Davis Bill specified that only when 50 percent of eligible voters in a state declared loyalty to the Union could the process of readmission begin. At that point the provisional governor could summon a state constitutional convention, whose delegates were to be elected by voters who had never borne arms against the United States. The new state constitutions would be required to abolish slavery, disenfranchise Confederate civil and military leaders, and repudiate debts accumulated by the state governments during the war. Only then would Congress formally readmit the states to the Union. Like the president's proposal, the Wade-Davis Bill left the question of political rights for Blacks up to the states.

Congress passed the Wade-Davis Bill a few days before it adjourned in 1864, but Lincoln disposed of it with a pocket veto, meaning that he killed the bill by keeping it "in his pocket" until it was too late for Congress to act on it during the legislative session. Predictably, Lincoln's pocket veto enraged Radical leaders and set up a showdown with the president over the future of Reconstruction policy. The debate over the proper course and objectives of Reconstruction was one that scholars would soon pick up. Indeed, historians were already struggling to make sense of its meaning in the latter stages of Reconstruction. They continue to do so today. (See "Debating the Past: Reconstruction.")

THE DEATH OF LINCOLN

What plan for Reconstruction the president might have ultimately approved no one can say. On the night of April 14, 1865, Lincoln and his wife attended a play at Ford's Theater in Washington. John Wilkes Booth, an actor fervently committed to the Southern cause, entered the presidential box from the rear and shot Lincoln in the head. Early the next morning, the president died.

The circumstances of Lincoln's death earned him immediate martyrdom. They also produced something close to hysteria throughout the North, especially when it soon became clear that Booth had been the leader of a conspiracy. One of his associates stabbed (but failed to kill) Secretary of State **William H. Seward** on the night of the assassination, and another abandoned at the last moment a plan to murder Vice President Andrew Johnson. Booth himself escaped on horseback into the Maryland countryside, where, on April 26, he was cornered by Union troops and shot to death in a blazing barn. Eight other people were convicted by a military tribunal of participating in the conspiracy. Four were hanged.

DEBATING THE PAST

RECONSTRUCTION

Debate over the nature of Reconstruction has been unusually intense. Indeed, few issues in American history have raised such deep and enduring passions.

Beginning in the late nineteenth century and continuing well into the twentieth, a relatively uniform and highly critical view of Reconstruction prevailed among historians. William A. Dunning's *Reconstruction, Political and Economic* (1907) was the principal scholarly expression of this view. Dunning portrayed Reconstruction as a corrupt and oppressive outrage imposed on a prostrate South by a vindictive group of Northern Republican Radicals. Unscrupulous carpetbaggers flooded the South and plundered the region. Ignorant and unfit African Americans were thrust into political offices. Reconstruction governments were awash in corruption and compiled enormous levels of debt. The Dunning interpretation dominated several generations of historical scholarship and helped shape such popular images of Reconstruction as those in the novel and film *Gone with the Wind*.

W. E. B. Du Bois, the great Black American scholar, offered one of the first alternative views in *Black Reconstruction* (1935). To Du Bois, Reconstruction was an effort by freed Blacks (and their white allies) to create a more democratic society in the South, and it was responsible for many valuable social innovations. In the early 1960s, John Hope Franklin and Kenneth Stampp, building on a generation of work by other scholars, published new histories of Reconstruction that also radically revised the Dunning interpretation. Reconstruction, they argued, was a genuine, if inadequate, effort to solve the problem of race in the South. Congressional Radicals were not saints, but they were genuinely concerned with protecting the rights of those who had been enslaved.

Reconstruction had brought important, if temporary, progress to the South and had created no more corruption there than was evident in the North at the same time. What was tragic about Reconstruction, the revisionists claimed, was not what it did to Southern whites but what it failed to do for Southern Blacks. It was, in the end, too weak and too short-lived to guarantee African Americans genuine equality.

In more recent decades, some historians have begun to question the conclusion of the first revisionists that Reconstruction had accomplished relatively little. Leon Litwack argued in *Been in the Storm So Long* (1979) that formerly enslaved men and women used the protections Reconstruction offered them to carve out a certain level of independence for themselves within Southern society: strengthening churches, reuniting families, and resisting the efforts of white planters to revive the gang labor system. Eric Foner's *Reconstruction: America's Unfinished Revolution* (1988) and *Forever Free* (2005) as well as Stephen Hahn's *A Nation Under Our Feet* (2003) strongly emphasized the rapid progress made by African Americans toward freedom and independence in a short time and the important role they played in shaping the execution of Reconstruction policies. Dylan Penningroth's *Claims of Kinfolk* (2003) and Susan Donvan's *Becoming Free in the Cotton South* (2007) focused on the particular experience of Blacks as a continuum stretching forth from slavery.

Gender historians have produced some of the most creative works in the field, focusing on the lives of Southern women, both Black and white. Notable are LeeAnn White's *The Civil War as a Crisis in Gender* (1995), Laura F. Edward's *Gendered Strife and*

A FREEDMEN'S BUREAU SCHOOL African American students and teachers stand outside a school for those formerly enslaved, one of many run by the Freedmen's Bureau throughout the defeated Confederacy in the first years after the war.

Confusion (1997), Tera Hunter's *To 'Joy My Freedom* (1997), and Leslie A. Schwalm's *A Hard Fight for We* (1997). A younger generation of gender scholars has focused on women's defense of their households and their bodies within a larger context of practices of femininity and masculinity: Diane Miller Sommerville's *Rape and Race in the Nineteenth-Century South* (2004), Hannah Rosen's *Terror in the Heart of Freedom* (2009), Kidada E. Williams's *The Left Great Marks on Me* (2012), and Carole Emberton's *Beyond Redemption* (2013). •

UNDERSTAND, ANALYZE, & EVALUATE

1. What are the popular interpretations of Reconstruction today? Do romanticized ideals of a benighted Old South, as seen in *Gone with the Wind*, still persist?
2. Why do you think the civil rights movement might have encouraged historians and others to reexamine Reconstruction?
3. How would a focus on gender history begin to ask new questions about Reconstruction?

To many Northerners, the murder of the president seemed evidence of an even darker conspiracy—an attempt by the unrepentant leaders of the defeated South to challenge the very authority of the nation's elected officials. Militant Republicans exploited such suspicions relentlessly in the ensuing months.

JOHNSON AND "RESTORATION"

Leadership fell immediately to Lincoln's successor, **Andrew Johnson** of Tennessee. A Democrat who had joined the Union ticket in 1864, he became president at a time of growing partisan passions.

Johnson revealed his plan for Reconstruction—or "Restoration," as he preferred to call it—soon after he took office and implemented it during the summer of 1865 when Congress was in recess. Like Lincoln, he offered some form of amnesty to Southerners who would take an oath of allegiance. In most other respects, however, his plan resembled the Wade-Davis Bill. The new president appointed a provisional governor in each state and charged

ABRAHAM LINCOLN This haunting photograph of Abraham Lincoln, showing clearly the weariness and aging that four years as a war president had created, was taken in Washington only four days before his assassination in 1865.

him with inviting qualified voters to elect delegates to a constitutional convention. To win readmission to Congress, a state had to revoke its ordinance of secession, abolish slavery and ratify the Thirteenth Amendment, and repudiate Confederate and state war debts.

By the end of 1865, all the seceded states had formed new governments—some under Lincoln's plan, others under Johnson's—and awaited congressional approval of them. But Radicals in Congress vowed not to recognize the Johnson governments for, by now, Northern opinion had become more hostile toward the South. Delegates to the Southern conventions had angered much of the North by their apparent reluctance to abolish slavery and their refusal to grant suffrage to Blacks. Southern states had also seemed to defy the North by electing prominent Confederate leaders to represent them in Congress, such as Alexander Stephens of Georgia, the former vice president of the Confederacy.

RADICAL RECONSTRUCTION

Reconstruction under Johnson's plan—often known as "presidential Reconstruction"—continued only until Congress reconvened in December 1865. At that point, Congress refused to seat the representatives of the "restored" states and created a new Joint Committee on Reconstruction to frame a policy of its own. The period of "congressional" or "Radical" Reconstruction had begun.

The Black Codes

Meanwhile, events in the South were driving Northern opinion in still more radical directions. Throughout the South in 1865 and early 1866, state legislatures enacted sets of laws known as the **Black Codes**, which authorized local officials to apprehend unemployed Blacks, fine them for vagrancy, and hire them out to private employers to satisfy the fines. Some codes forbade Blacks to own or lease farms or to take any jobs other than as plantation workers or domestic servants. (See "Consider the Source: Mississippi Black Codes.")

Those threatened by these codes called for swift intervention by federal troops and for legislation to protect them. Congress first responded by extending the life and expanding the powers of the Freedmen's Bureau so that it could nullify work agreements forced on freedmen under the Black Codes. Then, in April 1866, Congress passed the first Civil Rights Act, which declared Blacks to be full-fledged citizens of the United States and gave the federal government power to intervene in state affairs to protect the rights of citizens. Johnson vetoed both bills, but Congress overrode him each time.

The Fourteenth Amendment

In April 1866, the Joint Committee on Reconstruction proposed the **Fourteenth Amendment** to the Constitution. Without the support of Johnson, Congress approved it and sent it to the states for ratification. It offered the first constitutional definition of American citizenship. Everyone born in the United States, and everyone naturalized, was automatically a citizen and entitled to all the "privileges and immunities" guaranteed by the Constitution, including equal protection of the law under both state and national governments. For the first time in the nation's history, race or a prior condition of servitude was discarded as a barrier to full citizenship.

Significantly, the amendment extended the vote only to men, and specifically excluded Native Americans. The exclusion of women surfaced simmering tensions among those involved in the campaign for woman suffrage. Whereas some leaders like Elizabeth Cady Stanton insisted that the causes of civil rights for Blacks and for women should be pursued together, others wondered if both could be accomplished simultaneously. Frederick Douglass publicly split company with Stanton over the fate of the amendment. He decided that Black men had a stronger case for the vote and threw his full support to the amendment as written. Stanton and her allies lobbied Congress to have any reference to gender stricken from the amendment, but they stopped short of calling for the amendment never to be ratified.

While debate raged over the wording of the Fourteenth Amendment, congressional Radicals offered to readmit to the Union any state whose legislature ratified it. Only Tennessee did so. All other former Confederate states, along with Delaware and Kentucky, refused.

Radicals were undeterred and even began to grow in confidence as they anticipated gaining strength in the upcoming congressional election in the fall. And they were right. Bloody race riots in New Orleans and other Southern cities drew attention to the need for more vigorous legal and physical protection for African Americans, and to the failure of Johnson's policies. Johnson himself undercut his own and his party's popularity by delivering intemperate and mean-spirited speeches in which he openly disparaged the Radicals. He nearly fell off the stage delivering several speeches, which gave rise to rampant speculation that he was drinking too much. In the 1866 congressional elections, the voters returned an overwhelming majority of Republicans, most of them Radicals, to Congress. Congressional Republicans were now strong enough to enact a plan of their own, even over the president's objections.

CONSIDER THE SOURCE

MISSISSIPPI BLACK CODES (1865)

Mississippi enacted its first round of Black Codes only several months after the end of the Civil War. The codes relied heavily on the courts, the police, and the threat of violence to control Blacks.

Marriage and Employment

Section 3: . . . [I]t shall not be lawful for any freedman, free negro or mulatto to intermarry with any white person; nor for any person to intermarry with any freedman, free negro or mulatto; and any person who shall so intermarry shall be deemed guilty of felony, and on conviction thereof shall be confined in the State penitentiary for life. . . .

Section 6: . . . All contracts for labor made with freedmen, free negroes and mulattoes for a longer period than one month shall be in writing, and a duplicate, attested and read to said freedman, free negro or mulatto by a beat, city or county officer . . . and if the laborer shall quit the service of the employer before the expiration of his term of service, without good cause, he shall forfeit his wages for that year up to the time of quitting.

Section 7: . . . Every civil officer shall, and every person may, arrest and carry back to his or her legal employer any freedman, free negro, or mulatto who shall have quit the service of his or her employer before the expiration of his or her term of service without good cause . . .

Vagrancy and Mobility

Section 2: . . . All freedmen, free negroes and mulattoes in this State, over the age of eighteen years, found on the second Monday in January, 1866, or thereafter, with no lawful employment or business, or found unlawful assembling themselves together, either in the day or night time, and all white persons assembling themselves with freedmen, free negroes or mulattoes, or usually associating with freedmen, free negroes or mulattoes, on terms of equality, or living in adultery or fornication with a freed woman, freed negro or mulatto, shall be deemed vagrants, and on conviction thereof shall be fined in a sum not exceeding, in the case of a freedman, free negro or mulatto, fifty dollars, and a white man two hundred dollars, and imprisonment at the discretion of the court, the free negro not exceeding ten days, and the white man not exceeding six months.

Section 5: . . . All fines and forfeitures collected by the provisions of this act shall be paid into the county treasury of general county purposes, and in case of any freedman, free negro or mulatto shall fail for five days after the imposition of any or forfeiture upon him or her for violation of any of the provisions of this act to pay the same, that it shall be, and is hereby, made the duty of the sheriff of the proper county to hire out said freedman, free negro or mulatto, to any person who will, for the shortest period of service, pay said fine and forfeiture and all costs . . .

Weapons and Public Behavior

Section 1: . . . That no freedman, free negro or mulatto, not in the military service of the United States government, and not licensed so to do by the board of police of his or her county, shall keep or carry firearms of any kind, or any ammunition, dirk or bowie knife, and on conviction thereof in the county court shall be punished by fine . . .

Section 2: . . . Any freedman, free negro, or mulatto committing riots, routs, affrays, trespasses, malicious mischief, cruel treatment to animals, seditious speeches, insulting gestures, language, or acts, or assaults on any person, disturbance of the peace, exercising the function of a minister of the Gospel without a license from some regularly organized church, vending spirituous or intoxicating liquors, or committing any other misdemeanor, the punishment of which is not specifically provided for by law, shall, upon conviction thereof in the county court, be fined not less than ten dollars, and not more than one hundred dollars, and may be imprisoned at the discretion of the court, not exceeding thirty days.

Source: *Laws of the State of Mississippi, Passed at a Regular Session of the Mississippi Legislature, held in Jackson, October, November and December, 1965, Jackson, 1866,* pp. 82–93, 165–167.

UNDERSTAND, ANALYZE, & EVALUATE

1. How were the Black Codes similar to slavery?
2. What do the Black Codes suggest about the struggles the Black Americans would face after the Civil War and far beyond it?
3. What were the long term effects of the Black Codes on southern society?

THE CONGRESSIONAL PLAN

The Radicals passed three Reconstruction bills early in 1867 and overrode Johnson's vetoes of all of them. Nearly two years after the end of the war, these bills finally established a coherent plan for Reconstruction.

Under the congressional plan, Tennessee, which had ratified the Fourteenth Amendment, was promptly readmitted. But Congress rejected the Lincoln-Johnson governments of the other ten Confederate states and instead combined them into five military districts. A military commander governed each district and had orders to register qualified voters (defined as all adult Black males and those white males who had not participated in the war). These voters would elect conventions to prepare new state constitutions, which had to include provisions for Black suffrage. Once voters ratified the new constitutions, they could elect state governments. Congress had to approve a state's constitution, and the state legislature itself had to ratify the Fourteenth Amendment. Once enough states ratified the amendment to make it part of the Constitution, the former Confederate states could be restored to the Union.

By 1868, six of the ten remaining former Confederate states (Alabama, Arkansas, Florida, Louisiana, North Carolina, and South Carolina) had fulfilled these conditions and were readmitted to the Union. Conservatives whites held up the return of Georgia, Virginia, Texas, and Mississippi until 1870. By then, Congress had added an additional requirement for readmission—ratification of the **Fifteenth Amendment**, which forbade the states and the federal government to deny suffrage on account of "race, color, or previous condition of servitude." This opened the ballot box to all male citizens regardless of color, but kept it shut to all women. Ratification by the states was completed in 1870.

To stop Johnson from interfering with their plans, the congressional Radicals passed two laws in 1867. The Tenure of Office Act forbade the president to remove civil officials, including members of his own cabinet, without the consent of the Senate. The principal purpose of the law was to protect the job of Secretary of War Edwin M. Stanton, who was cooperating with the Radicals. The Command of the Army Act prohibited the president from issuing military orders except through the commanding general of the army (General Grant), who could not be relieved or assigned elsewhere without the consent of the Senate.

The congressional Radicals also took action to stop the Supreme Court from interfering with their plans. In 1866, the Court had declared in the case of *Ex parte Milligan* that military tribunals were unconstitutional in places where civil courts were functioning.

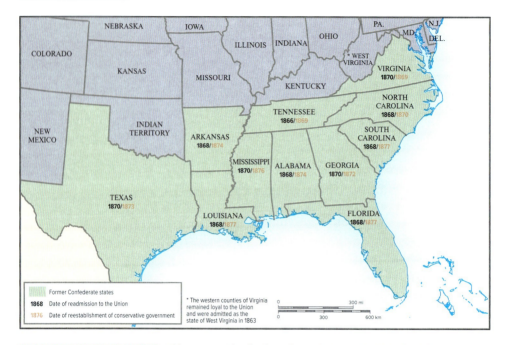

RECONSTRUCTION, 1866–1877 This map provides the date when each former Confederate state was readmitted by presidential order to the Union, as well as the date when a traditional white conservative elite took office as a majority in each state—an event white Southerners liked to call "redemption." • *What had to happen for a state to be readmitted to the Union? What had to happen before a state could experience "redemption"?*

Radicals in Congress immediately proposed several bills that would require two-thirds of the justices to support any decision overruling a law of Congress, would deny the Court jurisdiction in Reconstruction cases, would reduce its membership to three, and would even abolish it. The justices apparently took notice. Over the next two years, the Court refused to accept jurisdiction in any cases involving Reconstruction.

THE IMPEACHMENT OF ANDREW JOHNSON

President Johnson had long since ceased to be a serious obstacle to the passage of Radical legislation, but he was still the official charged with administering the Reconstruction programs. As such, the Radicals believed he remained a major impediment to their plans. Early in 1867, they began looking for reasons to begin formal efforts that would result in his **impeachment**. Republicans found it, they believed, when Johnson dismissed Secretary of War Stanton despite Congress's refusal to agree. Elated Radicals in the House quickly impeached the president for violating the recently passed Tenure of Office Act and sent the case to the Senate for trial.

The trial lasted throughout April and May 1868 and eventually broke Johnson's way. The Radicals put heavy pressure on all the Republican senators, but the Moderates vacillated. On the first three charges to come to a vote, seven Republicans joined the Democrats and independents to support acquittal. The Senate vote was 35 to 19, one vote short of the constitutionally required two-thirds majority needed to remove Johnson from office. After that, the Radicals dropped the impeachment effort.

THE SOUTH IN RECONSTRUCTION

Reconstruction may not have immediately accomplished what its most ambitious and hopeful framers intended, but it did have profound effects on the South.

POLITICS

One area of remarkable achievement during Reconstruction was Black politics. As hundreds of thousands of Black men began to vote for the very first time, they changed the South's political culture. They helped appoint or elect up to 2,000 Blacks to public offices, from small town mayors to congressional posts and the governorship of Louisiana. A total of 265 Blacks served as state legislators, 100 of whom had been born into slavery. At the federal level, the Mississippi Legislature appointed Hiram Revels to a seat in the Senate in 1870. An ordained minister with the African Methodist Episcopal

Library of Congress, Prints and Photographs Division
[LC-DIG-ppmsca-15783]

CRITICS' VIEW OF RECONSTRUCTION This Reconstruction-era cartoon expresses the view held by Southern white Democrats that they were being oppressed by Northern Republicans. President Grant (whose hat bears Abraham Lincoln's initials) rides in comfort in a giant carpetbag, guarded by bayonet-wielding soldiers, as the South staggers under the burden in chains. Evidence of military occupation is in the scarred background.

Church, he was the first Black to serve in Congress. That same year, Joseph Rainey of South Carolina broke the color line in the House of Representatives. Between 1870 and 1901, 22 Blacks took seats in Congress, 20 in the U.S. House of Representatives and 2 in the Senate.

Most white Southerners had no patience for Reconstruction governments led by Blacks or their supporters. They branded the few southern white Republicans with the derogatory term **scalawags**, slang for "scoundrels." These "scalawags" were typically former Whigs who had never felt comfortable in the Democratic Party or farmers who lived in remote areas where there had been little or no slavery. There was also resentment toward what were termed **carpetbaggers**, white Northerners, most veterans of the Union army, who sought their fortunes in the South as planters, businessmen, or professionals after the war's end. The term refers to their use of a cheap travel bag made from carpeting material.

EDUCATION

Among the most important accomplishments of Reconstruction was a dramatic change in Southern education. The impetus for educational reform in the South came from outside groups—the Freedmen's Bureau, Northern religious and private philanthropic organizations, the many Northern white women who traveled to the South to teach in freedmen's schools—and most importantly from Black Americans themselves.

Over the opposition of many Southern whites, who feared that education would give Blacks "false notions of equality," these reformers established a large network of schools for those freed from slavery—4,000 schools by 1870, staffed by 9,000 teachers (half of them Black), teaching 200,000 students. In the 1870s, Reconstruction governments began to build a comprehensive public school system. By 1876, more than half of all white children and about 40 percent of all Black children were attending schools in the South. Almost all were strictly segregated.

In addition to witnessing the emergence of primary and secondary education, Reconstruction also saw the growth of Black higher education and the formation of a network of what we now call Historically Black Colleges and Universities (HBCUs). Every southern government founded at least one HBCU during Reconstruction, but funded them unevenly. The Morrill Act of 1862 led to the creation of land-grant public universities dedicated to engineering, but these were segregated. Private HBCU's were founded in large part to alleviate the shortage of colleges and universities open to Blacks.

African Institute (now Cheyney University) was the first HBCU, founded in 1837 in Pennsylvania, by Quaker philanthropist Richard Humphries. Three others began before the Civil War, including Wilberforce in Ohio in 1856, which was the first university founded by a Black denomination, in this case the African Methodist Episcopal Church. Most HBCUs, though, opened their doors in the decades after the Civil War. Nearly all were in the South and many were begun by churches, especially Black but also white. The high point was 1867, when nine were founded.

LANDOWNERSHIP AND TENANCY

The most ambitious goal of the Freedmen's Bureau, shared by some Republican Radicals in Congress, was to reform land ownership in the South. By June 1865, the Bureau had settled nearly 10,000 Black families on their own land—most of it abandoned plantations in areas occupied by the Union army. Soon, however, the original plantation owners were demanding the restoration of their property. President Johnson, along with many Republicans, supported their demands, believing that openly confiscating land was going too far. The government eventually returned most of the land in question to the original white owners.

Even so, the distribution of land ownership in the South changed considerably in the postwar years. Among whites, there was a striking decline in land ownership, from 80 percent before the war to 67 percent by the end of Reconstruction. Some whites lost their land because of unpaid debt or increased taxes; others left the marginal lands they had owned to move to more fertile areas, where they rented. Among Blacks, during the same period, the proportion of landowners rose from virtually none to more than 20 percent.

Still, most Blacks did not own their own land and worked for others in one form or another. Many Black agricultural laborers—perhaps 25 percent—worked for wages. Most, however, became tenants of white landowners—working rented plots of land and paying their landlords either in cash or a share of their crops (hence the term **sharecropping**). As sharecroppers, Blacks enjoyed at least some degree of physical independence and had the sense of working their own land, even if they could rarely hope to buy it. But tenantry greatly benefited white landlords, relieving them of responsibility for the physical well-being of their workers and giving them control over their land and some power over Black laborers.

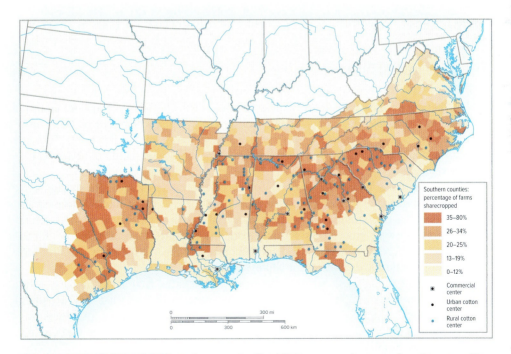

THE CROP-LIEN SYSTEM IN 1880 In the years after the Civil War, more and more southern farmers—white and Black—became tenants or sharecroppers on land owned by others. This map shows the percentage of farms that were within the so-called crop-lien system. Note the high density of sharecropping and tenant farming in the most fertile areas of the Deep South, the same areas where slaveholding had been most dominant before the Civil War. • *How did the crop-lien system contribute to the shift in southern agriculture toward one-crop farming?*

Many Blacks and some poor whites found themselves virtually imprisoned by the **crop-lien system.** Few of the traditional institutions of credit in the South returned after the war. In their stead emerged a new credit system, centered in large part on local country stores—some of them owned by planters, others by independent merchants. Blacks and whites, landowners and tenants—all depended on these stores. Since most who made their livelihood from the land did not have a steady cash flow, they typically purchased on credit from these merchants, who had no competition and thus could set interest rates as high as 50 or 60 percent. Poor farmers had to give the merchants a lien (or claim) on their crops as collateral for the loans (thus the term *crop-lien*). Those who suffered a few bad years in a row, as many did, could become trapped in a cycle of debt from which they could never escape.

Some Blacks who had acquired land during the early years of Reconstruction, and many poor whites who had owned land for years, gradually lost it as they fell into debt. Southern farmers became almost wholly dependent on cash crops—and most of all on cotton—because only such marketable commodities seemed to offer any possibility of escape from debt. The relentless planting of cotton ultimately contributed to soil exhaustion, which further undermined the Southern agricultural economy.

THE GRANT ADMINISTRATION

American voters in 1868 yearned for a strong, stable figure to guide them through the troubled years of Reconstruction. They turned trustingly to General **Ulysses S. Grant.**

The Soldier President

Grant could have had the nomination of either party in 1868. But believing that Republican Reconstruction policies were more popular in the North, he accepted the Republican nomination. The Democrats nominated former governor Horatio Seymour of New York. The campaign was a bitter one, and Grant's triumph was surprisingly narrow. Indeed, without the 500,000 new Black Republican voters in the South, he would not have won the popular vote.

Grant entered the White House with no formal political experience. He steadily supported the citizenship of formerly enslaved people and the policies of Radical Reconstruction. Refusing to compromise on matters of racial politics, he alienated some white Northerners who were growing disillusioned with the prolonged federal presence in the South and the effort to protect and extend democracy to Black Americans. And he was certainly not viewed favorably by most white Southerners.

By the end of Grant's first term, members of a substantial faction of the party—who referred to themselves as Liberal Republicans—had come to oppose what they derisively called "Grantism." Some Republicans suspected, correctly as it turned out, that there was a level of hidden corruption in the Grant administration. In 1872, hoping to prevent Grant's reelection, they nominated their own presidential candidate: Horace Greeley, veteran editor and publisher of the *New York Tribune.* The Democrats, somewhat reluctantly, named Greeley their candidate as well, hoping that the alliance with the Liberals would enable them to defeat Grant. But the effort was in vain. Grant won a substantial victory, polling 286 electoral votes to Greeley's 66.

The Grant Scandals

During the 1872 campaign, a series of scandals came to light that would plague Grant and the Republicans for the next four years. They became political fodder for white Southerners eager to discredit the Republican administration and its policies of racial equality. The first involved the French-owned Crédit Mobilier construction company, which had helped build the Union Pacific Railroad. The heads of Crédit Mobilier had used their positions as Union Pacific stockholders to steer large fraudulent contracts to their construction company, thus bilking the Union Pacific of millions. To prevent investigations, the directors had given Crédit Mobilier stock to key members of Congress. In 1872, a Congressional investigation revealed that some highly placed Republicans—including Grant's vice president, Schuyler Colfax—had accepted stock.

One dreary episode of malfeasance followed another in Grant's second term. Benjamin H. Bristow, Grant's third Treasury secretary, discovered that some of his officials and a group of distillers operating as a "whiskey ring" were cheating the government out of taxes by filing false reports. Then a House investigation revealed that William W. Belknap, secretary of war, had accepted bribes to retain an Indian-post trader in office (the so-called Indian ring). Other, lesser scandals also added to the growing impression that "Grantism" had brought rampant corruption to government.

The Greenback Question

Compounding Grant's problems was a financial crisis, known as the Panic of 1873. It began with the failure of a leading investment banking firm, Jay Cooke and Company, which had invested too heavily in postwar railroad building. There had been panics before—in 1819, 1837, and 1857—but this was the worst one yet.

Debtors now pressured the government to redeem federal war bonds with greenbacks, which would increase the amount of money in circulation. But Grant and most Republicans wanted a "sound" currency—based solidly on gold reserves—that would favor the interests of banks and other creditors. There was approximately $356 million in paper currency issued during the Civil War that was still in circulation. In 1873, the Treasury issued more currency in response to the panic. Banks were going under, unemployment rising, personal bankruptcies mounting, and farms being foreclosed on. In 1875, Republican leaders in Congress passed the Specie Resumption Act, which provided that after January 1, 1879, greenback dollars would be redeemed by the government and replaced with new certificates, firmly pegged to the price of gold. The law satisfied creditors, who had worried that debts would be repaid in paper currency of uncertain value. But "resumption" made things more difficult for debtors because the gold-based money supply could not easily expand.

In 1875, the "Greenbackers" formed their own political organization: the National Greenback Party. It failed to gain widespread support, but the money issue was to remain one of the most enduring issues in late-nineteenth-century American politics.

Republican Diplomacy

The Johnson and Grant administrations achieved their most significant successes in foreign affairs. These successes were engineered by two outstanding secretaries of state: William H. Seward and Hamilton Fish.

An ardent expansionist, Seward acted with as much daring as the demands of Reconstruction politics and the Republican hatred of President Johnson would permit. He accepted a Russian offer to buy Alaska for $7.2 million, despite criticism from many who derided the purchase as "Seward's Folly." In 1867, Seward also engineered the American annexation of the tiny Midway Islands, west of Hawaii.

Hamilton Fish's first major challenge was resolving the long-standing controversy over the American claims that Britain had violated neutrality laws during the Civil War by permitting English shipyards to build ships (among them the *Alabama*) for the Confederacy. American demands that England pay for the damage these vessels had caused became known as the "*Alabama* claims." In 1871, after a number of failed efforts, Fish forged an agreement, the Treaty of Washington, under which the British paid $15.5 million in damages and formally apologized. It marked the start of the long-running alliance between two nations that still exists today

THE ABANDONMENT OF RECONSTRUCTION

As the North grew increasingly preoccupied with its own political and economic problems, interest in Reconstruction began to wane. By the time Grant left office, Democrats had taken back eight of the governments of the former Confederate states. For three other states—South Carolina, Louisiana, and Florida—the end of Reconstruction had to wait for the withdrawal of the last federal troops in 1877.

THE SOUTHERN STATES FOR SOUTHERN WHITES

In the states where whites constituted a majority—the states of the upper South—overthrowing Republican control was relatively simple. By 1872, all but a handful of Southern whites had regained suffrage. Now a clear majority, they needed only to organize and elect their candidates—and control Black voters. Indeed, the key for whites seeking to regain political power was to manipulate the Black population through any means necessary.

In states where Blacks were a majority or where the populations of the two races were almost equal, whites used outright intimidation to undermine the Reconstruction regimes. Though these tactics were most pronounced in these states, they were universally deployed across the former Confederacy. Some planters refused to rent land to Republican Blacks; storekeepers refused to extend them credit; and employers refused to give them work, or when they did, pay them fairly. More significantly, secret societies—the **Ku Klux Klan**, the Knights of the White Camellia, the White League, the Red Shirts, and others—used terrorism. On October 25, 1870, in Eutaw, Alabama, the Klan first harassed and then began shooting into a political rally attended by about 2,000 Blacks, killing 4 and wounding 54. Black residents remembered it as the Eutaw Massacre. On election eve of 1874, in Eufaula, Alabama, members of the White League killed 7 Black men trying to vote and injured 70 more. This level of violence against Blacks, tragically, was common across the South.

The Republican Congress responded to this wave of repression with the **Enforcement Acts** of 1870 and 1871 (better known as the Ku Klux Klan Acts), which prohibited states from discriminating against voters on the basis of race and gave the national government the authority to prosecute crimes by individuals under federal law. The laws also authorized the president to use federal troops to protect civil rights—a provision President Grant used in 1871 in nine counties of South Carolina. The Enforcement Acts, although seldom enforced, discouraged Klan violence, which declined by 1872.

Waning Northern Commitment

The Northern commitment to civil rights did not last long. After the adoption of the Fifteenth Amendment in 1870, some reformers convinced themselves that their long campaign on behalf of Black people was now over, that with the vote Blacks ought to be able to take care of themselves. Former Radical leaders such as Charles Sumner and Horace Greeley now began calling themselves Liberals and cooperating with the Democrats. Within the South itself, many white Republicans moved into the Democratic Party as voters rejected Republican politicians whom they blamed for the financial crisis.

The Panic of 1873 further undermined support for Reconstruction. In the congressional elections of 1874, the Democrats won control of the House of Representatives for the first time since 1861. To appeal to Southern white voters, Grant even reduced the use of military force to prop up the Republican regimes in the South.

The Compromise of 1877

Grant had hoped to run for another term in 1876, but most Republican leaders resisted, shaken by recent Democratic successes and scandals in the White House. Instead, they settled on Rutherford B. Hayes, three-time governor of Ohio and a champion of civil service reform. The Democrats united behind Samuel J. Tilden, the reform governor of New York, who had been instrumental in overthrowing the corrupt Tweed Ring of New York City's Tammany Hall.

Although the campaign was bitter, few differences of principle distinguished the candidates. The election produced an apparent Democratic victory. Tilden carried the South and several large Northern states, and his popular margin over Hayes was nearly 300,000 votes.

Candidate (Party)	Electoral Vote	Popular Vote (%)
Rutherford B. Hayes (Republican)	185	4,036,298 (48)
Samuel J. Tilden (Democratic)	184	4,300,590 (51)

81.8% of electorate voting

THE ELECTION OF 1876 The election of 1876 was one of the most controversial in American history. As in the elections of 1824, 1888, 2000, and 2016, the winner of the popular vote—Samuel J. Tilden—was not the winner of the electoral vote. The final decision as to who would be president was not made until the day before the official inauguration in March. • *How did the Republicans turn this apparent defeat into a victory?*

But disputed returns from Louisiana, South Carolina, Florida, and Oregon, whose electoral votes totaled 20, threw the election in doubt. Hayes could still win if he managed to receive all 20 votes.

The Constitution had established no method to determine the validity of disputed returns. The decision clearly lay with Congress, but it was not obvious with which house or through what method. (The Senate was Republican, and the House was Democratic.) Members of each party naturally supported a solution that would yield them the victory. Finally, late in January 1877, Congress tried to break the deadlock by creating a special electoral commission composed of five senators, five representatives, and five justices of the Supreme Court. The congressional delegation consisted of five Republicans and five Democrats. The Court delegation included two Republicans, two Democrats, and the only independent, Justice David Davis. But when the Illinois legislature elected Davis to the U.S. Senate, the justice resigned from the commission. His seat went instead to a Republican justice. The commission voted along straight party lines, 8 to 7, awarding every disputed vote to Hayes.

Behind this seemingly partisan victory lay a series of elaborate and sneaky compromises among leaders of both parties. When a Democratic filibuster threatened to derail the electoral commission's report, Republican Senate leaders met secretly with Southern Democratic leaders. As the price of their cooperation, the Southern Democrats exacted several pledges from the Republicans, which became known as the **Compromise of 1877**: the appointment of at least one Southerner to the Hayes cabinet, control of federal patronage in their areas, generous internal improvements, federal aid for the Texas and Pacific Railroad, and most important, withdrawal of the remaining federal troops from the South.

In his inaugural address, Hayes announced that the South's most pressing need was the restoration of "wise, honest, and peaceful local self-government," and he soon withdrew the troops and let white Democrats take over the remaining Southern state governments. The outcome of the election created such bitterness that not even Hayes's promise to serve only one term could mollify his critics.

The president and his party hoped to build up a "new Republican" organization in the South committed to modest support for Black rights. Although many white Southern leaders sympathized with Republican economic policies, resentment of Reconstruction was so deep that supporting the party became politically impossible. The "solid" Democratic South, which would survive until the mid-twentieth century, was taking shape.

THE LEGACY OF RECONSTRUCTION

Reconstruction was a time when the United States made great strides toward the promise of democracy for all. Never before had the civil rights of citizens been so clearly enumerated or extended to those other than landholding white men; never before had the power of the federal government been marshaled so vigorously to protect these rights. In particular, Black Americans achieved a new level of dignity and power unimaginable only a few years earlier. Perhaps most important, Black Americans themselves managed to carve out a society and culture of their own and to create or strengthen their own institutions.

Still, Reconstruction did not bring long-lasting equality. Within little more than a decade after a devastating war, the white South had largely regained control of its own institutions and, to a great extent, restored its traditional ruling class to power. It would soon dismantle many of the legal freedoms won by African Americans and reinforce its prevailing ideology of white supremacy. Reconstruction was notable for its limitations. The United States had largely failed in its first serious effort to resolve its oldest and deepest social problem—the problem of racial injustice.

Given the odds confronting them, however, Black Americans had reason for considerable pride in the gains they were able to make during Reconstruction. And future generations would be grateful for the two great charters of freedom—the Fourteenth and Fifteenth Amendments to the Constitution—which, although widely ignored at the time, would one day serve as the basis for a "Second Reconstruction" that would renew the fight to bring freedom to all Americans.

THE NEW SOUTH

The Compromise of 1877 was supposed to be the first step toward developing a stable, permanent Republican Party in the South. In that respect, at least, it failed. In the years following the end of Reconstruction, white Southerners established the Democratic Party as the only viable political organization for the region's whites. Even so, the South did change in some of the ways that the framers of the Compromise had hoped.

THE "REDEEMERS"

Many white Southerners rejoiced at the restoration of what they liked to call "home rule." But in reality, political power in the region was soon more restricted than at any time since the Civil War. Once again, most of the South fell under the control of a powerful, conservative white ruling class, whose members smugly called themselves "**Redeemers**" for their efforts to "redeem" the South from the clutches of Republican rule.

In some places, this post-Reconstruction ruling class was much the same as it had been in the antebellum period. In Alabama, for example, the old planter elite retained much of its former power. In other areas, however, the Redeemers constituted a genuinely new ruling class of merchants, industrialists, railroad developers, and financiers. Some were former planters, some northern immigrants, some ambitious, upwardly mobile white Southerners from the region's lower social tiers. They combined a defense of "home rule" and social conservatism with a commitment to economic development.

The various Redeemer governments of the New South behaved in many respects quite similarly. Virtually all the new Democratic regimes lowered taxes, reduced public spending, and drastically diminished state services. One state after another eliminated or cut its support for public school systems. All ignored Black freedoms and violently suppressed it.

INDUSTRIALIZATION AND THE NEW SOUTH

Many white southern leaders in the post-Reconstruction era hoped to see their region develop a vigorous industrial economy, a "**New South**." Henry Grady, editor of the *Atlanta Constitution,* and other New South advocates seldom challenged white supremacy, but they did promote the virtues of thrift, industry, and progress—qualities that prewar Southerners had often denounced in northern society.

Southern industry did expand dramatically in the years after Reconstruction, most visibly in textile manufacturing. In the past, southern planters had usually shipped their cotton to manufacturers in the North or in Europe. Now textile factories appeared in the South itself—many of them drawn to the region from New England by the abundance of water power, the ready supply of cheap labor, the low taxes, and the accommodating conservative governments. The tobacco-processing industry similarly established an important foothold in the region. In the lower South, and particularly in Birmingham, Alabama, the iron (and, later, steel) industry grew rapidly.

Railroad development also increased substantially in the post-Reconstruction years. Between 1880 and 1890, trackage in the South more than doubled. And in 1886, the South changed the gauge (width) of its trackage to correspond with the standards of the North. No longer would it be necessary for cargoes heading into the South to be transferred from one train to another at the borders of the region.

Yet southern industry developed within strict limits, and its effects on the region were never even remotely comparable to the effects of industrialization on the North. The southern share of national manufacturing doubled in the last twenty years of the century, but it was still only 10 percent of the total. Similarly, the region's per capita income increased 21 percent in the same period, but average income in the South was still only 40 percent of that in the North; in 1860 it had been more than 60 percent. And even in those industries where development had been most rapid—textiles, iron, railroads—much of the capital had come from, and many of the profits thus flowed to, the North.

The growth of southern industry required the region to recruit a substantial industrial workforce for the first time. From the beginning, a high percentage of factory workers were women. Heavy male casualties in the Civil War had created a large population of unmarried women who desperately needed employment. Hours were long (often as much as twelve hours a day), and wages were far below the northern equivalent; indeed, one of the greatest attractions of the South to industrialists was that employers were able to pay workers there as little as one-half of what northern workers received. Life in most mill towns was rigidly controlled by the owners and managers of the factories, who rigorously suppressed attempts at protest or union organization. Company stores sold goods to workers at inflated prices and issued credit at exorbitant rates (much as country stores did in agrarian areas), and mill owners ensured that no competing merchants were able to establish themselves in the community.

Although some industries, such as textiles, did not offer many opportunities to Black workers, a number of others, such as tobacco, iron, and lumber, did. In these industries, and the towns that hosted them, Black and white cultures came into close contact. This contact fueled the determination of local white leaders to protect white supremacy.

Black Americans and the New South

The "New South creed" was not popular with whites alone. Many Black Americans were attracted to the vision of progress and self-improvement as well. Some formerly enslaved people (and, as the decades passed, their offspring) succeeded in elevating themselves into the middle class: acquiring property, establishing small businesses, or entering professions. The Black middle class heavily funded and supported the expansion of Black colleges and institutes.

The chief spokesman for this commitment to education was **Booker T. Washington**, founder and president of the Tuskegee Institute in Alabama. Born into slavery, Washington had worked his way out of poverty after acquiring an education (at Virginia's Hampton Institute). He urged others to follow the same road to self-improvement. Washington's message was both cautious and hopeful. Black Americans should attend school, learn skills, and establish a solid footing in agriculture and the trades. Industrial, not classical, education should be their goal. Blacks should, moreover, refine their speech, improve their dress, and adopt habits of thrift and personal cleanliness; they should, in short, adopt the standards of the white middle class. Only thus, Washington claimed, could they win the respect of the white population.

In a famous speech in Georgia in 1895, Washington outlined a controversial approach that became widely known as the **Atlanta Compromise**. Blacks, he said, should forgo agitation for political rights and concentrate on self-improvement and preparation for equality. Washington offered a powerful challenge to those whites who wanted to discourage Black Americans from acquiring an education or winning economic gains. But his message also assured whites that Blacks would not overtly challenge the emerging system of segregation.

The Lost Cause

Almost immediately after the end of the Civil War, and gaining momentum during Reconstruction, white Southerners started to frame their failed fight in ways that depicted themselves in the best possible light. The effort formally began when Edward Pollard published the *Lost Cause: A New Southern History of the War of the Confederates,* in 1866. His account of the war pitted invading Yankees against gallant Confederates seeking to protect their homes and their rights to happiness and property. The "Lost Cause" quickly became a vital part of white southern culture.

White women emerged as leaders of the Lost Cause. They organized cemeteries for fallen Confederates, raised money for statues, and campaigned for local and state "Confederate Memorial Days." More importantly, they ensured that the "Lost Cause" depiction of the Civil War would be part of the curriculum in every public school for generations.

Lost Cause history depicted almost all Confederate leaders as saints, none more so than General Robert E. Lee, who came to symbolize the self-sacrifice of men defending a culture of honor from greedy Northerners. According to this interpretation, the war was not about slavery but northern avarice. Indeed, slavery was presented as a benighted institution that boosted the national economy and reflected a God-ordained racial hierarchy, good for both white and Blacks, who were content to live out their lives serving their moral and intellectual superiors. White women were depicted as defenders of the Confederacy and symbols of purity in need of protection from Black men, who invariably sought to sexually assault them.

Lost Cause mythology was perhaps most visible in public squares. Hundreds of southern towns and cities erected monuments to the Confederacy, accompanied by descriptions of the righteousness of the cause. Officials named streets, schools, and parks after Confederate generals. Formal organizations sprang up to preserve and extend this public culture of the Lost Cause: most notably, the United Daughters of the Confederacy in 1894 and the United Sons of the Confederate Veterans (now simply the Sons of Confederate Veterans) in 1896. Both are still active today.

Public debate about these Confederate monuments and Confederate namesakes has become widespread in the twenty-first century. The nation has begun grappling with questions about what public commemoration says about its values. Heroes are expected to represent exemplary ideas and behavior; if someone enshrined in a statue fought to preserve slavery and racial hierarchy, under what circumstances is it proper to leave it in place? These are difficult but pressing questions, calling Americans to think about how they express what they care about most.

The Birth of Jim Crow

As the popularity of Lost Cause mythology suggests, white Southerners strongly resisted the idea of racial equality. The federal support that had been crucial to protecting the rights of Black Southerners all but vanished when the last federal troops withdrew and the Supreme Court stripped the Fourteenth and Fifteenth Amendments of much of their significance. In the so-called civil rights cases of 1883, the Court ruled that while the Fourteenth Amendment prohibited state governments from discriminating against people because of race, it did not restrict private organizations or individuals from doing so.

Eventually, the Court also validated state legislation that institutionalized the separation of the races. In *Plessy v. Ferguson* (1896), a case involving a Louisiana law that required segregated seating on railroads, the Court held that separate accommodations did not deprive Blacks of equal rights if the accommodations were equal. In *Cumming v. County Board of Education* (1899), the Court ruled that communities could establish schools for whites only, even if there were no comparable schools for Blacks.

White Southerners were particularly determined to strip African Americans of the right to vote. In some states, disenfranchisement had begun almost as soon as Reconstruction ended. But in other areas, Black voting continued for some time after Reconstruction—largely because conservative whites believed they could control the Black electorate and use it to beat back the attempts of poor white farmers to take control of the Democratic Party. In the 1890s, franchise restrictions became much more rigid. During those years, some small white farmers, sensing that the Black vote was being used against them, began to demand complete Black disenfranchisement. At the same time, many members of the conservative elite began to fear that poor whites might unite politically with poor Blacks to challenge them.

In devising laws to disenfranchise Black voters, the southern states had to find ways to evade the Fifteenth Amendment. Two devices emerged to accomplish this goal: the poll tax, or some form of property qualification (few Blacks were prosperous enough to meet such requirements); and the "literacy" or "understanding" test, which required voters to demonstrate an ability to read and to interpret the Constitution. Even those African Americans who could read had a hard time passing the difficult test white officials gave them, which often required them to interpret an arcane part of the Constitution to the satisfaction of a white elected official. (The laws affected poor white voters as well as Blacks.) By the late 1890s, the Black vote had decreased by 62 percent, the white vote by 26 percent.

Popular culture reflected these political developments. The rise of **minstrel shows**—slapstick dramatic representations of Black culture—typically embodied racist ideas. "Corked-up" whites (a reference to black makeup made by burning cork) grossly caricatured African Americans as silly, unintelligent, sensual, and immoral. Late in the 1800s, Blacks founded their own minstrel shows in part to modify these stereotypes, though with only modest success. (See "Patterns of Popular Culture: The Minstrel Show.")

Laws restricting the franchise and segregating schools were only part of a network of state and local statutes—collectively known as the **Jim Crow laws**—that by the first years of the twentieth century had institutionalized an elaborate system of racial hierarchy reaching into almost every area of southern life. Blacks and whites could not ride together in the same railroad cars, sit in the same waiting rooms, use the same washrooms, eat in the same restaurants, or sit in the same theaters. They had to live in different sections of the same town or city. Blacks had no access to many public parks, beaches, or picnic areas; they could not be patients in many hospitals. The Jim Crow laws stripped Blacks of many of the modest social, economic, and political gains they had made in the late nineteenth century.

The 1890s witnessed a dramatic increase in white violence against Blacks, which served to inhibit Black protest against Jim Crow or agitation for equal rights. The worst of such violence was the lynching of Blacks by white mobs, which reached appalling levels. In the 1890s, there was an average of 187 lynchings each year, more than 80 percent of them in the South. The vast majority of victims were Black men accused of crimes they did not commit and who rarely enjoyed due process or a fair trial. Those who participated in lynchings often saw their actions as a legitimate form of law enforcement, but also as a way to deliver a stark message about the power of white supremacy.

The rise of lynchings shocked the conscience of many white Americans in a way that other forms of racial injustice did not. In 1892, **Ida B. Wells**, a committed Black journalist, published a series of impassioned articles after the lynching of three friends in Memphis, Tennessee; her articles launched what became an international antilynching movement. The movement gradually attracted substantial support from whites in the North and even the South, particularly from women. Its goal was a federal antilynching law, which would allow the national government to do what state and local governments in the South were generally unwilling to do: punish those responsible for lynchings. Although such a law was introduced to Congress in 1918 and reintroduced in the 1920s, 1930s, and 1940s, it never made it past southern senators.

Library of Congress Prints and Photographs Division Washington, D.C. 20540 USA [LC-USZ62-29285]

LYNCHING OF HENRY SMITH, PARIS, TEXAS, 1893 A large, almost festive crowd numbering about 10,000 gathered to watch the lynching of Henry Smith. Accused of murdering a four-year-old white girl who was the daughter of a law enforcement officer, Smith was mutilated and burned on this scaffold. Over 4,000 lynchings occurred between the end of slavery and the late 1930s, mostly in the South. They reached their peak in the 1890s and the first years of the twentieth century. Lynchings such as this one—publicized well in advance and attracting whole families who traveled great distances to see them—were relatively infrequent. Most lynchings were the work of smaller groups, operating with less visibility.

PATTERNS OF POPULAR CULTURE

The Minstrel Show

The minstrel show was one of the most popular forms of entertainment in the United States in the second half of the nineteenth century. It was also a testament to the high level of racism in American society both before and after the Civil War. Minstrel performers were mostly white, usually disguised as Black. But African American performers also formed their own minstrel shows, transforming them into vehicles for training Black entertainers and developing new forms of music and dance.

Before the Civil War, white minstrel show performers blackened their faces with cork and presented grotesque stereotypes of the slave culture of the American South. Among the most popular of the stumbling, ridiculous characters invented for these shows were such figures as "Zip Coon" and "Jim Crow" (whose name later resurfaced as a label for late-nineteenth-century segregation laws). A typical minstrel show featured a group of men seated in a semicircle facing the audience. The man in the center ran the show, played the straight man for the jokes of others, and led the music—lively dances and sentimental ballads played on banjos, castanets, and other instruments and sung by soloists or the entire group.

Library of Congress Prints & Photographs Division [LC-USZ62-2659]

MINSTRELSY AT HIGH TIDE The Primrose & West minstrel troupe—a lavish and expensive entertainment that drew large crowds in the late 1800s—was one of many companies to offer this brand of entertainment to eager audiences all over the country. It featured interracial members enacting idealized versions of elite white culture and lampooning Black culture and Black people. It was part of the wider world of minstrelsy, which built on white performers "blackening" their faces and performing skits and songs that mocked Black Americans. Black entertainers formed their own troupes as well, offering a range of comedy and satire for Black audiences, a well as versions that fed racial stereotypes to white audiences.

After the Civil War, white minstrels began to expand their repertoire. Drawing from the famous and successful freak shows of P. T. Barnum and other entertainment entrepreneurs, some began to include Siamese twins, bearded ladies, and even a supposedly eight-foot two-inch "Chinese giant" in their shows. They also incorporated sex, both by including women in some shows and, even more popularly, by recruiting female impersonators. One of the most successful minstrel performers of the 1870s was Francis Leon, who delighted crowds with his female portrayal of a flamboyant "prima donna."

One reason white minstrels began to move in these new directions was that they were facing competition from Black performers, who could provide more-authentic versions of Black music, dance, and humor and usually brought more talent to the task. The Georgia Minstrels, organized in 1865, was one of the first all-Black minstrel troupes and had great success in attracting white audiences in the Northeast. By the 1870s, touring African American minstrel groups were numerous. Black minstrels used many of the conventions of the white shows. There were dances, music, comic routines, and sentimental recitations. Some performers even chalked their faces to make themselves look as dark as the white blackface performers with whom they were competing. Black minstrels sometimes denounced slavery (at least indirectly) and did not often demean Black culture or intelligence. But they could not entirely escape caricaturing African American life.

The Black minstrel shows did help develop some important forms of African American entertainment and transform them into a part of the national culture. Black minstrels introduced new forms of dance, derived from the informal traditions of slavery and Black community life. They showed the "buck and wing," the "stop time," and the "Virginia essence," which established the foundations for the tap and jazz dancing of the early twentieth century. They also improvised musically and began experimenting with forms that over time contributed to the growth of ragtime, jazz, and rhythm and blues.

Eventually, Black minstrelsy—like its white counterpart—evolved into other forms of theater, including the beginnings of serious Black drama. At Ambrose Park in Brooklyn in the 1890s, for example, the celebrated Black comedian Sam Lucas (a veteran of the minstrel circuit) starred in the play *Darkest America,* which one Black newspaper later described as a "delineation of Negro life, carrying the race through all their historical phases from the plantation, into reconstruction days and finally painting our people as they are today, cultured and accomplished in the social graces, [holding] the mirror faithfully up to nature."

Interest in the minstrel show did not die altogether. In 1927, Hollywood released *The Jazz Singer,* the first feature film with sound. It was about the career of a white minstrel performer, and its star was one of the most popular singers of the twentieth century: Al Jolson, whose career had begun on the blackface minstrel circuit years before. •

UNDERSTAND, ANALYZE, & EVALUATE

1. How did minstrel shows performed by white minstrels reinforce prevailing attitudes toward African Americans?
2. Minstrel shows performed by Black minstrels often conformed to existing stereotypes of African Americans. Why?
3. Can you think of any popular entertainments today that carry remnants of the minstrel shows of the nineteenth century?

CONCLUSION

Reconstruction changed the nation. In the North, it solidified the power of the Republican Party. The rapid expansion of the northern economy accelerated, drawing more and more of its residents into a burgeoning commercial world. In the South, Reconstruction fundamentally rearranged the relationship between white and Black citizens. Black Americans managed to carve out a much larger sphere of social and cultural activity than they had ever been able to create under slavery. Black churches proliferated in great numbers. Black schools, printing presses, and colleges were founded and flourished. New South advocates pioneered new industrial advances in parts of the region.

But in many ways the South remained what it had always been: a largely rural society with a sharply defined class and racial structure. The great majority of Blacks farmed not as landowners but as tenants and sharecroppers. The result was a form of economic bondage, driven by debt, only scarcely less oppressive than the legal bondage of slavery. Black Americans initially participated actively and effectively in southern politics. After a few years of widespread Black voting and officeholding, however, the forces of white supremacy shoved Black Americans to the margins of the southern political world, where they would mostly remain until the 1960s. Indeed, the South maintained a deep commitment among its white citizens to the subordination of Black Americans—a commitment solidified in the 1890s and the early twentieth century when white Southerners erected an elaborate legal system of segregation known as the Jim Crow laws. Tragically, the promise of the great Reconstruction amendments to the Constitution—the Fourteenth and Fifteenth—remained largely unfulfilled in the South as the century drew to its close.

KEY TERMS/PEOPLE/PLACES/EVENTS

Andrew Johnson 357
Atlanta Compromise 373
Black Codes 359
Booker T. Washington 372
carpetbagger 364
Charles Sumner 353
Compromise of 1877 370
crop-lien system 366
Enforcement Acts 368
Fifteenth Amendment 361
Fourteenth Amendment 359
Freedmen's Bureau 352
Ida B. Wells 375
impeachment 362
Jim Crow laws 374
Ku Klux Klan 368
minstrel show 374
New South 371
Plessy v. Ferguson 374
Radical Republicans 353
Reconstruction 353
Redeemers 371
scalawag 364
sharecropping 365
Thaddeus Stevens 353
Ulysses S. Grant 366
Wade-Davis Bill 353
William H. Seward 355

RECALL AND REFLECT

1. What were the principal questions facing the nation at the end of the Civil War?
2. What were the achievements of Reconstruction? Where did it fail and why?
3. What new problems arose in the South as the North's interest in Reconstruction waned?
4. What was the Compromise of 1877, and how did it affect Reconstruction?
5. How did the "New" South differ from the South before the Civil War?

Design element: Stars and Stripes: McGraw Hill Education.

APPENDIX

The Declaration of Independence
The Constitution of the United States

THE DECLARATION OF INDEPENDENCE

In Congress, July 4, 1776,

THE UNANIMOUS DECLARATION OF THE THIRTEEN UNITED STATES OF AMERICA

When, in the course of human events, it becomes necessary for one people to dissolve the political bands which have connected them with another, and to assume, among the powers of the earth, the separate and equal station to which the laws of nature and of nature's God entitle them, a decent respect to the opinions of mankind requires that they should declare the causes which impel them to the separation.

We hold these truths to be self-evident, that all men are created equal; that they are endowed by their Creator with certain unalienable rights; that among these, are life, liberty, and the pursuit of happiness. That, to secure these rights, governments are instituted among men, deriving their just powers from the consent of the governed; that, whenever any form of government becomes destructive of these ends, it is the right of the people to alter or to abolish it, and to institute a new government, laying its foundation on such principles, and organizing its powers in such form, as to them shall seem most likely to effect their safety and happiness. Prudence, indeed, will dictate that governments long established, should not be changed for light and transient causes; and, accordingly, all experience hath shown, that mankind are more disposed to suffer, while evils are sufferable, than to right themselves by abolishing the forms to which they are accustomed. But, when a long train of abuses and usurpations, pursuing invariably the same object, evinces a design to reduce them under absolute despotism, it is their right, it is their duty, to throw off such government and to provide new guards for their future security. Such has been the patient sufferance of these colonies, and such is now the necessity which constrains them to alter their former systems of government. The history of the present King of Great Britain is a history of repeated injuries and usurpations, all having, in direct object, the establishment of an absolute tyranny over these States. To prove this, let facts be submitted to a candid world:

He has refused his assent to laws the most wholesome and necessary for the public good.

He has forbidden his governors to pass laws of immediate and pressing importance, unless suspended in their operation till his assent should be obtained; and, when so suspended, he has utterly neglected to attend to them.

He has refused to pass other laws for the accommodation of large districts of people, unless those people would relinquish the right of representation in the legislature; a right inestimable to them, and formidable to tyrants only.

He has called together legislative bodies at places unusual, uncomfortable, and distant from the depository of their public records, for the sole purpose of fatiguing them into compliance with his measures.

He has dissolved representative houses repeatedly for opposing, with manly firmness, his invasions on the rights of the people.

He has refused, for a long time after such dissolutions, to cause others to be elected; whereby the legislative powers, incapable of annihilation, have returned to the people at large for their exercise; the state remaining, in the meantime, exposed to all the danger of invasion from without, and compulsions within.

He has endeavored to prevent the population of these States; for that purpose, obstructing the laws for naturalization of foreigners, refusing to pass others to encourage their migration hither, and raising the conditions of new appropriations of lands.

He has obstructed the administration of justice, by refusing his assent to laws for establishing judiciary powers.

He has made judges dependent on his will alone, for the tenure of their offices, and the amount and payment of their salaries.

He has erected a multitude of new offices, and sent hither swarms of officers to harass our people, and eat out their substance.

He has kept among us, in time of peace, standing armies, without the consent of our legislatures.

He has affected to render the military independent of, and superior to, the civil power.

He has combined, with others, to subject us to a jurisdiction foreign to our Constitution, and unacknowledged by our laws; giving his assent to their acts of pretended legislation:

For quartering large bodies of armed troops among us:

For protecting them by a mock trial, from punishment, for any murders which they should commit on the inhabitants of these States:

For cutting off our trade with all parts of the world:

For imposing taxes on us without our consent:

For depriving us, in many cases, of the benefit of trial by jury:

For transporting us beyond seas to be tried for pretended offences:

For abolishing the free system of English laws in a neighboring province, establishing therein an arbitrary government, and enlarging its boundaries, so as to render it at once an example and fit instrument for introducing the same absolute rule into these colonies:

For taking away our charters, abolishing our most valuable laws, and altering, fundamentally, the powers of our governments:

For suspending our own legislatures, and declaring themselves invested with power to legislate for us in all cases whatsoever.

He has abdicated government here, by declaring us out of his protection, and waging war against us.

He has plundered our seas, ravaged our coasts, burnt our towns, and destroyed the lives of our people.

He is, at this time, transporting large armies of foreign mercenaries to complete the works of death, desolation, and tyranny, already begun, with circumstances of cruelty and perfidy scarcely paralleled in the most barbarous ages, and totally unworthy the head of a civilized nation.

He has constrained our fellow citizens, taken captive on the high seas, to bear arms against their country, to become the executioners of their friends, and brethren, or to fall themselves by their hands.

He has excited domestic insurrections amongst us, and has endeavored to bring on the inhabitants of our frontiers, the merciless Indian savages, whose known rule of warfare is an undistinguished destruction of all ages, sexes, and conditions.

In every stage of these oppressions, we have petitioned for redress, in the most humble terms; our repeated petitions have been answered only by repeated injury. A prince, whose character is thus marked by every act which may define a tyrant, is unfit to be the ruler of a free people.

Nor have we been wanting in attention to our British brethren. We have warned them, from time to time, of attempts made by their legislature to extend an unwarrantable jurisdiction over us. We have reminded them of the circumstances of our emigration and settlement here. We have appealed to their native justice and magnanimity, and we have conjured them, by the ties of our common kindred, to disavow these usurpations, which would inevitably interrupt our connections and correspondence. They, too, have been deaf to the voice of justice and consanguinity. We must, therefore, acquiesce in the necessity which denounces our separation, and hold them as we hold the rest of mankind, enemies in war, in peace, friends.

We, therefore, the representatives of the United States of America, in general Congress assembled, appealing to the Supreme Judge of the world for the rectitude of our intentions, do, in the name, and by the authority of the good people of these colonies, solemnly publish and declare, that these united colonies are, and of right ought to be, free and independent states: that they are absolved from all allegiance to the British Crown, and that all political connection between them and the state of Great Britain is, and ought to be, totally dissolved; and that, as free and independent states, they have full power to levy war, conclude peace, contract alliances, establish commerce, and to do all other acts and things which independent states may of right do. And, for the support of this declaration, with a firm reliance on the protection of Divine Providence, we mutually pledge to each other our lives, our fortunes, and our sacred honor.

The foregoing Declaration was, by order of Congress, engrossed, and signed by the following members:

John Hancock

NEW HAMPSHIRE
Josiah Bartlett
William Whipple
Matthew Thornton

CONNECTICUT
Roger Sherman
Samuel Huntington
William Williams
Oliver Wolcott

NEW YORK
William Floyd
Philip Livingston
Francis Lewis
Lewis Morris

NEW JERSEY
Richard Stockton
John Witherspoon
Francis Hopkinson
John Hart
Abraham Clark

MASSACHUSETTS BAY
Samuel Adams
John Adams
Robert Treat Paine
Elbridge Gerry

PENNSYLVANIA
Robert Morris
Benjamin Rush
Benjamin Franklin
John Morton
George Clymer
James Smith
George Taylor
James Wilson
George Ross

DELAWARE
Caesar Rodney
George Read
Thomas M'Kean

MARYLAND
Samuel Chase
William Paca
Thomas Stone
Charles Carroll, of Carrollton

RHODE ISLAND
Stephen Hopkins
William Ellery

VIRGINIA
George Wythe
Richard Henry Lee
Thomas Jefferson
Benjamin Harrison
Thomas Nelson Jr.
Francis Lightfoot Lee
Carter Braxton

NORTH CAROLINA
William Hooper
Joseph Hewes
John Penn

SOUTH CAROLINA
Edward Rutledge
Thomas Heyward Jr.
Thomas Lynch Jr.
Arthur Middleton

GEORGIA
Button Gwinnett
Lyman Hall
George Walton

Resolved, That copies of the Declaration be sent to the several assemblies, conventions, and committees, or councils of safety, and to the several commanding officers of the continental troops; that it be proclaimed in each of the United States, at the head of the army.

THE CONSTITUTION OF THE UNITED STATES[1]

We the People of the United States, in Order to form a more perfect Union, establish Justice, insure domestic Tranquility, provide for the common defence, promote the general Welfare, and secure the Blessings of Liberty to ourselves and our Posterity, do ordain and establish this CONSTITUTION for the United States of America.

Article I

Section 1.
All legislative Powers herein granted shall be vested in a Congress of the United States, which shall consist of a Senate and House of Representatives.

Section 2.
The House of Representatives shall be composed of Members chosen every second Year by the People of the several States, and the Electors in each State shall have the Qualifications requisite for Electors of the most numerous Branch of the State Legislature.

No Person shall be a Representative who shall not have attained to the Age of twenty-five Years, and been seven Years a Citizen of the United States, and who shall not, when elected, be an Inhabitant of that State in which he shall be chosen.

[Representatives and direct Taxes[2] shall be apportioned among the several States which may be included within this Union, according to their respective Numbers, which shall be determined by adding to the whole Number of free Persons, including those bound to Service for a Term of Years, and excluding Indians not taxed, three fifths of all other Persons.][3] The actual Enumeration shall be made within three Years after the first Meeting of the Congress of the United States, and within every subsequent Term of ten Years, in such Manner as they shall by Law direct. The Number of Representatives shall not exceed one for every thirty Thousand, but each State shall have at Least one Representative; and until such enumeration shall be made, the State of New Hampshire shall be entitled to chuse three, Massachusetts eight, Rhode-Island and Providence Plantations one, Connecticut five, New York six, New Jersey four, Pennsylvania eight, Delaware one, Maryland six, Virginia ten, North Carolina five, South Carolina five, and Georgia three.

When vacancies happen in the Representation from any State, the Executive Authority thereof shall issue Writs of Election to fill such Vacancies.

The House of Representatives shall chuse their Speaker and other Officers; and shall have the sole Power of Impeachment.

Section 3.
The Senate of the United States shall be composed of two Senators from each State, chosen by the Legislature thereof, for six Years; and each Senator shall have one Vote.

Immediately after they shall be assembled in Consequence of the first Election, they shall be divided as equally as may be into three Classes. The Seats of the Senators of the first Class shall be vacated at the Expiration of the second Year, of the second Class at the Expiration of the fourth Year, and of the third Class at the Expiration of the sixth Year,

---

1 This version, which follows the original Constitution in capitalization and spelling, was published by the United States Department of the Interior, Office of Education, in 1935.
2 Altered by the Sixteenth Amendment.
3 Negated by the Fourteenth Amendment.

so that one-third may be chosen every second Year; and if Vacancies happen by Resignation, or otherwise, during the Recess of the Legislature of any State, the Executive thereof may make temporary Appointments until the next Meeting of the Legislature, which shall then fill such Vacancies.

No Person shall be a Senator who shall not have attained to the Age of thirty Years, and been nine Years a Citizen of the United States, and who shall not, when elected, be an Inhabitant of that State for which he shall be chosen.

The Vice President of the United States shall be President of the Senate, but shall have no vote, unless they be equally divided.

The Senate shall chuse their other Officers, and also a President pro tempore, in the absence of the Vice President, or when he shall exercise the Office of President of the United States.

The Senate shall have the sole Power to try all Impeachments. When sitting for that purpose they shall be on Oath or Affirmation. When the President of the United States is tried, the Chief Justice shall preside: And no person shall be convicted without the Concurrence of two thirds of the Members present.

Judgment in Cases of Impeachment shall not extend further than to removal from Office, and disqualification to hold and enjoy any Office of honor, Trust, or Profit under the United States: but the Party convicted shall nevertheless be liable and subject to Indictment, Trial, Judgment, and Punishment, according to Law.

Section 4.
The Times, Places and Manner of holding Elections for Senators and Representatives, shall be prescribed in each State by the Legislature thereof; but the Congress may at any time by Law make or alter such Regulations, except as to the Places of Chusing Senators.

The Congress shall assemble at least once in every Year, and such Meeting shall be on the first Monday in December, unless they shall by Law appoint a different Day.

Section 5.
Each House shall be the Judge of the Elections, Returns and Qualifications of its own Members, and a Majority of each shall constitute a Quorum to do Business; but a smaller number may adjourn from day to day, and may be authorized to compel the Attendance of absent Members, in such Manner, and under such Penalties, as each House may provide.

Each House may determine the Rules of its Proceedings, punish its Members for disorderly Behaviour, and, with the Concurrence of two thirds, expel a Member.

Each House shall keep a Journal of its Proceedings, and from time to time publish the same, excepting such Parts as may in their Judgment require Secrecy; and the Yeas and Nays of the Members of either House on any question shall, at the Desire of one fifth of those Present, be entered on the Journal.

Neither House, during the Session of Congress, shall, without the Consent of the other, adjourn for more than three days, nor to any other Place than that in which the two Houses shall be sitting.

Section 6.
The Senators and Representatives shall receive a Compensation for their Services, to be ascertained by Law, and paid out of the Treasury of the United States. They shall in all Cases, except Treason, Felony, and Breach of the Peace, be privileged from Arrest during their Attendance at the Session of their respective Houses, and in going to and returning from the same; and for any Speech or Debate in either House, they shall not be questioned in any other Place.

No Senator or Representative shall, during the Time for which he was elected, be appointed to any civil Office under the Authority of the United States, which shall have been created, or the Emoluments whereof shall have been increased, during such time; and no Person holding any Office under the United States shall be a Member of either House during his continuance in Office.

Section 7.
All Bills for raising Revenue shall originate in the House of Representatives; but the Senate may propose or concur with Amendments as on other bills.

Every Bill which shall have passed the House of Representatives and the Senate, shall, before it become a Law, be presented to the President of the United States; If he approve he shall sign it, but if not he shall return it, with his Objections, to that House in which it shall have originated, who shall enter the Objections at large on their Journal, and proceed to reconsider it. If after such Reconsideration two thirds of that House shall agree to pass the bill, it shall be sent, together with the objections, to the other House, by which it shall likewise be reconsidered, and if approved by two thirds of that House, it shall become a Law. But in all such Cases the Votes of both Houses shall be determined by Yeas and Nays, and the Names of the Persons voting for and against the Bill shall be entered on the Journal of each House respectively. If any Bill shall not be returned by the President within ten Days (Sundays excepted) after it shall have been presented to him, the Same shall be a Law, in like Manner as if he had signed it, unless the Congress by their Adjournment prevent its Return, in which Case it shall not be a Law.

Every Order, Resolution, or Vote to which the Concurrence of the Senate and House of Representatives may be necessary (except on a question of Adjournment) shall be presented to the President of the United States; and before the Same shall take Effect, shall be approved by him, or being disapproved by him, shall be repassed by two thirds of the Senate and House of Representatives, according to the Rules and Limitations prescribed in the Case of a Bill.

Section 8.
The Congress shall have Power To lay and collect Taxes, Duties, Imposts and Excises, to pay the Debts and provide for the common Defence and general Welfare of the United States; but all Duties, Imposts and Excises shall be uniform throughout the United States;

To borrow money on the credit of the United States;

To regulate Commerce with foreign Nations, and among the several States, and with the Indian Tribes;

To establish an uniform rule of Naturalization, and uniform Laws on the subject of Bankruptcies throughout the United States;

To coin Money, regulate the Value thereof, and of foreign Coin, and fix the Standard of Weights and Measures;

To provide for the Punishment of counterfeiting the Securities and current Coin of the United States;

To establish Post Offices and post Roads;

To promote the Progress of Science and useful Arts, by securing for limited Times to Authors and Inventors the exclusive Right to their respective Writings and Discoveries;

To constitute Tribunals inferior to the Supreme Court;

To define and punish Piracies and Felonies committed on the high Seas, and Offenses against the Law of Nations;

To declare War, grant Letters of Marque and Reprisal, and make Rules concerning Captures on Land and Water;

To raise and support Armies, but no Appropriation of Money to that Use shall be for a longer Term than two Years;

To provide and maintain a Navy;

To make Rules for the Government and Regulation of the land and naval forces;

To provide for calling forth the Militia to execute the Laws of the Union, suppress Insurrections and repel Invasions;

To provide for organizing, arming, and disciplining the Militia, and for governing such Part of them as may be employed in the Service of the United States, reserving to the States respectively, the Appointment of the Officers, and the Authority of training the Militia according to the discipline prescribed by Congress;

To exercise exclusive Legislation in all Cases whatsoever, over such District (not exceeding ten Miles square) as may, by Cession of particular States, and the acceptance of Congress, become the Seat of the Government of the United States, and to exercise like Authority over all Places purchased by the Consent of the Legislature of the State in which the Same shall be, for the Erection of Forts, Magazines, Arsenals, Dock-yards, and other needful Buildings;—And

To make all Laws which shall be necessary and proper for carrying into Execution the foregoing Powers, and all other Powers vested by this Constitution in the Government of the United States, or in any Department or Officer thereof.

Section 9.

The Migration or Importation of such Persons as any of the States now existing shall think proper to admit, shall not be prohibited by the Congress prior to the Year one thousand eight hundred and eight, but a tax or duty may be imposed on such Importation, not exceeding ten dollars for each Person.

The privilege of the Writ of Habeas Corpus shall not be suspended, unless when in Cases of Rebellion or Invasion the public Safety may require it.

No bill of Attainder or ex post facto Law shall be passed.

No capitation, or other direct, Tax shall be laid unless in Proportion to the Census or Enumeration herein before directed to be taken.

No Tax or Duty shall be laid on Articles exported from any State.

No Preference shall be given by any Regulation of Commerce or Revenue to the Ports of one State over those of another: nor shall Vessels bound to, or from, one State, be obliged to enter, clear, or pay Duties in another.

No Money shall be drawn from the Treasury, but in Consequence of Appropriations made by Law; and a regular Statement and Account of the Receipts and Expenditures of all public Money shall be published from time to time.

No Title of Nobility shall be granted by the United States: And no Person holding any Office of Profit or Trust under them, shall, without the Consent of the Congress, accept of any present, Emolument, Office, or Title, of any kind whatever, from any King, Prince, or foreign State.

Section 10.

No State shall enter into any Treaty, Alliance, or Confederation; grant Letters of Marque and Reprisal; coin Money; emit Bills of Credit; make any Thing but gold and silver Coin a Tender in Payment of Debts; pass any Bill of Attainder, ex post facto Law, or Law impairing the Obligation of Contracts, or grant any Title of Nobility.

No State shall, without the Consent of the Congress, lay any Imposts or Duties on Imports or Exports, except what may be absolutely necessary for executing its inspection

Laws; and the net Produce of all Duties and Imposts, laid by any State on Imports or Exports, shall be for the use of the Treasury of the United States; and all such Laws shall be subject to the Revision and Control of the Congress.

No state shall, without the Consent of Congress, lay any duty of Tonnage, keep Troops, or Ships of War in time of Peace, enter into any Agreement or Compact with another State, or with a foreign Power, or engage in War, unless actually invaded, or in such imminent Danger as will not admit of delay.

Article II

Section 1.
The executive Power shall be vested in a President of the United States of America. He shall hold his Office during the Term of four years, and, together with the Vice President, chosen for the same Term, be elected, as follows:

Each State shall appoint, in such Manner as the Legislature thereof may direct, a Number of Electors, equal to the whole Number of Senators and Representatives to which the State may be entitled in the Congress: but no Senator or Representative, or Person holding an Office of Trust or Profit under the United States, shall be appointed an Elector.

[The Electors shall meet in their respective States, and vote by Ballot for two persons, of whom one at least shall not be an Inhabitant of the same State with themselves. And they shall make a List of all the Persons voted for, and of the Number of Votes for each; which List they shall sign and certify, and transmit sealed to the Seat of the Government of the United States, directed to the President of the Senate. The President of the Senate shall, in the Presence of the Senate and House of Representatives, open all the Certificates, and the Votes shall then be counted. The Person having the greatest Number of Votes shall be the President, if such Number be a Majority of the whole Number of Electors appointed; and if there be more than one who have such Majority, and have an equal Number of Votes, then the House of Representatives shall immediately chuse by Ballot one of them for President; and if no Person have a Majority, then from the five highest on the List the said House shall in like Manner chuse the President. But in chusing the President, the Votes shall be taken by States, the Representation from each State having one Vote; a quorum for this Purpose shall consist of a Member or Members from two-thirds of the States, and a Majority of all the States shall be necessary to a Choice. In every Case, after the Choice of the President, the Person having the greatest Number of Votes of the Electors shall be the Vice President. But if there should remain two or more who have equal votes, the Senate shall chuse from them by Ballot the Vice President.][4]

The Congress may determine the Time of chusing the Electors, and the Day on which they shall give their Votes; which Day shall be the same throughout the United States.

No person except a natural-born Citizen, or a Citizen of the United States, at the time of the Adoption of this Constitution, shall be eligible to the Office of President; neither shall any Person be eligible to that Office who shall not have attained to the Age of thirty-five years, and been fourteen Years a Resident within the United States.

In Case of the Removal of the President from Office, or of his Death, Resignation, or Inability to discharge the Powers and Duties of the said Office, the same shall devolve on the Vice President, and the Congress may by Law provide for the Case of Removal, Death, Resignation, or Inability, both of the President and Vice President, declaring what Officer

4 Revised by the Twelfth Amendment.

shall then act as President, and such Officer shall act accordingly, until the disability be removed, or a President shall be elected.

The President shall, at stated Times, receive for his Services a Compensation, which shall neither be increased nor diminished during the Period for which he shall have been elected, and he shall not receive within that Period any other Emolument from the United States, or any of them.

Before he enter on the execution of his Office, he shall take the following Oath or Affirmation:—"I do solemnly swear (or affirm) that I will faithfully execute the Office of President of the United States, and will, to the best of my Ability, preserve, protect, and defend the Constitution of the United States."

Section 2.
The President shall be Commander in Chief of the Army and Navy of the United States, and of the Militia of the several States, when called into the actual Service of the United States; he may require the Opinion, in writing, of the principal Officer in each of the executive Departments, upon any subject relating to the Duties of their respective Offices, and he shall have Power to Grant Reprieves and Pardons for Offenses against the United States, except in Cases of Impeachment.

He shall have Power, by and with the Advice and Consent of the Senate, to make Treaties, provided two-thirds of the Senators present concur; and he shall nominate, and by and with the Advice and Consent of the Senate, shall appoint Ambassadors, other public Ministers and Consuls, Judges of the supreme Court, and all other Officers of the United States, whose Appointments are not herein otherwise provided for, and which shall be established by Law: but the Congress may by Law vest the Appointment of such inferior Officers, as they think proper, in the President alone, in the Courts of Law, or in the Heads of Departments.

The President shall have Power to fill up all Vacancies that may happen during the Recess of the Senate, by granting Commissions which shall expire at the End of their next Session.

Section 3.
He shall from time to time give to the Congress Information of the State of the Union, and recommend to their Consideration such Measures as he shall judge necessary and expedient; he may, on extraordinary occasions, convene both Houses, or either of them, and in Case of Disagreement between them, with respect to the Time of Adjournment, he may adjourn them to such Time as he shall think proper; he shall receive Ambassadors and other public Ministers; he shall take care that the Laws be faithfully executed, and shall Commission all the Officers of the United States.

Section 4.
The President, Vice President and all civil Officers of the United States, shall be removed from Office on Impeachment for, and Conviction of, Treason, Bribery, or other high Crimes and Misdemeanors.

Article III

Section 1.
The judicial Power of the United States, shall be vested in one supreme Court, and in such inferior Courts as the Congress may from time to time ordain and establish. The Judges, both of the supreme and inferior Courts, shall hold their Offices during good Behaviour, and shall, at stated Times, receive for their Services, a Compensation, which shall not be diminished during their Continuance in Office.

Section 2.
The judicial Power shall extend to all Cases, in Law and Equity, arising under this Constitution, the Laws of the United States, and Treaties made, or which shall be made, under their Authority;—to all Cases affecting ambassadors, other public ministers and consuls;—to all cases of admiralty and maritime Jurisdiction;—to Controversies to which the United States shall be a Party;—to Controversies between two or more States;—between a State and Citizens of another State;[5]—between Citizens of different States—between Citizens of the same State claiming Lands under Grants of different States, and between a State, or the Citizens thereof, and foreign States, Citizens, or Subjects.

In all Cases affecting Ambassadors, other public Ministers and Consuls, and those in which a State shall be Party, the supreme Court shall have original Jurisdiction. In all the other Cases before mentioned, the supreme Court shall have appellate Jurisdiction, both as to Law and Fact, with such Exceptions, and under such Regulations as the Congress shall make.

The trial of all Crimes, except in Cases of Impeachment, shall be by Jury; and such Trial shall be held in the State where the said Crimes shall have been committed; but when not committed within any State, the Trial shall be at such Place or Places as the Congress may by Law have directed.

Section 3.
Treason against the United States, shall consist only in levying War against them, or in adhering to their Enemies, giving them Aid and Comfort. No Person shall be convicted of Treason unless on the Testimony of two Witnesses to the same overt Act, or on Confession in open Court.

The Congress shall have power to declare the Punishment of Treason, but no Attainder of Treason shall work Corruption of Blood, or Forfeiture except during the Life of the Person attained.

Article IV

Section 1.
Full Faith and Credit shall be given in each State to the public Acts, Records, and judicial Proceedings of every other State. And the Congress may by general Laws prescribe the Manner in which such Acts, Records and Proceedings shall be proved, and the Effect thereof.

Section 2.
The Citizens of each State shall be entitled to all Privileges and Immunities of Citizens in the several States.

A Person charged in any State with Treason, Felony, or other Crime, who shall flee from Justice, and be found in another State, shall on demand of the executive Authority of the State from which he fled, be delivered up, to be removed to the State having Jurisdiction of the crime.

No Person held to Service or Labour in one State, under the Laws thereof, escaping into another, shall, in Consequence of any Law or Regulation therein, be discharged from such Service or Labour, but shall be delivered up on Claim of the Party to whom such Service or Labour may be due.

---

5 Qualified by the Eleventh Amendment.

Section 3.

New States may be admitted by the Congress into this Union; but no new State shall be formed or erected within the Jurisdiction of any other State; nor any State be formed by the Junction of two or more States, or parts of States, without the Consent of the Legislatures of the States concerned as well as of the Congress.

The Congress shall have Power to dispose of and make all needful Rules and Regulations respecting the Territory or other Property belonging to the United States; and nothing in this Constitution shall be so construed as to Prejudice any Claims of the United States, or of any particular State.

Section 4.

The United States shall guarantee to every State in this Union a Republican Form of Government, and shall protect each of them against Invasion; and on Application of the Legislature, or of the Executive (when the Legislature cannot be convened) against domestic Violence.

Article V

The Congress, whenever two-thirds of both Houses shall deem it necessary, shall propose Amendments to this Constitution, or, on the Application of the Legislatures of two-thirds of the several States, shall call a Convention for proposing Amendments, which, in either Case, shall be valid to all Intents and Purposes, as part of this Constitution, when ratified by the Legislatures of three-fourths of the several States, or by Conventions in three-fourths thereof, as the one or the other Mode of Ratification may be proposed by the Congress; Provided that no Amendment which may be made prior to the Year One thousand eight hundred and eight shall in any Manner affect the first and fourth Clauses in the Ninth Section of the first Article; and that no State, without its Consent, shall be deprived of its equal Suffrage in the Senate.

Article VI

All Debts contracted and Engagements entered into, before the Adoption of this Constitution, shall be as valid against the United States under this Constitution, as under the Confederation.

This Constitution, and the Laws of the United States which shall be made in Pursuance thereof; and all Treaties made, or which shall be made, under the Authority of the United States, shall be the supreme Law of the Land; and the Judges in every State shall be bound thereby, any Thing in the Constitution or Laws of any State to the Contrary notwithstanding.

The Senators and Representatives before mentioned, and the Members of the several State Legislatures, and all executive and judicial Officers, both of the United States and of the several States, shall be bound by Oath or Affirmation to support this Constitution; but no religious Tests shall ever be required as a qualification to any Office or public Trust under the United States.

Article VII

The Ratification of the Conventions of nine States shall be sufficient for the Establishment of this Constitution between the States so ratifying the same.

Done in Convention by the Unanimous Consent of the States present the Seventeenth Day of September in the Year of our Lord one thousand seven hundred and Eighty seven, and of the Independence of the United States of America the Twelfth. In Witness whereof We have hereunto subscribed our Names.[6]

George Washington
President and deputy from Virginia

NEW HAMPSHIRE
John Langdon
Nicholas Gilman

MASSACHUSETTS
Nathaniel Gorham
Rufus King

CONNECTICUT
William Samuel Johnson
Roger Sherman

NEW YORK
Alexander Hamilton

NEW JERSEY
William Livingston
David Brearley
William Paterson
Jonathan Dayton

PENNSYLVANIA
Benjamin Franklin
Thomas Mifflin
Robert Morris
George Clymer
Thomas FitzSimons
Jared Ingersoll
James Wilson
Gouverneur Morris

DELAWARE
George Read
Gunning Bedford Jr.
John Dickinson
Richard Bassett
Jacob Broom

MARYLAND
James McHenry
Daniel of St. Thomas Jenifer
Daniel Carroll

VIRGINIA
John Blair
James Madison Jr.

NORTH CAROLINA
William Blount
Richard Dobbs Spaight
Hugh Williamson

SOUTH CAROLINA
John Rutledge
Charles Cotesworth Pinckney
Charles Pinckney
Pierce Butler

GEORGIA
William Few
Abraham Baldwin

Articles in Addition to, and Amendment of, the Constitution of the United States of America, Proposed by Congress, and Ratified by the Legislatures of the Several States, Pursuant to the Fifth Article of the Original Constitution.[7]

[Article I]

Congress shall make no law respecting an establishment of religion, or prohibiting the free exercise thereof; or abridging the freedom of speech, or of the press; or the right of the people peaceably to assemble, and to petition the Government for a redress of grievances.

[Article II]

A well regulated Militia, being necessary to the security of a free State, the right of the people to keep and bear Arms shall not be infringed.

---

6 These are the full names of the signers, which in some cases are not the signatures on the document.
7 This heading appears only in the joint resolution submitting the first ten amendments.

[Article III]

No Soldier shall, in time of peace, be quartered in any house, without the consent of the Owner, nor in time of war, but in a manner to be prescribed by law.

[Article IV]

The right of the people to be secure in their persons, houses, papers, and effects, against unreasonable searches and seizures, shall not be violated, and no Warrants shall issue, but upon probable cause, supported by Oath or affirmation, and particularly describing the place to be searched, and the persons or things to be seized.

[Article V]

No person shall be held to answer for a capital or otherwise infamous crime, unless on a presentment or indictment of a Grand Jury, except in cases arising in the land or naval forces, or in the Militia, when in actual service in time of War or public danger; nor shall any person be subject for the same offence to be twice put in jeopardy of life or limb; nor shall be compelled in any criminal case to be a witness against himself, nor be deprived of life, liberty, or property, without due process of law; nor shall private property be taken for public use, without just compensation.

[Article VI]

In all criminal prosecutions, the accused shall enjoy the right to a speedy and public trial, by an impartial jury of the State and district wherein the crime shall have been committed, which district shall have been previously ascertained by law, and to be informed of the nature and cause of the accusation; to be confronted with the witnesses against him; to have compulsory process for obtaining witnesses in his favour, and to have the Assistance of Counsel for his defense.

[Article VII]

In suits at common law, where the value in controversy shall exceed twenty dollars, the right of trial by jury shall be preserved, and no fact tried by a jury, shall be otherwise reexamined in any Court of the United States, than according to the rules of the common law.

[Article VIII]

Excessive bail shall not be required, nor excessive fines imposed, nor cruel and unusual punishments inflicted.

[Article IX]

The enumeration of the Constitution, of certain rights, shall not be construed to deny or disparage others retained by the people.

[Article X]

The powers not delegated to the United States by the Constitution, nor prohibited by it to the States, are reserved to the States respectively, or to the people.

[Amendments I-X, in force 1791.]

[Article XI][8]

The Judicial power of the United States shall not be construed to extend to any suit in law or equity, commenced or prosecuted against one of the United States by Citizens of another State, or by Citizens or Subjects of any Foreign State.

[Article XII][9]

The Electors shall meet in their respective States and vote by ballot for President and Vice-President, one of whom, at least, shall not be an inhabitant of the same State with themselves; they shall name in their ballots the person voted for as President, and in distinct ballots the person voted for as Vice-President, and they shall make distinct lists of all persons voted for as President, and of all persons voted for as Vice-President, and of the number of votes for each, which lists they shall sign and certify, and transmit sealed to the seat of the government of the United States, directed to the President of the Senate;— The President of the Senate shall, in the presence of the Senate and House of Representatives, open all the certificates and the votes shall then be counted;—The person having the greatest number of votes for President, shall be the President, if such number be a majority of the whole number of Electors appointed; and if no person have such majority, then from the persons having the highest numbers not exceeding three on the list of those voted for as President, the House of Representatives shall choose immediately, by ballot, the President. But in choosing the President, the votes shall be taken by states, the representation from each state having one vote; a quorum for this purpose shall consist of a member or members from two-thirds of the states, and a majority of all the states shall be necessary to a choice. And if the House of Representatives shall not choose a President whenever the right of choice shall devolve upon them, before the fourth day of March next following, then the Vice-President shall act as President, as in the case of the death or other constitutional disability of the President.—The person having the greatest number of votes as Vice-President, shall be the Vice-President, if such number be a majority of the whole number of Electors appointed, and if no person have a majority, then from the two highest numbers on the list, the Senate shall choose the Vice-President; a quorum for the purpose shall consist of two-thirds of the whole number of Senators, and a majority of the whole number shall be necessary to a choice. But no person constitutionally ineligible to the office of President shall be eligible to that of Vice-President of the United States.

[Article XIII][10]

Section 1.
Neither slavery nor involuntary servitude, except as a punishment for crime whereof the party shall have been duly convicted, shall exist within the United States, or any place subject to their jurisdiction.

Section 2.
Congress shall have power to enforce this article by appropriate legislation.

---

8 Adopted in 1798.
9 Adopted in 1804.
10 Adopted in 1865.

[Article XIV][11]

Section 1.
All persons born or naturalized in the United States, and subject to the jurisdiction thereof, are citizens of the United States and of the State wherein they reside. No State shall make or enforce any law which shall abridge the privileges or immunities of citizens of the United States; nor shall any State deprive any person of life, liberty, or property, without due process of law; nor deny to any person within its jurisdiction the equal protection of the laws.

Section 2.
Representatives shall be apportioned among the several States according to their respective numbers, counting the whole number of persons in each State, excluding Indians not taxed. But when the right to vote at any election for the choice of electors for President and Vice-President of the United States, Representatives in Congress, the Executive and Judicial officers of a State, or the members of the Legislature thereof, is denied to any of the male inhabitants of such State, being twenty-one years of age, and citizens of the United States, or in any way abridged, except for participation in rebellion, or other crime, the basis of representation therein shall be reduced in the proportion which the number of such male citizens shall bear to the whole number of male citizens twenty-one years of age in such State.

Section 3.
No person shall be a Senator or Representative in Congress, or elector of President and Vice-President, or hold any office, civil or military, under the United States, or under any State, who, having previously taken an oath, as a member of Congress, or as an officer of the United States, or as a member of any State legislature, or as an executive or judicial officer of any State, to support the Constitution of the United States, shall have engaged in insurrection or rebellion against the same, or given aid or comfort to the enemies thereof. But Congress may by a vote of two-thirds of each House, remove such disability.

Section 4.
The validity of the public debt of the United States, authorized by law, including debts incurred for payment of pensions and bounties for services in suppressing insurrection or rebellion, shall not be questioned. But neither the United States nor any State shall assume or pay any debts or obligation incurred in aid of insurrection or rebellion against the United States, or any claim for the loss or emancipation of any slave; but all such debts, obligations, and claims shall be held illegal and void.

Section 5.
The Congress shall have the power to enforce, by appropriate legislation, the provisions of this article.

[Article XV][12]

Section 1.
The right of citizens of the United States to vote shall not be denied or abridged by the United States or by any State on account of race, color, or previous condition of servitude—

Section 2.
The Congress shall have power to enforce this article by appropriate legislation.

11 Adopted in 1868.
12 Adopted in 1870.

[Article XVI][13]

The Congress shall have power to lay and collect taxes on incomes, from whatever source derived, without apportionment among the several States, and without regard to any census or enumeration.

[Article XVII][14]

The Senate of the United States shall be composed of two Senators from each State, elected by the people thereof, for six years; and each Senator shall have one vote. The electors in each State shall have the qualifications requisite for electors of the most numerous branch of the State legislatures.

When vacancies happen in the representation of any State in the Senate, the executive authority of such State shall issue writs of election to fill such vacancies: *Provided,* That the legislature of any State may empower the executive thereof to make temporary appointments until the people fill the vacancies by election as the legislature may direct.

This amendment shall not be so construed as to affect the election or term of any Senator chosen before it becomes valid as part of the Constitution.

[Article XVIII][15]

Section 1.
After one year from the ratification of this article the manufacture, sale, or transportation of intoxicating liquors within, the importation thereof into, or the exportation thereof from the United States and all territory subject to the jurisdiction thereof for beverage purposes is hereby prohibited.

Section 2.
The Congress and the several States shall have concurrent power to enforce this article by appropriate legislation.

Section 3.
This article shall be inoperative unless it shall have been ratified as an amendment to the Constitution by the legislatures of the several States, as provided in the Constitution, within seven years from the date of the submission hereof to the States by the Congress.

[Article XIX][16]

The right of citizens of the United States to vote shall not be denied or abridged by the United States or by any State on account of sex.

Congress shall have power to enforce this article by appropriate legislation.

[Article XX][17]

Section 1.
The terms of the President and Vice-President shall end at noon on the 20th day of January, and the terms of Senators and Representatives at noon on the 3d day of January, of the

---

13 Adopted in 1913.
14 Adopted in 1913.
15 Adopted in 1918.
16 Adopted in 1920.
17 Adopted in 1933.

years in which such terms would have ended if this article had not been ratified; and the terms of their successors shall then begin.

Section 2.
The Congress shall assemble at least once in every year, and such meeting shall begin at noon on the 3d day of January, unless they shall by law appoint a different day.

Section 3.
If, at the time fixed for the beginning of the term of the President, the President elect shall have died, the Vice-President elect shall become President. If a President shall not have been chosen before the time fixed for the beginning of his term or if the President elect shall have failed to qualify, then the Vice-President elect shall act as President until a President shall have qualified; and the Congress may by law provide for the case wherein neither a President elect nor a Vice-President elect shall have qualified, declaring who shall then act as President, or the manner in which one who is to act shall be selected, and such person shall act accordingly until a President or Vice-President shall have qualified.

Section 4.
The Congress may by law provide for the case of the death of any of the persons from whom the House of Representatives may choose a President whenever the right of choice shall have devolved upon them, and for the case of the death of any of the persons from whom the Senate may choose a Vice-President whenever the right of choice shall have devolved upon them.

Section 5.
Sections 1 and 2 shall take effect on the 15th day of October following the ratification of this article.

Section 6.
This article shall be inoperative unless it shall have been ratified as an amendment to the Constitution by the legislatures of three-fourths of the several States within seven years from the date of its submission.

[Article XXI][18]

Section 1.
The eighteenth article of amendment to the Constitution of the United States is hereby repealed.

Section 2.
The transportation or importation into any State, Territory, or possession of the United States for delivery or use therein of intoxicating liquors, in violation of the laws thereof, is hereby prohibited.

Section 3.
This article shall be inoperative unless it shall have been ratified as an amendment to the Constitution by conventions in the several States, as provided in the Constitution, within seven years from the date of the submission hereof to the States by the Congress.

18 Adopted in 1933.

[Article XXII][19]

No person shall be elected to the office of the President more than twice, and no person who has held the office of President, or acted as President, for more than two years of a term to which some other person was elected President shall be elected to the office of the President more than once.

But this Article shall not apply to any person holding the office of President when this Article was proposed by the Congress, and shall not prevent any person who may be holding the office of President, or acting as President, during the term within which this Article becomes operative from holding the office of President or acting as President during the remainder of such term.

This article shall be inoperative unless it shall have been ratified as an amendment to the Constitution by the legislatures of three-fourths of the several states within seven years from the date of its submission to the states by the Congress.

[Article XXIII][20]

Section 1.

The District constituting the seat of Government of the United States shall appoint in such manner as the Congress may direct:

A number of electors of President and Vice-President equal to the whole number of Senators and Representatives in Congress to which the District would be entitled if it were a State, but in no event more than the least populous State; they shall be in addition to those appointed by the States, but they shall be considered, for the purposes of the election of President and Vice-President, to be electors appointed by a State; and they shall meet in the District and perform such duties as provided by the twelfth article of amendment.

Section 2.
The Congress shall have power to enforce this article by appropriate legislation.

[Article XXIV][21]

Section 1.
The right of citizens of the United States to vote in any primary or other election for President or Vice President, for electors for President or Vice President, or for Senator or Representative in Congress, shall not be denied or abridged by the United States or any state by reason of failure to pay any poll tax or other tax.

Section 2.
The Congress shall have the power to enforce this article by appropriate legislation.

[Article XXV][22]

Section 1.
In case of the removal of the President from office or of his death or resignation, the Vice President shall become President.

---

19 Adopted in 1951.
20 Adopted in 1961.
21 Adopted in 1964.
22 Adopted in 1967.

Section 2.
Whenever there is a vacancy in the office of the Vice President, the President shall nominate a Vice President who shall take office upon confirmation by a majority vote of both Houses of Congress.

Section 3.
Whenever the President transmits to the President Pro Tempore of the Senate and the Speaker of the House of Representatives his written declaration that he is unable to discharge the powers and duties of his office, and until he transmits to them a written declaration to the contrary, such powers and duties shall be discharged by the Vice President as Acting President.

Section 4.
Whenever the Vice President and a majority of either the principal officers of the executive departments or of such other body as Congress may by law provide, transmit to the President Pro Tempore of the Senate and the Speaker of the House of Representatives their written declaration that the President is unable to discharge the powers and duties of his office, the Vice President shall immediately assume the powers and duties of the office as Acting President.

Thereafter, when the President transmits to the President Pro Tempore of the Senate and the Speaker of the House of Representatives his written declaration that no inability exists, he shall resume the powers and duties of his office unless the Vice President and a majority of either the principal officers of the executive departments or of such other body as Congress may by law provide, transmit within four days to the President Pro Tempore of the Senate and the Speaker of the House of Representatives their written declaration that the President is unable to discharge the powers and duties of his office. Thereupon Congress shall decide the issue, assembling within forty-eight hours for that purpose if not in session. If the Congress, within twenty-one days after receipt of the latter written declaration, or, if Congress is not in session, within twenty-one days after Congress is required to assemble, determines by two-thirds vote of both Houses that the President is unable to discharge the powers and duties of his office, the Vice President shall continue to discharge the same as Acting President; otherwise, the President shall resume the powers and duties of his office.

[Article XXVI][23]

Section 1.
The right of citizens of the United States, who are eighteen years of age or older, to vote shall not be denied or abridged by the United States or by any State on account of age.

Section 2.
The Congress shall have power to enforce this article by appropriate legislation.

[Article XXVII][24]

No law varying the compensation for the services of the Senators and Representatives shall take effect until an election of Representatives shall have intervened.

---

23 Adopted in 1971.
24 Adopted in 1992.

GLOSSARY

9/11 attacks A series of coordinated attacks on September 11, 2001, in which Al Qaeda operatives hijacked four planes, destroyed the World Trade towers in New York, damaged the Pentagon outside Washington, and caused the death of all aboard United Flight 93, altogether killing 3,000 people.

A. Philip Randolph Leader of the Brotherhood of Sleeping Car Porters and activist in the American labor and civil rights movements.

Abigail Adams Supporter of expanding women's rights and protections in the new United States; wife of John Adams.

abolitionist An advocate for the end of a state-approved practice or institution; the term is used most often in connection with the eradication of slavery.

Abraham Lincoln Lawyer and diplomat originally from Kentucky who served as the 16th president of the United States during the Civil War.

Abraham Lincoln Brigade A group of roughly 3,000 young Americans who traveled to Spain to join a fight against the fascists there.

Adams-Onís Treaty Agreement between the United States and Spain in 1819 that gave Florida to the United States in exchange for dropping its claims to Texas.

Adolf Hitler Leader of the German Nazi Party who took power in 1933; fascist whose assumption of dictatorial powers and belief in Aryan racial superiority resulted in the deaths of millions of innocent people including six million Jews in the Holocaust.

affirmative action A policy that grants special consideration in the hiring and promoting of members of groups that historically have faced discrimination.

Affordable Care Act Also known as Obamacare, health-care legislation meant to expand Americans' access to insurance.

Afghan war Post-9/11 military campaign against the Taliban in Afghanistan who were believed to aid Al Qaeda and harbor Osama bin Laden.

AFL-CIO New name of the 1955 merger of the labor groups American Federation of Labor and the Congress of Industrial Organizations.

Agricultural Adjustment Act Act that created a federal agency empowered to achieve parity by controlling the production of seven basic commodities.

Agricultural Marketing Act Proposed by President Hoover in April 1929, it established the first major government program to help farmers maintain prices.

AIDS Acquired immune deficiency syndrome, the set of symptoms and illnesses caused by the HIV virus that killed many Americans in the late twentieth century.

Al Qaeda Network of Islamic extremists responsible for the 9/11 attacks as well as other terrorist acts.

Alamo A Catholic mission in San Antonio that was the site of a major battle in the Texas Revolution in which Mexican forces put down insurgents seeking Texas independence.

Albany Plan Benjamin Franklin's 1754 proposal for a "general government" to manage relations between the colonies and Native Americans; rejected by the colonies at the beginning of the French and Indian War.

Alexander H. Stephens Vice-president of the Confederacy.

Alexander Hamilton One of the country's founders, Hamilton championed a strong central government as a Federalist and was influential in Washington's cabinet.

Alexis de Tocqueville French aristocrat who toured the United States in the early 1830s and wrote *Democracy in America*.

Alger Hiss A high-ranking member of the State Department accused in 1948 of passing classified documents to a communist agent; eventually convicted of perjury.

Alice Paul Head of the National Woman's Party.

Alien and Sedition Acts A group of laws passed under President John Adams that limited new immigrants' access to citizenship and gave the federal government broad powers to limit criticism of the government.

Allies One of the two sides in the First World War, comprised of Britain, France, Russia, and in 1917, the United States.

GLOSSARY · G-1

American Federation of Labor (AFL) Union of skilled workers, formed in 1881 and led by Samuel Gompers, that used strikes to gain concessions from management.

American Indian Movement (AIM) Organization of Native American activists formed in 1968 to promote native self-determination.

American Patriots Term for supporters of American independence during the Revolutionary War.

American Plan A euphemism for the open shop; the crusade for this became a pretext for a harsh campaign of union-busting.

American Socialist Party Political party for economic reform created in 1901 that was closely aligned with organized labor.

American System Henry Clay's economic plan to bolster and unify the American economy by raising protective tariffs, developing the transportation system, and establishing a strong national bank.

Amistad Ship at the center of an 1841 Supreme Court case over the foreign slave trade; the Court decided that the Africans who had commandeered the ship had been illegally captured and sold and were granted freedom.

Andrew Carnegie Scottish immigrant who became a steel magnate and then philanthropist during the Gilded Age.

Andrew Jackson Seventh president of the United States; had distinguished himself at the Battle of New Orleans in the War of 1812 and as a Native American fighter.

Andrew Johnson Democrat from Tennessee who served as Lincoln's vice president and, upon Lincoln's assassination, became the seventeenth president; opposed Radical Republican policies on Reconstruction.

Anne Hutchinson Critic of the clerical doctrine of grace who sparked the Antinomian heresy that challenged the spiritual authority of established clergy.

antebellum The period before a war; in U.S. history, the term is commonly used to describe the pre–Civil War period.

Antietam Site of a Union victory on September 17, 1862, that stands as the single bloodiest day in American military history, blunted Confederate progress northward, and prompted Lincoln to issue the preliminary Emancipation Proclamation.

Antifederalists Term used by Federalists to describe those who were against ratification of the Constitution.

Antinomianism A belief that salvation comes from God's grace alone and not from good works.

Antonio Lopez de Santa Anna Politician and general who served multiple stints as president of Mexico and led Mexico in the war with the United States, 1846–1848.

Apollo program A NASA program to land an astronaut on the moon in the 1960s.

appeasement A foreign policy that accepts (rather than opposes) the aggressive moves of another state or actor.

Appomattox Court House Site of Lee's surrender to Grant.

Armory Show An event in New York City that displayed works of the French postimpressionists and of some American modern artists; supported by the Ashcan artists.

Army–McCarthy hearings One of the earliest televised hearings of Congress, the spectacle of Joseph McCarthy demeaning and bullying witnesses whom he accused of communist sympathies led to his loss of public support and official censure for unbecoming conduct.

Articles of Confederation The first adopted plan for union by the states that established a federal Congress with the power to tax and issue money; each state had one vote.

artisan An independent, skilled craftsperson.

Ashcan school Art movement whose members produced work startling in its naturalism and stark in its portrayal of the social realities of the era.

Atlanta Compromise This term describes Booker T. Washington's philosophy, stated in an 1895 speech, that Blacks should forgo agitation for political rights and concentrate on self-improvement and preparation for equality.

Atlantic Charter Statement of shared aims issued by America and Britain in August 1941; the two nations called for a new world order based on self-determination, economic cooperation, and anti-militarism.

Atlantic World The peoples and empires around the Atlantic Ocean rim that became interconnected in the sixteenth century.

Ayatollah Ruhollah Khomeini Anti-American religious leader of Iran who took control in 1979 after the previous leader was deposed.

Baby Boom A period of increased birthrate; the term is used most often to describe such a demographic trend from 1946 to 1964.

Bacon's Rebellion A major conflict in Virginia pitting the ruling gentry class against Black and white laborers and enslaved Blacks seeking greater freedoms and opportunities that resulted in a sharper definition between Native American and white spheres of influence.

Baltimore and Ohio Railroad The oldest railroad in the United States, originally built to help Maryland compete with canals in other states.

Bank War Term used for President Andrew Jackson's fight against Nicholas Biddle and supporters of the Bank of the United States; Jackson ultimately succeeded in eliminating the Bank.

Barack Obama Former U.S. senator from Illinois and first Black American elected president, who set a progressive agenda during his eight years in office.

Bartolomé de Las Casas Dominican friar who fought for fairer treatment of indigenous people in Spanish colonies.

Battle of Fallen Timbers The 1794 defeat of Native American group the Miami in the Ohio Valley, which forced the defeated Miami to agree to a treaty that ceded indigenous lands to the United States.

Battle of the Bulge The last major battle on the Western Front during World War II, as the Germans were finally stopped at Bastogne.

Bay of Pigs Failed invasion of Cuban exiles supported by the United States to overthrow the Castro regime in 1961.

Beats Term used to describe artists and authors, like Jack Kerouac, who were critics of middle-class society and conformity.

Benedict Arnold Military hero early in the Revolution; he lost hope as the war progressed and ultimately conspired with the British.

Benito Mussolini Leader of Italy's Fascist Party before and during World War II.

Benjamin Franklin Inventor, author, diplomat, and one of the most famous people of the 1700s; served as a colonial agent in England during the early part of the conflict between the colonies and England.

Benjamin Harrison Republican senator who was elected president in 1888 in one of the most corrupt elections in American history.

Betty Friedan Author of the 1963 *The Feminine Mystique* who described the frustration of many women who found themselves limited socially, economically, and intellectually in postwar America.

Bill Clinton Former governor of Arkansas and forty-second president during a period of economic prosperity in the 1990s; impeached but acquitted on charges of obstructing justice and lying about an extramarital affair.

Bill of Rights First ten amendments to the U.S. Constitution; limited the new government's ability to infringe upon certain fundamental rights.

Black Codes State laws that developed after the Civil War in the former Confederate states to limit the political power and mobility of Black Americans.

Black Hawk War Attack by Sauk (or Sac) and Fox against white settlers from 1831 to 1832 that ended after a brutal response by American forces.

Black Power A philosophy of racial empowerment and distinctiveness as opposed to assimilation into white culture.

Bonus Army The group of more than 20,000 World War I veterans who marched in 1932 into Washington demanding payment of owed monies from the federal government.

Booker T. Washington The chief spokesman for a commitment to Black education and the founder and president of Tuskegee University.

Boston Massacre Inflammatory description of a deadly clash between a mob and British soldiers on March 5, 1770, that became a symbol of British oppression for many colonists.

Boston Tea Party Dramatic attempt by Boston leaders to show colonial contempt for the Tea Act; they dumped British tea into Boston harbor and triggered similar acts of resistance in colonial cities.

Boxer Rebellion A revolt begun by Chinese nationalists against foreigners in China.

braceros Contract laborers from Mexico allowed into the United States during World War II in response to wartime labor shortages.

brinksmanship The attempt to gain a negotiating advantage by pushing a situation to the edge of war or other disaster.

Brotherhood of Sleeping Car Porters An important union dominated and led by African Americans, including A. Philip Randolph.

Browder v. Gayle The 1956 district court decision affirmed by the Supreme Court that ruled that Montgomery's bus segregation laws were unconstitutional.

Brown v. Board of Education of Topeka The 1954 Supreme Court Decision that overturned "separate but equal" opinion of *Plessy v. Ferguson* and provided federal support for the civil rights movement.

Bull Moose Party Also known as the Progressive Party, launched by Theodore Roosevelt ahead of the 1912 presidential election.

Cahokia Major trading center in the Mississippi River valley near modern-day St. Louis, Missouri, from the seventh to the thirteenth centuries.

Californios Hispanic residents of California.

Calvin Coolidge Former governor of Massachusetts and vice president under Warren Harding; became president when Harding died in office; elected to the office of president in 1924 but did not run again in 1928.

Camp David Accords Peace treaty between Israel and Egypt brokered by President Carter in 1978.

Cane Ridge Kentucky site of the 1801 religious revival that lasted a number of days with thousands of attendees.

capitalist Owner of material or financial assets useful for the accumulation of additional wealth.

carpetbaggers Slang term used by white Southern Democrats to describe white men from the North, many of them veterans, who settled in the South as hopeful planters, businessmen, and professionals and supported Republican policies.

Cecilius Calvert The second Lord Baltimore who, with his father George Calvert, was instrumental in the founding of Baltimore.

Central Intelligence Agency (CIA) Replaced the World War II-era Office of Strategic Services in 1947; tasked with the collection of information related to national security from around the world through open and covert methods.

Central Powers One of the sides in the First World War, comprised of Germany, Austria-Hungary, and the Ottoman Empire.

Cesar Chavez Leader of the mostly Hispanic United Farm Workers (UFW) in the 1970s.

Charles E. Coughlin A Catholic priest famous for his national radio broadcasts; he proposed a series of monetary reforms that he insisted would restore prosperity and ensure economic justice.

Charles Sumner United States Senator from Massachusetts who was a leading voice against slavery and for Black liberties.

charter A formal order from a governmental leader or body, like the king of a court, often granting the recipient power over a body of land, a business, or a people.

checks and balances A system that grants the various branches of government the power to oversee or constrain other branches, so that no part grows too powerful.

Chester A. Arthur Became president when Garfield was assassinated.

Chief Joseph Leader of the Nez Percé tribe in the Pacific Northwest during the late 1870s who fought efforts to force the tribe onto a reservation in the Idaho territory.

Chinese Exclusion Act The federal law of 1882 that blocked Chinese immigration and prevented those Chinese already living in America from becoming citizens for ten years.

Christian Coalition A religious political coalition formed in the 1970s to elect candidates supportive of its evangelical values.

Christopher Columbus Sea captain working for the Spanish crown whose trans-Atlantic voyages helped introduce the "New World" to Europeans.

citizenship The legal recognition of a person's inclusion in a body politic by the extension of various rights and privileges and the expectation of various duties and obligations.

city beautiful movement Led by architect Daniel Burnham, the movement sought to impose order and symmetry on the disordered life of American cities.

Civilian Conservation Corps (CCC) Agency that created camps in national parks and forests for young unemployed men from the cities to work in a semi-military environment; projects included planting trees, building reservoirs, and improving agricultural irrigation.

Claudette Colvin An early leader in the civil rights movement who was arrested at the age of fifteen for not giving up her seat on a Montgomery bus nine months before Rosa Parks.

Clovis people Term used for the oldest inhabitants of the Americas most probably from modern-day Siberia who would have traveled the Bering Strait some 11,000 years ago.

Coercive Acts Parliament's retaliation against the Boston Tea Party that was meant to coerce Boston colonists by reducing the colony's self-government.

Cold War A simmering conflict between the United States and Soviet Union that emerged at the end of World War II, expressed in ideological terms of difference between capitalism and communism but often executed as a competition for power and security; tensions took many forms, including espionage, an arms race, and proxy wars around the world.

colonization A process by which a country or territory falls, usually by force, under the control of a hostile country or territory.

colony A geographic area in one nation under control by another nation and typically occupied at least partly by settlers from that other nation.

committees of correspondence First called for by Samuel Adams, the committees formed in Boston and other parts of the colonies to share information about British abuses of power.

Common Sense Thomas Paine's popular pamphlet that encouraged independence from England by arguing that colonists could never be truly free under the English constitution.

Commonwealth v. Hunt Massachusetts Supreme Court decision (1842) that established the legality of unions in Massachusetts.

Community Action programs Part of Johnson's "war on poverty," programs that employed members of poor communities in designing and administering local services.

Compromise of 1850 Seeking to diffuse tensions over slavery, this series of bills admitted California as a free state, abolished slavery in Washington, D.C., and established the Fugitive Slave Act.

Compromise of 1877 Rutherford B. Hayes's promise to withdraw federal troops from the South, effectively ending Reconstruction, in exchange for the support of Southern delegates in the disputed election of 1876 presidential election.

concentration policy U.S. government policy introduced in 1851 that forced Native American tribes to live in specific regions, thereby opening up new areas for settlement.

Coney Island The famous and popular amusement park located on a Brooklyn beach.

Confederate States of America Also known as the Confederacy, those slave states that seceded from the Union and declared an independent nation.

Confiscation Acts Two laws passed by the federal government during the Civil War, in 1861 and 1862, designed to free enslaved people held by Confederates.

Congress of Industrial Organizations (CIO) Led by John L. Lewis, this committee expanded the constituency of the American labor movement.

Congress of Racial Equality (CORE) Established in 1942, Black organization that mobilized popular resistance to discrimination in new ways such as sit-ins and "Freedom Rides"

conquistador A European (especially Spanish and Portuguese) conqueror of the Americas (particularly Mexico and Peru) during the fifteenth and sixteenth centuries.

conscription The practice of requiring citizens to serve in the military or other national service; the draft.

conservationist A proponent of the protection of land for carefully managed development, as opposed to a preservationist, who seeks to protect nature from development altogether.

Constitution The legal framework of the United States created to resolve limitations of the Articles of Confederation.

consumerism An increased focus on purchasing goods for personal use; the protection or promotion of consumer interests.

containment The Cold War strategy that called for preventing the spread of communism, by force or by other means.

cotton kingdom Term used for the lower South, referring to an economy and culture built on cotton production.

Cotton Mather Puritan theologian who, drawing from the knowledge of a man he held in slavery, Onesimus, helped introduce smallpox immunizations to America.

counterculture A way of life opposed to the prevailing culture; the term typically refers to the revolution in lifestyles, values, and behavior among some young people of the 1960s.

Court-packing plan Derogatory nickname for one aspect of President Roosevelt's proposal to overhaul the federal court system.

covenant A Puritan belief that an individual's relationship with God and with others rested on mutual respect, duty, and consent.

Coxey's Army A group of unemployed who marched on Washington, led by Ohio Populist Jacob S. Coxey, to demand relief from the depression.

Creole A person of European or African ancestry born in the Americas; also, a person of mixed European and African ancestry.

crop-lien system A credit system widely used in the South after the Civil War in which farmers promised a portion of their future crops in exchange for supplies from local merchants.

Cuban missile crisis A thirteen-day standoff between the USSR and the United States over Soviet nuclear missiles in Cuba that ended with the Kennedy-Khrushchev Pact.

cult of domesticity The early-nineteenth-century belief that women were the guardians of family and religious virtue within the home.

D-Day The Allied attack of June 6, 1944, across the English Channel against Hitler's forces in France.

Dale Carnegie Author of the best-selling self-help manual *How to Win Friends and Influence People*, published in 1936.

Daniel Webster Prominent diplomat and politician of the early republic, serving as senator, representative, and secretary of state; regularly championed issues in defense of the Union.

Darwinism The argument that the human species had evolved from earlier forms of life through a process of "natural selection."

Daughters of Liberty Organization of women in the colonies that led the boycott against the Tea Act.

David Walker Black abolitionist who encouraged Blacks to unite and take any necessary measures to fight slavery and other forms of discrimination.

Dawes Plan A 1924 agreement in which American banks would provide loans to Germany that would then be used to pay war reparations to Britain and France.

Dawes Severalty Act Legislation that provided for the gradual elimination of most tribal ownership of land and the allotment of tracts to individual owners.

Declaration of Independence A founding document of the United States, the declaration explained why the colonies were breaking away from Britain. Drafted by Thomas Jefferson, who borrowed concepts from other works, and edited by the Second Continental Congress, the declaration also appealed to foreign countries and spurred colonies to reform themselves as states.

Defense of Marriage Act (DOMA) A 1996 federal law that allowed states to decide whether or not to recognize same-sex marriages.

deism The belief that God created but does not actively control the universe.

Denmark Vesey A freed Black living in Charleston, South Carolina, who in 1822 planned a thwarted slave rebellion that may have numbered over 9,000.

deregulation The process of removing government controls over industries such as airlines, trucking, electricity supply, and banking, with the intention of stimulating competition and innovation.

Dollar Diplomacy Foreign policies, especially those of the Taft administration in Latin America, that privileged American economic interests.

Dominion of New England Colonial entity formed when James II combined the government of Massachusetts with the government of the rest of the New England colonies and, later, also included New York and New Jersey.

Donald Trump New York Republican businessman and billionaire elected forty-fifth president in a surprising victory over Democrat Hillary Clinton.

Dorr Rebellion Named after the leader, Thomas L. Dorr, the Dorr Rebellion was a failed attempt by a group in Rhode Island to set up a new state government with expanded voting rights.

Douglas MacArthur American general who headed the occupation of a conquered Japan at the end of World War II; also led UN forces against the North Koreans in the Korean War until relieved by President Truman for making insubordinate statements.

Dred Scott decision The 1857 Supreme Court ruling that effectively stated that enslaved people were not citizens but, instead, property and therefore could not bring a suit in the federal courts.

Dust Bowl A region that stretched north from Texas into the Dakotas and experienced a decade-long drought that began in 1930.

Dwight D. Eisenhower U.S. general in charge of the invasion of France across the English Channel, later the supreme Allied commander and then president of the United States elected in 1952.

détente The easing of hostilities between countries, used especially in connection with the Cold War in the 1970s.

Earth Day First started in 1970, a day of events meant to heighten public awareness of environmental issues.

Echo Park A national park on the border between Utah and Colorado that was threatened by development in the 1950s; environmental organizations rallied to block the dam project.

Edward Bellamy Author of the utopian novel *Looking Backward* (1888) in which government monopolies created an equitable society.

Eleanor Roosevelt Outspoken supporter of racial justice and source of continuous pressure on the federal government to ease discrimination against Blacks; wife of Franklin Roosevelt.

Eli Whitney American inventor best known for developing the cotton gin.

Elizabeth Cady Stanton Abolitionist and women's rights advocate who co-organized the Seneca Falls Convention.

Elizabeth I Protestant Queen of England for the latter half of the 1500s who presided over the beginnings of English colonial enterprises in America.

Elizabeth Keckley Personal seamstress of Mary Todd Lincoln who bought freedom for herself and her son through sewing.

Emancipation Proclamation Lincoln's executive order of 1863 declaring that those held in slavery in the Confederate states were forever free.

embargo A ban on trade with another country, especially the refusal to allow foreign ships to unload goods at port.

encomienda The right to extract tribute and labor from the native peoples on large tracts of land in Spanish America; also the name given to the land and village in such tracts.

Enforcement Acts Also known as the Ku Klux Klan Acts, these congressional acts in 1870 and 1871 prohibited states from discriminating against voters on the basis of race and gave the national government authority to prosecute crimes by individuals under federal law.

Enlightenment An intellectual movement that stressed the importance of science and reason in the pursuit of truth.

Erie Canal A constructed waterway that connected the Great Lakes to the Hudson River and transformed New York into an economic powerhouse of the young United States.

Eugene V. Debs Leader of the American Railway Union in the Pullman strike of 1894; presidential candidate for the Socialist Party.

eugenics The pseudo-scientific movement that attributed genetic weakness to various races and ethnicities; also describes efforts to control or isolate supposed hereditary traits through selective breeding, sterilization, immigration restriction, and other forms of social engineering.

evangelist A devout person who aims to convert others to the faith through preaching and missionary work.

Factory Girls Association The 1834 union originally formed by Lowell workers to protest pay cuts.

factory system A method of manufacturing involving powered machinery, usually run by water, that allowed the use of unskilled labor and yielded greater output than in the artisan tradition.

Fair Deal Harry Truman's twenty-one-point domestic program supporting expansion of Social Security, an increase of the minimum wage, public housing, and environmental/public works planning.

Farm Security Administration Created in 1937, this agency provided loans to help famers cultivating submarginal soil to relocate to better lands; in the end, it moved no more than a few thousand.

Farmers' Alliances Began among southern farmers in 1875 but spread nationwide; formed cooperatives and other marketing mechanisms.

fascism A term originating with Mussolini's Fascist Party and applying to any antidemocratic regime with a supreme leader, intolerance of dissent, faith in militarism over diplomacy, and a belief in national or ethnic superiority.

Federal Deposit Insurance Corporation (FDIC) Established by the Glass-Steagall Act of June 1933, this guaranteed all bank deposits up to $2,500.

Federal Highway Act of 1956 Massive ten-year federal project to build over 40,000 miles of interstate highways initiated under President Eisenhower.

federalism A political system dividing powers between state and federal governments that together constitute a federation.

Federalists Term for supporters of the Constitution and later a political party that favored a strong central government.

Fidel Castro One of the leaders against Fulgencio Batista's dictatorship of Cuba, Fidel Castro took control of Cuba in 1959 and turned Cuba into a communist state with support from the Soviet Union.

Fifteenth Amendment An 1870 constitutional amendment that forbade the states and the federal government from denying suffrage to any male citizen on account of race, color, or previous condition of servitude.

fireside chats President Franklin Roosevelt's regular radio addresses during which he explained to the people in simple terms his programs and plans, helping build public confidence in the administration.

First Battle of Bull Run The first major battle of the Civil War; also known as First Battle of Manassas.

First Continental Congress Early gathering of colonial delegates in 1774 that called for the repeal of all oppressive laws of Parliament since 1763.

Five Civilized Tribes Term used for the five native societies of the American South that had adopted some Euro-American social structures and institutions by the 1830s.

flappers Young women who challenged traditional expectations in the mid-1920s.

Fort Necessity Site of the opening skirmish in the French and Indian War, this stockade in the Ohio Valley was unsuccessfully protected by Militia Colonel George Washington.

Fort Sumter Fort in Charleston, South Carolina, that was shelled by Confederate forces on April 12, 1861, forcing its surrender and marking the start of armed conflict in the Civil War.

Forty-niners A slang term for people who flocked to California in 1849 in search of gold.

Fourteen Points President Woodrow Wilson's list of principles for which he believed the nation should be fighting during the First World War.

Fourteenth Amendment An 1868 constitutional amendment that granted citizenship to all persons born in the United States and prohibited states from denying "life, liberty, or property, without due process of law" or equal protection under the law.

Frances Perkins First female member of the cabinet; appointed by Roosevelt as secretary of labor.

Francis Cabot Lowell Pioneer of American textile manufacturing who created one of the first complete mills in Waltham, Massachusetts.

Frank Capra Italian-born director whose films presented social messages; films included *Mr. Deeds Goes to Town*.

Franklin Delano Roosevelt Assistant secretary of the navy, governor of New York, and president from 1933 until his death in office in 1945; his campaign and subsequent administrations offered Americans "a new deal"

Frederick Douglass African American abolitionist and reformer who was a major voice against slavery in both his writings and in public speeches throughout America and Europe.

free silver Economic philosophy that advocated for the coining of silver; farmers and others believed that expanding the money supply in this way would increase prices for their products and ease their debt payments.

Free-Soil Party The antislavery party that emerged during the 1848 presidential and congressional elections.

Freedmen's Bureau U.S. bureau established in 1858 that aimed to help former enslaved people forge independent lives.

freedom rides Civil rights initiative in which groups of interracial students traveled by bus through segregated states.

Freedom Summer Attempt by civil rights activists in the summer of 1964 to encourage Black voter registration mostly in segregated states like Alabama and Mississippi.

French and Indian War Colonists' name given to the Seven Years' War in the colonies that strained the relationship of England to its colonies and marked the decline of relationships between Native Americans and Europeans.

Gabriel Prosser Leader of a thwarted large-scale slave revolt outside Richmond, Virginia, in 1800.

Gadsden Purchase The 1853 American acquisition from Mexico of nearly 30,000 acres of land that now form modern southern Arizona and southwestern New Mexico.

George B. McClennan Union general who ran unsuccessfully as a Northern Democrat in the 1864 presidential election against Lincoln.

George Calvert The first Lord Baltimore who, with his son Cecilius, the second Lord Baltimore, was instrumental in the founding of Maryland.

George Grenville Prime Minister to King George III who increased troops and taxes in the colonies after the French and Indian War and made many colonists believe colonial self-rule was under attack.

George H. W. Bush Vice president under Reagan who served as the forty-first president during the first Gulf War and several years of economic downturn.

George III King of England in 1760 who wanted to reassert the crown's authority; he was mentally unstable for most of his reign.

George W. Bush Son of the forty-first president and former governor of Texas, he pushed through broad tax cuts and led the country after the 9/11 attacks, initiating an aggressive war against terror.

George Washington Military leader and one of the founders of the United States; served as first president.

Gerald R. Ford Vice president appointed by Nixon in 1973 who took over the presidency in 1974 after Nixon's resignation.

Gerald R. Ford Vice president to Richard Nixon who assumed the presidency upon Nixon's resignation.

Geronimo Apache chief and medicine man who led the fight against resettlement efforts by Mexico and then the United States.

Gettysburg Pennsylvania town that was the site of a major Civil War battle on July 1-3, 1863, in which the Union Army turned back the Confederate march northward.

GI Bill Officially known as the Servicemen's Readjustment Act of 1944; provided housing, education, and job-training subsidies to veterans.

Gibbons v. Ogden Supreme Court case of 1824 that strengthened federal authority over interstate commerce.

globalization The process of interaction and exchange between peoples and ideas from different parts of the globe.

Good Neighbor Policy Franklin Roosevelt's position regarding the countries of Latin America, which marked a departure over prior approaches by offering commercial relationships and cooperation rather than military interventions to promote hemispheric stability and United States interests.

gospel of wealth Term popularized by Andrew Carnegie to argue that those with immense wealth carry a greater burden to use that wealth for social progress.

Grangers Founded in 1867, the first major farm organization in the country to mobilize against railroads and other special interests; predecessor to the farmers' alliances of the late nineteenth century.

Great Awakening The first major American religious revival, begun in earnest in the 1730s.

Great Depression The major economic downturn of the 1930s that ended with American entrance into World War II.

Great Migration The movement of nearly half a million Black people from the rural South to industrial cities in the North in the era of the First World War.

Great Recession of 2008 Economic crisis fueled by the collapse of the housing market and poor regulation of financial industries.

Great Society LBJ's legislative initiatives that focused on addressing the social problems of poverty, decaying cities, and poor schools.

greenbacks Paper currency not backed by gold or silver.

Grover Cleveland Reform governor of New York who was elected president in 1884 and again in 1892.

Guadalcanal One of the southern Solomon Islands, American forces assaulted a Japanese garrison here for six months in 1942-1943 before successfully driving them out.

Guantanamo U.S. naval base in Cuba where suspected terrorists from the war on terror have been interrogated and detained since 2002.

Gulf of Tonkin Resolution Act of Congress in 1964 that gave President Johnson the authority to escalate the conflict in Vietnam based on questionable accounts of attacks made on American ships by the North Vietnamese.

Half-Breeds Political group within the Republican Party led by James G. Blaine of Maine, who favored reform.

Handsome Lake Seneca native and revivalist of the Second Great Awakening who called for a return to native traditions among the Iroquois.

Harlem Renaissance Term used to describe the flourishing artistic life created by a new generation of Black intellectuals in New York who focused on the richness of their own racial heritage.

Harpers Ferry Site of the federal arsenal in Virginia raided by John Brown in 1859.

Harriet Beecher Stowe Abolitionist, best known for her novel *Uncle Tom's Cabin*.

Harry Hopkins Former director of New York State relief agency; appointed by Roosevelt to administer New Deal agencies FERA and the WPA.

Harry S. Truman Democratic senator from Missouri, then Franklin Roosevelt's vice presidential candidate in the 1944 election, and president of the United States from 1945 to 1953.

Hartford Convention Meeting of New Englanders, many of them Federalists, that denounced the War of 1812; they had just adjourned when news of Jackson's victory in New Orleans and the peace treaty became known.

Hawley-Smoot Tariff This 1930 act established the highest import duties in history and stifled global commerce; it contributed to the dramatic contraction of international trade leading up to the Great Depression.

Haymarket bombing In a clash between striking laborers and police in Chicago on May 1, 1886, an unknown person threw a bomb into a crowd killing seven police and injuring nearly seventy others.

headright system A grant system that allowed new settlers to acquire fifty acres of land in a variety of ways.

Henry Cabot Lodge The Republican senator and chair of the Foreign Relations Committee who obstructed and opposed the ratification of the Treaty of Versailles after the First World War.

Henry Clay Prominent politician from Kentucky, serving as Speaker of the House, senator, and secretary of state, and running unsuccessfully for president three times; one of the founders of the Whig Party; helped bring about the Missouri Compromise of 1820 while Speaker.

Henry David Thoreau Transcendentalist who urged Americans to resist both social conformity and unjust laws.

Henry Ford Early leader of the automobile industry who stressed the standardization of parts and assembly lines.

Henry George Author of *Progress and Poverty* (1879), which argued for tax reform on land as a way to break the power of monopolies.

Herbert Hoover Elected president in 1928, he personified the modern, prosperous, middle-class society of the New Era, but also came to be associated with the failure to adequately respond to the Great Depression.

Herman Melville Author of the 1851 classic *Moby Dick* that captured harsh aspects of nineteenth-century American culture.

Hessians German mercenaries hired by England during the American Revolutionary War.

Hetch Hetchy Valley that was the object of a 1906 controversy over a proposed dam, which was ultimately built; brought the contending views of the early conservation movement to a head.

Hideki Tojo General and leader of the war party in Japan; replaced more moderate prime minister in 1941.

Hillary Rodham Clinton Former first lady, senator of New York, secretary of state, and in 2016 the first woman to earn the presidential nomination of a major political party.

Hindenburg German dirigible that crashed in flames in Lakehurst, New Jersey, in 1937, broadcast live over the radio.

Hiroshima Japanese site of the first detonation of an atomic bomb against an enemy nation, dropped by the United States in 1945.

Ho Chi Minh Nationalist and communist leader of the Vietnamese independence movement against the French from 1946 to 1954, and then leader of North Vietnam until his death in 1969.

Holocaust Systematic Nazi campaign of the 1930s and 1940s to exterminate Jews and other "undesirable" groups of Europe.

Homestead Act Federal legislation permitting any citizen or prospective citizen, including those who had once been enslaved, to purchase 160 acres of public land in the western United States for a small fee after living on it for five years.

Homestead Strike A strike of the steel mill union in 1892 that led to armed conflict and the involvement of state militia.

Hoovervilles The term used to describe the shantytowns that unemployed people established on the outskirts of cities during the Depression.

Horace Mann Educational reformer who promoted education as essential to a strong democracy.

Horatio Alger Author of Gilded Age books whose hardworking heroes go from "rags to riches"

Horizontal integration A corporate combination where a group of businesses that do the same thing are consolidated.

House Un-American Activities Committee (HUAC) Congressional committee that held widely publicized investigations into communist subversion within the American government.

Hudson River school New York landscape painters known for their depictions of spectacular vistas.

Huey P. Long Senator from Louisiana known as "the Kingfish" who initially supported Roosevelt but broke with him; champion of the Share-Our-Wealth Plan, which would have guaranteed every American a home and income.

Hull House The most famous of the settlement houses, opened in Chicago in 1889.

Ida B. Wells African American journalist whose reporting in the late nineteenth century on racial violence launched what became an international antilynching movement.

Immigration and Nationality Act of 1965 Legislation that revised laws from the 1920s, allowing for greater immigration from most areas and removing preferences for northern European immigrants.

impeachment The process of charging a public official with misconduct, with the potential for punishment including loss of office.

imperialism The process whereby an empire or nation pursues military, political, or economic advantage by extending its rule over external territories and peoples.

impressment The act of forcing people to serve in a navy or other military operation; the term is most commonly used in connection with the actions of British fleets against American sailors in the early 1800s.

indentured servitude The condition of being bound to an employer for a specific period of time, usually in exchange for the cost of passage to a new land. The labor practice was most commonly used in Britain's American colonies.

Indian Civil Rights Act Law passed by Congress in 1968 that extended Bill of Rights protection to residents of reservations but also accepted native legal authority inside reservations.

Indian Removal Act An 1830 act that allowed the federal government to negotiate with Native Americans to relocate them west of the Mississippi River.

Indian Territory Present-day Oklahoma, the land designated for Native Americans forced to relocate west of the Mississippi River by the Indian Removal Act.

Industrial Revolution The transformation of the economy from manual to mechanized forms of production; started in Britain in the eighteenth century and later spread to other places, including in the United States in the nineteenth century.

Industrial Workers of the World (IWW) Known to its opponents as the "Wobblies," the radical labor organization led by "Big Bill" Haywood; advocated a single union for all workers.

Industrialization The process of a society changing from predominately agricultural production to a society based on factory production.

intercontinental ballistic missile (ICBM) Missile capable of traveling over oceans and into space to deliver nuclear strikes.

Interstate Commerce Act The first effective federal railroad regulation, passed in 1887; administered by a five-person agency.

Iran-Contra scandal Covert and illegal operation by members of the Reagan administration who funneled money gained from selling arms to the anti-American government of Iran to the anticommunist rebels or "Contras" in Nicaragua.

Iraq War An armed conflict (2003–2011) of Western countries led by the United States to depose Iraq leader Saddam Hussein, set up a more democratic government, and defend it against local insurgents.

Iroquois Confederacy Organization of five Native American nations that traded regularly with the French and English in the early colonial period, but their relationships with the colonists deteriorated during the mid to late 1700s.

isolationism A foreign policy that avoids forging alliances or lending support to other nations, especially in wartime.

Issei Japanese immigrants.

J. P. Morgan Banker and creator of U.S. Steel.

Jacob Leisler Colonist who raised a militia in 1689 and proclaimed himself the head of government in New York.

Jacob Riis New York newspaper photographer who wrote *How the Other Half Lives,* which used photos and words to expose the harshness of tenement life.

James A. Garfield Veteran Republican congressman from Ohio and a Half-Breed; won the presidency in the 1880 election; assassinated in 1881.

James Henry Hammond South Carolina senator who coined the expression "cotton is king."

James K. Polk North Carolinian and 11th president of the United States who served from 1845 to 1849.

James Madison Fourth president of the United States; instrumental in the creation of the U.S. Constitution.

James Oglethorpe Veteran British general who spearheaded the founding of Georgia.

Jamestown First colonial settlement of the London Company in North America.

Jane Addams Influential social worker and advocate of the settlement house movement.

Jarena Lee An African American woman who preached in public in contrast to rules and customs that prohibited her from doing so during the first half of the nineteenth century.

Jay's Treaty Crafted in response to continued British seizure of American ships in 1794 by the chief justice of the Supreme Court, John Jay; resolved the dispute by acknowledging American supremacy over the Northwest territory and producing a commercial relationship with Britain.

Jefferson Davis President of the Confederacy.

jeremiad A sermon of despair at society's lost moral virtue, usually warning about dire consequences in the world and the afterlife.

Jim Crow laws A dense network of state and local statutes that institutionalized an elaborate system of racial hierarchy.

Jimmy Carter Former governor of Georgia who served as the thirty-ninth president of the United States during a period of rising oil prices, economic recession, and the Iran hostage crisis.

jingoes A term coined in the late nineteenth century to refer to advocates for expanded U.S. economic, political, and military power in the world.

John Adams One of the country's founders; first vice president of the United States and the second president of the United States.

John Birch Society Ultra-conservative organization led by Robert Welch, who was convinced communism had infiltrated all levels of the American government.

John Brown Radical abolitionist who aimed to foment a slave insurrection in the South.

John Burgoyne General for British northern forces, defeated at Saratoga.

John C. Calhoun Prominent South Carolinian politician serving as senator, secretary of state, vice president, and secretary of war; supporter of slavery and of nullification, or the states' rights to nullify federal laws if they found them unconstitutional.

John Collier Commissioner of Indian affairs whose goal was to reverse the pressures of assimilation and instead champion Native American rights.

John D. Rockefeller Founder of Standard Oil, famous for horizontal and vertical integration, and the wealthiest man of the Gilded Age.

John Dos Passos Author of the *U.S.A.* trilogy in the 1930s, which attacked materialistic American culture.

John Foster Dulles Secretary of state under Eisenhower who advocated for aggressive action against communism, as well as brinksmanship as a strategy to gain concessions from foreign powers.

John J. Pershing The American general who led the expedition chasing Pancho Villa in 1916 and later commanded the American forces in the First World War.

John Kennedy First Catholic to be elected president after defeating Richard Nixon in 1960; assassinated in 1963.

John L. Lewis Leader of the United Mine Workers, champion of industrial unionism, and first president of the CIO.

John Marshall Chief Justice of the United States Supreme Court for over thirty years and a Federalist, Marshall most famously rendered the opinion in *Marbury v. Madison*.

John Muir The leading preservationist in the United States and founder of the Sierra Club.

John Peter Zenger New York publisher tried for libel whose case expanded free speech.

John Quincy Adams Son of John Adams, he served as secretary of state under James Monroe, president from 1825 to 1829, and later as a member of the House of Representatives.

John Smith World traveler and writer whose leadership helped the Jamestown colony survive.

John Steinbeck Author of *The Grapes of Wrath*, perhaps the best-known depiction of Depression-era American life.

John Tyler Virginian who became the tenth president after the death of William Henry Harrison only a month into his administration; regularly clashed with his fellow Whigs and was not nominated for reelection by his party.

John Winthrop Governor of the Massachusetts Bay Company who dominated colonial politics.

Jonathan Edwards New England Congregationalist preacher who was famous for vivid descriptions of hell and damnation.

Josef Stalin Communist dictator of the Soviet Union from the 1920s until his death in 1953; responsible for the death or exile of millions of Soviet citizens.

Joseph and Mary Brant Mohawk brother and sister who allied with the British, thus harming the unity of the Iroquois Confederacy's neutrality during the Revolutionary War.

Joseph Smith Founder of the Mormon faith.

Judith Sargent Murray Essayist of the early republic who argued for a larger role for women in the new country.

Julius and Ethel Rosenberg New York couple accused, convicted, and executed for passing secret information regarding America's atomic bomb to the Soviets; Julius was likely guilty, Ethel likely innocent.

Kansas-Nebraska Act Passed by Congress in May 1854, it allowed residents of the territories of Kansas and Nebraska to decide whether slavery would be permitted there.

Kate Chopin A southern writer who explored the oppressive features of traditional marriage; known for her shocking novel *The Awakening*.

King Philip's War The most prolonged and deadly encounter between whites and Native Americans in the seventeenth century.

Knights of Labor Short-lived early national labor union that championed eight-hour workdays and the end of child labor, open to almost all workers.

Know-Nothings Name used for the anti-immigrant, anti-Catholic group that formed in 1850 and organized as the American Party.

Korematsu v. U.S. U.S. Supreme Court case that upheld the internment of more than 100,000 Japanese Americans to detention camps.

Ku Klux Klan One of many secret societies that used terrorism and physical violence to intimidate those freed from slavery and undercut their constitutional rights, especially the right to vote.

Langston Hughes A leading writer of the Harlem Renaissance.

League of Nations The organization President Woodrow Wilson hoped would implement the principles of the Fourteen Points after the First World War; it ultimately came into being without the United States as a member.

lend-lease A system that allowed the Franklin Roosevelt administration to lend or lease arms to the British without explicitly violating the Neutrality Acts.

Levittowns Named after developer William Levitt, Levittowns were inexpensive suburban developments of similarly built homes.

Lewis and Clark On the direction of President Jefferson, Meriwether Lewis and William Clark led an expedition from Missouri to the Pacific in order to gather information on the lands acquired in the Louisiana Purchase.

Liberia African nation established in 1830 by freed Blacks whose manumission and voyage was sponsored by the American Colonization Society (ACS).

Liberty League An organization founded by conservative business leaders, led by the Du Pont family, and focused on attacking the New Deal.

***Life* magazine** New and enormously popular photographic journal, first published in 1936, it had the largest readership of any publication in the United States other than *Reader's Digest*.

Little Bighorn Site of the 1876 battle in which Colonel George Custer and his men were surprised and killed by a large army of Sioux warriors.

long drive A journey over grasslands that allowed western cattle ranchers to deliver their animals to railroad centers.

Lord Cornwallis British officer with early successes as leader of the Southern British forces but who was forced to surrender at Yorktown in 1781.

Lord Dunmore's Proclamation British promise of 1775 to grant freedom to people enslaved by Patriots in exchange for joining their military forces against the rebelling colonists.

Lost Generation Author Gertrude Stein used this term to describe the young Americans emerging from World War I.

Louis D. Brandeis Lawyer and Supreme Court justice; author of *Other People's Money*, which was about the "curse of bigness."

Louisa May Alcott Author of the *Little Women* series about an ambitious girl who fought conventional society to become a writer.

Lowell or Waltham system A factory system used to mass-produce textiles, primarily in New England, that relied on young women workers who lived in factory communities.

Loyalists (Tories) Supporters of England and the king, they may have represented a third of the white colonial population; many left America after the Revolution.

Lucretia Mott Abolitionist and women's rights advocate who co-organized the Seneca Falls Convention.

Lusitania The British passenger liner sunk by Germany in May 1915, killing almost 1,200 people, including 128 Americans.

Lyndon Johnson Often called LBJ, the president from 1963 to 1969 who promoted major social reform in his Great Society legislation and expanded America's military role in Vietnam.

Malcolm X Leader of the civil rights movement who promoted Black Power; assassinated in 1965.

Manhattan Project The American military's secret operation to develop an atomic bomb.

Manifest Destiny An ideology holding that God or fate intended the United States to expand its dominion across the North American continent.

manumission The act of freeing enslaved people.

Mao Zedong Leader of the communist armies of China in ongoing conflict with the nationalist government of Chiang Kai-shek in the 1930s and 1940s; won that battle and declared a communist China in 1949.

Marbury v. Madison Important decision of the United States Supreme Court that established the court's authority over the constitutionality of laws.

Marcus Garvey The African American leader who encouraged Black people to reject assimilation into white society and develop pride in their own race and culture; he founded the Universal Negro Improvement Association (UNIA).

Margaret Sanger The pioneer of the American birth-control movement.

Mark Twain Pen name of Samuel Langhorne, nineteenth-century American author and humorist who wrote *The Adventures of Tom Sawyer* (1876) and *The Adventures of Huckleberry Finn* (1885).

Marshall Plan American program after World War II meant to spark economic recovery in Europe and thus cultivate economic ties and broader alliances between the Continent and the United States; eventually channeled $13 billion into the economies of sixteen participating countries.

Martin Luther King Jr. Baptist minister who came to prominence during the Montgomery bus boycott and went on to be the voice of the civil rights movement until his assassination in 1968.

Martin Van Buren Eighth president of the United States after serving as Andrew Jackson's vice president; nicknamed the little magician, but struggled as president during a difficult depression.

Massachusetts Bay Company Group of Puritan merchants in England who organized a new colonial venture in America.

Mayflower Compact Document that the Pilgrims signed to establish a government for themselves.

McCarthyism Name given to the anticommunist crusade of Senator Joseph McCarthy in the early 1950s, during which he recklessly persecuted alleged communists, often without evidence.

McCulloch v. Maryland Supreme Court case that confirmed the implied powers of Congress by upholding the constitutionality of the Bank of the United States.

Medicaid Social welfare program created during the Johnson administration that extended medical care to all ages in need.

Medicare Social welfare program created under the Johnson administration to provide health care to elderly Americans.

mercantilism An economic theory popular in Europe from the sixteenth through eighteenth centuries holding that nations were in competition with one another for wealth, and that the state should maximize its wealth by limiting imports and establishing new colonies that would provide access to precious minerals, spices, and enslaved labor.

Mesoamerica Land area of the Archaic period including the lower portion of modern-day Mexico and the rest of Central America where many native societies flourished.

mestizo A person of mixed European and American descent, traditionally in Spanish-speaking territories and nations.

Metacom Leader of an attempt by Native Americans in seventeenth-century New England to drive out English settlers and resist encroachment on their lands.

middle grounds Places where European and indigenous cultures interacted and where neither side had a military advantage.

middle passage The name given to the route used by slave ships between Africa and the Americas.

Mikhail Gorbachev Soviet leader who initiated broad economic reforms, government restructuring, and changes in military policy including a reduced Soviet presence in Eastern Europe and who resigned when the Soviet Union ceased to exist.

minstrel show Form of popular theater and entertainment from the early 1800s to the early 1900s that openly mocked and degraded African American culture.

Missouri Compromise Agreement of 1820 that defused sectional conflict by agreeing to admit Maine as a free state and Missouri as a slave state, and to henceforth prohibit slavery in the Louisiana Purchase territory in the regions north of the southern border of Missouri.

Molly Maguires A secret society of Irish-born coal miners willing to use violence to deal with management.

monopoly A business entity that controls an industry or market sector without competition.

Monroe Doctrine Articulated in 1823, the policy of the United States that warned against European interference in the American continents and promised the United States would stay out of European affairs.

Morse code System, designed by Samuel Morse, of long and short electrical bursts that made the telegraph system a viable, long-distance communication system.

muckraker A journalist who exposes scandal, corruption, and injustice; the term was especially popular during the progressive era.

My Lai massacre The deliberate American killing of hundreds of Vietnamese villagers in the hamlet of My Lai in early 1968.

Nat Turner Leader of a slave revolt in 1831 in Southampton County, Virginia, that killed 60 white men, women, and children before being crushed.

Nathaniel Hawthorne Novelist, best known for *The Scarlet Letter* and *The House of the Seven Gables*, who wrote about the misery caused by egotism.

National Association for the Advancement of Colored People (NAACP) Founded in 1909 when the Niagara Movement joined with sympathetic white progressives; the goal was equal rights.

National Consumers League (NCL) Formed in the 1890s under the leadership of Florence Kelley; the goal was to force retailers and manufacturers to improve wages and working conditions.

National Labor Relations Board (NLRB) Part of the Wagner Act, this board was an enforcement mechanism that compelled employers to recognize and bargain with legitimate unions.

National Organization for Women (NOW) A leading advocacy organization for women's rights, formed in 1966.

National Origins Act of 1924 Law that banned immigration from East Asia entirely and reduced the quota for European immigrants; the result was an immigration system that greatly favored northwestern Europeans.

National Recovery Administration (NRA) A federal agency that called on businesses to accept the regulation of wages, hours, prices, and other labor practices with the goal of stabilizing the economy, maintaining the workforce, and reducing competition; invalidated by the Supreme Court in 1935.

National Socialist (Nazi) Party Germany's National Socialist organization, headed by Adolf Hitler, that came to power in 1933.

nativism A belief in the superiority of native-born inhabitants over immigrants; in particular, an anti-immigrant movement that began in the early 1800s in the United States and crested with the passage of immigration restriction in 1924.

Navigation Acts Three acts that Parliament passed to regulate colonial commerce.

Nelson Mandela Leader of the African National Congress and force against apartheid who became the first Black president of South Africa in 1994.

neoconservatives A small but influential group of conservative intellectuals who rejected liberalism after the turmoil of the 1960s.

Neutrality Acts Series of laws between 1935 and 1937 that created a mandatory arms embargo against both sides in any military conflict and legislated other inhibitors of American involvement in another foreign war.

New Deal The broad array of reform initiatives launched during the administration of Franklin Roosevelt in the 1930s that dramatically increased the impact of the federal government on economic life and the personal welfare of citizens.

New Freedom Presidential candidate Woodrow Wilson's 1912 program that supported a progressive agenda, claiming that corporate combinations threatened economic freedom.

New Frontier JFK's campaign plan for progressive domestic reforms.

New Jersey Plan Plan presented by William Paterson of New Jersey during the Constitutional Convention to have a single legislative body with equal representation for all states regardless of population.

New Nationalism Presidential candidate Theodore Roosevelt's 1912 program supporting economic concentration and using government to regulate and control it.

New Right Conservative movement that began in the 1970s and culminated in the election of Ronald Reagan.

New South A term referring to the economic modernization and industrialization of the South after Reconstruction.

Nicholas Biddle Ran the Second Bank of the United States; fought ultimately unsuccessfully against President Andrew Jackson for the survival of the institution.

Nicola Sacco and Bartolomeo Vanzetti The two anarchists accused of murder who were eventually executed in the 1927 amid widespread nativist prejudices and fears.

Nineteenth Amendment The amendment to the constitution ratified in 1920 that guaranteed women the right to vote.

Nisei The American-born children of Japanese immigrants.

Nixon Doctrine Foreign policy plan under President Nixon to continue to support allies' military defense needs while cutting back on the commitment of American forces to those needs.

Noah Webster Author, teacher, and promoter of the new American nation, best known for his dictionaries and spellers that helped standardize the American language.

North Atlantic Treaty Organization (NATO) The postwar alliance of the United States and many of the countries of Western Europe, unified against invasion by the Soviet Union.

Northwest Ordinance A 1787 decree that created a single political territory out of the land north of the Ohio River.

NSC-68 A 1950 report commissioned by the Truman administration and issued by the National Security Council; called for a major expansion of American military power to combat the threat of communism and the Soviet Union.

nullification The theory that individual states, as the original creators of the federal government, possess the right to invalidate federal laws if they find them unconstitutional.

Occupy Wall Street (OWS) Protest movement of 2011 against economic inequality that began as an encampment in a Wall Street area park.

Office of Price Administration (OPA) Unpopular federal organization during World War II tasked with fighting economic inflation.

Office of War Information (OWI) Agency charged with disseminating the official U.S. viewpoint and encouraging domestic war efforts during World War II.

Okies Collective term for families from the Dust Bowl (though not all came from Oklahoma) traveling to California in search of jobs.

Okinawa Fierce battle in the Pacific on an island 370 miles south of Japan; saw the use of kamikaze planes and the loss of over 100,000 Japanese troops.

Oneida "Perfectionists" Members of a Utopian experiment in upstate New York who rejected traditional notions of family and marriage in favor of communal bonds.

Open Door The metaphor Secretary of State John Hay used in 1898 to characterize the access to Chinese markets he desired for the United States; it was later expanded to refer to a policy of granting equal trade access to all countries.

Oregon Trail A 2,170-mile wagon route that linked the Missouri River to western Oregon and was a main passageway for white western settlers between the 1830s and early 1870s.

Osama bin Laden Leader of Al Qaeda from 1987 until 2011, when he was killed in Pakistan by U.S. military special forces.

Palmer Raids Led by U.S. Attorney General A. Mitchell Palmer, these 1919 and 1920 police actions targeted alleged radical centers throughout the country, often using extralegal means.

Panama Canal The canal finished in 1914 that linked the Atlantic and the Pacific by creating a channel through Central America.

Pancho Villa The Mexican revolutionary whom Gen. John Pershing unsuccessfully pursued into Mexico after Villa killed Americans along the border.

Panic of 1819 Six-year depression that began with a price collapse of American trade goods.

Panic of 1837 Triggered by an executive order that the government would accept payment for land only in gold or silver or a currency backed by one of the metals, this economic crisis included business failures, a spike in unemployment, falling prices, and even bread riots.

Panic of 1893 The beginning of the most severe depression the United States had experienced at the time; triggered by the Philadelphia and Reading Railroad bankruptcy.

parity A complicated formula for setting an adequate price for farm goods and ensuring that farmers would earn back at least their production costs no matter the fluctuations of national or world agricultural markets.

Patrick Henry Virginia politician who lead the fight against the Stamp Act and declared supporters of Parliament taxes were enemies of the colonies.

Paxton Boys A group of Pennsylvania frontiersmen who demanded tax relief and massacred a number of Conestoga.

Pearl Harbor American naval base in Hawaii and headquarters of the Pacific Fleet; attacked by Japanese on December 7, 1941.

peculiar institution Southern term for slavery depicting it as a special institution of the South.

Pendleton Act First national civil service measure, passed in 1883, that tested applicants' qualifications for federal jobs rather than assigning them through patronage connections, among other reforms; largely symbolic at first but grew in reach over time.

Pequot War War in Connecticut during 1637 between English settlers and Native Americans of the region.

Pinckney's Treaty Agreement between the United States and Spain that guaranteed access to the Mississippi River for American trade and protection from Native Americans in Spain's territories.

planter class Wealthy slaveholding planters of the South who dominated southern society despite their limited number.

Plessy v. Ferguson An 1896 Supreme Court decision that ruled that separate accommodations for Blacks and whites were legal so long as they were equal.

Plymouth Plantation First Pilgrim settlement in Massachusetts.

Pontiac Ottawa chief who led a coalition of native nations to war against the British from 1763 to 1766; achieved some gains including pressuring the British to restrain their settlers from the trans-Appalachian west, but eventually was undermined by internal divisions, disease, and the brutal violence of settlers and the British military.

Popular Front This was a broad coalition of antifascist groups on the left; communism was its driving force.

popular sovereignty A term coined by Stephen A. Douglas to adjudicate the expansion of slavery in the western territories by allowing settlers to decide the status of slavery for their territory.

Populism A reform movement of the 1890s that promoted federal government policies to redistribute wealth and power from national elites to common people; more generally, refers to a political doctrine that supports the rights of the people over the elite.

Popé Native American religious leader who led a successful uprising against the Spanish in 1680.

Powhatan Chief of the Powhatan Confederacy and father of Pocahontas.

preservationist Activist who believes in the protection of natural environments rather than their managed development.

Proclamation of 1763 Attempt by England to reduce violence between Native Americans and English colonists by legally barring settlement beyond the Appalachian Mountains.

progressivism The reform movement of the late nineteenth and early twentieth centuries that attempted to impose order on a society rapidly changing amidst industrialization, immigration, and urbanization.

prohibition Complete ban on the sale and manufacture of alcoholic beverages; 1920 amendment put it into effect on a national level.

Protestant Reformation The schism in the Catholic Church that began in 1517 with Martin Luther and led to new forms of Christian denominations still recognizable today.

Public Health Service Organization created in 1912; goal was to prevent occupational diseases and create common health standards.

Puerto Rico Part of the Spanish Empire from 1508 until 1898, when it fell under the control of the United States; became an American territory in 1917.

Pullman strike An 1894 railroad strike that escalated to twenty-seven states and territories, ultimately broken by federal troops and resulting in management's victory.

Puritans A sect of Protestants of England that wished to "purify" the Church of England of its Catholic ceremonies and practices.

Quakers A Protestant sect that called themselves the Society of Friends and believed that all could attain salvation by cultivating their inherent divinity.

quasi war The name given to the undeclared war between the United States and France of 1798–1799.

Queen Liliuokalani Nationalist leader of Hawaii elevated to the throne in 1891.

Rachel Carson Author of the 1962 book *Silent Spring* who argued that overuse of pesticides was destroying the environment.

Radical Republicans A wing of the Republican Party in the mid-nineteenth century that aggressively opposed slavery and, after the Civil War, fought to expand and protect African American civil rights.

Ralph Waldo Emerson Transcendentalist philosopher who urged individuals to find fulfillment and self-improvement in nature.

range wars Conflicts between sheepmen and cattlemen, ranchers, and farmers.

Reagan Doctrine President Reagan's policy to combat communism by supporting anticommunist regimes and revolutionaries with money and military aid.

Reaganomics Supply-side economic policy embraced by Reagan and based on the idea that reducing taxes on corporations and the wealthy would encourage new investment and social well-being.

Rebecca Cox Jackson Radical African American religious figure who broke with the free Black church movement in Philadelphia and ultimately joined the Shaker movement during the mid-nineteenth century.

Reconstruction The process by which the federal government, between 1865 and 1877, controlled the former Confederate states and set the conditions for their readmission to the Union.

Reconstruction Finance Corporation (RFC) The bill established a government agency to provide federal loans to troubled banks, railroads, and other businesses.

Red Scare A period of intense popular fear and government repression of real or imagined leftist radicalism; usually associated with the years immediately following World War I.

Redeemers Coalition of white southern landowners, business interests, and professionals who sought to "redeem" the South after the Civil War by limiting the influence of the Republican Party and violently overthrowing federal reconstruction policies and Black American citizenship rights.

relocation centers Euphemistic name for areas of detention for more than 100,000 Japanese Americans during World War II, rounded up and evacuated against their will.

republicanism A system of governance in which power derives from the people, rather than from a ruling family, aristocratic class, or some other supreme authority.

Republicans Name for those who wished to limit the new government's power, in opposition to the Federalists.

Revolution of 1800 Thomas Jefferson's term for his election in 1800 which saw the peaceful transfer of power between ideologically opposed parties.

Richard Nixon Vice President under Dwight D. Eisenhower and president from 1969 until his resignation in 1974 as a result of the Watergate scandal.

Richard Wright Author of *Native Son,* a story of a young African American man broken by the system of racial oppression.

Roanoke The first English colony attempted in the North America.

Robert E. Lee Superintendent of West Point and later general of the Northern Army of Virginia, he won a string of early victories during the Civil War and later presided over the official surrender of the Confederacy.

Robert Fulton American inventor of the first commercially successful steamboat, the *Clermont.*

Robert M. La Follette Wisconsin governor, senator, and nationally known progressive.

Rocky Mountain school Group of late-nineteenth-century painters known for large-scale depictions of western landscapes.

Roe v. Wade Supreme Court decision of 1973 that made abortion legal in all states based on privacy rights during the early period of pregnancy and made it difficult for states to create legislation against abortion until the second and third trimesters.

Roger B. Taney Appointed secretary of the treasury under President Jackson to remove federal deposits from the Bank of the United States; he was later appointed by Jackson as chief justice of the Supreme Court and wrote the *Dred Scott v. Sanford* decision in 1857.

Roger Williams Controversial minister who established the Rhode Island colony where people of different faiths could worship without interference.

Ronald Reagan Former governor of California elected as fortieth president in 1982 who led a conservative reform of American politics during his two terms.

Roosevelt Corollary President Theodore Roosevelt's amendment to the Monroe Doctrine, stating that the United States had the right not only to oppose European intervention in the Western Hemisphere but also to intervene in the domestic affairs of its neighbors should they be unable to maintain order and sovereignty.

Rosa Parks Civil rights activist who was arrested for not giving up her seat to a white passenger on a bus in Montgomery, Alabama, in 1955; her arrest spurred the Montgomery bus boycott.

Rosie the Riveter Popular image of a woman performing industrial work during World War II in America; became a symbol of female contributions to the war effort.

Rutherford B. Hayes Elected president in 1876 but largely ineffectual in office; the deals that led to his election are often described as marking the end of Reconstruction.

Saddam Hussein Iraqi leader whose invasion of Kuwait led to the first Gulf War; deposed by the United States in the second Gulf War.

Sagebrush Rebellion A late 1970s movement by western conservatives who opposed federal environmental regulations and restrictions on development.

Salem witchcraft trials An instance of the widespread hysteria in the 1680s regarding supposed satanic influences in New England that often targeted women.

Sam Houston A soldier, lawyer, congressman, and governor of Tennessee, he eventually moved to Texas where he commanded the Texas armies in their successful battle for independence from Mexico in 1836.

Samuel Gompers Union organizer under whose leadership the American Federation of Labor (AFL) grew by combining similar skilled unions together.

Sarah Bagley Creator of the Female Labor Reform Association that promoted better working conditions and shorter workdays in the mid-1840s.

Saratoga Site in New York where, with the help of Benedict Arnold, General Horatio Gates surrounded British General John Burgoyne and forced his surrender.

scalawags Slang term referring to Southern whites who supported the Republican Party and federal Reconstruction policies after the Civil War.

Scopes trial Tennessee case that attracted intense national attention to the debate over whether to teach evolution or creationism in the schools.

Scotch-Irish Scottish Presbyterians who had settled in northern Ireland in the early seventeenth century.

Scottsboro case A 1931 case that was among the most notorious examples of racism in the United States; an all-white jury convicted nine Black teenagers of raping two white women despite overwhelming evidence of their innocence.

secession The act of asserting independence by withdrawing membership from a political state; it refers in particular to the South's departure from the United States in 1861.

Second Continental Congress Body of colonial representatives formed after the battles of Lexington and Concord to help resolve the conflict with Great Britain.

Second Great Awakening A wave of Protestant revival in the early 1800s signified by large congregations and dynamic sermons.

second middle passage Term coined by historian Ira Berlin to describe the forced movement of enslaved people within the United States, primarily from the upper South to the cotton states.

Second New Deal Launched in the spring of 1935 in response both to growing political pressures and continuing economic crisis, these new proposals represented a shift toward more openly anticorporate initiatives; this wave also included more long-term reforms including Social Security.

Securities and Exchange Commission (SEC) A federal government agency established in 1934 to police the stock market.

Selective Service Act The 1917 law that created a national draft to provide men to fight the First World War.

Seminole Wars Term used to describe a series of conflicts between United States forces and the Seminole, a mixture of Native Americans and Black settlers in Florida; the first one, in 1816–1819, saw Andrew Jackson lead an invasion of Spanish Florida, during which Jackson's troops chased Seminole raiders and seized Spanish forts.

Seneca Falls Convention The 1848 meeting that produced the Declaration of Sentiments and Resolutions arguing for women's inalienable rights.

separation of powers The partitioning of authority to distinct branches of a government.

Seven Years' War Called the French and Indian War in the American colonies, the Seven Years' War was a global conflict between England and France ultimately won by England in 1763.

Shakers Utopian religious society committed to complete celibacy, equality of the sexes, and a simple ordered life.

sharecropping A farming system in which large landowners rent their fields to farmers, usually families, in return for a share of the crop's production.

Shays's Rebellion A 1786 uprising of poor Massachusetts farmers demanding relief from their debts.

Sherman Antitrust Act Legislation, passed in 1890, aimed at prohibiting corporate combinations that restrained competition; it was largely ineffective.

Shiloh Name of battle site in southwestern Tennessee where, on April 6–7, 1862, Union forces won control of the upper Mississippi River.

silent majority Phrase coined by Richard Nixon to refer to conservative voters, as opposed to the more vocal counterculture.

sit-down strike A planned labor stoppage in which workers assume their positions in a factory or other workplace but refuse to perform their duties, thus preventing the use of strikebreakers.

slave codes Laws passed in the British colonies or in American states granting white slaveholders absolute authority over the enslaved; these included laws depriving the enslaved of property, free movement, and legal defenses.

Social Darwinism The belief that societies are subject to the laws of natural selection and that some societies or peoples are innately superior to others.

Social Gospel The effort to make faith a tool of social reform.

social justice A movement that seeks justice for whole groups or societies rather than individuals.

Social Security Act Passed in 1935; established several distinct programs including a pension program for workers (though many were excluded);

also established a system of unemployment insurance and aid to people with disabilities as well as dependent children.

socialism A political theory that advocates government (rather than private) ownership and management of the means of production and distribution.

Sojourner Truth Formerly enslaved women who lectured extensively on behalf of equal rights for Blacks and women.

Sons of Liberty Groups of male colonists who organized against England's enforcement of the Stamp Act and who terrorized British officials.

sovereignty The authority to govern; popular sovereignty refers to the idea that the source of this authority is the people, who confer authority through elections.

Spanish Civil War The 1936–1939 war between Spain's liberal republican government and the conservative forces of General Francisco Franco.

Spanish-American War War of 1898 between the United States and Spain; took place in Cuba and the Philippines and resulted in American possession of or great influence over those areas and others.

specie circular Executive order issued by President Jackson that required gold or silver to buy public lands.

spoils system The process whereby elected officials give out government jobs as reward for political favors.

Sputnik The first earth-orbiting satellite launched by the Soviet Union in 1957.

stagflation An economic condition in which inflation is high, unemployment is high, and growth is low.

Stalwarts Political group within the Republican Party led by Roscoe Conkling of New York that favored traditional, professional machine politics.

Stamp Act Hated act passed by Prime Minister Grenville of England that required an official stamp on all paper documents in the colonies and united the colonies against England.

Stephen A. Douglas Democratic senator from Illinois who brokered the Compromise of 1850 and ran unsuccessfully against Lincoln and others in the presidential election of 1860.

Stephen F. Austin Immigrant from Missouri who established the first legal American settlement in the Mexican territory of Texas in 1822.

Stephen H. Long Dispatched by the United States government, he explored the Platte and South Platte Rivers in present-day Nebraska and Colorado.

Stonewall Riot Landmark 1969 event in the gay liberation movement that began as patrons of the Stonewall Inn, a gay nightclub, reacted violently against a police raid.

Stono Rebellion A revolt of enslaved workers in South Carolina during the colonial period.

Strategic Defense Initiative (SDI) President Reagan's missile defense plan, which included ground and space-based antimissile defense weapons.

Students for a Democratic Society (SDS) Leading organization of student radicalism in the 1960s.

Sugar Act British act of 1764 designed to stop sugar smuggling in the colonies by lowering taxes on molasses but enforcing their payment and forcing compliance with trade laws.

Sunbelt The southeastern and southwestern regions of the United States.

Susan B. Anthony Abolitionist and one of the most iconic and active leaders of the early women's right movement.

Syngman Rhee Pro-Western though not reliably democratic leader of South Korea in 1950.

Taft-Hartley Act Officially known as the Labor-Management Relations Act of 1947; made it illegal to operate a business in which no worker could be hired without first being a member of a union.

Tallmadge Amendment Proposed amendment to Missouri's admission to statehood that would have gradually turned it from a slave state to a free state.

Tammany Hall Urban machine led by famously corrupt city boss William M. Tweed.

Taylorism Named for Frederick Winslow Taylor, an attempt to use scientific management to improve factory production.

Tea Act A 1773 act passed by England that gave the British East India Company the right to export tea to the colonies without paying the same taxes that were imposed on colonial merchants; the act enraged American merchants and colonists boycotted tea.

Tea Party movement Conservative political movement begun in 2009 to reduce taxes and government regulations

Teapot Dome The location of rich naval oil reserves in Wyoming and part of a national scandal during the Harding administration.

Tecumseh A chief of the Shawnees who worked to unite native peoples against the threat of white expansion; died fighting for Britain in the War of 1812.

Tejanos The Mexican residents of Texas.

Teller Amendment The 1898 amendment to the war declaration against Spain that promised no American intention to occupy or control Cuba in the wake of an American victory.

temperance Self-restraint, especially concerning drink; the temperance movement pushed for bans on the sale and consumption of alcohol.

tenements By the late nineteenth century, this was a descriptor used for slum dwellings.

Tennessee Valley Authority (TVA) A regional planning program focused on water resources as a source of cheap electric power.

terrorism The use of violence as a form of intimidation against peoples and governments.

Tet offensive Large coordinated attack on January 31, 1968, of American strongholds in South Vietnam by communists forces that led to lost support for the Vietnam War in America.

Thaddeus Stevens U.S. Representative from Pennsylvania who was an abolitionist and a leader of the Radical Republicans.

The Federalist Papers A collection of essays written by Alexander Hamilton, James Madison, and John Jay that supported ratification of the Constitution.

The Jazz Singer The first feature-length film with spoken dialogue, or "talkie."

The Other America Title of Michael Harrington's 1962 about the chronic problem of poverty in America.

the Prophet Also known as Tenskwatawa, the Shawnee prophet was a charismatic speaker, leader, and younger brother of Tecumseh; the Battle of Tippecanoe disillusioned many of his followers.

theocracy A form of government in which political power is believed to derive from a deity, and in which religious and government structures are intertwined.

Theodore Roosevelt Rough Rider during the Spanish-American War, governor of New York, Republican Party and progressive leader, vice president, and president of the United States from 1901 to 1909.

Thirteenth Amendment Passed on January 31, 1865, this constitutional amendment formally abolished slavery.

Thomas J. ("Stonewall") Jackson Confederate general who commanded troops in major Civil War engagements in the first half of war before being mortally wounded in the Battle of Chancellorsville in May 1863.

Thomas Jefferson One of the founders of the United States, he wrote most of the Declaration of Independence and served in all levels of government, both locally and nationally, in his long career.

Tiananmen Square Location in China of the 1989 student uprising that was brutally put down by Chinese authorities.

Townsend Plan Created by a California physician, this proposal focused on federal pensions for older adults.

Townshend Duties External taxes passed by England's Charles Townshend that taxed goods imported to the colonies; hated by the colonists, the taxes were later repealed after a colonial boycott of English goods.

Trail of Tears Term for the forced journey made by Native Americans to the Indian Territory that began in the winter of 1838, killing perhaps a quarter or more of the migrants.

transcendentalism A philosophical and literary movement of the early nineteenth century that sought beauty and truth in nature and the individual, rather than in formalized education, politics, or religion.

Treaty of Guadalupe Hidalgo Agreement to end the Mexican War in which the United States gained California and New Mexico and the Texas boundary was drawn at the Rio Grande.

trench warfare A common form of fighting in the First World War, whereby armies sought cover below ground from artillery bombardment, machine guns, and other military technologies.

Triangle Shirtwaist Company fire This disastrous event in New York was influential in finally passing a series of pioneering labor laws.

triangular trade A simplified description of the complex trade networks of the Atlantic World; the triangle metaphor refers to the trade in rum, enslaved people, and sugar among New England, Africa, and the West Indies.

Truman Doctrine Expressed policy of the United States, articulated in 1947, to support groups or governments fighting against communists around the world.

Turner thesis The theory articulated by Frederick Jackson Turner in 1893 that westward expansion into the frontier had defined and continually renewed American ideas about democracy and individualism.

U-2 Crisis A 1960 diplomatic incident between the United States and the Soviet Union when a U.S. U-2 spy plane was shot down by the Soviets after crossing Soviet air space.

U.S.S. *Maine* American vessel sunk in Havana harbor in February 1898; explosion was blamed on a Spanish mine and used by popular press to urge war, though it was likely caused by mechanical error.

Ulysses S. Grant Chief of the Union armies (at the beginning of 1864) and eighteenth president who supervised much of Reconstruction.

United Nations An international organization, established in the peace talks after World War II, that contained a General Assembly and a Security Council of the five major powers.

United Service Organization (USO) Organization that recruited thousands of young women during World War II to serve as hostesses and sustain the morale of servicemen.

United States Sanitary Commission An organization of Northern civilian volunteers who raised money and support for the care of the sick and wounded during the Civil War; organized large numbers of female nurses to serve in field hospitals.

vaudeville A form of theater adapted from French models; the most popular urban entertainment into the first decades of the twentieth century.

vertical integration The arrangement by which a company takes ownership of businesses in various stages of production and distribution within the same industry.

Viet Cong Military arm of the National Liberation Front (NLF); communists in South Vietnam who fought against South Vietnamese rule and the American military with the support of the North Vietnamese.

Vietnamization Term used by President Nixon to describe transferring the responsibility for fighting the Vietnam War to the South Vietnamese.

Virginia and Kentucky Resolutions Written by Jefferson and Madison, respectively, in response to the Alien and Sedition Acts, the resolutions argued that states had the right to nullify federal laws.

Virginia House of Burgesses The first elected legislature in what would become the United States.

Virginia Plan James Madison's proposal during the Constitutional Convention for a two-house legislature where states would be represented in both bodies in proportion to their population.

Virginia Resolves Term used for a group of resolutions passed by the Virginia legislature declaring only the colonies' governments had the right to tax colonists.

virtual representation British political theory holding that members of Parliament represented all British subjects, not just those from the specific region that had elected them.

Vladimir Putin Political leader of Russia since 1999.

Voting Rights Act of 1965 Legislation that expanded the right to vote by providing federal protections to African Americans who had previously been barred by local and state regulations.

W. E. B. Du Bois Sociologist, historian, one of the first African Americans to receive a degree from Harvard, author of *The Souls of Black Folk*, chief spokesman for fighting for Black civil rights, founding member of the Niagara Movement and the NAACP.

Wade-Davis Bill The 1864 bill stipulating that all Confederate states seeking readmission to the Union have a majority of its voters take a loyalty oath to the federal government; it never passed because Lincoln refused to sign it.

Walt Whitman Writer who helped define American literature with his book of poems, *Leaves of Grass*, and his focus on individual freedom.

War Hawks Term given to a group of congressmen who argued for war with Britain in 1812.

War Industries Board The 1917 agency created to coordinate government purchases of military supplies.

Warren G. Harding Senator from Ohio who was elected president in 1920; generally remembered as ill-suited for the office.

Warsaw Pact A Soviet-led alliance of its Eastern European satellite nations created in 1955.

Washington Irving Successful author in the early 1800s of historical works and short stories including "The Legend of Sleepy Hollow."

Washington, D.C. The capital of the United States designed by French engineer Pierre L'Enfant; took many years to develop into a major city.

Watergate Named after the site where Nixon's operatives were first arrested, the scandal involving Nixon's use of illegal campaign tactics and attempts to cover up those tactics.

Webster-Ashburton Treaty This 1842 treaty helped define the northern border between British colonies (now Canada) and the United States and soothed American anger at British actions in the Caroline and Creole affairs.

Webster-Hayne debate Senate exchange between Daniel Webster and Robert Hayne that focused on the issue of states' rights versus national power, pitting the nullification advocate Hayne against the nationalist Webster.

welfare capitalism A corporate strategy for discouraging labor unrest by improving working conditions, hours, wages, and other elements of workers' lives.

Western Union Telegraph Company The dominant telegraph company by 1860, formed by a combination of smaller telegraph companies.

Whigs Political party formed during the presidency of Andrew Jackson to oppose Jackson and to support a more active federal government; favored industrial, commercial, and infrastructure development to promote economic growth.

Whiskey Rebellion A 1794 uprising of western Pennsylvania farmers opposed to a new federal whiskey tax; put down by troops led by President Washington.

Wilbur and Orville Wright Builders of the first self-powered airplane, successfully flown in 1903.

William Berkeley The royal governor of Virginia during Bacon's Rebellion.

William H. Seward Secretary of state in both President Lincoln's and President Johnson's administrations who negotiated purchase of Alaska from Russia in 1867.

William Henry Harrison Ninth president of the United States who died shortly after taking office; experienced veteran of combat against Native Americans at an early age and later governor of Indiana Territory; defeated Tecumseh at the Battle of Tippecanoe.

William Howard Taft Theodore Roosevelt's most trusted lieutenant and his handpicked successor, who was elected president in 1908.

William Howe British commander who led troops in capturing New York in 1776 and Philadelphia in 1777, but who was largely seen as ineffective until replaced in 1778.

William James Harvard psychologist and most prominent publicist of pragmatism.

William Jennings Bryan Congressman from Nebraska, tireless 1896 presidential candidate, and author of the "Cross of Gold" speech; also later secretary of state under President Woodrow Wilson and a Christian fundamentalist witness in the Scopes trial.

William Lloyd Garrison Founder of *The Liberator*, a newspaper that focused on the harsh truths of slavery and argued for the immediate release of the enslaved and extension of citizenship to all.

William M. Tweed The famously corrupt boss of New York's political machine Tammany Hall.

William McKinley Governor of Ohio and former congressman who was elected president in 1896 and 1900; assassinated in 1901.

William Penn Outspoken Quaker who led the colony of Pennsylvania after receiving a royal land grant.

William Pitt Leading English secretary of state and prime minister who ran England's war effort during the Seven Years' War.

William Randolph Hearst The most powerful U.S. newspaper chain owner; by 1914, he controlled nine newspapers and two magazines.

William T. Sherman Union general who captured Atlanta and in late 1864 marched his troops across Georgia and the Carolinas, burning crops and buildings as he went and crippling the Confederacy.

Wilmot Proviso Failed congressional plan to prohibit slavery in any territory acquired from Mexico.

Winston Churchill Outspoken British prime minister in power during most of World War II.

Women's Christian Temperance Union (WCTU) Advocate for abstinence from the consumption of alcohol, led by Frances Willard after 1879; single largest women's organization in American history by 1911.

Woodrow Wilson Native Virginian, president of Princeton University, governor of New Jersey, Democratic Party and progressive leader, and president of the United States from 1913 to 1921.

Woodstock Music festival held in upstate New York in August of 1969 that became a symbol of the counterculture.

Worcester v. Georgia Supreme Court case of 1832 that affirmed federal authority over individual states' authority concerning the affairs of Native Americans.

Works Progress Administration (WPA) Established in 1935, this was a system of work relief for the unemployed, but on a larger scale than earlier, similar endeavors; it included such programs as the Federal Art Project, the Federal Music Project, and the Federal Writers' Project.

Wounded Knee Located on the Pine Ridge Indian Reservation in South Dakota, it was the site of a massacre of between 150 and 300 Sioux, including women and children, by the U.S. Army on December 29, 1890.

XYZ Affair Name given to an international incident between U.S. and French diplomats that sparked the quasi war between France and the United States.

Yalta Conference A 1945 meeting between Roosevelt, Churchill, and Stalin that laid the groundwork for the United Nations but otherwise established only murky or soon-violated agreements on the partition of Germany and the question of who would rule Poland.

yellow journalism Sensationalist reporting, particularly in newspapers of the late nineteenth and early twentieth centuries, so named for the color of a character in one of the papers' comic strips.

yeomen farmer Small farmer who worked his own soil and held no enslaved laborers.

Yorktown Virginia site of the last major battle of the American Revolution, where Lord Cornwallis surrendered to George Washington and French forces in 1781.

Young America A political movement supporting free trade and territorial expansion in America during the late 1840s and 1850s.

Zachary Taylor American hero of the Mexican War who was elected president in 1848.

Zimmermann Telegram The communication intercepted in early 1917 from the German foreign secretary to the German ambassador in Mexico outlining a deal to draw the Mexicans into the war against the United States.

zoot suits Style of dress among Mexican American youths during World War II that featured padded shoulders and baggy pants; led to attacks by whites for flaunting wartime protocols to conserve clothing materials.

INDEX

Abolitionism. *See also* Emancipation
   Amistad incident and, 267, 291
   anti-abolitionism, 290-291
   arguments for, 262
   in Civil War, 327
   divisions within, 291, 293
   free Black leaders of, 286-287, 290
   as global movement, 288-289
   in North, 121, 286-287, 290, 295
   women and, 283, 291
ACS (American Colonization Society), 286
Adams, Abigail, 123, 124
Adams, Charles Francis, Jr., 336
Adams, John
   Boston Massacre and, 97
   correspondence with Abigail, 123, 124
   Declaration of Independence and, 111
   in election of 1800, 152-153, 166
   independence supported by, 107
   as minister to London, 128
   peace negotiations of, 120
   political ideology of, 98, 126
   as president, 150-153
   on taverns, 101
   as vice president, 142
Adams, John Quincy, 183, 191-192, 198-201, 267
Adams, Samuel, 97, 100-103, 107, 142
Adams-Onís Treaty (1819), 192
Advertising, in colonial period, 68
Africa. *See also specific countries*
   economy in, 17-18
   European exploration of, 7
   gold in, 17
   religion in, 18, 75
   resettlement of free Blacks in, 286
   slavery in, 16-18, 62
   trade and, 16-18, 62, 66-67
   women in, 18
African Americans. *See also* Free Blacks; Slavery and enslaved persons
   as abolitionists, 286-287, 290
   Black Codes and, 359-361
   citizenship for, 141, 359, 366
   civil rights movement and, 276, 350
   in Civil War, 327, 329-330, 337
   in Congress, 363-364

   demand for federal aid, 354
   discrimination against, 243, 281, 359, 368
   education for, 281, 353, 364, 372-373
   employment of, 359, 372
   families and, 270
   in horse racing, 164
   land ownership by, 352-353, 364-366
   lynchings of, 290, 325, 375
   in middle class, 372
   music and, 268
   in New South, 372-373
   population trends among, 228
   poverty of, 266, 286
   in reform movements, 283, 286
   religion of, 17, 159-160, 283, 285-286
   voting rights for, 141, 204, 353, 361, 369, 374
   wages for, 365
   women, 283, 285-286
African Methodist Episcopal Church, 268, 283, 363-364
Age of globalization, 16
Age of revolutions, 118-119
Agriculture. *See also* Plantations; Rural areas; *specific crops*
   agrarian ideal and, 146, 155, 166
   in antebellum period, 229, 248-251, 253-257
   in Chesapeake region, 27-30
   crop-lien system, 365, 366
   indigenous peoples and, 4-5, 15, 27
   in Mayan region, 4
   in New England, 65, 232
   in North, 5, 188, 232, 248-251
   settlement formation and, 3
   sharecropping, 365
   in South, 65, 69-70, 160-161, 189, 229, 253-257
   technology for, 160-161, 249-250
   tenant farmers, 365
   in West, 229
   yeoman farmers, 125
Air pollution, 238
Alabama
   indigenous lands in, 211
   secession of, 320
   statehood for, 189
Alamo, siege of (1836), 297

Alaska, 2, 368
Albany Plan, 84
Alcott, Louisa May, 331
Alexander, Francis, 146
Algonquians, 6, 22, 26, 50
Alien and Sedition Acts of 1798, 151-152
Allen, Ethan, 96, 112
Alliance, Treaty of (1778), 151
Almanacs, 77, 78
Amendments to U.S. Constitution. *See* Bill of Rights; *specific amendments*
American Antislavery Society, 290, 291
American Colonization Society (ACS), 286
American Fur Company, 189
American Indians. *See* Indigenous peoples
American Medical Association (AMA), 281
American Party, 230
American Patriots, 111-116, 120-123
American Revolution. *See also* Colonies and colonial period
   beginnings of, 102-104, 107
   British surrender in, 109, 117
   economic effects of, 125, 130-131
   events leading up to, 92-102
   France and, 111, 115-117, 120
   global influence of, 118-119
   government following, 125-128
   historical debates on, 108-110
   indigenous peoples and, 114, 122
   maps of, 103, 113, 115, 117
   in mid-Atlantic colonies, 113-115
   mobilization for war, 111
   in New England, 112-113
   objectives of, 107, 110
   peace negotiations, 117, 120, 128
   people of color in, 75, 109, 111, 117
   as political and military conflict, 106
   religion influenced by, 120-121, 158
   Seven Years' War and, 87
   slavery and, 107, 121-122
   social effects of, 120-125
   in South, 116-117
   women in, 109, 123, 124
American System, 199, 200, 219

Americas. *See also* Latin America; North America; South America
 biological and cultural exchanges in, 11-13, 15
 civilization growth in, 3-6
 diseases spread by Europeans in, 1, 9, 13
 Dutch settlements in, 23
 early history in international context, 16-17
 English settlements in, 18-22, 25-37, 39-53
 Enlightenment in, 76, 79, 108
 French settlements in, 22-23, 44-45, 47, 84
 migration to, 2-3, 16
 missionaries in, 9, 13, 23, 46, 84, 157
 origins of name, 7
 precontact period, 2-3
 Spanish Empire in, 9-13, 15, 46-47
Amherst, Jeffrey, 87, 88
Amistad incident (1839), 267, 291
Anderson, Robert, 321
Andros, Edmund, 52
Anesthetics, 281
Anglican Church. *See* Church of England
Antebellum period. *See also* Civil War; Slavery and enslaved persons
 agriculture in, 229, 248-251, 253-257
 birth and mortality rates in, 228, 246, 260
 class divisions in, 242-245, 259-261
 commerce and industry in, 236-238
 communications technology in, 234-236
 culture during, 244-248
 economy in, 236-238, 242-244
 education during, 246, 281-282
 families in, 245-246
 free Blacks in, 243-244, 265-266, 268
 gender relations in, 246
 immigration during, 228-229
 industry in, 236-238, 257
 labor force in, 238-242
 leisure activities in, 246-248
 literature in, 246, 273-274
 nativism in, 230
 population trends in, 228-229
 reform movements during, 279-286
 social patterns in, 242-248
 technological advances during, 237-238
 transportation during, 231-234, 257
Anthony, Susan B., 283, 331
Anti-abolitionism, 290-291
Antietam, Battle of (1862), 329, 341
Antifederalists, 135, 138-139, 142
Anti-Mason Party, 208, 218
Anti-Masonry movement, 218
Anti-Nebraska Democrats, 311
Anti-Nebraska Whigs, 311
Antinomian controversy, 36
Anti-Semitism, 229
Antislavery movement. *See* Abolitionism
Appomattox Court House, 347
Aptheker, Herbert, 262
Archaic period, 3
Argentina, 11
Arizona, 46, 310
Arkansas, 321, 353
Army, U.S. *See also* Confederate Army; Union Army
 expansion of, 171
 indigenous peoples and, 210, 213-214
 scaling down of, 168
 size of, 325
 in War of 1812, 181
Arnold, Benedict, 112-115
Aroostook War, 224
Art, Romanticism and, 272-273
Articles of Confederation, 111, 127-128, 136, 138, 139, 141
Artisans, 57, 65, 238, 241-242
Ashburton, Lord, 224
Ashley, Andrew and William, 190
Asia. *See also specific countries*
 migration to Americas from, 2
 trade with, 7, 125
Assimilation of indigenous peoples, 11, 29, 178, 198, 282
Associated Press, 236
Astor, John Jacob, 189
Atlanta (Georgia), 327, 345, 347
Atlanta Compromise, 373
Atlantic World, 16-17, 118, 288
Attucks, Crispus, 98
Austin, Moses, 189
Austin, Stephen F., 297
Austro-Hungarian Empire, 86
Aztecs, 4, 9, 12, 13

Bacon, Francis, 76
Bacon, Nathaniel and Bacon's Rebellion, 30-31
Bacterial infections, 279. *See also specific diseases*
Bagley, Sarah, 239
Baillie, James, 307
Bailyn, Bernard, 108
Balboa, Vasco de, 7-8
Baldwin, Joseph G., 274
Baltimore, Lord, 32-33, 40, 41, 53
Baltimore and Ohio Railroad, 233
Bancroft, George, 108
Bank of the United States, 145, 186, 193, 196-197, 215-216, 224
Banks, national, 144-145, 186, 193, 218-219, 325
Bank War, 215-216
Baptist, Ed, 263
Baptists, 36, 74, 121, 158, 160, 268
Barbados, 23, 42, 45, 46
Barbary states, 168
Bard, James, 187
Barlow, Francis Channing, 348
Barnum, P. T., 247, 248, 377
Barron, James, 176
Barton, Clara, 331
Baseball, 248
Bastille, storming of (1789), 119
Battles. *See specific names of battles*
Beard, Charles A., 138
Bear Flag Revolt, 302
Beauregard, P. G. T., 321, 338, 339
Becker, Carl, 108
Beckert, Sven, 263
Becknell, William, 189
Belknap, William W., 367
Benjamin, Judah P., 336
Bennett, Gordon, 222
Benson, Lee, 206
Berkeley, John, 43
Berkeley, William, 30, 31
Berlin, Ira, 256, 262-263
Berry, Diane, 263
Bett, Mum, 121
Biddle, Nicholas, 215, 216
Bierstadt, Albert, 273
Bill of Rights, 139, 142-143, 147
Bingham, George Caleb, 204
Birney, James G., 291
Birth rates, 57, 228, 246, 260
Black Americans. *See* African Americans
Black Belt, 189
Black Codes, 359-361
Black Death, 6, 13
Black Hawk and Black Hawk War, 210, 211
Blassingame, John, 262

"Bleeding Kansas," 311, 313
Bolívar, Simón, 119, 200, 288
Bonaparte, Napoleon. See
    Napoleon Bonaparte
Bonds
    Confederate, 332, 351
    Federalist program for, 144–145
    U.S. Treasury, 325
Books. See Literature
Booth, John Wilkes, 355
Borderlands of British North
    America, 25, 44–51
Boston, John, 326
Boston (Massachusetts)
    American Revolution in, 112
    Perkins School for the Blind
        in, 282
    population of, 73, 229
    settlement of, 35, 37
    smuggling in, 96
    taverns in, 101
Boston Massacre, 97, 98
Boston Tea Party, 101
Boyer, Paul, 72
Bradford, William, 34
Brady, Mathew, 326
Bragg, Braxton, 334, 339, 345
Brant, Joseph and Mary, 114
Brazil, 11, 18, 23, 62, 289
Breckinridge, John C., 316
Breen, T. H., 110
Bristow, Benjamin H., 367
British East India Company,
    99, 101
British Empire. See also American
    Revolution; England
    abolition of slavery in, 288
    administration of, 83, 89–90
    in Caribbean, 44–46
    imperialism and, 82, 89–92
    India in, 86, 169
Brook Farm community, 276
Brooks, Preston, 312
Brown, John, 311, 312, 316
Brown, Joseph, 332
Brown, Robert E., 138
Bubonic plague, 6, 13
Buchanan, James, 313–315,
    320, 321
Bucktails faction, 207
Bullocke, James, 164
Bull Run, Battles of, 338, 341
Bunker Hill, Battle of (1775), 112
Burgoyne, John, 114
Burnside, Ambrose E., 341
Burr, Aaron, 152, 153, 175

Business. See also Economy; Industry
    in antebellum period, 236
    corporations, 236
    monopolies, 19, 62
Butler, Andrew P., 311–312

Cabot, John, 18
Cahokia, 5
Calhoun, John C.
    on nullification, 208–210
    as secretary of state, 224
    as secretary of war, 191
    on slavery, 313
    on transportation
        improvements, 188
    as vice president, 208, 209
    as War Hawk, 180
    in Whig Party, 218, 219
California. See also specific cities
    American settlers in, 302
    Bear Flag Revolt in, 302
    Chinese Americans in, 305
    gold rush in, 305, 307
    Hispanic residents of, 302
    indigenous peoples in, 302, 305
    Mexican War and, 302, 303
    migration to, 302, 305, 307
    Spanish colony in, 46
    statehood for, 307, 308
    U.S. annexation of, 303
Calloway, Colin, 110
Calvert family, 32–33. See also
    Baltimore, Lord
Calvin, John, 19
Calvinists, 63, 74, 158
Camp, Stephanie, 262
Canada. See also specific cities
    American invasions of, 112–113, 180
    anti-British factions in, 224
    boundary with Maine, 224
    Caroline affair and, 224
    French settlements in, 22–23
    indigenous peoples and, 178, 179
    movement for U.S. annexation
        of, 310
    Oregon boundary with, 300, 301
    Seven Years' War and, 86–88
Canals, 231, 232, 257
Cane Ridge camp meeting, 158
Canvassing for a Vote (Bingham), 204
Capitalism, 238, 276
Capitalists, 163, 236, 238, 242, 259
Caribbean region. See also
    specific locations
    British Empire in, 44–46
    colonies in, 7, 11, 23, 44–46

Dutch settlement in, 23
French possessions in, 88
resettlement of free Blacks in, 286
slavery in, 18, 45–46, 62, 288–289
Spanish Empire in, 11, 44–45
sugar production in, 18, 23, 45, 254
trade in, 44–46, 66–67, 92, 125
Carnegie, Andrew, 338
Carolina colony, 41–42.
    See also North Carolina;
    South Carolina
Caroline affair (1837), 224
Carpetbaggers, 356, 364
Carteret, George, 43
Cass, Lewis, 305
Catholic Church
    in Africa, 18, 75
    American Revolution and, 121
    in colonial period, 32–33,
        46–48, 74
    in England, 20, 41, 52
    French Revolution and, 118
    immigrants and, 230
    indigenous peoples and, 11,
        23, 46, 84
    political influence of, 86–87
Catlin, George, 211
Cavaliers, 40
Central America. See also
    specific countries
    Catholic Church in, 9
    early civilizations of, 3, 4
    independent provinces of, 119
Central Pacific Railroad Company,
    324–325
Central Park (New York City), 243
Champlain, Samuel de, 22–23
Chancellorsville, Battle of (1863), 342
Charles I (England), 35, 40, 99
Charles II (England), 41–43, 52
*Charles River Bridge v. Warren
    Bridge* (1837), 217
Charleston (South Carolina)
    American Revolution in, 116
    establishment of, 41
    Federalist support in, 146
    Fort Sumter in, 321
    free Blacks in, 269
    Jewish populations in, 75
    population of, 73
    slavery in, 65
Charter of Liberties
    (Pennsylvania), 44
Charters for trading companies, 19
Chattanooga, Battle of (1863), 345
Checks and balances, 137, 140

I-4 · INDEX

Cherokee people, 90–91, 122, 130, 197–198, 211–214
Chesapeake-Leopard incident (1807), 176–178
Chesapeake region. *See also* Maryland; Virginia
   Bacon's Rebellion in, 30–31
   indentured servants in, 29–31, 33, 56
   indigenous peoples in, 26–31, 33
   life expectancy in, 57
   reorganization and expansion in, 27–29, 32
   slavery in, 29–33
   women in, 29, 61
Chicago (Illinois), 233, 249
Chickamauga, Battle of (1863), 345
Chickasaw people, 130, 211, 214
Childbirth, 60, 61, 157
Children. *See also* Education; Families
   labor of, 242
   mortality rates for, 61, 260
   of enslaved persons, 18, 30, 42, 63, 69
Chile, 2, 11, 12
China, 225, 305
Chinese Americans, 305
Choctaw people, 130, 211, 214
Christianity. *See also* Catholic Church; Protestantism; *specific denominations*
   enslaved persons and, 267–268
   evangelical, 76, 158–159, 279
Churchill, Winston, 86
Church of England, 17, 20, 33, 36, 41, 53, 74, 120
Church of Jesus Christ of Latter-day Saints (Mormons), 277–279
Cities and towns. *See also specific cities and towns*
   in colonial period, 73–74
   crime in, 74, 163
   in early civilizations, 4, 5
   food availability in, 245
   growth of, 166, 229
   inequality, 242–243
   leisure activities in, 247–248
   middle class in, 244
   migration to, 162–163
   pollution in, 163
   poverty in, 243
   slavery in, 265
Citizenship
   for African Americans, 141, 359, 366
   Fourteenth Amendment and, 141, 359, 361, 371, 374

   for immigrants, 140, 151
   for indigenous peoples, 141
Civil Rights Act of 1866, 359
Civil rights movement, 276, 350
Civil War (1861–1865). *See also* Antebellum period; Confederate Army; Reconstruction; Union Army
   African Americans in, 327, 329–330, 337
   campaigns and battles of, 337–347
   casualties of, 337, 351
   causes of, 261, 320, 324
   economic and social effects of, 333
   financing of, 325, 332
   foreign powers and, 323, 336, 368
   historical debates on, 320, 321
   Lost Cause history of, 373
   maps of, 340, 342–344, 346, 347
   mobilization for, 324–327, 329–333
   opposing sides in, 321, 323
   politics during, 326–327, 329
   sea power in, 334–336
   secession crisis and, 320–323
   strategy and diplomacy in, 333–336
   technologies in, 337–338
   western theater of, 336, 338–339
   women in, 330–331, 333
Clark, George Rogers, 114
Clark, William, 174
Class divisions. *See also* Inequality; Social mobility; *specific classes*
   in African societies, 18
   in antebellum period, 242–245, 259–261
   in colonial period, 33, 41, 68, 96
   horse racing and, 164–165
   industrialization and, 163
   in Jacksonian period, 207
   in South, 259–261
Clay, Henry
   American System of, 199, 200, 219
   Compromise of 1850 and, 308
   as House Speaker, 180, 191, 195, 199, 200
   Missouri Compromise and, 195
   in nullification crisis, 210
   peace negotiations by, 183
   as presidential candidate, 199, 200, 216, 219, 300
   as secretary of state, 200
   as War Hawk, 180
   in Whig Party, 218–219, 300
Clinton, DeWitt, 207
Clinton, Henry, 114, 116, 117
Clovis people, 2

Coal industry, 162, 237
Coercive (Intolerable) Acts of 1774, 101, 102
Cohens v. Virginia (1821), 196
Colfax, Schuyler, 367
Colleges and universities. *See also specific institutions*
   for African Americans, 364, 372
   in colonial period, 78
   HBCUs, 364
   land-grant institutions, 324, 364
   private, 157
   women in, 246
Colonies and colonial period. *See also* American Revolution; Imperialism; *specific colonies*
   birth and mortality rates in, 30, 57, 60, 61
   borderlands in, 25, 44–51
   in Caribbean, 7, 11, 23, 44–46
   cities and towns in, 73–74
   class divisions in, 33, 41, 68, 96
   culture during, 45–46, 50
   Dutch, 23, 42–43
   economy in, 28, 29, 65–69
   education in, 74, 76–78
   English, 18–22, 25–37, 39–53
   families in, 60–61
   French, 22–23, 44–45, 47, 84
   government in, 29, 30, 34–36, 43, 52–53, 79–80
   immigrants in, 30, 55–59, 61, 63
   incentives for, 19–20
   industry in, 65–66
   medicine in, 57, 60
   middle grounds in, 44, 49–51
   mortality rates in, 30
   politics in, 30, 33, 53, 79–80
   population of, 30, 56–57, 60–64, 71, 73
   Portuguese, 15, 18, 118–119
   religion in, 32–36, 41–44, 46–48, 53, 74–76
   restoration colonies, 40–44
   sites of resistance in, 99–101
   slavery in, 29–33, 41–43, 55, 57, 62–63, 65, 69–71
   social patterns in, 69–71, 73–74
   Spanish, 7, 11, 15, 44–45, 118–119, 189
   taverns in, 74, 99–102
   taxation in, 90, 92–97, 99, 101
   trade in, 27, 34, 42, 48, 51, 66–67, 92
   women in, 29, 36, 57, 60–61, 101–102

Colorado, 174
Columbia University, 78
Columbus, Christopher, 1, 2, 7, 8, 15, 44
Comanche people, 51, 297
Command of the Army Act of 1867, 361
Commerce. See Economy; Trade
Committee on the Conduct of the War, 334
Committees of correspondence, 97, 102
Common Sense (Paine), 110
Commonwealth v. Hunt (1842), 242
Compromise of 1850, 308-309
Compromise of 1877, 370, 371
Conciliatory Propositions, 102
Concord, Battle of (1775), 103-104, 112
Conestoga people, 91
Confederacy. See also Confederate Army
  constitution of, 331
  diplomacy of, 336
  establishment of, 320
  Fort Sumter and, 321
  government of, 331-332
  Lost Cause history of, 373
  mobilization of, 331-333
  monuments to, 373
  naval blockade of, 326, 333-336
  railroad system of, 321
  war financing in, 332
Confederate Army
  advantages of, 321, 323
  campaigns and battles, 337-347
  commanders of, 334
  mobilization of, 332-333
  surrender of, 347
  uniforms worn by, 337
Confederate States of America. See Confederacy
Confederation Congress, 127-129, 135
Confederation period, 127-131, 134-136
Confiscation Acts, 327
Congregationalists, 76, 78, 121
Congress, U.S. See also House of Representatives, U.S.; Senate, U.S.
  African Americans in, 363-364
  Bill of Rights approved by, 142
  checks and balances by, 137
  Committee on the Conduct of the War, 334

funding of internal improvements and, 188
  Reconstruction and, 353, 355, 358, 359
  Supreme Court authority to nullify acts of, 169
Connecticut. See also specific cities
  emancipation in, 121
  government of, 35-36
  slavery in, 65
Conquistadores, 9
Conscription
  in American Revolution, 111
  in Civil War, 325, 332-333
Conscription Act of 1862, 332
Conservation, 276
Conservatives, 327, 353
Constitution, Confederate, 331
Constitution, U.S. See also specific amendments
  Antifederalists on, 135, 138-139, 142
  Bill of Rights, 139, 142-143, 147
  drafting of, 136-137, 140
  Federalists on, 134-135, 139, 141-142
  historical views of, 108-109
  meaning of, 138-140
  necessary and proper clause of, 142, 196
  ratification of, 135, 141-142
  slavery and, 136, 137, 139-140
  Supreme Court interpretation of, 169, 195-197
Constitutional Convention, 136-138, 142
Consumerism, in colonial period, 68-69
Continental army, 111, 116
Continental Association, 102
Continental Congress, 102, 107, 110-111, 114, 127
Continental currency, 111
Coode, John, 53
Cooper, Anthony Ashley, 41
Cooper, James Fenimore, 273
Cooper, Peter, 233
Cornwallis, Lord, 109, 116-117
Corporations, 236
Corruption
  in Grant administration, 366, 367
  in presidential elections, 200
  in Reconstruction, 356
  of Tammany Hall, 369
Cortés, Hernando, 9
Cotton gin, 160-161, 256, 260

Cotton kingdom, 256, 258-259
Cotton production, 160-161, 189, 254-259, 333
Countryman, Edward, 110
Covenants, 70
Coverture, 60
Craftspeople. See Artisans
Crawford, William H., 199, 200
Crédit Mobilier scandal, 367
Credit systems, 366, 372
Creek people, 181, 211, 214
Creole incident (1841), 224
Creoles, 84, 119
Crime
  in cities, 74, 163
  in colonial period, 74, 79
  slave codes on, 261
Criminal justice system, 282
Crittenden Compromise, 321
Crockett, Davy, 297
Cromwell, Oliver, 40-41
Crop-lien system, 365, 366
Crops. See Agriculture
Cuba
  abolition of slavery in, 289
  Columbus in, 7
  Ostend Manifesto and, 309
  as Spanish colony, 44
  Western Design expedition and, 40
Cult of domesticity, 246
Cultural nationalism, 156-160
Culture. See also Literature; Popular culture
  in antebellum period, 244-248
  art, 272-273
  in colonial period, 45-46, 50
  in early nineteenth century, 156-160
  exchanges in Americas, 11-13, 15
  music, 268, 376, 377
  of enslaved persons, 262, 267-270
Cumming v. County Board of Education (1899), 374
Currency
  in colonial period, 66, 92
  Confederate, 332, 351
  Continental, 111
  gold-based, 215, 220, 367
  greenbacks, 325, 367
  silver-based, 215, 220
  U.S. Treasury notes as, 325
Currency Act of 1764, 92
Currier, Nathaniel, 280
Custer, George A., 334

I-6 · INDEX

Dame schools, 78
Dartmouth College v. Woodward (1819), 195–196
Daughters of Liberty, 101
Davis, David, 370
Davis, Henry, 353
Davis, Jefferson, 310, 331–332, 334, 345, 347
Dawes, William, 103
Day, Benjamin, 222
Death rates. *See* Mortality rates
De Bow, James D. B., 257
De Bry, Theodor, 21
Debt
 from American Revolution, 130–131
 from Civil War, 355, 358
 in England, 89–90
 federal, 144–145, 168, 219
 state, 144, 145
Decentralization, 6, 83, 111, 131, 138–139, 199
Declaration of Independence, U.S., 110–111, 122, 125–126
Declaration of Sentiments and Resolutions, 283–285
Declaratory Act of 1766, 93, 96
Deism, 158
Delaware, 44, 142
Democracy in America (Tocqueville), 205
Democratic Party
 in Jacksonian period, 207, 218
 in South, 369–371
 western expansion and, 300, 310–311
 Young America movement in, 309–310
Democratic-Republican Party. *See* Republicans
Democratization of government, 202–205
Demos, John, 72
Depression of 1837, 220, 224
Descartes, René, 76
Detroit (Michigan), 90
Dew, Thomas R., 313
Diallo, Ayuba Suleiman, 75
Dias, Bartholomeu, 7
Dickinson, John, 107
Diet. *See* Foods
Diplomacy. *See also* Treaties
 with Barbary states, 168
 with China, 225
 in Civil War, 333–336
 in Europe, 86

failures of, 128, 150
 with Latin America, 198–199
 with Mexico, 302, 303
 Republican, 367–368
 of Whigs, 224–225
Diplomatic revolution, 86
Disarmament, 183
Discrimination. *See also* Racism; Segregation
 African Americans and, 243, 281, 359, 368
 in education, 281
 in employment, 243, 359
 immigrants and, 243
 voting rights and, 368
Diseases. *See also* Epidemics; Health and health care; Medicine; *specific diseases*
 bacterial infections, 279
 in colonial period, 57, 60
 European spread in Americas, 1, 9, 13
 germ theory of, 281
 indigenous peoples and, 1, 9, 13, 34, 46, 171
 westward migration and, 299
District of Columbia. *See* Washington, D.C.
Diversity. *See* Race and ethnicity
Dix, Dorothea, 331
Doctors. *See* Medicine
Domesticity, cult of, 246
Dominion of New England, 52
Donelson, Fort, 339
Donovan, Susan, 356
Dorr, Thomas L. and Dorr Rebellion, 204
Douglas, Stephen A., 308–311, 315, 316
Douglass, Frederick, 139, 287, 289–291, 293, 319, 324, 359
Draft. *See* Conscription
Drake, Francis, 20
Dred Scott decision (1857), 314, 315
The Drunkard's Progress (Currier), 280
Du Bois, W. E. B., 16, 121, 356
Dunning, William A., 356
Duquesne, Fort, 85, 87
Dutch. *See* Netherlands
Dutch Reformed, 74
Dutch West India Company, 42
Duval, Kathleen, 50–51

Eating habits. *See* Foods
Eaton, John H., 209
Economic nationalism, 216, 324–325

Economy. *See also* Business; Panics; Trade
 in Africa, 17–18
 American Revolution and, 125, 130–131
 in antebellum period, 236–238, 242–244
 capitalism and, 238, 276
 in Civil War, 333
 in colonial period, 28, 29, 65–69
 depressions and, 220, 224
 in England, 19
 Federalist policies for, 144–145
 growth of, 186–187, 197
 inflation and, 111, 123, 332
 mercantilist view of, 19
 recessions and, 216
Edict of Nantes (1598), 63
Education. *See also* Colleges and universities
 for African Americans, 281, 353, 364, 372–373
 in antebellum period, 246, 281–282
 in colonial period, 74, 76–78
 for persons with disabilities, 282
 in early civilizations, 4
 for indigenous peoples, 78, 157, 282
 medical, 157
 nationalism and, 158
 private schools, 156–157
 public schools, 78, 156–157, 281, 364, 371
 Reconstruction and, 364
 reform of, 281–282
 segregation in, 364, 374
 for enslaved persons, 78
 in South, 260, 281, 364, 371
 in West, 281, 282
 for women, 78, 156–157, 246, 260
Edward, Laura F., 356–357
Edwards, Jonathan, 76, 78
Egerton, Douglas, 110
Elections. *See also* Voting rights
 1789, 142
 1792, 146
 1796, 150
 1800, 152–153, 166, 205
 1804, 168, 175
 1808, 177
 1816, 191
 1820, 191, 199
 1824, 199–200, 205
 1828, 200–201, 205, 207
 1832, 216
 1836, 219
 1840, 220–221, 224

1844, 225, 300
1848, 305
1852, 309
1856, 313-314
1860, 316-317, 319
1864, 327
1868, 366
1872, 366
1876, 369-370
turnout in, 205
Electoral college, 140, 168, 191, 205
Electricity, 17, 79
Elizabeth I (England), 20
Elkins, Stanley, 262
Emancipation. *See also* Abolitionism
as Civil War aim, 324
as global movement, 288-289
in North, 121, 256
politics of, 327, 329
Tallmadge Amendment on, 195
Thirteenth Amendment and, 288, 329, 358
Emancipation Proclamation, 328, 329, 336
Embargoes, 177-178
Emberton, Carole, 357
Emerson, Ralph Waldo, 103, 274-275
Employment. *See also* Labor force; Unemployment; Women in workforce
of African Americans, 359, 372
Black Codes on, 360
discrimination in, 243, 359
Encomiendas, 11, 46
Enforcement Acts of 1870 and 1871, 368
England. *See also* British Empire
American colonies of, 18-22, 25-37, 39-53
anti-British factions in Canada, 224
Civil War in, 40-41
economy in, 19
exploration by, 8
French conflicts with, 84-89, 147
Industrial Revolution in, 68, 160, 162
Ireland and, 18-19
Jay's Treaty and, 147, 149, 150, 178
naval power of, 112, 147
Navigation Acts and, 51, 52, 67, 99
Parliament of, 36, 40-41, 51-52, 83, 93-95, 98-102
political philosophy in, 97-99
railroads in, 232
religion in, 20, 41, 52

restoration of monarchy in, 41
slave trade and, 18, 62, 288
textile industry in, 19, 162, 187, 323, 336
trade and, 19, 66-67, 176, 178, 183, 238
U.S. Civil War and, 323, 336, 368
War of 1812 and, 180-183, 188
English Civil War, 40-41
English Reformation, 20
Enlightenment, 17, 60, 76, 79, 108, 118, 288
Entertainment, 246-248.
See also specific activities
Environmentalism and environmental issues, 163, 238, 276
Epidemics. *See also* Diseases
bubonic plague, 6, 13
lack of medical knowledge regarding, 171
measles, 298
smallpox, 9, 34, 79
Equality. *See also* Inequality
in colonial period, 63
Enlightenment idea of, 118
in gender relations, 283
Reconstruction and, 370
in republican ideology, 125-126
Equiano, Olaadah, 62
Era of Good Feelings, 185, 191-193
Erie Canal, 231, 232
Eriksson, Leif, 6
Essex Junto, 175
Ethnicity. *See* Race and ethnicity
Europe. *See also specific regions and countries*
balance of power in, 86
diplomacy in, 86
Enlightenment in, 17, 76, 118
immigrants from, 160, 162, 229
imperialism of, 16
Industrial Revolution in, 68, 160, 162
Napoleonic Wars in, 87, 175-176, 180
population growth in, 6
Protestantism in, 19, 63, 86-87
trade and, 6, 7, 66-67, 253
Eutaw Massacre (1870), 368
Evangelicals, 76, 158-159, 279
Excise tax, 144, 145, 147
Executive branch. *See also* Presidents, U.S.
checks and balances by, 137, 140
constitutional provisions for, 143

judiciary power and, 169
patronage within, 167-168
separation of powers and, 126, 137
spoils system and, 208
Virginia Plan and, 136
Ex parte Milligan (1866), 361-362

Factory Girls Association, 239
Factory system, 162-163, 236-242
Fair Oaks, Battle of (1862), 340
Fallen Timbers, Battle of (1794), 130, 178
Families. *See also* Children; Marriage
in Africa, 18
African American, 270
in antebellum period, 245-246
in colonial period, 60-61
middle class, 245, 246
in Old Northwest, 188
of enslaved persons, 46, 69, 268
Fanon, Frantz, 16
Farming. *See* Agriculture
Farragut, David G., 339
Far West. *See* West
Federal government. *See also*
Constitution, U.S.; Executive branch; Judicial branch; Legislative branch
advocates for reform of, 135-136
checks and balances within, 137, 140
in Confederation period, 127-131, 134-136
debt of, 144, 168, 219
economic growth and, 186-187, 197
in Jacksonian period, 208-210
primacy over state governments, 197
separation of powers within, 126, 137
surplus funds of, 219
Federalism, defined, 137
The Federalist Papers, 142
Federalists
on Constitution, 134-135, 139, 141-142
decline of, 150-153, 191, 199
economic policies of, 144-145
Essex Junto group of, 175
geographic areas of support, 146, 182
judicial branch and, 153, 168
Republican opposition to, 145-146
Fell, Margaret, 43

Female Labor Reform Association, 239
Feminism. *See also* Women
  in antebellum period, 282–285
  literature and, 275
Ferdinand of Aragon, 7
Fifteenth Amendment, 361, 369, 371, 374
Fifth Amendment, 314
Fillmore, Millard, 308, 309, 314
First Battle of Bull Run (1861), 338, 341
First Continental Congress, 102
First Great Awakening, 75–76
Fish, Hamilton, 367, 368
Fishing, 3–5, 66
Fiske, John, 138
Five Civilized Tribes, 211, 213–214
Fletcher v. Peck (1810), 195
Flintlock rifles, 39
Florida. *See also specific cities*
  French settlement in, 22
  indigenous lands in, 211, 214
  rice production in, 254
  secession of, 320
  Seminole Wars in, 192, 214
  Spanish colony in, 11, 46, 47
  U.S. acquisition of, 179–180, 191–192
  War of 1812 in, 181
Foner, Eric, 356
Foods
  in antebellum period, 245
  in cultural exchanges, 15
  draft in Confederacy, 332
  health improvement with, 280
  for enslaved persons, 264
Foote, Julia, 283
Foreign relations. *See* Diplomacy; Imperialism; Trade
Forrest, Nathan Bedford, 330
Fort Pillow Massacre (1864), 330
Forts. *See specific names of forts*
Forty-niners, 305
Founding Fathers, 136, 326
Fourteenth Amendment, 141, 359, 361, 371, 374
Fowler, Orson and Lorenzo, 280
Fox, George, 43
France. *See also* Napoleon Bonaparte
  abolition of slavery in, 288
  American Revolution and, 111, 115–117, 120
  Bastille Day in, 119
  colonies of, 22–23, 44–45, 47, 84

  English conflicts with, 84–89, 147
  exploration by, 8
  Huguenots in, 63
  indigenous peoples and, 84
  industrialization in, 162
  quasi war with U.S., 150–151
  revolution in, 87, 118, 119, 146
  trade and, 49, 84, 176, 178
  U.S. Civil War and, 323, 336
Francis I (France), 22
Franklin, Benjamin
  in American Revolution, 115–116
  Declaration of Independence and, 111
  deism and, 158
  education supported by, 78
  Enlightenment thought of, 76
  Federalist support by, 142
  on indigenous peoples, 84
  parliamentary testimony of, 94–95
  peace negotiations by, 120
  Pennsylvania Gazette, 89
  Poor Richard's Almanac, 77
  scientific experiments by, 79
  on Stamp Act, 93–95
Franklin, John Hope, 356
Free Blacks. *See also* African Americans
  as abolitionists, 286–287, 290
  in antebellum period, 243–244, 265–266, 268
  citizenship for, 141
  religion and, 268
  resettlement in Liberia, 286
  voting rights for, 141, 204
Freedmen's Bureau, 352–353, 357, 359, 364
Freeman, Elizabeth, 121
Freemasons, 218
Free-Soil Party, 291, 305, 309, 312–313
Free-staters, 311, 315
Frémont, John C., 302, 314
French and Indian War. *See* Seven Years' War
French Revolution, 87, 118, 119, 146
Frey, Sylvia R., 109
Fugitive slave laws, 140, 287, 291, 307–309
Fuller, Margaret, 275
Fulton, Robert, 165, 187, 197
Fundamental Articles of New Haven, 35–36

Fundamental Constitution for Carolina, 41
Fundamental Orders of Connecticut, 35
Fur trade, 22–23, 31, 34, 66, 84, 92, 178, 189–190

Gadsden, James and Gadsden Purchase, 310
Gage, Thomas, 102–103
Galen, 60
Gallatin, Albert, 168, 183
Gama, Vasco da, 7
Gambling, 164–165
Garrison, William Lloyd, 286, 290, 291, 293
Gaspée sinking (1772), 99
Gates, Horatio, 114, 116
Gathering and hunting societies, 3–5
Gender relations. *See also* Women
  in American Revolution, 123
  in antebellum period, 246
  in colonial period, 60–61
  cult of domesticity and, 246
  feminism on, 283
  indigenous peoples and, 6, 160
  in utopian communities, 277
Genovese, Eugene, 262
Geographical mobility, 243–244
George I (England), 83
George II (England), 48, 83
George III (England), 90, 93
Georgia. *See also specific cities*
  American Revolution in, 116
  Constitution ratified by, 142
  English colony in, 47–49
  indigenous lands in, 178, 211, 213
  rice production in, 65, 254
  secession of, 320
  slavery in, 65
  Spanish colony in, 47
German Americans, 229, 337
Germany
  immigrants from, 229
  industrialization in, 162
  Protestantism in, 19, 63
Germ theory of disease, 281
Gettysburg, Battle of (1863), 325, 344–345
Ghana, kingdom of, 18
Ghent, Treaty of (1814), 183
Gibbons v. Ogden (1824), 197
Gilbert, Humphrey, 20, 22
Globalization, 16, 162–163
Glorious Revolution, 52–53

INDEX · I-9

Gold
  in Africa, 17
  currency based on, 215, 220, 367
  in Spanish America, 9
Gold rushes, 305, 307
Goodyear, Charles, 237
Government. *See also* Federal government; Politics; State governments
  in colonial period, 29, 30, 34-36, 43, 52-53, 79-80
  of Confederacy, 331-332
  democratization of, 202-205
  in Europe, 6
  in Inca Empire, 4
  republican, 125-126
Grady, Henry, 371
Graham, Sylvester, 280
Grant, Ulysses S.
  in Civil War battles, 339, 340, 342-343, 345-347
  portrait of, 335
  as president, 366-369
  Union Army command by, 334, 345, 346
Great Awakenings, 75-76, 158-159, 271, 279
Great Compromise, 137
Great Lakes region
  agriculture in, 249
  British troops in, 128, 183
  cities of, 229, 249
  indigenous peoples in, 50, 51
  War of 1812 in, 180
Greece, 119
Greeley, Horace, 222, 366, 369
Greenbacks, 325, 367
Greene, Nathanael, 116
Greenville, Treaty of (1795), 130
Grenville, George, 90, 92, 93
Grenville, Richard, 20
Grimké, Sarah and Angelina, 282-283
Griswold, Robert, 152
Guadalupe Hidalgo, Treaty of (1848), 303
Guerrilla warfare, 214
Gutman, Herbert, 262
Guttierez, Ramon, 50

Hahn, Stephen, 356
Haitian Revolution, 118, 171, 288
Hale, John P., 309
Halleck, Henry W., 334
Hamalainen, Pekka, 51
Hamilton, Alexander
  Burr and, 153, 175
  in Confederation period, 130, 135-136
  on Constitution, 139
  economic proposals of, 144-145
  Federalist Papers and, 142
  as treasury secretary, 143
  on Whiskey Rebellion, 147
Hamilton Manufacturing Company, 240-241
Hammond, Bray, 206
Hammond, James Henry, 258-259
Hancock, John, 103
Hancock, Winfield Scott, 334
Handsome Lake (revivalist), 160
Harpers Ferry raid (1859), 316
Harrison, William Henry, 178-180, 220-221, 224
Harrison Land Law of 1800, 178
Hartford (Connecticut), 35, 36
Hartford Convention, 182-183
Harvard University, 78
Hasenclever, Peter, 66
Hawaii, U.S. annexation of, 310
Hawthorne, Nathaniel, 276, 292
Hayes, Rutherford B., 369-370
Hayne, Robert Y., 209, 210
HBCUs (Historically Black Colleges and Universities), 364
Headright system, 29, 33, 41
Health and health care.
  See also Diseases; Medicine
  antebellum theories of, 279-280
  for enslaved persons, 264
Heidler, David and Jeanne, 207
Hemings, Sally, 141
Henry, Fort, 339
Henry VII (England), 18
Henry VIII (England), 20
Henry, Patrick, 93, 142
Henry the Navigator (Portugal), 7
Herskovits, Melville J., 262
Hervieu, Auguste, 265
Hessians, 107, 113-114
Higher education. *See* Colleges and universities
Hispanics, 298, 302
Hispaniola, 7, 10, 13, 40, 44
Historians
  on American Revolution, 108-110
  on Civil War causes, 320, 321
  on Constitution, 138-140
  on indigenous peoples, 50-51
  interpretational differences among, 14-15
  on Jacksonian period, 206-207
  on Reconstruction, 356-357
  on slavery, 262-263
  on witchcraft trials, 72-73
Historically Black Colleges and Universities (HBCUs), 364
Hoe, Richard, 236
Hofstadter, Richard, 206
Holiday celebrations, 246-247
Holland (Netherlands), 23, 33
Holmes, Oliver Wendell, 276, 281
Holton, Woody, 110, 139
Homer, Winslow, 348
Homestead Act of 1862, 324
Hood, John Bell, 334, 345
Hooker, Joseph, 341-342, 344
Hooker, Thomas, 35
Hopis, 4, 6
Horne, Gerald, 263
Horse racing, 164-166, 248
Horseshoe Bend, Battle of (1814), 181
House of Burgesses (Virginia), 29, 30, 93
House of Delegates (Maryland), 33
House of Representatives, U.S.
  African Americans in, 364
  in presidential elections, 153, 200
  representation in, 137
  selection of members, 140
  violence within, 152
  War Hawks in, 180
Housing
  middle class, 244, 245
  in Old Northwest, 188
  on plantations, 15, 189
  for enslaved persons, 264
  working class, 244
Houston, Sam, 297-298
Howard, Oliver O., 353
Howe, Elias, 237
Howe, William, 113, 114
Hudson, Henry, 23
Hudson River school, 272-273
Huguenots, 63
Humoralism, 60
Humphries, Richard, 364
Hunter, Tera, 357
Hunting and gathering societies, 3-5
The Hurly-Burly Pot (Baillie), 307
Huron people, 22, 23
Hutchinson, Anne, 36
Hutchinson, Thomas, 93

Illinois. *See also* specific cities
  American Revolution in, 114
  indigenous lands in, 178, 197, 210
  statehood for, 189

Immigrants and immigration
  Alien Act and, 151
  in antebellum period, 228–229
  Catholic, 230
  Chinese, 305
  citizenship for, 140, 151
  in colonial period, 30, 55–59, 61, 63
  discrimination against, 243
  European, 160, 162, 229
  as indentured servants, 30, 56, 58–59
  Jewish, 229
  in labor force, 239, 241, 242, 305
  nativism and, 230, 279
  poverty of, 229
  in West, 305
Immunizations, 79, 281
Impeachment of Johnson, 362
Imperialism, 16, 82, 89–92. *See also* Colonies and colonial period
Impressment, 87, 89, 176–177, 183
Inca Empire, 3–4, 9, 12
Income. *See also* Wages
  in antebellum period, 242–243
  in New South, 372
  taxation of, 325, 332
Indentured servants
  in Caribbean region, 45
  in Chesapeake region, 29–31, 33, 56
  as immigrants, 30, 56, 58–59
  passage to Americas, 56, 58–59
  women as, 56–57, 61
India, 7, 86, 87, 169
Indiana
  indigenous lands in, 178
  New Harmony community in, 277
  statehood for, 189
Indian Removal Act of 1830, 211
Indians. *See* Indigenous peoples
Indian Territory, 213–214
Indigenous peoples.
  See also specific groups
  agriculture and, 4–5, 15, 27
  American Revolution and, 114, 122
  assimilation of, 11, 29, 178, 198, 282
  biological and cultural exchanges with Europeans, 11–13, 15
  in California, 302, 305
  in Chesapeake region, 26–31, 33
  citizenship for, 141
  diseases and, 1, 9, 13, 34, 46, 171
  education for, 78, 157, 282
  enslavement of, 15
  Five Civilized Tribes, 211, 213–214

  gender relations among, 6, 160
  in Jacksonian period, 210–215
  lands of, 89–91, 122, 130, 147, 171, 178, 188, 197, 210–215
  Lewis and Clark expedition and, 174
  middle grounds and, 50–51
  migrant conflicts with, 300
  in New England, 34, 37, 39
  Penn's relationship with, 44
  in precontact Americas, 2–3
  religion and, 11, 23, 46, 84, 160, 179
  on reservations, 213–215, 282
  sovereignty of, 147, 197, 211
  in Spanish Empire, 10, 11, 46
  trade and, 48, 91, 171, 189–190
  treaties with, 30–31, 50, 84, 122, 130, 178, 188, 212–213
  wars involving, 22, 29, 37, 39, 130, 179–181, 210
  in West, 130, 171, 178, 211, 213–215
  white attitudes toward, 210, 215
Individualism, 273, 300
Industrialization
  agriculture and, 249
  class divisions and, 163
  environmental costs of, 238
  global process of, 162–163
  in New South, 371–372
  obstacles to, 66
  technological advances and, 237
Industrial Revolution, 68, 160, 162–163, 227–228
Industry. *See also* Business; *specific industries*
  in antebellum period, 236–238, 257
  in colonial period, 65–66
  factory system, 162–163, 236–242
  globalization of, 162–163
  growth of, 144
  labor force and, 162–163, 238–242
  in South, 257, 260, 371–372
  technology and, 161, 237
Inequality. *See also* Class divisions; Poverty
  in antebellum period, 242–243
  in Jacksonian period, 202
  in urban areas, 242–243
Inflation, 111, 123, 332
Influenza, 13
Innovations. *See* Technology
Inoculation. *See* Vaccinations
Interchangeable parts, 161, 237
Intolerable (Coercive) Acts of 1774, 101, 102
Inuit people, 4

Inventions. *See* Technology
Ireland, 18–19, 63, 99, 229
Irish Americans, 229, 230, 241, 244, 325, 337
Iron industry, 65–66, 162, 257, 372
Iroquois Confederacy
  American Revolution and, 114, 122
  fur trapping by, 189
  gender relations in, 160
  lands of, 6, 130
  in middle grounds, 50
  relationship with Europeans, 22, 82, 84–85, 89
Irving, Washington, 158
Isabella of Castile, 7
Islam and Muslims, 18, 75

Jackson, Andrew. *See also* Jacksonian period
  background of, 202, 207
  corrupt bargain and, 200
  historical views of, 206–207
  horse racing and, 164
  on indigenous peoples, 210–215
  Kitchen Cabinet of, 209
  portrait of, 217
  as president, 202–220
  Seminole Wars and, 192
  as slaveholder, 207
  on Texas statehood, 298
  in War of 1812, 181, 182, 201
Jackson, Rebecca Cox, 285–286
Jackson, Thomas J. "Stonewall," 334, 339–342
Jackson, William Henry, 278
Jacksonian period, 202–220
  bank war in, 215–216
  democratization of government in, 202–205
  federal government in, 208–210
  historical debates on, 206–207
  indigenous peoples in, 210–215
  mass politics during, 203–205, 207–208
  political changes in, 217–219
  specie circular in, 220
Jamaica, 40, 41, 45
James I (England), 20, 22, 26, 27, 30, 35
James II (England), 41–43, 52, 53
James, C. L. R., 16
Jameson, J. Franklin, 108
Jamestown settlement, 20, 26–29, 31, 47
Japan, industrialization in, 162
Jay, John, 120, 142, 147

Jay Cooke and Company, 367
Jay's Treaty (1794), 147, 149, 150, 178
The Jazz Singer (film), 377
Jefferson, Thomas
  agrarian ideal of, 146, 155, 166
  Declaration of Independence
    and, 111, 122
  deism and, 158
  Enlightenment thought of, 76
  on indigenous peoples, 122, 141,
    157, 178, 210
  Louisiana Purchase under, 171, 174
  on Missouri Compromise, 194, 195
  on national bank, 145
  on Northwest Territory, 128
  political ideology of, 125
  portrait of, 167
  as president, 152-153, 166-178, 183
  on religious tolerance, 75
  Republicans and, 143, 146, 149,
    151, 155, 167-168
  as secretary of state, 143, 191
  as slaveholder, 122, 141
  Statute of Religious Liberty
    and, 127
  as vice president, 150
  western exploration under,
    171-174
Jeremiads, 75
Jesuit missionaries, 23, 84
Jews and Judaism
  anti-Semitism and, 229
  in colonial period, 36, 49, 75
  immigration by, 229
Jim Crow laws, 374-375
Jobs. *See* Employment; Labor force
Johnson, Andrew
  impeachment of, 362
  as president, 351, 357-359, 361,
    365, 367-368
  as vice president, 327, 355
Johnson, Walter, 262-263
Johnson v. McIntosh (1823), 197
Johnston, Albert Sidney, 334, 339
Johnston, Joseph E., 340, 345-347
Joint Committee on
  Reconstruction, 358, 359
Jolson, Al, 377
Journalism. *See* Newspapers
Judaism. *See* Jews and Judaism
Judicial branch. *See also* Supreme
    Court, U.S.
  checks and balances by, 137, 140
  constitutional provisions for,
    143, 169
  Federalist control of, 153, 168

separation of powers and, 137
Virginia Plan and, 136
Judiciary Act of 1789, 143, 169
Judiciary Act of 1801, 153, 168

Kansas
  "Bleeding Kansas," 311, 313
  Lecompton constitution in, 315
  statehood for, 315
  territory of, 311
Kansas-Nebraska Act of 1854, 311
Karlsen, Carol, 72
Kearny, Stephen W., 302
Keckley, Elizabeth, 265-266
Kendall, Amos, 203
Kennesaw Mountain, Battle of
    (1864), 345
Kentucky
  Cane Ridge camp meeting in, 158
  horse racing in, 164
  Maysville Road in, 215
  statehood for, 147
Kerber, Linda, 109
Key, Francis Scott, 182
King, Martin Luther, Jr., 276
King George's War, 85
King Philip's War, 37, 39
King William's War, 73, 85
Kitchen Cabinet, of Jackson, 209
Know-Nothings, 230, 311, 314
Knox, Henry, 143
Kongo, Kingdom of, 18, 75
Ku Klux Klan, 368

Labor force. *See also* Employment;
    Indentured servants; Slavery
    and enslaved persons; Wages;
    Women in workforce
  in antebellum period, 238-242
  children in, 242
  in colonial period, 29
  factory system and, 162-163,
    236-242
  immigrant, 239, 241, 242, 305
  industrial, 162-163, 238-242
  native workers, 238-239
  working conditions for, 241, 242
Labor unions, 239, 242
Lafayette, Marquis de, 111, 117
Land and landownership.
  See also Plantations
  by African Americans, 352-353
  in colonial period, 29, 33, 96
  government sales of, 219, 220
  Homestead Act and, 324
  indigenous, 89-91, 122, 130, 147,
    171, 178, 188, 197, 210-215

in Northwest Territory, 128-129
in Reconstruction, 352-353,
    364-366
reservations, 213-215, 282
Land-grant institutions, 324, 364
Language of enslaved persons, 268
Las Casas, Bartolomé de, 9, 10
Latin America. *See also* South
    America; *specific countries*
  abolition of slavery in, 288
  Monroe Doctrine and, 198-199
  revolutions in, 119, 198-199
  U.S. foreign policy toward,
    198-199
Latinos/Latinas, 298, 302
Latter-day Saints (Mormons),
    277-279
Laudonniere, Rene Goulaine de, 22
Laws and regulations. *See also*
    specific laws and regulations
  Black Codes, 359-361
  fugitive slave laws, 140, 287, 291,
    307-309
  Jim Crow, 374-375
  personal liberty laws, 291, 307
  slave codes, 30-32, 45, 63,
    261, 333
Lecompton constitution, 315
Lee, Jarena, 283, 285, 286
Lee, "Mother" Ann, 277
Lee, Richard Henry, 107
Lee, Robert E., 316, 334, 335,
    340-347, 373
Legal system. *See* Judicial branch;
    Laws and regulations
Legislative branch. *See also*
    Congress, U.S.
  checks and balances by,
    137, 140
  separation of powers and,
    126, 137
  Virginia Plan and, 136
Leisler, Jacob, 52-53
Leisure activities, 246-248.
  See also specific activities
L'Enfant, Pierre, 166
Leon, Francis, 377
Lewis, Meriwether, 172-174
Lexington, Battle of (1775),
    103-104, 112
Liberalism, in education, 78
Liberia, resettlement of free Blacks
    in, 286
Liberty Party, 291
Life expectancy, 57
Limited liability corporations, 236

INDEX

Lincoln, Abraham
  assassination of, 351, 355
  Civil War and, 325-329, 333-334, 339-341, 345-347, 351
  debates with Douglas, 315
  on economy, 250
  election and reelection of, 317, 327
  Emancipation Proclamation of, 328, 329, 336
  inaugural address of, 321
  portrait of, 358
  on Reconstruction, 353, 355
  on slavery, 315-316, 319
Lincoln, Mary Todd, 265-266
Literacy
  in colonial period, 76-78
  national rates of, 282
  voting rights and, 374
Literature. *See also specific authors and works*
  in antebellum period, 246, 273-274
  cultural nationalism and, 158
  Romanticism and, 246, 273-276
  sentimental novels, 291-293
  of transcendentalists, 274-276
  by women, 275, 291-293
Litwack, Leon, 356
Livingston, Robert R., 165, 170, 171, 197
Locke, John, 41, 75, 76, 111, 118, 151
Loco Focos, 218
London Company, 22, 26, 27, 34
Long, Stephen H., 190
Longstreet, Augustus B., 274
Lord Dunmore's Proclamation, 107, 121, 122
Lords of Trade, 52
Los Angeles (California), 46
Lost Cause history, 373
Louis XIV (France), 41, 47, 84
Louis XVI (France), 86
Louisbourg (Cape Breton Island), 85, 87
Louisiana. *See also specific cities*
  Reconstruction in, 353
  secession of, 320
  statehood for, 171
  sugar production in, 256
Louisiana Purchase, 171, 174, 195, 310
Louisiana Territory, 47, 84, 170-175
Lovejoy, Elijah, 290
Lowell, Francis Cabot, 187
Lowell system, 239-241
Loyalists (Tories), 112, 114, 116, 120, 121

Loyalty oaths, 353, 355, 357
Lucas, Sam, 377
Lundy, Benjamin, 290
Luther, Martin, 19
Lynchings, 290, 325, 375
Lyon, Matthew, 152

Machine tools, 161, 237
Macon's Bill No. 2 (1810), 178
Madison, James
  on Bill of Rights, 142
  in Confederation period, 130, 136
  on Constitution, 136, 137, 139
  Enlightenment thought of, 76
  Federalist Papers and, 142
  on national bank, 145
  as president, 177, 178, 180, 188
  Republicans and, 143, 146, 151
  as secretary of state, 169, 177, 191
Magellan, Ferdinand, 8-9
Maier, Pauline, 108
Main, Jackson Turner, 138-139
Maine, 36, 195, 224
Malaria, 26-27, 57
Mali, empire of, 18
Manifest Destiny, 296
Mann, Horace, 281
Manufacturing. *See* Industry
Manumission, 42, 121, 266
Marbury v. Madison (1803), 169
Marcy, William L., 208
Marion, Francis, 116
Marriage. *See also* Families
  Black Codes on, 360
  in colonial period, 60-61
  polygamy, 278, 279
  of enslaved persons, 268
  in Spanish America, 11, 15
Marshall, John, 153, 169, 175, 195-197, 216-217
Mary (England), 20
Mary II (England), 52, 53
Maryland
  Albany Plan and, 84
  Catholicism in, 32-33, 53, 74
  slavery in, 33, 63
  tobacco production in, 33, 69, 254
Maryland Toleration Act of 1649, 33
Mason, James M., 336
Massachusetts. *See also specific cities*
  Brook Farm community in, 276
  in Dominion of New England, 52
  emancipation in, 121
  establishment of, 35
  public schools in, 78, 281
  religion in, 35, 36

  slavery in, 65
  taverns in, 99-101
  textile industry in, 187, 239
  witchcraft trials in, 72-73
Massachusetts Bay Company, 35, 36
Mass politics, 203-205, 207-208
Mather, Cotton, 38-39, 79, 281
Matrilineal descent, 6, 18
Maurer, Louis, 165
Mayans, 4, 13
Mayflower Compact, 34
Maysville Road (Kentucky), 215
McClellan, George B., 327, 334, 338-341
McCormick, Cyrus H., 249-250
McCulloch v. Maryland (1819), 196-197
McDonald, Forrest, 138
McDowell, Irvin, 338, 339
McHenry, Fort, 182
Meacham, Jon, 207
Meade, George C., 334, 344, 347
Measles, 13, 298
Medicine. *See also* Diseases; Health and health care
  anesthetics, 281
  in colonial period, 57, 60
  in early civilizations, 4
  education in, 157
  nursing profession, 331
  science and, 157, 280-281
  vaccinations, 79, 281
Meiji reforms (Japan), 162
Melville, Herman, 273
Men. *See* Gender relations
Mental health reform, 331
Mercantilism, 19
Merrell, James, 50
Mesoamerican civilizations, 4
Mestizos, 15
Metacom "King Philip" (Wampanoag chief), 39
Methodists, 76, 158, 159
Mexican War (1846-1848), 302-304
Mexico
  border with Texas, 301, 303
  Catholic Church in, 9
  early civilizations in, 3, 4, 9
  independence of, 119, 189
  Spanish rule in, 9, 11, 12
  Texas territory and, 297-298
Meyers, Marvin, 206
Miami people, 130
Michigan, 178. *See also specific cities*
Mid-Atlantic colonies, 57, 113-115

Middle class
  African Americans in, 372
  in antebellum period, 244-245, 260
  in cities, 244
  cult of domesticity of, 246
  families and, 245, 246
  horse racing and, 165
  industrialization and, 163
  women in, 244, 246
Middle grounds, 44, 49-51
Middle passage, 62, 67, 256
Midwives, 60, 157
Migration. *See also* Immigrants and immigration
  to Americas, 2-3, 16
  to California, 302, 305, 307
  to cities, 162-163
  motivations for, 16-17, 188
  to Oregon, 298-299
  to South, 256
  to Texas, 297, 298
  westward, 188, 244, 298-300
Military. *See also* Army, U.S.; Confederate Army; Conscription; Navy, U.S.; Union Army; War and warfare; Weapons
  in American Revolution, 111-114, 116-117
  in Civil War, 325, 327, 329-330, 332-336, 337-347
  in early civilizations, 4
  in Mexican War, 302-303
  in War of 1812, 180-182
Military Academy at West Point, 168, 334
Military Telegraph Corps, U.S., 338
Miller, Arthur, 72
Miller, Perry, 72
Mining, 9, 66, 305, 307
Minorities. *See* Race and ethnicity
Minstrel shows, 247-248, 374, 376-377
Minutemen, 102, 103
Missionaries
  education by, 157, 282
  Jesuit, 23, 84
  in middle grounds, 49
  revivalism and, 160
  in Spanish Empire, 9, 13, 46, 47
Mississippi
  Black Codes in, 360-361
  indigenous lands in, 178, 211
  secession of, 320, 322
  slavery in, 259
  statehood for, 189
Mississippi River
  Civil War and, 338-339, 343-344
  French control of, 84, 170
  steamboats on, 187-188, 231
Missouri, statehood for, 185, 193, 195
Missouri Compromise, 193-195, 304, 310-311, 314, 321
Mittelberger, Gottlieb, 58-59
Mohawk and Hudson Railroad, 233
Mohawks, 39, 114, 122
Monitor (ironclad warship), 335
Monopolies, 19, 62
Monroe, James, 171, 177, 191, 192, 198-199
Monroe Doctrine, 198-199
Montagnais, 22
Montcalm, Marquis de, 87
Montesquieu, Baron de, 137
Montezuma (Aztec leader), 9
Montgomery, Richard, 113
Montreal, 23, 84, 88
Moran, Thomas, 273
Morgan, Edmund S., 108
Morgan, William, 218
Mormons (Church of Jesus Christ of Latter-day Saints), 277-279
Morrill Land Grant College Act of 1862, 324, 364
Morris, Robert, 130
Morse, Samuel F. B. and Morse code, 235
Mortality rates
  in antebellum period, 228, 260
  for children, 61, 260
  in colonial period, 30, 57, 60, 61
Morton, William, 281
Motherhood, 123
Mott, Lucretia, 283
Mount Holyoke College, 246
Murfreesboro, Battle of (1862), 339
Murphy, Isaac, 164
Murray, Judith Sargent, 123, 157
Music, 268, 376, 377
Muskogean people, 6
Muslims and Islam, 18, 75
Mutiny Act of 1765, 92, 96

Napoleon Bonaparte, 118-119, 151, 169-171, 176, 178
Napoleonic Wars, 87, 175-176, 180-181, 183, 186, 192
Nash, Gary, 50, 108
Nashville, Battle of (1864), 346
National Bank Acts of 1863-1864, 325
National banks, 144-145, 186, 193, 218-219, 325. *See also* Bank of the United States
National Greenback Party, 367
Nationalism
  art and, 272-273
  cultural, 156-160
  economic, 216, 324-325
  in Era of Good Feelings, 191
  Federalists and, 139
  in foreign policy, 198-199
  Manifest Destiny and, 296
  Missouri Compromise and, 195
  Supreme Court and, 197, 198, 217
  after War of 1812, 185, 191
National Republican Party. *See* Whig Party
National Road, 187
National sovereignty, 146-147, 149
National Trades' Union, 242
Native American Party, 230
Native Americans. *See* Indigenous peoples
Nativism, 230, 279
Naturalization Act of 1790, 140, 151
Navigation Acts, 51, 52, 67, 99
Navy, U.S.
  in Civil War, 326, 333-336, 338
  conflict with British ships, 176-177
  in Mexican War, 302-303
  in quasi war with France, 151
  scaling down of, 168
  in War of 1812, 180
Nebraska, 310-311
Necessary and proper clause (U.S. Constitution), 142, 196
Necessity, Fort, 85
Netherlands
  colonies of, 23, 42-43
  Puritans in, 33
  slave trade and, 18
New Amsterdam, 42
New England. *See also specific states*
  agriculture in, 65, 232
  Albany Plan and, 84
  American Revolution in, 112-113
  Dominion of New England, 52
  expansion of, 35-37
  indigenous peoples in, 34, 37, 39
  life expectancy in, 57
  Mather on, 38-39
  Puritans in, 20, 33-35, 38, 70-76
  revolt of, 182-183
  secession threat in, 175, 182
  settlement of, 33-36

New England—(Cont.)
  slavery in, 65
  social patterns in, 70-71, 73
  textile industry in, 161, 187, 237, 239
  water power in, 161
  women in, 36, 61
New England Antislavery Society, 290
New France, 74, 84-85
New Hampshire, 36, 52, 121
New Harmony community, 277
New Haven (Connecticut), 35-36, 78
New Jersey
  American Revolution in, 113-114
  colony of, 43, 52
  Constitution ratified by, 142
  emancipation in, 121
New Jersey Plan, 136
New Light revivalists, 76
New Mexico. *See also specific cities*
  American settlers in, 189, 302
  Gadsden Purchase and, 310
  indigenous peoples in, 4
  Mexican War in, 302, 303
  Spanish colony in, 11, 46
  statehood for, 307
  territory, 301-302
  U.S. annexation of, 303
New Netherland, 23, 42
New Ochota, Treaty of (1835), 212-213
New Orleans (Louisiana)
  British control of, 88
  Burr and, 175
  in Civil War, 338-339
  French control of, 84, 170
  race riots in, 359
  U.S. purchase of, 170, 171
  War of 1812 in, 182
Newport (Rhode Island), 73, 75
New South, 371-375
New Spain, 47, 189
Newspapers. *See also specific newspapers*
  abolitionist, 287, 290
  in colonial period, 68, 74, 78, 89
  mass circulation of, 236
  penny press, 221-223, 296
  telegraph and, 236
New World. *See* Americas
New York. *See also specific cities*
  Albany Plan and, 84
  American Revolution in, 112-114
  Anti-Masonry movement in, 218
  Bucktails faction in, 207

  colony of, 42-43, 52
  Constitution ratified by, 142
  emancipation in, 121
  Iroquois Confederacy in, 6
  Oneida Community in, 277
  Seneca Falls Convention in, 283-285
  slavery in, 42
  War of 1812 in, 182
New York City
  Central Park in, 243
  draft riots in, 325
  Jewish populations in, 75
  population of, 73, 166, 229
  slavery in, 256
  Tammany Hall in, 369
New York Herald, 222, 223, 236
New York Sun, 221, 222, 236
New York Times, 222
New York Tribune, 222, 314, 366
Nicholson, Francis, 52
Nicolls, Richard, 42
Nightingale, Florence, 331
Nineteenth Amendment, 141
Nissenbaum, Stephen, 72
Non-Intercourse Act of 1809, 178
North. *See also* Civil War; Sectionalism; Union Army; *specific regions and states*
  abolitionism in, 121, 286-287, 290, 295
  agriculture in, 5, 188, 232, 248-251
  American Revolution in, 112-114
  in colonial period, 70-71, 73
  economic nationalism in, 324-325
  free Blacks in, 204, 243, 286
  rural life in, 250-251
  slavery in, 65, 70, 121, 256
  transportation systems in, 231, 321
North, Lord, 97, 101, 102, 115, 117
North America. *See also specific countries*
  borderlands of, 25, 44-51
  Dutch colonies in, 23
  early civilizations of, 4-6
  English colonies in, 18-22, 25-37, 39-53
  European exploration of, 8, 9
  French colonies in, 22-23
  migration to, 2, 3
  Spanish Empire in, 11
North Carolina
  American Revolution in, 116
  formation of, 42
  indigenous lands in, 213

  Regulator revolt in, 92
  Roanoke colony in, 20-22
  secession of, 321
  statehood for, 147
Northeast. *See also specific states*
  agriculture in, 5, 232, 248-249
  European immigrants in, 229
  Federalist support in, 146
  impact of embargo in, 177
  industry in, 227, 229, 238
Northern confederacy, 175
Northwest Ordinance, 129, 146, 256
Northwest Territory, 128-129, 147, 171, 178, 256
Norton, Mary Beth, 72-73, 109
Noyes, John Humphrey, 277
Nullification, 208-210
Nursing profession, 331

Ogden, Aaron, 197
Oglethorpe, James, 47-49
Ohio, 187, 203-204
Old Light traditionalists, 76
Old Northwest, 188, 189, 249-250, 298, 310
Old Southwest, 189
Olmec people, 4
Oñate, Don Juan de, 11
O'Neale, Peggy, 209
Oneida "Perfectionists," 277
Onesimus, 79, 281
Onís, Luis de, 191, 192
Opechancanough, 29
Opie, Amelia, 63
Ordinance of 1784, 128
Ordinance of 1785, 128, 129
Oregon
  boundary with Canada, 300, 301
  European exploration of, 9
  migration to, 298-299
  statehood for, 307
Oregon Trail, 299
Osceola (Seminole chief), 214
Ostend Manifesto, 309
Otis, James, 93
Ottoman Empire, 119
Owen, Robert, 277

Pacifism, 43, 120, 291
Paine, Thomas, 76, 110
Palmer, F. F., 190
Panics
  of 1819, 192-193
  of 1837, 219-220
  of 1873, 367, 369
Parkman, Francis, 50

INDEX · I-15

Parliament (England), 36, 40–41, 51–52, 83, 93–95, 98–102
Paterson, William, 136
Patrilineal descent, 6
Patriots, 111–116, 120–123
Patronage, 167–168
Paxton Boys, 91, 92
Peace of Paris (1763), 88
Peale, Rembrandt, 167
Peculiar institution of slavery, 261–267
Penitentiaries, 282
Penn, William, 43–44
Penningroth, Dylan, 356
Pennsylvania. *See also specific cities*
   Albany Plan and, 84
   Charter of Liberties in, 44
   emancipation in, 121
   establishment of, 43
   immigrants in, 63
   turnpike in, 165
   Whiskey Rebellion in, 147
Pennsylvania Gazette (newspaper), 89
Penny press, 221–223, 296
Pequot War, 37, 40
Perkins School for the Blind, 282
Perry, Oliver Hazard, 180
Personal liberty laws, 291, 307
Peru
   Inca Empire in, 3–4, 9, 12
   independence of, 119
   migration to, 2
   Spanish rule in, 11, 12
Pessen, Edward, 206
Philadelphia (Pennsylvania)
   American Revolution in, 114
   Constitutional Convention in, 136–137
   Continental Congresses in, 102, 107, 110–111
   establishment of, 43
   Paxton Boys in, 92
   population of, 73, 166, 229
Philip II (Spain), 20
Philippines, European exploration of, 9
Phillips, Ulrich B., 262
Photography, 326–327
Phrenology, 280
Physicians. *See* Medicine
Pickens, Andrew, 116
Pierce, Franklin, 309–311, 313
Pike, Zebulon Montgomery, 174, 190
Pilgrims, 34, 35, 37
Pinckney, Charles C., 150, 168, 177

Pinckney, Thomas, 149
Pinckney's Treaty (1795), 149, 150, 170
Piracy, 47, 168
Pitcairn, Thomas, 103
Pitt, William, 87, 90, 96
Pizarro, Francisco, 9
Plantations. *See also* Agriculture; Slavery and enslaved persons
   commercial sector and, 257
   cotton, 189
   growth of, 188, 253
   headright system and, 29
   housing on, 15, 189
   sugar, 254
   tobacco, 65, 69
Planter class, 259, 371
Plattsburgh, Battle of (1814), 182
Plessy v. Ferguson (1896), 374
Plymouth Company, 22, 33
Plymouth Plantation, 33–34, 37
Pocahontas, 28–29, 50
Pocket veto, 355
Poe, Edgar Allan, 273–274
"Poem on the Rising Glory of America," 158
Political parties. *See also specific parties*
   in 1820s and 1830s, 199, 205, 207
   legitimization of, 205, 207
   nativist, 230
   nomination conventions held by, 208
   origins of party system, 145
   two-party system, 199, 200, 207, 217
Politics. *See also* Elections; Government; Political parties
   in African societies, 17
   Catholic Church and, 86–87
   in Civil War, 326–327, 329
   in colonial period, 30, 33, 53, 79–80
   in Confederacy, 332
   of emancipation, 327, 329
   in England, 97–98
   of indigenous peoples, 6
   in Jacksonian period, 217–219
   mass, 203–205, 207–208
   of Reconstruction, 353, 363–364
   tavern culture and, 99–101
Polk, James K., 225, 298, 300–305, 309
Pollard, Edward, 373
Pollution
   air, 238
   in cities, 163

   industrial, 238
   water, 238
Polo, Marco, 7
Polygamy, 278, 279
Pontiac (Ottawa chief), 89, 90
Poor Richard's Almanac (Franklin), 77
Pope, John, 340–341
Popé (Pueblo leader), 11
Popular culture
   horse racing in, 164–166, 248
   leisure activities and, 246–248
   minstrel shows in, 247–248, 374, 376–377
   penny press in, 221–223, 296
   sentimental novels in, 291–293
   taverns in, 100–101
Popular sovereignty, 118, 119, 304–305, 308, 316
Population growth
   in antebellum period, 228–229
   in colonial period, 30, 57, 71
   in England, 19
   in Europe, 6
   in fifteenth century, 6
   as motivation for migration, 188
   in nineteenth century, 163, 188
   in Spanish America, 11
Populism, 108
Portage des Sioux, Treaties of (1815), 188
Port Royal Experiment, 352
Portugal
   colonies of, 15, 18, 118–119
   exploration by, 1, 7
   French invasion of, 119
   slave trade and, 18
Positivists, 14
Pottawatomie Massacre (1856), 311
Poverty
   of African Americans, 266, 286
   in antebellum period, 243
   in cities, 243
   of immigrants, 229
Powhatan and Powhatan Confederacy, 26–29
Precontact Americas, 2–3
Prejudice. *See* Discrimination
Presbyterians, 63, 121, 158, 160
Presidents, U.S. *See also* Elections; *specific presidents*
   assassination of, 351, 355
   checks and balances by, 137
   impeachment of, 362
   Virginia Dynasty and, 191
   war powers of, 326, 329

Press. *See* Newspapers
Press freedom, 79, 143
Preston, Thomas, 97
Prigg v. Pennsylvania (1842), 291
Primogeniture, 70
Princeton University, 78
Printing technology, 78, 222, 236
Prison reform, 282, 331
Proclamation of 1763, 90-92
Prosser, Gabriel, 160, 269
Prostitution, 298
Protestantism. *See also specific denominations*
   American Revolution and, 121
   in colonial period, 33, 74
   in Europe, 19, 63, 86-87
   immigrant concerns in, 230
   revivalism and, 75-76, 158-160, 179, 279
Protestant Reformation, 19
Providence (Rhode Island), 36
Prussia, 86, 111
Publick Occurrences (newspaper), 78
Public schools, 78, 156-157, 281, 364, 371
Publius, 142
Pueblo people, 4, 6, 11
Puerto Rico
   abolition of slavery in, 289
   Spanish colony in, 44
   Western Design expedition and, 40
Punch, John, 30
Puritans, 20, 33-35, 38, 70-76, 100

Quakers, 36, 43-44, 78, 120, 283
Quasi war with France, 150-151
Quebec, 22-23, 84, 87-88, 113, 120
Queen Anne's War, 85

Race and ethnicity. *See also* Immigrants and immigration; *specific racial and ethnic groups*
   in California, 307
   in Spanish America, 15
Race riots, 359
Racism. *See also* Discrimination; White supremacy
   of anti-abolitionists, 290
   in minstrel shows, 247-248, 374, 376-377
   phrenology and, 280
Radical Reconstruction, 358-359, 361-362, 366

Radical Republicans
   on Committee on the Conduct of the War, 334
   on emancipation, 327, 329
   on Reconstruction, 353, 355, 356, 358, 359, 361-362
Railroads. *See also specific railroads*
   in Civil War, 324-325, 337-338
   Confederate, 321
   government funding for, 233
   history and growth of, 232-234
   industrialization and, 162
   in New South, 372
   plantation economy and, 257
   public land grants for, 233
   telegraph and, 235-236
   transcontinental, 310, 324-325
   in West, 310
Rainey, Joseph, 364
Rakove, Jack, 139
Raleigh Walter, 20, 22
Ralph Wheelock's Farm (Alexander), 146
Randolph, Edmund, 136, 143, 145, 147
Randolph, John, 266
Raymond, Henry, 222
Recessions in 1830s, 216
Reconstruction, 350-371
   abandonment of, 368-371
   Congress and, 353, 355, 358, 359
   education and, 364
   historical debates on, 356-357
   land redistribution in, 352-353, 364-366
   legacy of, 350, 370-371
   plans for, 353, 355, 361
   politics of, 353, 363-364
   Radical, 358-359, 361-362, 366
   readmission to Union in, 353, 355, 358, 359, 361, 362
   state governments in, 353, 358, 363-364
Redeemers, 371
Reform movements
   African Americans in, 283, 286
   in antebellum period, 279-286
   in education, 281-282
   in health and medicine, 279-281
   rehabilitation and, 282
   temperance, 271, 279
   women in, 279, 282-286
Regulations. *See* Laws and regulations
Regulator revolt, 92
Rehabilitation, 282

Religion. *See also specific religions and denominations*
   in Africa, 18, 75
   of African Americans, 17, 159-160, 283, 285-286
   American Revolution and, 120-121, 158
   in colonial period, 32-36, 41-44, 46-48, 53, 74-76
   in England, 20, 41, 52
   indigenous peoples and, 11, 23, 46, 84, 160, 179
   as motivation for migration, 16-17
   revivalism and, 75-76, 158-160, 179, 279
   in rural areas, 251
   of enslaved persons, 46, 69, 75, 159-160, 267-268
   women and, 159, 283, 285-286
Religious freedom, 33, 35, 36, 41, 63, 129, 142
Religious tolerance, 33, 35, 36, 41, 74, 75, 118
Remini, Robert V., 207
Republican government, 125-126
Republican Party. *See also* Radical Republicans
   conservatives, 327, 353
   diplomacy of, 367-368
   founding of, 145, 311
   ideology of, 313, 315
   Lincoln administration and, 326, 334
   in South, 366
Republicans
   fragmentation of, 199, 200
   geographic areas of support, 146
   Jeffersonian, 143, 146, 149, 151, 155, 167-168
   opposition to Federalists, 145-146
   presidential nomination convention held by, 208
   Revolution of 1800 and, 152-153
Reservations, 213-215, 282
Restoration colonies, 40-44
Restoration of English monarchy, 41
Revels, Hiram, 363-364
Revere, Paul, 97, 98, 103
Revivalism, 75-76, 158-160, 179, 279
Revolutionary War. *See* American Revolution
Revolution of 1800, 152-153, 183
Revolutions. *See also* American Revolution
   age of, 118-119
   diplomatic, 86

French, 87, 118, 119, 146
Haitian, 118, 171, 288
industrial, 68, 160, 162–163
in Latin America, 119, 198–199
Rhode Island. *See also specific cities*
  Dorr Rebellion in, 204
  emancipation in, 121
  religious diversity in, 36, 75
  statehood for, 147
  voting rights in, 204
Ribault, Jean, 22
Rice production, 42, 65, 254
Richmond (Virginia), 331, 337, 339–341, 345–347, 352
Richter, Daniel, 50
Rights. *See also* Voting rights
  Bill of Rights, 139, 142–143, 147
  civil, 276, 350, 359
  of indigenous peoples, 197
Riots
  in colonial period, 131
  draft laws and, 325
  free Blacks and, 243
  Loco Focos and, 218
  race, 359
Ripley, George, 276
Roads
  in Inca Empire, 4
  Maysville, 215
  National Road, 187
  in South, 257
  turnpikes, 165–166, 231
Roanoke colony, 20–22
Rockingham, Marquis of, 93, 96
Rocky Mountain Fur Company, 190
Rogin, Michael, 206
Rolfe, John, 27–29
Roman Catholic Church. *See* Catholic Church
Romanticism, 246, 272–276
Rosecrans, William, 345
Rosen, Hannah, 357
Ross, John, 212–213
Roundheads, 40
Rousseau, Jean-Jacques, 76, 118
Royal African Company, 62, 75
Royal Society of London, 79
Rural areas. *See also* Agriculture
  leisure activities in, 247
  lifestyle in, 250–251
  migration to cities from, 162–163
  Republican support in, 146
Rush-Bagot agreement (1817), 183

Russia
  Napoleonic Wars and, 180
  Seven Years' War and, 86
  U.S. acquisition of Alaska from, 368

Sacajawea, 174
St. Augustine (Florida), 11, 47, 49
St. Leger, Barry, 114
Salary. *See* Income; Wages
Salem witchcraft trials, 72–73
Salt Lake City (Utah), 279
Sampson, Deborah, 123
San Francisco (California), 46, 305
San Ildefonso, Treaty of (1800), 169–170
San Jacinto, Battle of (1836), 298
Santa Anna, Antonio López de, 297–298
Santa Fe (New Mexico), 11, 302
Santa Fe Trail, 189, 299
Saratoga, Battle of (1777), 114–116
Saugus Ironworks, 65–66
Sauk people, 210, 211
Savannah (Georgia), 48, 116, 346, 347
Saxton, Alexander, 206
Scalawags, 364
Schlesinger, Arthur M., 108, 206
Schools. *See* Education
Schwalm, Leslie A., 357
Science
  in Enlightenment, 17, 60, 76, 79
  medicine and, 157, 280–281
Scotch-Irish immigrants, 63
Scott, Dred, 314, 315
Scott, Thomas, 338
Scott, Winfield, 213, 302–303, 309, 334
Secession
  New England threats of, 175, 182
  of southern states, 320–323, 331
Second Battle of Bull Run (1862), 341
Second Continental Congress, 107, 110–111, 114, 127
Second Great Awakening, 158–159, 271, 279
Second middle passage, 256
Sectionalism
  agricultural markets and, 249
  "Bleeding Kansas" as symbol of, 311
  crises of 1850s, 309–317
  Dred Scott decision and, 314
  in horse racing, 164

slavery debates and, 193–195, 304–310
  tariffs and, 200
Sedition Act of 1798, 151–152
Segregation. *See also* Discrimination
  in education, 364, 374
  in horse racing, 164
  Jim Crow laws and, 374–375
Seminole people, 192, 211, 214
Seminole Wars, 192, 214
Semmelweis, Ignaz, 281
Senate, U.S.
  African Americans in, 363–364
  representation in, 137
  sectionalism in, 311–312
  selection of members, 140
Seneca Falls Convention, 283–285
Sentimental novels, 291–293
Separation of church and state, 36, 127
Separation of powers, 126, 137
Sequoyah (Cherokee chief), 198
Seven Days, Battle of the (1862), 340
Seven Years' War, 47, 84–89, 91, 169
Seward, William H., 336, 355, 367–368
Sex and sexuality
  enslaved persons, 69, 260, 264
  in colonial period, 61, 69
  in Oneida Community, 277
  prostitution, 298
Seymour, Horatio, 366
Shakers, 277, 286
Sharecropping, 365
Shattuck, Job, 131
Shaw, Robert Gould, 329, 330
Shays, Daniel and Shays's Rebellion, 131, 134, 136, 140
Shelburne, Lord, 117
Sherman, William Tecumseh, 334, 345–347, 352
Shiloh, Battle of (1862), 339
Silver, 9, 215, 220
Simms, William Gilmore, 274
Singer, Isaac, 237
1619 Project, 110
Slater, Samuel, 160
Slaughterhouse Offal Act of 1862, 238
Slave codes, 30–32, 45, 63, 261, 333
Slavery and enslaved persons. *See also* Abolitionism; African Americans; Antebellum period; Emancipation; Plantations; Slave trade

Slavery—(*Cont.*)
  in Africa, 16–18, 62
  American Revolution and, 107, 121–122
  in Brazil, 18, 62, 289
  in Caribbean region, 18, 45–46, 62, 288–289
  in cities, 265
  in colonial period, 29–33, 41–43, 55, 57, 62–63, 65, 69–71
  Compromise of 1850 and, 308–309
  Constitution and, 136, 137, 139–140
  cotton cultivation and, 160–161, 189, 255
  culture of, 262, 267–270
  debates regarding, 185, 193–195, 304–310, 313
  in early civilizations, 4
  education for, 78
  families of, 46, 69, 268
  fugitive slave laws, 140, 287, 291, 307–309
  historical debates on, 262–263
  indigenous peoples as, 15
  language and music of, 268
  legal and social status of, 63, 261
  life under, 261, 264
  Missouri Compromise and, 193–195, 304, 310–311, 314, 321
  in North, 65, 70, 121, 256
  as peculiar institution, 261–267
  population trends among, 228
  pro-slavery arguments, 313, 324
  punishment of, 261, 264
  rebellions by, 70, 118, 160, 266, 267, 269
  religion of, 46, 69, 75, 159–160, 267–268
  running away, 70, 214
  sexual abuse of, 69, 260, 264
  in territories, 304–310
  women as, 69, 260, 262, 264
Slave trade
  Africa and, 16–18, 62
  domestic, 266–267
  middle passage and, 62, 67, 256
  opposition to, 288
  prohibitions on, 121, 288, 308
  in triangular trade, 66–67
Slidell, John, 336
Smallpox, 1, 9, 13, 34, 79, 90, 171, 281
Smith, Henry Nash, 375
Smith, John, 27, 33, 34, 50
Smith, Joseph, 278, 279
Smith, Sydney, 272
Smith-Rosenberg, Carroll, 110
Smuggling, 92, 96
Snyder, Christina, 207
Social mobility, 69, 243–244. *See also* Class divisions
Society of Freemasons, 218
Society of Friends. *See* Quakers
Soldiers. *See* Military
Sommerville, Diane Miller, 357
Sons of Liberty, 93, 100, 111
South. *See also* Civil War; Confederacy; Confederate Army; Reconstruction; Sectionalism; Slavery and enslaved persons; *specific regions and states*
  agriculture in, 65, 69–70, 160–161, 189, 229, 253–257
  American Revolution in, 116–117
  class divisions in, 259–261
  in colonial period, 57, 62, 69–70
  cotton economy in, 160–161, 189, 254–259, 333
  Democratic Party in, 369–371
  early civilizations in, 5
  education in, 260, 281, 364, 371
  industry in, 257, 260, 371–372
  international trade and, 253
  Lost Cause history in, 373
  migration to, 256
  New South, 371–375
  plantations in, 65, 69–70, 188, 189, 253, 257
  Republican support in, 146, 366
  secession of states in, 320–323, 331
  War of 1812 in, 182
  white society in, 257, 259–261
  women in, 260, 333, 372
South America. *See also* Latin America; *specific countries*
  Catholic Church in, 9
  Dutch settlement in, 23
  early civilizations of, 3–4
  European exploration of, 7, 8
  migration to, 2
  slavery in, 18, 288
  Spanish Empire in, 9, 11, 12
  trade with, 125
South Carolina. *See also specific cities*
  American Revolution in, 116
  formation of, 42
  French settlement in, 22
  nullification doctrine in, 208–210
  rice production in, 65, 254
  secession of, 320–322
  slavery in, 63, 65, 121, 259
  Stono Rebellion in, 70
Southeast borderlands, 47
Southern Homestead Act of 1866, 353
Southwest. *See also specific states*
  borderlands of, 46–47
  expansion of, 301–302, 304
  horse racing in, 164
  indigenous peoples in, 4, 6
  plantation system in, 189
  Spanish colonies in, 11, 12
  white settlement in, 147, 178, 189
Sovereignty
  of colonies, 99
  federalism and, 137
  indigenous, 147, 197, 211
  national, 146–147, 149
  popular, 118, 119, 304–305, 308, 316
  of states, 134
Spain. *See also* Spanish Empire
  diplomacy with, 128
  exploration by, 1, 7–9, 11
  French invasion of, 119
  Pinckney's Treaty and, 149, 150, 170
  Seven Years' War and, 169
Spanish Armada, 20
Spanish Empire
  abolition of slavery in, 288–289
  in Americas, 9–13, 15, 46–47
  in Caribbean region, 11, 44–45
  colonies of, 7, 11, 15, 44–45, 118–119, 189
  indigenous peoples in, 10, 11, 46
  Latin American revolutions against, 198–199
  rebellions against, 9, 11
Specie circular, 220
Specie Resumption Act of 1875, 367
Speech, freedom of, 142
Spinoza, Baruch, 76
Spoils system, 208
Sports, 248
Spotsylvania Court House, Battle of (1864), 345
Stamp Act Congress, 93
Stamp Act of 1765, 78, 93–96, 100
Stampp, Kenneth, 262, 356
Stanton, Edwin M., 361, 362
Stanton, Elizabeth Cady, 283, 331, 359
"The Star-Spangled Banner" (Key), 182

State governments
  constitutions for, 126–127
  debt of, 144, 145
  federal surplus distributed to, 219
  nullification doctrine and, 208–210
  primacy of federal government over, 197
  in Reconstruction, 353, 358, 363–364
  republicanism and, 125–126
States' rights
  in Confederacy, 332
  Supreme Court cases on, 196, 197
  Webster-Hayne debate on, 209
Statute of Religious Liberty (Virginia), 127
Steamboats, 163, 165, 187–188, 197, 231, 257
Steam engines, 162, 232
Stephens, Alexander H., 324, 331, 358
Steuben, Baron von, 111
Stevens, John, 232
Stevens, Thaddeus, 327, 353
Stockton and Darlington Railroad, 232
Stone tools, 2, 3
Stono Rebellion, 70
Story, Joseph, 203
Stowe, Harriet Beecher, 283, 291–293, 313
Stratton, Charles "Tom Thumb," 247
Strikes, 239, 242. *See also* Unions
Stuyvesant, Peter, 42
Submarines, 338
Suffrage. *See* Voting rights
Sugar Act of 1764, 92, 93
Sugar production, 18, 23, 45, 254
Sullivan, John, 114
Sumner, Charles, 311–312, 327, 353, 369
Sumter, Fort, 321, 325
Sumter, Thomas, 116
Supreme Court, U.S. *See also specific cases and justices*
  Constitution as interpreted by, 169, 195–197
  on indigenous lands, 197
  nationalism and, 197, 198, 217
  powers of, 143, 169
  Reconstruction and, 361–362
  on segregation, 374
  states' rights cases heard by, 196, 197

Supreme Order of the Star-Spangled Banner. *See* Know-Nothings
Sutter, John, 305

Talleyrand (prince), 150
Tallmadge Amendment, 193, 195
Tammany Hall, 369
Tammany Society, 152
Tanaghrisson, 85
Taney, Roger B., 216–217, 314
Tariffs
  in Civil War, 324
  in colonial period, 92
  Federalists and, 144, 145
  nullification doctrine and, 208, 210
  sectionalism and, 200
  textile industry and, 187
Taverns, 74, 99–102, 247
Taxation
  of British colonies, 90, 92–97, 99, 101
  in Civil War, 325, 332
  in Confederation period, 130–131
  in England, 20, 40
  excise tax, 144, 145, 147
  of income, 325, 332
Taylor, Zachary, 301, 302, 305, 307, 308
Tea Act of 1773, 101
Technology. *See also* Weapons; *specific technologies*
  agricultural, 160–161, 249–250
  in antebellum period, 237–238
  in Atlantic World, 17
  in Civil War, 337–338
  in colonial period, 65–66, 78
  electricity, 17, 79
  household, 244
  industrial, 161, 237
  machine tools, 161, 237
  printing, 78, 222, 236
  stone tools, 2, 3
  in textile industry, 160, 162, 187
  transportation, 163, 165–166, 231–234
Tecumseh (Shawnee chief), 178–180, 183
Tejanos, 298
Telegraphs, 234–236, 338
Temperance movement, 271, 279
Tenant farmers, 365
Tennessee
  horse racing in, 164
  indigenous lands in, 178
  readmission to Union, 359, 361

  Reconstruction in, 353
  secession of, 321
  statehood for, 147
Ten Percent Plan, 353, 355
Tenskwatawa "the Prophet," 178–179, 183
Tenth Amendment, 151, 208
Tenure of Office Act of 1867, 361, 362
Texas
  American settlers in, 189, 297–298
  border with Mexico, 301, 303
  flag of republic of, 297
  Hispanic residents of, 298
  independence of, 297–298
  secession of, 320, 322
  Spanish colony in, 46, 47
  statehood for, 298, 300, 301
  sugar production in, 256
Textile industry
  in England, 19, 162, 187, 323, 336
  factory system in, 162, 237
  growth of, 186–187
  labor force in, 239–241
  Lowell or Waltham system for, 239–241
  in South, 257, 260, 371–372
  technological advances in, 160, 162, 187
  water power for, 161, 162, 237, 372
Thames, Battle of the (1813), 180
Theater, 247–248
Theocracy, 35
Thirteenth Amendment, 288, 329, 358
Thomas, Jesse B., 195
Thomas Amendment, 195
Thoreau, Henry David, 275–276
Three Fires Confederacy, 90
Ticonderoga, Fort, 112, 114
Tilden, Samuel J., 369
Tippecanoe, Battle of (1811), 179
Tobacco production, 27–30, 33, 65, 68–69, 254, 372
Tocqueville, Alexis de, 205
Tom Thumb (Charles Stratton), 247
Tool use, 2, 3
Tories. *See* Loyalists
Toussaint-Louverture (Haiti), 118
Towns. *See* Cities and towns
Townshend, Charles, 96–97
Townshend Duties, 96–97, 101
Trade. *See also* Economy; Slave trade; Tariffs
  Africa and, 16–18, 62, 66–67
  in Caribbean region, 44–46, 66–67, 92, 125

INDEX

Trade—(Cont.)
  in colonial period, 27, 34, 42, 48, 51, 66-67, 92
  in early civilizations, 4, 5
  embargoes on, 177-178
  England and, 19, 66-67, 176, 178, 183, 238
  European, 6, 7, 66-67, 253
  France and, 49, 84, 176, 178
  fur, 22-23, 31, 34, 66, 84, 92, 178, 189-190
  indigenous peoples and, 48, 91, 171, 189-190
  mercantilist view of, 19
  as motivation for migration, 16
  Navigation Acts and, 51, 52, 67, 99
  triangular, 66-67
  in West, 189-190
Trade unions, 239, 242
Trafalgar, Battle of (1805), 176
Trail of Tears, 213-214
Trains. See Railroads
Transcendentalists, 274-276
Transcontinental railroad, 310, 324-325
Transportation. See also Railroads; Roads
  canals, 231, 232, 257
  innovations in, 163, 165-166
  steamboats, 163, 165, 187-188, 197, 231, 257
Treaties. See also specific names of treaties
  in American Revolution, 120, 128, 135
  Barbary states and, 168
  indigenous peoples and, 30-31, 50, 84, 122, 130, 178, 188, 212-213
Trench warfare, 337
Trent affair (1861), 336
Triangular trade, 66-67
Trist, Nicholas, 303
Trollope, Frances, 265
"Trotting Cracks" on the Snow (Maurer), 165
Truck farming, 248-249
Truth, Sojourner, 283
Tubman, Harriet, 269
Turner, Frederick Jackson, 206
Turner, Nat, 266, 269, 313
Turnpikes, 165-166, 231
Twain, Mark, 274
Tweed, William M. and Tweed Ring, 369
Twelfth Amendment, 150, 200
Tyler, John, 224, 225, 298, 300

Uncle Tom's Cabin (Stowe), 291-293, 313
Underground railroad, 269-270
Unemployment
  of African Americans, 359
  in Panic of 1837, 220
  in Panic of 1873, 367
Union Army
  advantages of, 321
  African Americans in, 327, 329-330
  campaigns and battles, 337-347
  commanders of, 333-334, 341
  mobilization of, 325
  uniforms worn by, 337
Union Pacific Railroad Company, 324-325, 367
Unions, 239, 242. See also specific unions
Unitarianism, 158
United States Sanitary Commission, 331
United States Treasury notes, 325
Universalism, 158
Universities. See Colleges and universities
University of Pennsylvania, 78, 157
Urban areas. See Cities and towns
Utah. See also specific cities
  Mormon settlements in, 279
  statehood for, 307
Utopian communities, 276-277
Utrecht, Treaty of (1713), 85

Vaccinations, 79, 281
Vallandigham, Clement, 327
Van Buren, Martin, 207-209, 216, 218-221, 224, 300, 305
Vance, Zebulon M., 332
Van Cleve, George William, 139-140
Van de Passe, Simon, 28
Vatican, 18, 52, 86, 121. See also Catholic Church
Vergennes, Count de, 115-116, 120
Verger, Jean Baptist de, 112
Vermont
  American Revolution in, 114
  emancipation in, 121
  establishment of, 96
  statehood for, 147
Verrazzano, Giovanni da, 22
Vesey, Denmark, 269
Vespucci, Amerigo, 7
Vicksburg siege (1863), 343, 344

Violence. See also Riots; War and warfare
  against African Americans, 368, 375
  by anti-abolitionists, 290-291
Virginia. See also specific cities
  Bacon's Rebellion in, 30-31
  Civil War battles in, 338-342, 345, 346
  Constitution ratified by, 142
  Jamestown settlement in, 20, 26-29, 31, 47
  secession of, 321, 331
  slavery in, 29-32, 63, 269
  Statute of Religious Liberty in, 127
  tobacco production in, 27-30, 65, 69, 254
Virginia and Kentucky Resolutions, 151
Virginia Company, 27, 29, 30, 34
Virginia Dynasty, 191
Virginia (ironclad warship), 335
Virginia Plan, 136, 137
Virginia Resolves, 93
Virtual representation, 99
Voltaire, 118
Voting rights. See also Elections
  for African Americans, 141, 204, 353, 361, 369, 374
  in colonial period, 35, 98
  discrimination and, 368
  Fifteenth Amendment and, 361, 369, 371, 374
  in Jacksonian period, 203-205
  literacy requirements and, 374
  Nineteenth Amendment and, 141
  restrictions on, 374
  for women, 140, 141, 204, 283

Wade, Benjamin E., 327, 334, 353
Wade-Davis Bill, 353, 355, 357
Wages. See also Income
  for African Americans, 365
  for enslaved laborers, 42
  for textile factory workers, 239
Waldstreicher, David, 139-140
Walker, David, 287, 319
Waltham system, 239
Wampanoags, 34, 35, 39
Wang Hya, Treaty of (1844), 225
War and warfare. See also Military; Weapons; specific wars and battles
  debt from, 89-90, 130-131
  guerrilla, 214
  religious, 86

total war, 29
trench, 337
War for Independence. *See* American Revolution
War Hawks, 180
Warner, Susan, 292, 293
War of 1812
  economic growth following, 186
  England and, 180-183, 188
  events leading up to, 176, 180
  indigenous peoples and, 180-181
  Jackson in, 181, 182, 201
  nationalism after, 185, 191
  peace negotiations, 183
Washington, Booker T., 372-373
Washington, D.C., 166-167, 182, 327
Washington, George
  in American Revolution, 111, 113-114, 116, 117
  at Constitutional Convention, 136, 142
  death of, 157
  Farewell Address by, 148-149
  Federalist support by, 142, 146
  at Fort Necessity, 85
  Mount Vernon estate of, 68, 142, 143
  political ideology of, 98
  as president, 142-146
  on Proclamation of 1763, 92
  as slaveholder, 122, 141
  Whiskey Rebellion and, 147
Washington, Martha, 68
Washington, Treaty of (1871), 368
Water pollution, 238
Water power, 161, 162, 237, 372
Watt, James (inventor), 162
Wayne, Anthony, 130, 178
Wealth
  in antebellum period, 242-243, 259
  of capitalists, 163
  colonial possessions and, 86
  of fur traders, 43, 52
  in Jacksonian period, 203
  of planters, 42, 167, 181, 189
  support for Constitution and, 138
Weapons
  in Civil War, 337
  flintlock rifles, 39
  muskets, 9, 161, 337
Webster, Daniel, 182, 195-196, 209, 216, 218-219, 224
Webster, Noah, 158
Webster-Ashburton Treaty (1842), 224

Webster-Hayne Debate, 209
Wells-Barnett, Ida B., 375
Wesley, John and Charles, 76
Wesleyan College, 246
West. *See also specific regions and states*
  agriculture in, 229
  canal routes to, 232
  Civil War in, 340, 342-344
  early civilizations in, 4
  Eastern images of, 190
  education in, 281, 282
  expansion in, 188-190, 300-304
  exploration of, 171-174
  Hispanics in, 298
  Homestead Act and, 324
  indigenous peoples in, 130, 171, 178, 211, 213-215
  migration to, 188, 244, 298-300
  Mormon settlements in, 278, 279
  railroads in, 310
  Republican support in, 146
  securing, 146-147
  trade in, 189-190
  trails leading to, 189, 299
West, Benjamin, 88
Western Design expedition, 40
Western Union Telegraph Company, 235
West Point, Military Academy at, 168, 334
West Virginia, 338
Wheelwright, John, 36
Whig Party
  Anti-Masonry movement and, 218
  diplomacy of, 224-225
  fragmentation of, 311
  geographic areas of support, 218
  ideology of, 199, 218
  in Jacksonian period, 207, 216-218
  presidential candidates, 216, 219-221, 224, 300, 305, 309
Whiskey Rebellion, 147
White, Deborah Gray, 262
White, John, 22
White, LeeAnn, 356
White, Richard, 50
Whitefield, George, 76
White League, 368
Whites
  as abolitionists, 286, 290
  African society as viewed by, 17
  attitudes toward indigenous peoples, 210, 215
  as indentured servants, 30, 61

  in Old Northwest, 188, 310
  in southern society, 257, 259-261
  in Southwest, 147, 178, 189
  wars with indigenous peoples, 22, 29, 37, 39, 130, 179-181, 210
White supremacy
  American Revolution and, 110
  Civil War and, 320
  ideology of, 122, 371
  in Jacksonian period, 202
  Ku Klux Klan and, 368
  lynchings and, 375
  phrenology and, 280
  in South, 121, 371, 372
Whitman, Walt, 273
Whitney, Eli, 160-161
Whittier, John Greenleaf, 289
Wilberforce, William, 288, 289
Wilentz, Sean, 139, 206
Wilkes, Charles, 336
Wilkinson, James, 171, 175
William and Mary College, 78
William of Orange, 52-53
Williams, Eric, 16
Williams, Kidada E., 357
Williams, Roger, 36
Wilmot, David and Wilmot Proviso, 304, 306
Winthrop, John, 35
Witchcraft trials, 72-73
Witgen, Michael, 51
Wolfe, James, 87, 88
Women. *See also* Feminism; Gender relations; Marriage; Women in workforce
  as abolitionists, 283, 291
  African American, 283, 285-286
  in African society, 18
  in American Revolution, 109, 123, 124
  childbirth and, 60, 61, 157
  in Civil War, 330-331, 333
  in colonial period, 29, 36, 57, 60-61, 101-102
  cult of domesticity and, 246
  education for, 78, 156-157, 246, 260
  as indentured servants, 56-57, 61
  literacy rate for, 76
  middle class, 244, 246
  motherhood and, 123
  as preachers, 283, 285-286
  in prostitution, 298
  in reform movements, 279, 282-286
  religious activity of, 159, 283, 285-286

Women—(*Cont.*)
  enslavement of, 69, 260, 262, 264
  in South, 260, 333, 372
  voting rights for, 140, 141, 204, 283
  witchcraft trials and, 72–73
  working class, 246
  as writers, 275, 291–293
Women in workforce
  in antebellum period, 239
  in Civil War, 330–331, 333
  in New South, 372
  as nurses, 331
  textile industry and, 239–241
  as union members, 239, 242
Women's National Loyal League, 331
Wood, Gordon, 109–110, 139
Woodland people, 4
Wool. *See* Textile industry
Worcester v. Georgia (1832), 197, 211
Workforce. *See* Labor force
Working class
  horse racing and, 165
  housing for, 244
  revolts in Europe, 243
  standard of living for, 163
  women in, 246
World War I (1914–1918), 337
World War II (1939–1945), 337

XYZ Affair, 150

Yale University, 78
Yancey, William L., 267
Yellow fever, 171
Yeoman farmers, 125
Yorktown, Battle of (1781), 109, 117, 123
Young, Brigham, 279
Young America movement, 309–310

Zagarri, Rosemarie, 110
Zenger, John Peter, 79
Zunis, 4